Classics in Mathematics

Saunders Mac Lane Homology

Saunders Mac Lane was born on August 4, 1909 in Connecticut. He studied at Yale University and then at the University of Chicago and at Göttingen, where he received the D. Phil. in 1934. He has taught at Harvard, Cornell and the University of Chicago.

Mac Lane's initial research was in logic and in algebraic number theory (valuation theory). With Samuel Eilenberg he published fifteen papers on algebraic topology. A number of them involved the initial steps in the cohomology of groups and in other aspects of homological algebra – as well as the discovery of category theory. His famous undergraduate textbook *Survey of modern algebra,* written jointly with G. Birkhoff, has remained in print for over 50 years. Mac Lane is also the author of several other highly successful books.

Saunders Mac Lane

Homology

Reprint of the 1975 Edition

Springer

Saunders Mac Lane
Department of Mathematics, University of Chicago
Chicago, IL 60637-1514
USA

Originally published as Vol. 114 of the
Grundlehren der mathematischen Wissenschaften

Mathematics Subject Classification (1991): 18-02, 18AXX, 18CXX, 18GXX

ISBN 3-540-58662-8 Springer-Verlag Berlin Heidelberg New York

CIP data applied for

41/3111 – 5 4 3 2 1 Printed on acid-free paper

Saunders Mac Lane

Homology

Springer-Verlag
Berlin Heidelberg New York

Saunders Mac Lane
University of Chicago
Department of Mathematics
5734 University Avenue
Chicago, IL 60637
USA

Fourth Printing 1994

With 7 Figures

AMS Subject Classification (1991): 18–02, 18Axx, 18Cxx, 18Gxx

ISBN 3-540-03823-X Springer-Verlag Berlin Heidelberg New York
ISBN 0-387-03823-X Springer-Verlag New York Berlin Heidelberg

Library of Congress Catalog Card Number 75-509640

Printed in Germany

Typesetting: Universitätsdruckerei H. Stürtz AG, Würzburg
41/3111 – 5 4 3 2 1 Printed on acid-free paper

To Dorothy

Preface

In presenting this treatment of homological algebra, it is a pleasure to acknowledge the help and encouragement which I have had from all sides. Homological algebra arose from many sources in algebra and topology. Decisive examples came from the study of group extensions and their factor sets, a subject I learned in joint work with OTTO SCHILLING. A further development of homological ideas, with a view to their topological applications, came in my long collaboration with SAMUEL EILENBERG; to both collaborators, especial thanks. For many years the Air Force Office of Scientific Research supported my research projects on various subjects now summarized here; it is a pleasure to acknowledge their lively understanding of basic science.

Both REINHOLD BAER and JOSEF SCHMID read and commented on my entire manuscript; their advice has led to many improvements. ANDERS KOCK and JACQUES RIGUET have read the entire galley proof and caught many slips and obscurities. Among the others whose suggestions have served me well, I note FRANK ADAMS, LOUIS AUSLANDER, WILFRED COCKCROFT, ALBRECHT DOLD, GEOFFREY HORROCKS, FRIEDRICH KASCH, JOHANN LEICHT, ARUNAS LIULEVICIUS, JOHN MOORE, DIETER PUPPE, JOSEPH YAO, and a number of my current students at the University of Chicago — not to mention the auditors of my lectures at Chicago, Heidelberg, Bonn, Frankfurt, and Aarhus. My wife, DOROTHY, has cheerfully typed more versions of more chapters than she would like to count. Messrs. SPRINGER have been unfailingly courteous in the preparation of the book; in particular, I am grateful to F. K. SCHMIDT, the Editor of this series, for his support. To all these and others who have helped me, I express my best thanks.

Chicago, 17. February 1963

SAUNDERS MAC LANE

Added Preface

In the third printing, several errors have been corrected. In particular, the previous erroneous construction of the splitting in the Homology classification theorem (Theorem III, 4.3) has been replaced by a correct proof, due essentially to DOLD (A. DOLD, *Lectures on Algebraic Topology,* Grundlehren der mathematischen Wissenschaften, vol. 200, Springer 1972). Also, the axioms on page 260 for allowable short exact sequences have been modified, so that they will actually apply where they are used on page 376.

SAUNDERS MAC LANE

Table of Contents

Introduction

Our subject starts with homology, homomorphisms, and tensors.

Homology provides an algebraic "picture" of topological spaces, assigning to each space X a family of abelian groups $H_0(X), \ldots, H_n(X), \ldots$, to each continuous map $f: X \to Y$ a family of group homomorphisms $f_n: H_n(X) \to H_n(Y)$. Properties of the space or the map can often be effectively found from properties of the groups H_n or the homomorphisms f_n. A similar process associates homology groups to other Mathematical objects; for example, to a group Π or to an associative algebra Λ. Homology in all such cases is our concern.

Complexes provide a means of calculating homology. Each n-dimensional "singular" simplex T in a topological space X has a boundary consisting of singular simplices of dimension $n-1$. If K_n is the free abelian group generated by all these n-simplices, the function ∂ assigning to each T the alternating sum ∂T of its boundary simplices determines a homomorphism $\partial: K_n \to K_{n-1}$. This yields (Chap. II) a "complex" which consists of abelian groups K_n and boundary homomorphisms ∂, in the form

$$0 \leftarrow K_0 \overset{\partial}{\leftarrow} K_1 \overset{\partial}{\leftarrow} K_2 \overset{\partial}{\leftarrow} K_3 \overset{\partial}{\leftarrow} \cdots.$$

Moreover, $\partial\partial = 0$, so the kernel C_n of $\partial: K_n \to K_{n-1}$ contains the image ∂K_{n+1}. The factor group $H_n(K) = C_n / \partial K_{n+1}$ is the n-th homology group of the complex K or of the underlying space X. Often a smaller or simpler complex will suffice to compute the same homology groups for X. Given a group Π, there is a corresponding complex whose homology is that appropriate to the group. For example, the one dimensional homology of Π is its factor commutator group $\Pi/[\Pi, \Pi]$.

Homomorphisms of appropriate type are associated with each type of algebraic system; under composition of homomorphisms the systems and their homomorphisms constitute a "category" (Chap. I). If C and A are abelian groups, the set $\operatorname{Hom}(C, A)$ of all group homomorphisms $f: C \to A$ is also an abelian group. For C fixed, it is a covariant "functor" on the category of all abelian groups A; each homomorphism $\alpha: A \to A'$ induces the map $\alpha_*: \operatorname{Hom}(C, A) \to \operatorname{Hom}(C, A')$ which carries each f into its composite αf with α. For A fixed, Hom is contravariant: Each $\gamma: C' \to C$ induces the map γ^* in the opposite direction, $\operatorname{Hom}(C, A) \to \operatorname{Hom}(C', A)$, sending f to the composite $f\gamma$. Thus $\operatorname{Hom}(?, A)$ applied to a complex $K = ?$ turns the arrows around to give a complex

$$\mathrm{Hom}(K_0, A) \xrightarrow{\partial^*} \mathrm{Hom}(K_1, A) \xrightarrow{\partial^*} \mathrm{Hom}(K_2, A) \to \cdots.$$

Here the factor group (Kernel ∂^*)/(Image ∂^*) is the cohomology $H^n(K, A)$ of K with coefficients A. According to the provenance of K, it yields the cohomology of a space X or of a group Π.

An extension of a group A by a group C is a group $B \supset A$ with $B/A \cong C$; in diagramatic language, an extension is just a sequence

$$E: 0 \to A \to B \to C \to 0$$

of abelian groups and homomorphisms which is exact in the sense that the kernel of each homomorphism is exactly the image of the preceding one. The set $\mathrm{Ext}^1(C, A)$ of all extensions of A by C turns out to be an abelian group and a functor of A and C, covariant in A and contravariant in C.

Question: Does the homology of a complex K determine its cohomology? The answer is almost yes, provided each K_n is a free abelian group. In this case $H^n(K, A)$ is determined "up to a group extension" by $H_n(K)$, $H_{n-1}(K)$, and A; specifically, the "universal coefficient theorem" (Chap. III) gives an exact sequence

$$0 \to \mathrm{Ext}^1(H_{n-1}(K), A) \to H^n(K, A) \to \mathrm{Hom}(H_n(K), A) \to 0$$

involving the functor Ext^1 just introduced. If the K_n are not free groups, there is a more complex answer, involving the spectral sequences to be described in Chap. XI.

Tensors arise from vector spaces U, V, and W and bilinear functions $B(u, v)$ on $U \times V$ to W. Manufacture the vector space $U \otimes V$ generated by symbols $u \otimes v$ which are bilinear in $u \in U$ and $v \in V$ and nothing more. Then $u \otimes v$ is a universal bilinear function; to any bilinear B there is a unique linear transformation $T: U \otimes V \to W$ with $B(u, v) = T(u \otimes v)$. The elements of $V \otimes V$ turn out to be just the classical tensors (in two indices) associated with the vector space V. Two abelian groups A and G have a tensor product $A \otimes G$ generated by bilinear symbols $a \otimes g$; it is an abelian group, and a functor covariant in A and G. In particular, if K is a complex, so is $A \otimes K: A \otimes K_0 \leftarrow A \otimes K_1 \leftarrow \cdots$.

Question: Does the homology of K determine that of $A \otimes K$? Answer: Almost yes; if each K is free, there is an exact sequence

$$0 \to A \otimes H_n(K) \to H_n(A \otimes K) \to \mathrm{Tor}_1(A, H_{n-1}(K)) \to 0.$$

Here $\mathrm{Tor}_1(A, G)$ is a new covariant functor of the abelian groups A and G, called the "torsion product"; it depends (Chap. V) on the elements of finite order in A and G and is generated, subject to suitable relations, by pairs of elements $a \in A$ and $g \in G$ for which there is an integer m with $ma = 0 = mg$.

Take the cartesian product $X \times Y$ of two spaces. Can we calculate its homology from that of X and Y? A study of complexes constructed from simplices (Chap. VIII) reduces this question to the calculation of the homology of a tensor product $K \otimes L$ of two complexes. This calculation again involves the torsion product, via an exact sequence (the Künneth Thm, Chap. V)

$$0 \to \sum_{p+q=n} H_p(K) \otimes H_q(L) \to H_n(K \otimes L) \to \sum_{p+q=n-1} \operatorname{Tor}_1(H_p(K), H_q(L)) \to 0.$$

But woe, if A is a subgroup of B, $A \otimes G$ is not usually a subgroup of $B \otimes G$; in other words, if $E: 0 \to A \to B \to C \to 0$ is exact, the sequence of tensor products

$$0 \to A \otimes G \to B \otimes G \to C \otimes G \to 0,$$

is exact, *except* possibly at $A \otimes G$. Happily, the torsion product repairs the trouble; the given sequence E defines a homomorphism $E_*: \operatorname{Tor}_1(C, G) \to A \otimes G$ with image exactly the kernel of $A \otimes G \to B \otimes G$, and the sequence

$$0 \to \operatorname{Tor}_1(A, G) \to \operatorname{Tor}_1(B, G) \to \operatorname{Tor}_1(C, G) \xrightarrow{E_*} A \otimes G \to B \otimes G$$

is exact. Call E_* the connecting homomorphism for Tor_1 and $\otimes$.

But again woe, if A is a subgroup of B, a homomorphism $f: A \to G$ may not be extendable to a homomorphism $B \to G$; in other words, the exact sequence $0 \to A \to B \to C \to 0$ induces a sequence (opposite direction by contravariance!)

$$0 \to \operatorname{Hom}(C, G) \to \operatorname{Hom}(B, G) \to \operatorname{Hom}(A, G) \to 0$$

which may not be exact at $\operatorname{Hom}(A, G)$. Ext^1 to the rescue: There is a "connecting" homomorphism E^* which produces a longer exact sequence

$$0 \to \operatorname{Hom}(C, G) \to \operatorname{Hom}(B, G) \to \operatorname{Hom}(A, G) \xrightarrow{E^*}$$
$$\xrightarrow{E^*} \operatorname{Ext}^1(C, G) \to \operatorname{Ext}^1(B, G) \to \operatorname{Ext}^1(A, G) \to 0.$$

Now generalize; replace abelian groups by modules over any commutative ring R. Then $\operatorname{Ext}^1(A, G)$ ist still defined as an R-module, but the longer sequence may now fail of exactness at $\operatorname{Ext}^1(A, G)$. There is a new functor $\operatorname{Ext}^2(A, G)$, a new connecting homomorphism $E^*: \operatorname{Ext}^1(A, G) \to \operatorname{Ext}^2(C, G)$, and an exact sequence extending indefinitely to the right as

$$\cdots \to \operatorname{Ext}^n(C, G) \to \operatorname{Ext}^n(B, G) \to \operatorname{Ext}^n(A, G) \xrightarrow{E^*} \operatorname{Ext}^{n+1}(C, G) \to \cdots.$$

The elements of $\mathrm{Ext}^n(C, G)$ are suitable equivalence classes of long exact sequences

$$0 \to G \to B_{n-1} \to \cdots \to B_0 \to C \to 0$$

running from G to C through n intermediate modules. Similarly for the tensor product; there are functors $\mathrm{Tor}_n(A, G)$, described via suitable generators and relations, which enter into a long exact sequence

$$\cdots \to \mathrm{Tor}_{n+1}(C, G) \xrightarrow{E_*} \mathrm{Tor}_n(A, G) \to \mathrm{Tor}_n(B, G) \to \mathrm{Tor}_n(C, G) \to \cdots$$

induced by each $E\colon 0 \to A \to B \to C \to 0$. They apply also if the ring is not commutative – and A, B, and C are right R-modules, G a left R-module.

These functors Tor_n and Ext^n are the subject of homological algebra. They give the cohomology of various algebraic systems. If Π is a group, take R to be the group ring generated by Π over the integers. Then the group Z of integers is (trivially) an R-module; if A is any other R-module, the groups $\mathrm{Ext}^n_R(Z, A)$ are the cohomology groups $H^n(\Pi, A)$ of the group Π with coefficients in A. If $n=2$, $H^2(\Pi, A)$ turns out, as it should, to be the group of all extensions B of the abelian group A by the (non-abelian) group Π, where the structure of A as a Π-module specifies how A is a normal subgroup of B. If $n=3$, $H^3(\Pi, A)$ is a group whose elements are "obstructions" to an extension problem. Similarly, $\mathrm{Tor}_n(Z, A)$ gives the homology groups of Π. Again, if Λ is an algebra over the field F, construct Ext^n by long exact sequences of two-sided Λ-modules A. The algebra Λ is itself such a module, and $\mathrm{Ext}^n(\Lambda, A)$ is the cohomology of Λ with coefficients A; again Ext^2 and Ext^3 correspond to extension problems for algebras.

A module P is projective if every homomorphism $P \to B/A$ lifts to a homomorphism $P \to B$. Any free module is projective; write any module in terms of generators; this expresses it as a quotient of a free module, and hence of a projective module.

How can Tor_n and Ext^n be calculated? Write A as a quotient of a projective module P_0; that is, write an exact sequence $0 \leftarrow A \leftarrow P_0$. The kernel of $P_0 \to A$ is again a quotient of a projective P_1. This process continues to give an exact sequence $0 \leftarrow A \leftarrow P_0 \leftarrow P_1 \leftarrow \cdots$. The complex P is called a "projective resolution" of A. It is by no means unique; compare two such

$$\begin{array}{ccccccccc} 0 \leftarrow & A & \leftarrow & P_0 & \xleftarrow{\partial} & P_1 & \leftarrow & P_2 & \leftarrow \cdots \\ & \| & & \downarrow{\scriptstyle f_0} & & \downarrow{\scriptstyle f_1} & & \downarrow & \\ 0 \leftarrow & A & \leftarrow & P_0' & \xleftarrow{\partial} & P_1' & \leftarrow & P_2' & \leftarrow \cdots. \end{array}$$

Since P_0 is projective, the map $P_0 \to A$ lifts to $f_0\colon P_0 \to P_0'$. The composite map $P_1 \to P_0'$ lifts in turn to an $f_1\colon P_1 \to P_1'$ with $\partial f_1 = f_0 \partial$, and so on by recursion. The resulting comparison $f_n : P_n \to P'_n$ of complexes induces a homomorphism $H_n(P \otimes G) \to H_n(P' \otimes G)$. Reversing the roles of P and P' and deforming $P \to P' \to P$ to the identity (deformations are called homotopies) shows this an isomorphism $H_n(P \otimes G) \cong H_n(P' \otimes G)$. Therefore the homology groups $H_n(P \otimes G)$ do not depend on the choice of the projective resolution P, but only on A and G. They turn out to be the groups $\mathrm{Tor}_n(A, G)$. Similarly, the cohomology groups $H^n(P, G)$ are the groups $\mathrm{Ext}^n(A, G)$, while the requisite connecting homomorphisms E^* may be obtained from a basic exact homology sequence for complexes (Chap. II). Thus Tor and Ext may be calculated from projective resolutions. For example, if Π is a group, the module Z has a standard "bar resolution" (Chap. IX) whose cohomology is that of Π. For particular groups, particular resolutions are more efficient.

Qualitative considerations ask for the minimum length of a projective resolution of an R-module A. If there is a projective resolution of A stopping with $P_{n+1} = 0$, A is said to have homological dimension at most n. These dimensions enter into the arithmetic structure of the ring R; for example, if R is the ring Z of integers, every module has dimension at most 1; again for example, the Hilbert Syzygy Theorem (Chap. VII) deals with dimensions of graded modules over a polynomial ring.

Two exact sequences $0 \to A \to B \to C \to 0$ and $0 \to C \to D \to F \to 0$ may be "spliced" at C to give a longer exact sequence

$$\begin{array}{c} 0 \to A \to B \dashrightarrow D \to F \to 0; \\ \searrow C \nearrow \end{array}$$

in other words, an element of $\mathrm{Ext}^1(C, A)$ and an element of $\mathrm{Ext}^1(F, C)$ determine a two-fold extension which is an element of $\mathrm{Ext}^2(F, A)$, called their product (Chap. III). These and similar products for Tor can be computed from resolutions (Chap. VIII).

Every R-module is also an abelian group; that is, a module over the ring Z of integers. Call an extension $E: A \to B \to C$ of R-modules Z-split if the middle module B, regarded just as an abelian group, is the direct sum of A and C. Construct the group $\mathrm{Ext}^1_{(R,Z)}(C, A)$ using only such Z-split extensions. This functor has connecting homomorphisms E^* for those E which are Z-split. With the corresponding torsion functors and their connecting homomorphisms, it is the subject of relative homological algebra (Chap. IX). The cohomology of a group is such a relative functor. Again, if Λ is an algebra over the commutative ring K, all appropriate concepts are relative to K; in particular, the cohomology of Λ arises from exact sequences of Λ-bimodules which are split as sequences of K-modules.

Modules appear to be the essential object of study. But the exactness of a resolution and the definition of a projective are properties of homomorphisms; all the arguments work if the modules and the homomorphisms are replaced by any objects $A, B, \ldots$ with "morphisms" $\alpha: A \to B$ which can be added, compounded, and have suitable kernels, cokernels $(B/\alpha A)$, and images. Technically, this amounts to developing homological algebra in an abelian category (Chap. IX). From the functor $T_0(A) = A \otimes G$ we constructed a sequence of functors $T_n(A) = \mathrm{Tor}_n(A, G)$. More generally, let T_0 be any covariant functor which is additive $[T_0(\alpha_1 + \alpha_2) = T_0\alpha_1 + T_0\alpha_2]$ and which carries each exact sequence $0 \to A \to B \to C \to 0$ into a right exact sequence $T_0(A) \to T_0(B) \to T_0(C) \to 0$. We again investigate the kernel of $T_0(A) \to T_0(B)$ and construct new functors to describe it. If the category has "enough" projectives, each A has a projective resolution P, and $H_n(T_0(P))$ is independent of the choice of P and defines a functor $T_n(A)$ which enters into a long exact sequence

$$\cdots \to T_n(A) \to T_n(B) \to T_n(C) \xrightarrow{E_*} T_{n-1}(A) \to \cdots.$$

Thus T_0 determines a whole sequence of derived functors T_n and of connecting homomorphisms $E_*: T_n(C) \to T_{n-1}(A)$. These "derived" functors can be characterized conceptually by three basic properties (Chap. XII):

(i) The long sequence above is exact,

(ii) If P is projective and $n > 0$, $T_n(P) = 0$,

(iii) If $E \to E'$ is a homomorphism of exact sequences, the diagram of connecting homomorphisms commutes (naturality!):

$$\begin{array}{ccc} T_n(C) & \to & T_{n-1}(A) \\ \downarrow & & \downarrow \\ T_n(C') & \to & T_{n-1}(A'). \end{array}$$

In particular, given $T_0(A) = A \otimes G$, these axioms characterize $\mathrm{Tor}_n(A, G)$ as functors of A. There is a similar characterization of the functors $\mathrm{Ext}^n(C, A)$ (Chap. III). Alternatively, each derived functor T_n can be characterized just in terms of the preceding T_{n-1}: If $E: S_n(C) \to S_{n-1}(A)$ is another natural connecting homomorphism between additive functors, each "natural" map of S_{n-1} into T_{n-1} extends to a unique natural map of S_n into T_n. This "universal" property of T_n describes it as the left satellite of T_{n-1}; it may be used to construct products.

Successive and interlocking layers of generalizations appear throughout homological algebra. We go from abelian groups to modules to bimodules to objects in an abelian category; from rings to groups to algebras to Hopf algebras (Chap. VI); from exact sequences to Z-split exact sequences to a "proper" class of exact sequences characterized by axioms (Chap. XII). The subject is in process of rapid expansion; the most general formulation is yet to come. Hence this book will proceed from the special to the general, subsuming earlier results in the concluding treatment (Chap. XII) of additive functors in an abelian category relative to a proper class of exact sequences.

As each concept is developed, we take time out to stress its applications. Thus Chap. IV on the cohomology of groups includes the topological interpretation of the cohomology groups of Π as the cohomology of an aspherical space with fundamental group Π, as well as SCHUR's Theorem that every extension of a finite group by another finite group of relatively prime order must split. Chap. VII, on dimension, studies syzygies and separable algebras. Chap. X on the cohomology of algebraic systems includes the Wedderburn principal theorem for algebras and the cohomology (at various levels) of abelian groups. Chap. XI includes the standard construction of the spectral sequences of a filtration and of a bicomplex, used to construct the spectral sequence of a covering and of a group extension. (The latter is due to LYNDON and not, as often thought, to the subsequent work of HOCHSCHILD-SERRE). Much of the general development of homological algebra in the other chapters can be read independently of these results.

For the expert we note a few special features. The basic functors Ext and Tor are described directly: Ext, following YONEDA, by long exact sequences, Tor by an improved set of generators and relations. Resolutions are relegated to their proper place as a means of computation. All the varieties of algebras (coalgebras, Hopf Algebras, graded algebras, differential graded algebras) are described uniformly by commutative diagrams for the product maps. Relative homological algebra is treated at two levels of generality: First, by a "forgetful" functor, say one which regards an R-module just as an abelian group, later by a suitable proper class of short exact sequences in an abelian category. The cohomology of groups is defined functorially by the bar construction. This construction later appears in conceptual form: For a pair of categories with a forgetful functor and a functor constructing relative projectives (Chap. IX, § 7). The proper definition of connecting homomorphisms by additive relations (correspondences) is indicated; these relations are used to describe the transgression in a spectral sequence. This gives a convenient treatment of the transgression in LYNDON's spectral sequence. Diagram chasing works in an abelian category with subobjects or quotient objects replacing elements (XII.3).

Notations are standard, with the following few exceptions. A complex is X (latin), a commutative ring is K (greek). A "graded" module M is a family $M_0, M_1, \ldots$ of modules and *not* their direct sum $\sum M_n$, while a family ..., M_{-1}, M_0, M_1, ... is said to be "Z-graded". A monomorphism is written $\varkappa: A \rightarrowtail B$, an epimorphism $\sigma: B \twoheadrightarrow C$, while $\varkappa \| \sigma$ states that $0 \to A \to B \to C \to 0$ is exact. A dotted arrow $A \cdots\!\!\to B$ is a homomorphism to be constructed, a dashed arrow $A \dashrightarrow B$ is a group homomorphism between modules, a half arrow $A \rightharpoondown B$ is an additive relation. We distinguish between a bicomplex (XI.6) and a complex of complexes (X.9); we "augment" but do not "supplement" an algebra. The dual of a resolution is a "coresolution". If u is a cycle in the homology class h of $H_n(X)$, $u \in\in H_n$ is short for $u \in h \in H_n$, while h is written $h = \operatorname{cls} u$. The coboundary of an n-cochain f is $\delta f = (-1)^{n+1} f\, \partial$, *with* a sign (II.3).

A reference to Thm V.4.3 is to Theorem 3 of section 4 of Chap. V; if the chapter number is omitted, it is to a theorem in the chapter at hand. A reference such as Bourbaki [1999] is to that author's article, as listed in the bibliography at the back of our book and published in the year cited; [1999b] is to the second article by the same author, same year. The influential treatise by H. Cartan and S. Eilenberg on *Homological Algebra* is honored by omitting its date. The bibliography makes no pretense at completeness, but is intended to provide a guide to further reading, as suggested in the notes at the ends of some chapters or sections. These notes also contain occasional historical comments which give positive—and perhaps prejudiced—views of the development of our subject. The exercises are designed both to give elementary practice in the concepts presented and to formulate additional results not included in the text.

Chapter one

Modules, Diagrams, and Functors

Homology theory deals repeatedly with the formal properties of functions and their composites. The functions concerned are usually homomorphisms of modules or of related algebraic systems. The formal properties are subsumed in the statement that the homomorphisms constitute a category. This chapter will examine the notions of module and category.

1. The Arrow Notation

If X and Y are sets, the *cartesian product* $X \times Y$ is the set of all ordered pairs (x, y) for $x \in X$ and $y \in Y$.

The notation $f: X \to Y$ states that f is a *function on X to Y*. Formally, such a function may be described as an ordered triple $f = (X, F, Y)$, with F a subset of $X \times Y$ containing for each $x \in X$ exactly one pair (x, y).

Actually we write $f(x) = y$, as usual, for the value of f at the argument x. Notice that we normally write the function f to the left of its argument, as in $f(x)$. Notice also that each function f carries with it a definite set X as *domain* and a definite set Y as *range*.

If $f: X \to Y$ and $g: Y \to Z$ are functions, the *composite function* gf, sometimes written $g \circ f$, is the function on X to Z with the value $(gf)(x) = g(f(x))$ for each $x \in X$. Since functions are written on the left, gf means first apply f, then apply g. This composite is defined *only* when Range (f) = Domain (g); in particular, we do not define the composite when Range (f) is a proper subset of Domain (g).

For any set X, the identity 1 or 1_X is the function $1: X \to X$ with $1(x) = x$ for all x. If S is a subset of X the function $j: S \to X$ with values $j(s) = s$ for all $s \in S$ is called the ("identity") *injection* of S into X. For any $f: X \to Y$ the composite $fj: S \to Y$ (sometimes written $f|S$) is the function f "cut down" to the subset S of its domain. Similarly, when Y is a subset of W and $k: Y \to W$ is the injection (with $k(y) = y$), the composite $kf: X \to W$ is the function f with its range expanded from Y to W. Notice that the functions f and kf have the same *values* for each argument x, but they are different functions, since the range is different. This distinction, apparently pedantic, will pay off. (See Example 3 in II. 1.).

We use the usual notations of set theory, with $X \cap Y$ denoting the intersection of the sets X and Y and with $\emptyset$ the empty set.

2. Modules

Let R be a ring with identity $1 \neq 0$. A left *R-module* A is an additive abelian group together with a function $p: R \times A \to A$, written $p(r, a) = ra$, such that always

$$(r + r')\,a = ra + r'a, \qquad (rr')\,a = r(r'a),$$
$$r(a + a') = ra + ra', \qquad 1a = a.$$

It follows that $0a = 0$ and $(-1)a = -a$. Some authors define an R-module without requiring that $1a = a$, and call a module with this property unitary. In this book, ***every ring has an identity and every module is unitary.***

Our treatment of left R-modules will apply, *mutatis mutandis*, to *right R-modules*. They are abelian groups A with $ar \in A$ defined so as to satisfy the corresponding four identities; for example $a(rr') = (ar)r'$.

Modules appear in many connections. In case R is a field or a skew field, a left R-module is a left vector space over R. If F is a field and $R = F[x]$ the polynomial ring in one indeterminate x with coefficients in F, then an R-module is simply a vector space V over F together with a fixed linear transformation $T: V \to V$; namely, T is the transformation given by left multiplication by $x \in R$. Consider Z-modules, where Z denotes the ring of integers. For each positive integer m, $ma = a + \cdots + a$ (m times); hence a Z-module A is just an abelian group, with the usual meaning for integral multiples ma, $m \in Z$. If Z_k is the ring of integers modulo k, a Z_k-module A is an abelian group in which every element has order a divisor of k. Finally, take R to be a commutative ring generated by 1 and by an element d with $d^2 = 0$, so that R consists of all $m + nd$ for integer coefficients m and n; an R-module is then an abelian group A together with a homomorphism $d: A \to A$ such that $d^2 = 0$; such a pair (A, d) is called a "differential group" (II.1).

A subset S of an R-module A is a *submodule* (in symbols, $S \subset A$), if S is closed under addition and if $r \in R$, $s \in S$ imply $rs \in S$; then S itself is an R-module. The ring R is itself a left R-module. A submodule of R is a subset L of R closed under addition and with each $rL \subset L$; such a subset is also called a *left ideal* in R. If L is a left ideal in R and A a left R-module, the set

$$LA = \{\text{all finite sums } \textstyle\sum l_i a_i, \text{ for } l_i \in L,\ a_i \in A\}$$

is a submodule of A, called the *product* of the ideal L by the module A. In particular, the product LL' of two left ideals is a left ideal, and $(LL')A = L(L'A)$.

If A and B are both R-modules, the notation $\alpha: A \to B$ or $A \xrightarrow{\alpha} B$ states that α is an *R-module homomorphism* of A to B; that is, a function on A to B such that always

$$\alpha(a + a') = \alpha a + \alpha a', \qquad \alpha(ra) = r(\alpha a).$$

When $\alpha: A \to B$, call A the *domain* and B the *range* of α. The *image* $\operatorname{Im}(\alpha) = \alpha A$ consists of all elements αa for $a \in A$; it is a submodule of the range B; the *kernel* $\operatorname{Ker}(\alpha)$ consists of all a in A with $\alpha a = 0$; it is a submodule of the domain A. If $\alpha A = B$, we say that α is an *epimorphism* and write $\alpha: A \twoheadrightarrow B$, while if $\operatorname{Ker}\alpha = 0$ we say that α is a *monomorphism* and write $\alpha: A \rightarrowtail B$. Finally, α is an *isomorphism* if and only if α is both a monomorphism and an epimorphism. For each module A, the identity function $1_A: A \to A$ is an isomorphism. For any A and B, the zero or "trivial" function 0 with every $0(a) = 0$ is a homomorphism $0: A \to B$. A homomorphism $\omega: A \to A$ with range and domain equal is called an *endomorphism*.

If $\alpha_1, \alpha_2: A \to B$ are homomorphisms with the same domain A and the same range B, their sum $\alpha_1 + \alpha_2$, defined by $(\alpha_1 + \alpha_2)a = \alpha_1 a + \alpha_2 a$, is an R-module homomorphism $\alpha_1 + \alpha_2: A \to B$.

If $\alpha: A \to B$ and $\beta: B \to C$ are R-module homomorphisms, the *composite* function $\beta\alpha$ is also an R-module homomorphism $\beta\alpha: A \to C$; but note that this composite is defined only when Range α = Domain β. The composition of homomorphisms is associative when defined. A (two-sided) *inverse* of $\alpha: A \to B$ is a homomorphism $\alpha^{-1}: B \to A$ such that both $\alpha\alpha^{-1} = 1_B$ and $\alpha^{-1}\alpha = 1_A$. Moreover, α has an inverse if and only if it is an isomorphism, and the inverse is then unique. We write $\alpha: A \cong B$ when α is an isomorphism. A *left inverse* of α is any homomorphism γ: $B \to A$ with $\gamma\alpha = 1_A$; it need not exist or be unique.

A pair of homomorphisms (α, β) with Range α = Domain $\beta = B$,

$$A \xrightarrow{\alpha} B \xrightarrow{\beta} C,$$

is *exact* at B if $\operatorname{Ker} \beta = \operatorname{Im} \alpha$. A longer sequence of homomorphisms:

$$A_1 \xrightarrow{\alpha_1} A_2 \xrightarrow{\alpha_2} A_3 \to \cdots \to A_{n-1} \xrightarrow{\alpha_{n-1}} A_n$$

is said to be *exact* if (α_{i-1}, α_i) is exact at A_i, for each $i = 2, \ldots, n-1$.

For each submodule $T \subset B$ the injection is a monomorphism $j: T \to B$. For each $b \in B$ the set $b + T$ of all sums $b + t$ with $t \in T$ is a *coset* of T in B; two cosets $b_1 + T$ and $b_2 + T$ are either disjoint or equal (the latter when $b_1 - b_2 \in T$). Recall that the *quotient group* (factor group or difference group) B/T has as its elements the cosets of T in B, with $(b_1 + T) + (b_2 + T) = (b_1 + b_2) + T$ as addition. Since T is a submodule, the abelian group B/T becomes an R-module when the product of any $r \in R$ with a coset is defined by $r(b + T) = rb + T$; we call B/T a *quotient module*. The function η which sends each element $b \in B$ into its coset $\eta b = b + T$ is an epimorphism $\eta: B \twoheadrightarrow B/T$, called the *canonical* map or *projection* of B on B/T.

Proposition 2.1. *If $\beta: B \to B'$ with $T \subset \operatorname{Ker} \beta$, there is a unique module homomorphism $\beta': B/T \to B'$ with $\beta'\eta = \beta$; that is, the diagram*

$$\begin{array}{ccc} B & \xrightarrow{\eta} & B/T \\ & {\scriptstyle\beta}\searrow & \vdots{\scriptstyle\beta'} \\ & & B' \end{array}, \qquad \beta(T) = 0,$$

can be "filled in" by a unique β' so as to be commutative ($\beta'\eta = \beta$).

Proof. Set $\beta'(b + T) = \beta b$; since $T \subset \operatorname{Ker} \beta$, this is well defined. In particular, if $\beta: B \to B'$ is an epimorphism with kernel T, $\beta': B/T \cong B'$.

This result may be worded: Each β with $\beta(T) = 0$ *factors uniquely through* the projection η. This property characterizes $\eta: B \to B/T$ up to an isomorphism of B/T, in the following sense:

Proposition 2.2. *If $T \subset B$ and $\zeta: B \to D$ is such that $\zeta(T) = 0$ and each $\beta: B \to B'$ with $\beta(T) = 0$ factors uniquely through ζ, there is an isomorphism $\theta: B/T \cong D$ with $\zeta = \theta\eta$.*

Proof. Factor ζ through η and η through ζ, so $\zeta=\zeta'\eta$, $\eta=\eta'\zeta$. Hence $\zeta=(\zeta'\eta')\,\zeta=1\zeta$. But ζ factors *uniquely* through ζ, so $\zeta'\eta'=1$. Symmetrically, $\eta'\zeta'=1$. Hence $\eta'=(\zeta')^{-1}$ and ζ' is the desired isomorphism θ.

For any $T\subset B$ the injection j and the projection η yield an exact sequence.

$$0\to T\xrightarrow{j}B\xrightarrow{\eta}B/T\to 0.$$

Conversely, let

$$(\varkappa,\sigma): 0\to A\xrightarrow{\varkappa}B\xrightarrow{\sigma}C\to 0$$

be any *short exact sequence*; that is, an exact sequence of five R-modules with the two outside modules zero (and hence the two outside maps trivial). Exactness at A means that $\varkappa$ is a monomorphism, at B means that $\varkappa A=\operatorname{Ker}\sigma$, at C that σ is an epimorphism. Thus the short exact sequence may be written as $A\rightarrowtail B\twoheadrightarrow C$, with exactness at B. Now $\varkappa$ induces an isomorphism $\varkappa': A\cong\varkappa A$ and σ an isomorphism $\sigma': B/\varkappa A\cong C$; together these provide an isomorphism of short exact sequences, in the form of a commutative diagram

$$\begin{array}{ccccccccc} 0 & \longrightarrow & A & \xrightarrow{\varkappa} & B & \xrightarrow{\sigma} & C & \longrightarrow & 0\\ & & \downarrow{\scriptstyle \varkappa'} & & \| & & \downarrow{\scriptstyle (\sigma')^{-1}} & & \\ 0 & \to & \varkappa A & \xrightarrow{j} & B & \to & B/\varkappa A & \to & 0. \end{array} \tag{2.1}$$

In brief, a short exact sequence is but another name for a submodule and its quotient.

Each homomorphism $\alpha: A\to B$ determines two quotient modules

$$\operatorname{Coim}\alpha=A/\operatorname{Ker}\alpha,\qquad \operatorname{Coker}\alpha=B/\operatorname{Im}\alpha,$$

called the *coimage* and the *cokernel* of α. This definition gives two short exact sequences

$$\operatorname{Ker}\alpha\rightarrowtail A\twoheadrightarrow\operatorname{Coim}\alpha,\qquad \operatorname{Im}\alpha\rightarrowtail B\twoheadrightarrow\operatorname{Coker}\alpha, \tag{2.2}$$

an isomorphism $\operatorname{Coim}\alpha\cong\operatorname{Im}\alpha$, and a longer exact sequence

$$0\to\operatorname{Ker}\alpha\xrightarrow{j}A\xrightarrow{\alpha}B\xrightarrow{\eta}\operatorname{Coker}\alpha\to 0. \tag{2.3}$$

By Prop. 2.1, $\beta\alpha=0$ implies that β factors uniquely through η as $\beta=\beta'\eta$. Dually, if some $\gamma: A'\to A$ has $\alpha\gamma=0$, then γ factors through j as $\gamma=j\gamma'$ for a unique $\gamma': A'\to\operatorname{Ker}\alpha$. This property characterizes $j:\operatorname{Ker}\alpha\to A$ up to an isomorphism of $\operatorname{Ker}\alpha$. Observe the dual statements: α is a monomorphism if and only if $\operatorname{Ker}\alpha=0$, and is an epimorphism if and only if $\operatorname{Coker}\alpha=0$. This duality will be discussed in § 8.

If $\alpha: A\to B$ and $S\subset A$, the set αS of all elements αs for $s\in S$ is a submodule of B called the *image* of S under α. Similarly, if $T\subset B$, the set $\alpha^{-1}T$ of all $s\in A$ with $\alpha s\in T$ is a submodule of A, called the (complete) *inverse image* of T. In particular, $\operatorname{Ker}\alpha = \alpha^{-1}0$, where 0 denotes the submodule of B consisting only of the zero element.

For $K \subset S \subset A$ the module S/K is called a *subquotient* of A; it is a quotient module of the submodule S of A, and simultaneously a submodule of the quotient module A/K. Furthermore, if $K \subset K' \subset S' \subset S \subset A$, then K'/K is a submodule of S'/K and the composite projection $S' \to S'/K \to (S'/K)/(K'/K)$ has kernel K', hence the familiar isomorphism $(S'/K)/(K'/K) \cong S'/K'$. This allows us to write each subquotient $(S'/K)/(K'/K)$ of a subquotient S/K directly as a subquotient of A.

Let S/K be a subquotient of A, S'/K' one of A'. If $\alpha: A \to A'$ has $\alpha S \subset S'$ and $\alpha K \subset K'$, then $\alpha s + K'$ is a coset of S'/K' uniquely determined by the coset $s + K$ of S/K. Hence $\alpha_*(s+K) = \alpha s + K'$ defines a homomorphism

$$\alpha_*: S/K \to S'/K' \qquad (\alpha S \subset S',\ \alpha K \subset K') \tag{2.4}$$

called the homomorphism *induced* by α on the given subquotients.

If S and T are submodules of A, their *intersection* $S \cap T$ (as sets) is also a submodule, as is their *union* $S \cup T$, consisting of all sums $s + t$ for $s \in S$, $t \in T$. The *Noether isomorphism theorem* asserts that 1_A induces an isomorphism

$$1_*: S/(S \cap T) \cong (S \cup T)/T. \tag{2.5}$$

3. Diagrams

The diagram of R-modules and homomorphisms

$$\begin{array}{ccccccccc} 0 & \to & A & \xrightarrow{\varkappa} & B & \xrightarrow{\sigma} & C & \to & 0 \\ & & \downarrow{\scriptstyle\alpha} & & \downarrow{\scriptstyle\beta} & & \downarrow{\scriptstyle\gamma} & & \\ 0 & \to & A' & \xrightarrow{\varkappa'} & B' & \xrightarrow{\sigma'} & C' & \to & 0 \end{array} \tag{3.1}$$

is said to be *commutative* if $\varkappa'\alpha = \beta\varkappa: A \to B'$ (left square commutative!) and $\sigma'\beta = \gamma\sigma: B \to C'$ (right square commutative!). In general, a diagram of homomorphisms is commutative if any two paths along directed arrows from one module to another module yield the same composite homomorphism.

Lemma 3.1. *(The Short Five Lemma.) If the commutative diagram* (3.1) *of R-modules has both rows exact, then*

(i) *If α and γ are isomorphisms, so is β;*

(ii) *If α and γ are monomorphisms, so is β;*

(iii) *If α and γ are epimorphisms, so is β.*

The same conclusions hold for a diagram of (not necessarily abelian) groups.

Proof. Clearly (ii) and (iii) together yield (i). To prove (ii), take $b \in \operatorname{Ker} \beta$. The right square is commutative, so $\gamma\sigma b = \sigma'\beta b = 0$; as γ is a monomorphism, this means that $\sigma b = 0$. Since the top row is exact, there is an element a with $\varkappa a = b$. Now the left square is commutative, so $\varkappa'\alpha a = \beta\varkappa a = \beta b = 0$. But the bottom row is exact at A', so $\alpha a = 0$. Since α is a monomorphism, $a = 0$, and hence $b = \varkappa a = 0$. This proves β a monomorphism.

To prove (iii), consider any b' in B'. Since γ is an epimorphism there is a $c \in C$ with $\gamma c = \sigma' b'$; since the top row is exact, there is a $b \in B$ with $\sigma b = c$. Then $\sigma'(\beta b - b') = 0$ in C'. The exactness of the bottom row yields an $a' \in A'$ with $\varkappa' a' = \beta b - b'$. Since α is an epimorphism, there is an $a \in A$ with $\alpha a = a'$ and hence with $\beta\varkappa a = \varkappa'\alpha a = \beta b - b'$. Then $b' = \beta(b - \varkappa a)$ is in the image of β, q.e.d.

This type of proof is called "diagram chasing". Inspection shows that the chase succeeds just as well if the groups are non-abelian (multiplicative) groups.

By the same method, the reader should verify the following more general results (as formulated by J. Leicht):

Lemma 3.2. *(The Strong Four Lemma.) Let a commutative diagram*

$$\begin{array}{ccccccc} \cdot & \to & \cdot & \xrightarrow{\xi} & \cdot & \to & \cdot \\ \downarrow{\scriptstyle\tau} & & \downarrow{\scriptstyle\alpha} & & \downarrow{\scriptstyle\beta} & & \downarrow{\scriptstyle\nu} \\ \cdot & \to & \cdot & \xrightarrow{\eta} & \cdot & \to & \cdot \end{array} \tag{3.2}$$

have exact rows, τ an epimorphism, and ν a monomorphism. Then

$$\operatorname{Ker} \beta = \xi(\operatorname{Ker} \alpha), \qquad \operatorname{Im} \alpha = \eta^{-1}(\operatorname{Im} \beta).$$

Here the dots in the diagram stand for modules or for not necessarily abelian groups.

A simpler version (the Weak Four Lemma) states, for the same commutative diagram with exact rows, that β is a monomorphism if α and ν are monomorphisms and τ an epimorphism, while α is an epimorphism if τ and β are epimorphisms and ν a monomorphism. A more frequently used consequence is

Lemma 3.3. *(The Five Lemma.) Let a commutative diagram*

$$\begin{array}{ccccccccc} \cdot & \to & \cdot & \to & \cdot & \to & \cdot & \to & \cdot \\ \downarrow{\scriptstyle\alpha_1} & & \downarrow{\scriptstyle\alpha_2} & & \downarrow{\scriptstyle\alpha_3} & & \downarrow{\scriptstyle\alpha_4} & & \downarrow{\scriptstyle\alpha_5} \\ \cdot & \to & \cdot & \to & \cdot & \to & \cdot & \to & \cdot \end{array} \tag{3.3}$$

have exact rows. If $\alpha_1, \alpha_2, \alpha_4, \alpha_5$ are isomorphims, so is α_3. In more detail,

(i) *If α_1 is an epimorphism and α_2 and α_4 monomorphisms, then α_3 is a monomorphism,*

(ii) *If α_5 is a monomorphism and α_2 and α_4 epimorphisms, then α_3 is an epimorphism.*

Proof. Chase the diagram, or apply Lemma 3.2 twice to the left-hand and right-hand portions.

4. Direct Sums

The *external direct sum* $A_1 \oplus A_2$ of two R-modules A_1 and A_2 is the R-module consisting of all ordered pairs (a_1, a_2), for $a_i \in A_i$, with module operations defined by

$$(a_1, a_2) + (a_1', a_2') = (a_1 + a_1', a_2 + a_2'), \qquad r(a_1, a_2) = (r a_1, r a_2).$$

The functions ι and π defined by $\iota_1 a_1 = (a_1, 0)$, $\iota_2 a_2 = (0, a_2)$, $\pi_1(a_1, a_2) = a_1$, $\pi_2(a_1, a_2) = a_2$ are homomorphisms

$$A_1 \underset{\pi_1}{\overset{\iota_1}{\rightleftarrows}} A_1 \oplus A_2 \underset{\pi_2}{\overset{\iota_2}{\leftrightarrows}} A_2 \tag{4.1}$$

which satisfy the identities

$$\left.\begin{array}{ll} \pi_1 \iota_1 = 1_{A_1}, & \pi_1 \iota_2 = 0, \\ \pi_2 \iota_1 = 0, & \pi_2 \iota_2 = 1_{A_2}, \\ \iota_1 \pi_1 + \iota_2 \pi_2 = 1_{A_1 \oplus A_2}. & \end{array}\right\} \tag{4.2}$$

Call ι_1 and ι_2 the *injections* and π_1, π_2 the *projections* of the direct sum. The diagram (4.1) contains partial diagrams, to wit:

Injective direct sum diagram: $A \xrightarrow{\iota_1} A_1 \oplus A_2 \xleftarrow{\iota_2} A_2$,
Projective direct sum diagram: $A_1 \xleftarrow{\pi_1} A_1 \oplus A_2 \xrightarrow{\pi_2} A_2$,
One-sided direct sum diagram: $A_1 \oplus A_2 \leftrightarrows A_2$,
Sequential direct sum diagram: $A_1 \xrightarrow{\iota_1} A_1 \oplus A_2 \xrightarrow{\pi_2} A_2$;

in particular, the last diagram is a short exact sequence. Instead of defining the direct sum via elements, we can characterize each of these diagrams by conceptual properties. With a view to later generalizations (Chap. IX), our proofs of these properties will be so cast as to use only the diagram (4.1), the identities (4.2), and formal properties of the addition and composition of homomorphisms; in particular, the distributive laws $\beta(\alpha_1 + \alpha_2) = \beta\alpha_1 + \beta\alpha_2$ and $(\alpha_1 + \alpha_2)\gamma = \alpha_1\gamma + \alpha_2\gamma$.

Proposition 4.1. *For given modules A_1 and A_2 any diagram*

$$A_1 \underset{\pi_1'}{\overset{\iota_1'}{\rightleftarrows}} B \underset{\pi_2'}{\overset{\iota_2'}{\leftrightarrows}} A_2$$

of the form (4.1) *and satisfying the five identities like* (4.2) *is isomorphic to the direct sum diagram. In more detail, there is exactly one isomorphism*

$\theta: B \to A_1 \oplus A_2$ *such that*

$$\pi_j \theta = \pi_j', \qquad \theta \iota_j' = \iota_j, \qquad \text{for } j = 1, 2. \tag{4.3}$$

Proof. Define θ as $\theta = \iota_1 \pi_1' + \iota_2 \pi_2'$ and the analogue $\theta': A_1 \oplus A_2 \to B$ by $\theta' = \iota_1' \pi_1 + \iota_2' \pi_2$. The identities (4.2) show that θ' is a two-sided inverse for θ and thus that θ is an isomorphism; the properties (4.3) follow directly from (4.2). Also if θ satisfies (4.3), then $\theta = (\iota_1 \pi_1 + \iota_2 \pi_2)\theta = \iota_1 \pi_1' + \iota_2 \pi_2'$, so θ is indeed uniquely determined.

Next we characterize the one-sided direct sum diagram.

Proposition 4.2. *Any diagram* $A_2 \xrightarrow{\iota''} B \xrightarrow{\pi''} A_2$ *with* $\pi'' \iota'' = 1_{A_2}$ *is isomorphic to a "one-sided" direct sum diagram* $A_1 \oplus A_2 \leftrightarrows A_2$ *with* $A_1 = \operatorname{Ker} \pi''$.

The proof requires an isomorphism $\theta: B \to A_1 \oplus A_2$ with $\theta \iota'' = \iota_2$, $\pi_2 \theta = \pi''$. Define θ by $\theta b = (b - \iota'' \pi'' b, \pi'' b)$ and θ^{-1} by $\theta^{-1}(a_1, a_2) = a_1 + \iota'' a_2$.

To prove this without using elements, consider the diagram

$$\operatorname{Ker} \pi'' \xrightarrow{\iota'} B \underset{\pi''}{\overset{\iota''}{\leftrightarrows}} A_2,$$

with ι' the injection. Since $\pi''(1_B - \iota'' \pi'') = 0$, $1_B - \iota'' \pi''$ factors through ι' as $1_B - \iota'' \pi'' = \iota' \pi'$ for some $\pi': B \to \operatorname{Ker} \pi''$. Now $\pi'' \iota' = 0$ and $\iota' \pi' \iota' = \iota'$ give $\pi' \iota' = 1$, so we have identities like (4.2) and can apply Prop. 4.1.

Now write the direct sum as a short exact sequence (ι_1, π_2). Here ι_2 is a right inverse of π_2, while $\pi_1 \iota_1 = 1$ shows π_1 a left inverse of ι_1.

Proposition 4.3. *The following properties of a short exact sequence* $(\iota', \pi''): A_1 \rightarrowtail B \twoheadrightarrow A_2$ *are equivalent:*

(i) π'' *has a right inverse* $\iota'': A_2 \to B$, *with* $\pi'' \iota'' = 1$;
(ii) ι' *has a left inverse* $\pi': B \to A_1$, *with* $\pi' \iota' = 1$;
(iii) *The sequence is isomorphic (with identities on* A_1 *and* A_2*) to*

$$0 \to A_1 \xrightarrow{\iota_1} A_1 \oplus A_2 \xrightarrow{\pi_2} A_2 \to 0.$$

A short exact sequence with one (and hence all) of these properties is said to *split* (some authors say instead that the sequence is inessential).

Proof. We just observed that (iii) implies (i) and (ii). Conversely, exactness shows that ι' gives the isomorphism $A_1 \cong \operatorname{Ker} \pi''$, so (i) implies (iii) by Prop. 4.2. Similarly, (ii) implies (iii).

Now consider pairs of *coterminal* homomorphisms α_1, α_2, as in the diagram

$$D: A_1 \xrightarrow{\alpha_1} B \xleftarrow{\alpha_2} A_2. \tag{4.4}$$

Such a diagram is said to be *universal* with ends A_1 and A_2 if to every diagram $D': A_1 \to B' \leftarrow A_2$ with the same ends there exists a unique homomorphism of D to D' which is the identity on each A_j. In other words, D is universal if to each rectangular diagram

$$\begin{array}{ccccc} A_1 & \xrightarrow{\alpha_1} & B & \xleftarrow{\alpha_2} & A_2 \\ \| & & \downarrow{\scriptstyle\beta} & & \| \\ A_1 & \xrightarrow{\alpha_1'} & B' & \xleftarrow{\alpha_2'} & A_2, \end{array} \tag{4.5}$$

with D as first row and end maps the identities, there is a unique way of inserting the middle dotted arrow so that the whole diagram becomes commutative ($\beta\alpha_1=\alpha_1'$, $\beta\alpha_2=\alpha_2'$).

Proposition 4.4. *The (injective) direct sum diagram $A_1 \to A_1 \oplus A_2 \leftarrow A_2$ is universal with ends A_1 and A_2. Conversely, any diagram* (4.4) *which is universal with ends A_j is isomorphic to this direct sum diagram (with identities on A_1 and A_2).*

Proof. To show $A_1 \oplus A_2$ universal, define the homomorphism β needed for (4.5) as $\beta(a_1, a_2)=\alpha_1' a_1+\alpha_2' a_2$; that is, as $\beta=\alpha_1'\pi_1+\alpha_2'\pi_2$; this is the only choice for β. To prove the converse, it will suffice to show that any two diagrams universal with ends A_1 and A_2 are isomorphic (with identities on A_j). Suppose then that both rows in (4.5) are universal. Since the top row is universal, there is a $\beta: B \to B'$ with $\beta\alpha_j=\alpha_j'$; since the bottom row is universal, there is a β': $B' \to B$ with $\beta'\alpha_j'=\alpha_j$. Then $(\beta'\beta)\alpha_j=\alpha_j$, for $j=1, 2$. Since also $1_B \alpha_j=\alpha_j$, the uniqueness property for the top row gives $\beta'\beta=1_B$. Similarly the uniqueness for the bottom row gives $1_{B'}=\beta\beta'$. Hence β and β' are mutually inverse isomorphisms, q.e.d.

Since the universal diagram is unique up to an isomorphism, it follows that the maps α_j in any universal diagram with ends A_1 and A_2 are always monomorphisms, since they are such for the external direct sum diagram.

Notice that the proof of the converse part of the proposition did not use elements of the modules, but only formal arguments with homomorphisms. This proof is thus valid in any category, in the sense soon (§ 7) to be explained.

Dually, a pair of *coinitial* maps forming a diagram $D: A_1 \leftarrow C \to A_2$ is *couniversal* with ends A_1 and A_2 if to each rectangular diagram

$$\begin{array}{ccccc} A_1 & \xleftarrow{\gamma_1} & C & \xrightarrow{\gamma_2} & A_2 \\ \| & & \uparrow & & \| \\ A_1 & \xleftarrow{\gamma_1'} & C' & \xrightarrow{\gamma_2'} & A_2, \end{array} \tag{4.6}$$

with D as first row and with vertical maps 1 on each A_j, there is a unique way of inserting the middle dotted arrow to make the diagram commutative. The reader should prove

Proposition 4.5. *The (projective) direct sum diagram*

$$A_1 \xleftarrow{\pi_1} A_1 \oplus A_2 \xrightarrow{\pi_2} A_2$$

is couniversal with ends A_1 and A_2. Conversely, any diagram couniversal with ends A_1 and A_2 is isomorphic (identities on each A_j) to this diagram.

Direct sums of more than two modules work similarly. For example, in a direct sum $A_1 \oplus A_2 \oplus A_3$ an element may be regarded as an ordered triple (a_1, a_2, a_3) or as a function a on the set $\{1, 2, 3\}$ of indices with $a(i) \in A_i$. In general, given a family of modules $\{A_t\}$ indexed by an arbitrary set T, the *cartesian product* $\prod_t A_t$ is the set of all those functions f on T to the union of the sets A_t for which $f(t) \in A_t$ for each t. Define the module operations "termwise"; that is, define the functions $f+f'$ and rf for $r \in R$ by

$$(f+f')(t) = f(t) + f'(t), \qquad (rf)(t) = r(f(t)), \qquad t \in T.$$

Then $\prod_t A_t$ is an R-module. The homomorphisms $\pi_t\colon \prod_t A_t \to A$ defined by $\pi_t f = f(t)$ are called the *projections* of the cartesian product.

For given A_t, let $\{\gamma_t\colon B \to A_t\}$ be a diagram with one additional module B and one homomorphism γ_t for each $t \in T$. This diagram is *couniversal* with ends A_t if to each diagram $\{\gamma'_t\colon B' \to A_t \mid t \in T\}$ there exists a unique $\beta\colon B' \to B$ such that $\gamma'_t = \gamma_t \beta$ for all t. The projections of the cartesian product $\prod_t A_t$ yield such a couniversal diagram, and any two such diagrams are isomorphic, as before.

The *external direct sum* $\sum_t A_t$ of the same modules A_t is that submodule of $\prod_t A_t$ which consists of all those functions f with but a finite number of non-zero values. The homomorphisms $\iota_t\colon A_t \to \sum_t A_t$ are defined for each $a \in A_t$ by letting $\iota_t(a)$ be the function on T with $[\iota_t(a)](t) = a$, $[\iota_t(a)](s) = 0$ for $s \neq t$. These homomorphisms are called the *injections* of the direct sum. As in the case of two summands, the diagram $\{\iota_t\colon A_t \to \sum_t A_t\}$ is universal for given ends A_t, and is determined up to isomorphism by this fact.

For a *finite* number of summands the external direct sum is identical with the cartesian product. This implies that any finite universal diagram $\alpha_j\colon A_j \to B$, for $j = 1, \ldots, n$, yields a couniversal diagram $\{\gamma_j\colon B \to A_j\}$. More explicitly, each γ_j is that map which is uniquely determined (since B is universal) by the conditions $\gamma_j \alpha_j = 1_{A_j}$, $\gamma_j \alpha_k = 0$ for $j \neq k$. Dually, the reader should obtain a universal diagram from the couniversal one.

Direct sums may be treated in terms of submodules. If S_t is any family of submodules of B indexed by a set T, their *union* $\bigcup S_t$ is the set of all finite sums $s_1 + \cdots + s_n$ with each s_j in some S_t; it is a submodule of B containing all the S_t and contained in any submodule which contains all the S_t. Their *intersection* $\bigcap S_t$ is the intersection of the sets S_t;

it is a submodule of B contained in all the S_t which contains every submodule contained in every S_t. We also write $S_1 \cup S_2$ or $S_1 \cap S_2$ for the union or intersection of two submodules S_1, S_2.

Proposition 4.6. *For submodules $S_t \subset B$, $t \in T$, the following conditions are equivalent:*

(i) *The diagram $\{j_t: S_t \to B\}$, j_t the injection, is universal for ends S_t,*

(ii) $B = \cup S_t$ *and, for each $t_0 \in T$,* $S_{t_0} \cap (\bigcup_{t \neq t_0} S_t) = 0$.

Proof. Given (i), B is isomorphic to $\sum S_t$, which satisfies (ii). Conversely, given (ii), the condition $B = \cup S_t$ states that each $b \neq 0$ can be written as a finite sum $b = s_1 + \cdots + s_n$ of elements $s_i \neq 0$ belonging to different submodules S_{t_i}, $i = 1, \ldots, n$; the second condition of (ii) states that this representation is unique. For any other diagram $\{\alpha_t: S_t \to B'\}$ the homomorphism $\beta: B \to B'$ defined by $\beta(s_1 + \cdots + s_n) = \alpha_{t_1} s_1 + \cdots + \alpha_{t_n} s_n$ is the unique homomorphism with $\beta j_t = \alpha_t$; hence the universality.

When these conditions hold, B is called the ***internal direct sum*** of its submodules S_t. Therefore an internal direct sum is isomorphic to the external direct sum $\sum S_t$. In particular, B is the internal direct sum of two submodules S_1 and S_2 if and only if $S_1 \cap S_2 = 0$ and $S_1 \cup S_2 = B$; these conditions imply $B \cong S_1 \oplus S_2$.

Exercises

1. Show that a diagram (4.1) with $\pi_1 \iota_1 = 1$, $\pi_2 \iota_2 = 1$, $\pi_1 \iota_2 = 0$, and (ι_1, π_2) exact is a direct sum diagram.

2. If $\alpha: A \to A$ satisfies $\alpha^2 = \alpha$, then A is the direct sum of $\operatorname{Ker} \alpha$ and $\operatorname{Im} \alpha$.

3. Show that the diagram $\{\alpha_t: A_t \to B, t \in T\}$ is universal for given ends A_t if and only if (i) B is the union of its submodules $\alpha_t A_t$; (ii) there are homomorphisms $\pi_t: B \to A_t$ for $t \in T$ with $\pi_t \alpha_t = 1$ and $\pi_s \alpha_t = 0$ for $s \neq t$.

4. State and prove the dual of Prop. 4.6. (The dual of a submodule is a quotient module.)

5. If $\alpha_{ij}: A_i \to A'_j$ for $i, j = 1, 2$, show that there is a unique $\omega: A_1 \oplus A_2 \to A'_1 \oplus A'_2$ with $\pi'_j \omega \iota_i = \alpha_{ij}$ for $i, j = 1, 2$.

5. Free and Projective Modules

The ring R, as a left R-module, has the following characteristic property. If a is any element of an R-module A, there is a unique R-module homomorphism $\mu_a: R \to A$ with $\mu_a(1) = a$; namely, the function μ_a with $\mu_a(r) = ra$.

A free left R-module is any direct sum of isomorphic copies of the left R-module R. In view of the above property of R, we can say more explicitly that the left R-module F is *free on a subset* T of its elements if the homomorphisms $\mu_t: R \to F$ with $\mu_t(r) = rt$ form a universal diagram with ends R (one for each t). As each homomorphism $\nu: R \to A$ is uniquely determined by $\nu(1) \in A$, this universal property can be restated as follows.

Proposition 5.1. *The module F is free on a subset $T \subset F$ if and only if to each module A and each set function g on T to A there is a unique module homomorphism $\mu: F \to A$ with $\mu(t) = g(t)$ for every t.*

The isomorphism of internal and external direct sums gives.

Proposition 5.2. *The module F is free on a subset $T \subset F$ if and only if each element of F can be represented uniquely as a sum $\sum r_t t$ with coefficients $r_t \in R$ which are almost all zero (i. e., all but a finite number are zero).*

A module F free on T is determined up to isomorphism by T. Given R and any set T, we may construct an R-module free on T as $F = \sum_t Rt$, where Rt is the set of all rt for $r \in R$, with the obvious module structure.

The left module A is *generated* by a subset U of its elements if A is the only submodule of A containing all $u \in U$; that is, if every element of A can be written as a finite sum $\sum r_i u_i$ with each $r_i \in R$. A module free on T is generated by T.

Proposition 5.3. *Every R-module is isomorphic to a quotient of a free module.*

Proof. Given the module A, take a subset U generating A (e. g., take $U = A$). Form a free module F on U and the map $\mu: F \to A$ with $\mu(u) = u \in A$. Since U generates A, μ is an epimorphism, so $A \cong F/(\operatorname{Ker} \mu)$.

A module A is *finitely generated* (or, of *finite type*) if it is generated by a finite subset; that is, if it is isomorphic to a quotient module of a finite direct sum $R \oplus \cdots \oplus R$. A module C is *cyclic* (or monogenic) if it is generated by one element; then $C \cong R/L$, where L is a submodule of R (i. e., L is a left ideal in R). The main theorem of elementary divisor theory asserts that if R is a commutative integral domain in which every ideal is principal (i. e., monogenic), then any finitely generated R-module is isomorphic to a direct sum of cyclic modules. In particular ($R = Z$) any finitely generated abelian group is a direct sum of cyclic groups.

A module P is called *projective* if in each diagram

$$\begin{array}{ccc} & & P \\ & \swarrow{\scriptstyle\cdots} & \downarrow{\scriptstyle\gamma} \\ B & \xrightarrow[\sigma]{} & C \end{array} \tag{5.1}$$

with σ an epimorphism, the dotted arrow can be filled in to make the diagram commutative. In other words, given an epimorphism $\sigma: B \twoheadrightarrow C$, each map $\gamma: P \to C$ can be *lifted* to a $\beta: P \to B$ such that $\sigma\beta = \gamma$.

Lemma 5.4. *Every free module is projective.*

Proof. Let F be free on generators t. Since $\sigma B = C$, we can choose for each t an element $b_t \in B$ such that $\sigma b_t = \gamma t$. Then the unique $\beta: F \to B$ with $\beta t = b_t$ for each t lifts γ, as desired.

Projective modules will be repeatedly used. Note that a projective module need not be free. For example, take $R = Z \oplus Z$, the direct sum of the ring Z of integers with itself (with product $(m, n)(m', n') = (mm', nn')$). Then the first summand Z, as submodule of an R-module, is an R-module. It is clearly not free, but is projective according to

Proposition 5.5. *An R-module P is projective if and only if it is a direct summand of a free R-module.*

Proof. Suppose first that $\pi: F = P \oplus Q \to P$ with F free. Given any diagram (5.1), $\gamma\pi: F \to C$ lifts to $\beta: F \to B$ with $\sigma\beta = \gamma\pi$. The injection $\iota: P \to P \oplus Q$ has $\sigma(\beta\iota) = \gamma\pi\iota = \gamma$, so γ lifts to $\beta\iota$ and P is therefore projective.

Conversely, if P is projective, Prop. 5.3 gives an epimorphism $\varrho: F \to P$ with F free. Lift $1_P: P \to P$ to $\beta: P \to F$ with $\varrho\beta = 1$. By Prop. 4.2, F is the direct sum of $\beta P \cong P$ and $\operatorname{Ker} \varrho$.

Any subgroup of a free abelian group is free; hence every projective Z-module is free.

Exercises

1. Show that a direct summand of a projective module is projective.
2. For m, n relatively prime, show Z_m projective (but not free) over the ring Z_{mn} of integers modulo mn.
3. Prove: Any direct sum of projective modules is projective.

6. The Functor Hom

Let A and B be R-modules. The set

$$\operatorname{Hom}_R(A, B) = \{f \mid f: A \to B\}$$

of all R-module homomorphisms f of A into B is an abelian group, under the addition defined for $f, g: A \to B$ by $(f+g)a = fa + ga$. If $A = B$, $\operatorname{Hom}_R(A, A)$ is a ring under addition and composition of homomorphisms; this ring is called the ring of *R-endomorphisms* of A. In case the ring R is commutative, $\operatorname{Hom}_R(A, B)$ may be regarded not just as a group but as an R-module, when $tf: A \to B$ is defined for $t \in R$ and $f: A \to B$ by $(tf)(a) = t(fa)$ for all $a \in A$. That tf is still an R-module homomorphism follows from the calculation

$$(tf)(ra) = t(fra) = tr(fa) = rt(fa) = r[(tf)a]$$

which uses the commutativity of R.

This group Hom occurs frequently. If R is a field, $\operatorname{Hom}_R(A, B)$ is the vector space of all linear transformations of the vector space A into the vector space B. If G is an abelian group, and P the additive group of real numbers, modulo 1, both G and P can be regarded as Z-modules, and $\mathrm{Hom}_Z(G, P)$ is the character group of G. If $\varphi: R \to \mathrm{Hom}_Z(G, G)$ is any ring homomorphism, then the abelian group G becomes an R-module with left operators $rg = \varphi(r)g$. All left R-modules can be so obtained from such a G and φ.

Consider the effect of a fixed module homomorphism $\beta: B \to B'$ on $\mathrm{Hom}_R(A, B)$. Each $f: A \to B$ determines a composite $\beta f: A \to B'$, and $\beta(f+g) = \beta f + \beta g$. Hence the correspondence $f \to \beta f$ is a homomorphism

$$\beta_*: \mathrm{Hom}_R(A, B) \to \mathrm{Hom}_R(A, B') \tag{6.1}$$

of abelian groups, called the homomorphism "induced" by β. Explicitly, $\beta_* f = \beta \circ f$. If β is an identity, so is β_*; if β is a composite, so is β_*; in detail

$$(1_B)_* = 1_{\mathrm{Hom}(A,B)}, \qquad (\beta\beta')_* = \beta_* \beta'_*, \tag{6.2}$$

the latter whenever the composite $\beta\beta'$ is defined. We summarize (6.1) and (6.2) by the phrase: $\mathrm{Hom}_R(A, B)$ is a "covariant functor" of B (general definition in § 8).

For the first argument A a reverse in direction occurs. For a fixed module homomorphism $\alpha: A \to A'$, each $f': A' \to B$ determines a composite $f'\alpha: A \to B$ with $(f'+g')\alpha = f'\alpha + g'\alpha$. Hence $\alpha^* f' = f'\alpha$ defines an "induced" homomorphism

$$\alpha^*: \mathrm{Hom}_R(A', B) \to \mathrm{Hom}_R(A, B) \tag{6.3}$$

of abelian groups. Again $(1_A)^*$ is an identity map. If $\alpha: A \to A'$ and $\alpha': A' \to A''$, the composite $\alpha'\alpha$ is defined, and the induced maps are

$$\mathrm{Hom}_R(A'', B) \xrightarrow{\alpha'^*} \mathrm{Hom}_R(A', B) \xrightarrow{\alpha^*} \mathrm{Hom}_R(A, B);$$

one shows that $\alpha^*\alpha'^* = (\alpha'\alpha)^*$. This reversal of order generalizes the fact that the transpose of the product of two matrices is the product of their transposes in *opposite* order. Because of this reversal we shall say that $\mathrm{Hom}_R(A, B)$, for B fixed, is a *contravariant* functor of A.

Now vary both A and B. Given $\alpha: A \to A'$ and $\beta: B \to B'$, each $f: A' \to B$ determines a composite $\beta f \alpha: A \to B'$; the correspondence $f \to \beta f \alpha$ is a homomorphism

$$\mathrm{Hom}(\alpha, \beta): \mathrm{Hom}(A', B) \to \mathrm{Hom}(A, B')$$

of abelian groups, with $\alpha^*\beta_* = \mathrm{Hom}(\alpha, \beta) = \beta_*\alpha^*$. It has the properties

$$\mathrm{Hom}(1, 1') = \text{the identity},$$

$$\mathrm{Hom}(\alpha\alpha', \beta\beta') = \mathrm{Hom}(\alpha', \beta)\,\mathrm{Hom}(\alpha, \beta'),$$

whenever the composites $\alpha\alpha'$ and $\beta\beta'$ are defined. We say that Hom is a functor in two variables, contravariant in the first and covariant in the second, from R-modules to groups.

If $\alpha_1, \alpha_2 : A \to A'$ are two homomorphisms one shows that

$$\mathrm{Hom}(\alpha_1+\alpha_2, \beta) = \mathrm{Hom}(\alpha_1, \beta) + \mathrm{Hom}(\alpha_2, \beta). \tag{6.4}$$

Similarly, $\mathrm{Hom}(\alpha, \beta_1+\beta_2) = \mathrm{Hom}(\alpha, \beta_1) + \mathrm{Hom}(\alpha, \beta_2)$. These two properties state that Hom is an "additive" functor.

For B fixed, apply $\mathrm{Hom}(-, B)$ to a direct sum diagram (4.1). The result

$$\mathrm{Hom}(A_1, B) \underset{\pi_1^*}{\overset{\iota_1^*}{\leftrightarrows}} \mathrm{Hom}(A_1 \oplus A_2, B) \underset{\pi_2^*}{\overset{\iota_2^*}{\rightleftarrows}} \mathrm{Hom}(A_2, B)$$

changes injections ι_j to projections ι_j^*, but by (6.4) still satisfies the identities (4.2) for a direct sum diagram. Similarly, for A fixed, a direct sum diagram on modules B_1 and B_2 is carried by $\mathrm{Hom}(A, -)$ to a direct sum diagram (injections to injections). Thus

$$\left.\begin{aligned} \mathrm{Hom}(A_1 \oplus A_2, B) &\cong \mathrm{Hom}(A_1, B) \oplus \mathrm{Hom}(A_2, B), \\ \mathrm{Hom}(A, B_1 \oplus B_2) &\cong \mathrm{Hom}(A, B_1) \oplus \mathrm{Hom}(A, B_2). \end{aligned}\right\} \tag{6.5}$$

In particular, $\mathrm{Hom}(A, B_1) \rightarrowtail \mathrm{Hom}(A, B_1 \oplus B_2) \twoheadrightarrow \mathrm{Hom}(A, B_2)$ is exact.

Theorem 6.1. *For any module D and any sequence $0 \to A \xrightarrow{\varkappa} B \xrightarrow{\beta} L$ exact at A and B the induced sequence*

$$0 \to \mathrm{Hom}_R(D, A) \xrightarrow{\varkappa_*} \mathrm{Hom}_R(D, B) \xrightarrow{\beta_*} \mathrm{Hom}_R(D, L) \tag{6.6}$$

of abelian groups is exact.

Proof. To show $\varkappa_*$ a monomorphism, consider any $f: D \to A$ with $\varkappa_* f = 0$. For each $d \in D$, $\varkappa_* f d = \varkappa f d = 0$; since $\varkappa$ is a monomorphism, each $fd = 0$, so $f = 0$, and therefore $\varkappa_*$ is a monomorphism. Clearly, $\beta_* \varkappa_* = (\beta\varkappa)_* = 0_* = 0$, so $\mathrm{Im}\, \varkappa_* \subset \mathrm{Ker}\, \beta_*$. For the converse inclusion, consider $g: D \to B$ with $\beta_* g = 0$. Then $\beta g d = 0$ for each d. But $\mathrm{Ker}\beta = \varkappa A$, by the given exactness, so there is a unique a in A with $\varkappa a = gd$. Then $fd = a$ defines a homomorphism $f: D \to A$ with $\varkappa_* f = g$. Thus $\mathrm{Im}\, \varkappa_* \supset \mathrm{Ker}\beta_*$, which completes the proof of exactness.

By a corresponding argument, the reader should prove

Theorem 6.2. *If $M \xrightarrow{\alpha} B \xrightarrow{\sigma} C \to 0$ is exact, and D is any module, the induced sequence*

$$0 \to \mathrm{Hom}_R(C, D) \xrightarrow{\sigma^*} \mathrm{Hom}_R(B, D) \xrightarrow{\alpha^*} \mathrm{Hom}_R(M, D) \tag{6.7}$$

is exact.

A sequence $M \to B \to C \to 0$ exact at B and C is called a short *right exact* sequence. This theorem states that the functor $\mathrm{Hom}_R(-, D)$ for fixed D turns each short right exact sequence into a short left exact sequence; by the previous theorem, $\mathrm{Hom}_R(D, -)$ carries a short left exact sequence into a short left exact sequence. If $A \rightarrowtail B \overset{\sigma}{\twoheadrightarrow} C$ is a short exact sequence, we wish to have exact sequences

$$0 \to \operatorname{Hom}_R(D, A) \to \operatorname{Hom}_R(D, B) \xrightarrow{\sigma_*} \operatorname{Hom}_R(D, C) \to ?, \tag{6.6'}$$

$$0 \to \operatorname{Hom}_R(C, D) \to \operatorname{Hom}_R(B, D) \to \operatorname{Hom}_R(A, D) \to ?. \tag{6.7'}$$

By the two theorems above, each is exact except perhaps at the right end. With 0 for ? on the right, these would not usually be exact. For example, exactness of (6.6′) at $\operatorname{Hom}_R(D, C)$ would assert that each $h: D \to C$ has the form $h = \sigma h'$ for some $h': D \to B$; i.e., that each map h into the quotient $C = B/\varkappa A$ could be lifted to a map h' into B (as would be possible were D projective). To see that this need not be so, take $R = Z$ and $D = Z_m$ the cyclic group of order m. For the short exact sequence $Z \rightarrowtail Z \twoheadrightarrow Z_m$, with first map $\varkappa$ the operation of multiplication by m, the sequence (6.6′) becomes $0 \to 0 \to 0 \to \operatorname{Hom}(Z_m, Z_m) \to 0$, and is manifestly inexact. Similarly, (6.7′) can be inexact with a zero at ?, since for $A \subset B$ a homomorphism $f: A \to D$ cannot in general be extended to one of B into D. It will be possible to describe an object which is the "obstruction" to the problem of extending such an f. The group of these objects, placed at "?" in (6.7′), will restore exactness. This construction, done for both (6.6′) and (6.7′), is one of the objectives of homological algebra.

We now can formulate several characterizations of projective modules.

Theorem 6.3. *The following properties of a module D are equivalent:*

(i) *D is projective,*

(ii) *For each epimorphism $\sigma: B \twoheadrightarrow C$, $\sigma_*: \operatorname{Hom}_R(D, B) \to \operatorname{Hom}_R(D, C)$ is an epimorphism,*

(iii) *If $A \rightarrowtail B \twoheadrightarrow C$ is a short exact sequence, so is $0 \to \operatorname{Hom}_R(D, A) \to \operatorname{Hom}_R(D, B) \to \operatorname{Hom}_R(D, C) \to 0$,*

(iv) *Every short exact sequence $A \rightarrowtail B \twoheadrightarrow D$ splits.*

Proof. In (ii) the statement that σ_* is an epimorphism means that each $\gamma: D \to C$ can be factored as $\gamma = \sigma\beta$; this is exactly the statement that D is projective. Given the exactness of (6.6), (ii) is equivalent to (iii). Finally, if D is projective and $\sigma: B \twoheadrightarrow D$, the map $1_D: D \to D$ lifts to a $\beta: D \to B$ with $\sigma\beta = 1$, so the sequence of (iv) splits. Conversely, if every such sequence ending in D splits, write D as an image $\varrho: F \twoheadrightarrow D$ of some free module F. Since the sequence $\operatorname{Ker}\varrho \rightarrowtail F \twoheadrightarrow D$ splits, D is a direct summand of F, by Prop. 4.2, hence is projective, by Prop. 5.5.

Exercises

1. Any left ideal L in the ring R is an R-module, and $L \rightarrowtail R \twoheadrightarrow R/L$ is exact. Suppose $L^2 \neq L$.

(i) The sequence (6.6′) need not be exact with zero for ? on the right. Show this for $D = R/L$ by proving that $\operatorname{Hom}_R(R/L, R) \to \operatorname{Hom}_R(R/L, R/L)$ is not an epimorphism (1 is not an image!).

(ii) The sequence (6.7′) need not be exact with zero for ? on the right. Show this for $D=L$ by proving that $\mathrm{Hom}_R(R, L) \to \mathrm{Hom}_R(L, L)$ is not an epimorphism (1 not an image!).

2. For any set T of indices establish an isomorphism

$$\mathrm{Hom}_R(\textstyle\sum_t A_t, B) \cong \prod_t \mathrm{Hom}_R(A_t, B)$$

by mapping each $f\colon \sum A_t \to B$ into the collection of its restrictions $f_t : A_t \to B$.

3. For any set T of indices establish an isomorphism

$$\mathrm{Hom}_R(A, \prod_t B_t) \cong \prod_t \mathrm{Hom}_R(A, B_t).$$

7. Categories

A category consists of "objects" and "morphisms" which may sometimes be "composed". Formally, a category $\mathscr{C}$ is a class of *objects* $A, B, C, \ldots$ together with

(i) A family of disjoint sets $\hom(A, B)$, one for each pair of objects;

(ii) For each triple of objects A, B, C a function which assigns to $\alpha \in \hom(A, B)$ and $\beta \in \hom(B, C)$ an element $\beta\alpha \in \hom(A, C)$;

(iii) A function which assigns to each object A an element $1_A \in \hom(A, A)$;

all subject to the two axioms:

Associativity: If $\alpha \in \hom(A, B)$, $\beta \in \hom(B, C)$, and $\gamma \in \hom(C, D)$, then $\gamma(\beta\alpha) = (\gamma\beta)\alpha$;

Identity: If $\alpha \in \hom(A, B)$, then $\alpha 1_A = \alpha = 1_B \alpha$.

Write $\alpha\colon A \to B$ for $\alpha \in \hom(A, B)$ and call α a *morphism* of $\mathscr{C}$ with *domain* A and *range* B. By (ii), the composite $\beta\alpha$ is defined if and only if range $\alpha =$ domain β; the triple composite $\gamma\beta\alpha$ is associative whenever it is defined. Call a morphism $\varkappa$ an *identity* of $\mathscr{C}$ if both $\varkappa\alpha = \alpha$ whenever $\varkappa\alpha$ is defined and $\beta\varkappa = \beta$ whenever $\beta\varkappa$ is defined. Each 1_A is an identity. Conversely, if $\varkappa$ is an identity, then $\varkappa\colon A \to A$ for some object A, and $\varkappa = \varkappa 1_A = 1_A$: Each identity of $\mathscr{C}$ has the form 1_A for a unique object A. In other words, the identities of $\mathscr{C}$ determine the objects of $\mathscr{C}$. It is possible to describe a category simply as a class of morphisms, with a composite sometimes defined and subject to suitable axioms (Ex. 3 below).

A morphism $\theta\colon A \to B$ is called an *equivalence* in $\mathscr{C}$ if there is in $\mathscr{C}$ another morphism $\varphi\colon B \to A$ such that $\varphi\theta = 1_A$ and $\theta\varphi = 1_B$. Then φ is unique, for if also $\varphi'\theta = 1_A$, then $\varphi = 1_A\varphi = \varphi'\theta\varphi = \varphi' 1_B = \varphi'$. Call φ the *inverse* $\varphi = \theta^{-1}$ of the equivalence θ. The composite of two equivalences, when defined, is an equivalence.

A (multiplicative) group G is a category with one object G; let hom(G, G) be all elements of G. If a set M is closed under an associative multiplication with an identity, it is likewise a category with one object and composition given by multiplication.

A more typical example of a category is the category ${}_R\mathcal{M}$ of (left) modules over a fixed ring R. The objects of this category are all R-modules $A, B, C, \ldots$, the set hom(A, B) of morphisms is the set $\mathrm{Hom}_R(A, B)$ of all R-module homomorphisms of A to B, while the composite is the usual composite of homomorphisms. The axioms of associativity and existence of identities are obviously fulfilled. This category uses the *class* of all R-modules. We cannot say the *set* of all R-modules because this set would be an illegitimate totality in the usual axioms for set theory. If one adopts the Gödel-Bernays-von Neumann axioms for set theory [GÖDEL 1940], one has at hand larger totalities than sets, called classes, and one can legitimately speak of the class of all modules, or of all topological spaces. With this interpretation in view, we have defined a category to be a *class* of objects .... We call a category *small* if the class of its objects is a set.

To give other examples of categories it will suffice to specify the objects and the morphisms of the category; in most cases the range and domain of the morphisms, the composite, and the identities will have their standard meanings. We list a number of examples of categories which we shall meet.

The category of topological spaces. Objects, all topological spaces; morphisms, all continuous maps $f: X \to Y$ of one space into a second one.

The category of abelian groups. Objects, all abelian groups; morphisms, all homomorphisms of such.

The category of groups. Objects, all (not necessarily abelian) groups; morphisms, all homomorphisms of groups.

The category of sets. Objects, all sets; morphisms, all functions on a set to a set.

In the next examples R denotes a fixed ring.

The category of exact sequences of R-modules of length n. Objects, all exact sequences $S: A_1 \to A_2 \to \cdots \to A_{n-1} \to A_n$; morphisms $\Gamma: S \to S'$, all n-tuples $\Gamma = (\gamma_1, \gamma_2, \ldots, \gamma_n)$ of module homomorphisms $\gamma_i: A_i \to A'_i$ such that the diagram

$$\begin{array}{ccccccccc} A_1 & \to & A_2 & \to & \cdots & \to & A_{n-1} & \to & A_n \\ \downarrow{\scriptstyle \gamma_1} & & \downarrow{\scriptstyle \gamma_2} & & & & \downarrow{\scriptstyle \gamma_{n-1}} & & \downarrow{\scriptstyle \gamma_n} \\ A'_1 & \to & A'_2 & \to & \cdots & \to & A'_{n-1} & \to & A'_n \end{array}$$

is commutative. If $B = (\beta_1, \ldots, \beta_n)$: $S' \to S''$, the composite $B\Gamma$ is $(\beta_1 \gamma_1, \ldots, \beta_n \gamma_n)$.

One may also have the category of exact sequences infinite to the right or infinite to the left, or both. Another example is the category of short exact sequences $E: A \rightarrowtail B \twoheadrightarrow C$, with morphisms all triples (α, β, γ) of module homomorphisms for which the appropriate diagram (3.1) is commutative. It is by now amply clear how more examples can be constructed *ad libitum* — a category of sequences of exact sequences of....

It is also clear that a number of concepts applicable to modules will apply to the objects of any category — provided the definition of the concept makes reference not to the elements of the modules but only to modules and their homomorphisms. Thus, in any category $\mathscr{C}$, a diagram consisting of morphisms $\alpha_t: A_t \to C$ of $\mathscr{C}$, one for each t in a given set T, is *universal* for the given objects A_t (or, a *direct sum diagram* for the A_t) if to each diagram $\{\alpha'_t: A_t \to C' \mid t \in T\}$ on the same A_t there exists a unique morphism $\beta: C \to C'$ of $\mathscr{C}$ with $\beta\alpha_t = \alpha'_t$ for each $t \in T$. (For $T = \{1, 2\}$, this is exactly the property formulated in (4.5)). The previous uniqueness proof for the direct sum of two modules carries over verbatim to give

Proposition 7.1. *In any category $\mathscr{C}$ let $\{\alpha_t: A_t \to C\}$ and $\{\alpha'_t: A_t \to C'\}$ be two direct sum diagrams for the same family $\{A_t\}$ of objects. There is then a unique equivalence $\theta: C \to C'$ of $\mathscr{C}$ with $\theta\alpha_t = \alpha'_t$ for every t.*

An analogous uniqueness theorem holds for a direct product diagram, defined as a diagram $\{\gamma_t: B \to A_t \mid t \in T\}$ such that to each $\{\gamma'_t: B' \to A_t \mid t \in T\}$ there exists a unique morphism $\beta: B' \to B$ with $\gamma'_t = \gamma_t\beta$ for $t \in T$.

The definition of the direct product is exactly parallel to that for the direct sum, except that *all the arrows are reversed*. We say that the direct product is the "dual" of the direct sum. In general, the *dual* of any statement $\mathfrak{S}$ (in the first order propositional calculus) about a category $\mathscr{C}$ is the statement $\mathfrak{S}^*$ obtained by reversing the direction of all morphisms, replacing each composite $\alpha\beta$ of morphisms by $\beta\alpha$ and interchanging "domain" and "range". One observes at once that the dual of each axiom for a category is an axiom. It follows that the dual of a proof from these axioms of a statement $\mathfrak{S}$ about categories is a proof of the dual statement $\mathfrak{S}^*$. For example, the dual of Prop. 7.1 is the proposition which asserts the uniqueness (up to equivalence) of a direct product diagram with given ends. Since Prop. 7.1 has been proved from the axioms of a category, we have this dual proposition without further proof. However, a proposition $\mathfrak{S}$ whose statement involves only objects and morphisms may happen to be true in a particular category although the dual statement is false. For example, in the category of all denumerable abelian groups there exists a direct sum diagram with summands any denumerable list of denumerable groups $A_1, A_2, \ldots, A_n, \ldots$, but there does not exist a direct product of the same groups (essentially because the direct product which does exist in the category of all abelian groups is non-denumerable).

To each category one may construct an *opposite category* $\mathscr{C}^{\mathrm{op}}$. Take the objects of $\mathscr{C}^{\mathrm{op}}$ to be a class in $1-1$ correspondence $A^* \leftrightarrow A$ with the objects A of $\mathscr{C}$. Take the morphisms to be a class in $1-1$ correspondence $\alpha^* \leftrightarrow \alpha$ with the morphisms of $\mathscr{C}$. Decree that $\alpha^*: A^* \to B^*$ if and only if $\alpha: B \to A$, and that $\alpha^* \beta^*$ is defined and is $(\beta\alpha)^*$, exactly when $\beta\alpha$ is defined. Then $\mathscr{C}^{\mathrm{op}}$ is a category, and any statement $\mathfrak{S}^*$ about the category $\mathscr{C}$ is the same as the original statement $\mathfrak{S}$ about the category $\mathscr{C}^{\mathrm{op}}$. This, again, shows the dual $\mathfrak{S}^*$ of a provable statement $\mathfrak{S}$ provable. The $1-1$ function T with $T(A)=A^*$ and $T(\alpha)=\alpha^*$ is an "anti-isomorphism" of $\mathscr{C}$ to $\mathscr{C}^{\mathrm{op}}$, since $T(\beta\alpha)=T(\alpha)\,T(\beta)$.

Subsequently, we shall define a special sort of category, called an "abelian category", by requiring essentially that hom (A, B) be an abelian group and that kernels and cokernels exist, as in the case of the category of modules. It turns out that many theorems about modules remain true when the modules and their homomorphisms are replaced by the objects and the morphisms of any abelian category. The interested reader may turn at once to Chaps. IX and XII.

Exercises

1. In the category of topological spaces, show that the disjoint union of two spaces provides a direct sum diagram, and that the cartesian product $X \times Y$ of two spaces, with its usual topology and with the natural projections on X and Y, provides a direct product diagram.

2. Show that any two objects in the category of groups may be ends of a direct product diagram and of a direct sum diagram. (Note: The "direct sum" for not necessarily abelian groups is more often known as the "free product".)

3. Consider a class $\mathscr{M}$ of elements α, β, γ in which a product $\beta\alpha \in \mathscr{M}$ is sometimes defined. Call $\varkappa$ an identity of $\mathscr{M}$ if $\varkappa\beta = \beta$ whenever $\varkappa\beta$ is defined and $\alpha\varkappa = \alpha$ whenever $\alpha\varkappa$ is defined. Then $\mathscr{M}$ is called an *abstract category* if it satisfies the following axioms:

(i) The product $\gamma(\beta\alpha)$ is defined if and only if $(\gamma\beta)\alpha$ is defined. When either is defined, they are equal. This triple product will be written $\gamma\beta\alpha$.

(ii) The triple product $\gamma\beta\alpha$ is defined whenever both products $\gamma\beta$ and $\beta\alpha$ are defined.

(iii) For each α in $\mathscr{M}$ there exist identities $\varkappa, \varkappa'$ such that $\alpha\varkappa$ and $\varkappa'\alpha$ are defined.

Prove that the class of morphisms of a category is an abstract category, and conversely that the elements of any abstract category are the morphisms of a category $\mathscr{C}$ which is determined uniquely up to an isomorphism of categories.

8. Functors

Let $\mathscr{C}$ and $\mathscr{D}$ be categories. A *covariant functor* T on $\mathscr{C}$ to $\mathscr{D}$ is a pair of functions (both denoted by the same letter T): An "object function" which assigns to each object $C \in \mathscr{C}$ an object $T(C) \in \mathscr{D}$, and a "mapping function'' which assigns to each morphism $\gamma: C \to C'$ of $\mathscr{C}$ a morphism $T(\gamma): T(C) \to T(C')$ of $\mathscr{D}$. This pair of functions must satisfy the two following conditions:

$$T(1_C) = 1_{T(C)}, \qquad C \in \mathscr{C}, \tag{8.1}$$

$$T(\beta\gamma) = T(\beta)T(\gamma), \qquad \beta\gamma \text{ defined in } \mathscr{C}. \tag{8.2}$$

A covariant functor T on $\mathscr{C}$ to $\mathscr{D}$ is thus a mapping of $\mathscr{C}$ to $\mathscr{D}$ which preserves range, domain, identities, and composites.

For example, let R be a fixed ring. For any set T let $F(T) = \Sigma_t Rt$ be the free module on the set T. Then F is a covariant functor on the category of sets to the category of R-modules. Again, for example, let $\mathscr{G}$ be the category of all groups, with $G' = [G, G]$ the commutator subgroup of G; that is, the subgroup generated by all the "commutators" $g_1 g_2 g_1^{-1} g_2^{-1}$ for $g_i \in G$. Each homomorphism $\gamma: G \to H$ clearly maps G' to H' by a homomorphism γ'. The functions $T(G) = G'$ and $T(\gamma) = \gamma'$ make G' a covariant functor on $\mathscr{G}$ to $\mathscr{G}$. Similarly, the "factor-commutator" group $G/[G, G]$ may be regarded as a covariant functor on $\mathscr{G}$ to the category of abelian groups.

Let S and T be two covariant functors on $\mathscr{C}$ to $\mathscr{D}$. A *natural transformation* $h: S \to T$ is a function which assigns to each object $C \in \mathscr{C}$ a morphism $h(C): S(C) \to T(C)$ of $\mathscr{D}$ in such a fashion that for each morphism $\gamma: C \to C'$ of $\mathscr{C}$ the diagram

$$\begin{array}{ccc} S(C) & \xrightarrow{h(C)} & T(C) \\ \downarrow{\scriptstyle S(\gamma)} & & \downarrow{\scriptstyle T(\gamma)} \\ S(C') & \xrightarrow{h(C')} & T(C') \end{array} \tag{8.3}$$

is commutative in $\mathscr{D}$. When $h(C)$ satisfies this commutativity condition we say more briefly "h is natural". If in addition each $h(C)$ is an equivalence, we say that h is a *natural isomorphism.*

Intuitively, a "natural transformation" h is one which is defined in the same way or by the same formula for every object in the category in question. For instance, for each group G let $h(G): G \to G/[G, G]$ be the homomorphism which assigns to each element $g \in G$ its coset $g[G, G]$ in the factor-commutator group. The diagram like (8.3) is commutative, so h may be viewed as a natural transformation of the identity functor to the factor-commutator functor (both in the category of all groups). Other (and more incisive) examples of natural transformations will appear shortly (e. g., Prop. II.4.2 for relative homology).

A *contravariant functor* T on $\mathscr{C}$ to $\mathscr{D}$ consists of an *object function* T which assigns to each C a $T(C) \in \mathscr{D}$ and a *mapping function* T which assigns to each morphism $\gamma: C \to C'$ a morphism $T(\gamma): T(C') \to T(C)$ of $\mathscr{D}$, now in the opposite direction. This pair of functions must again satisfy two conditions:

$$T(1_C) = 1_{T(C)}, \qquad C \in \mathscr{C}, \tag{8.4}$$

$$T(\beta\gamma) = T(\gamma)\,T(\beta), \qquad \beta\gamma \text{ defined in } \mathscr{C}. \tag{8.5}$$

The reversed order of the factors in (8.5) is necessary to make sense, for $\beta\gamma$ defined means $\gamma: C \to C'$, $\beta: C' \to C''$, hence $T(\beta): T(C'') \to T(C')$ and $T(\gamma): T(C') \to T(C)$, so that $T(\gamma)\,T(\beta)$ is defined.

For a fixed R-module B we noted in § 6 that $\mathrm{Hom}_R(A, B)$ is a contravariant functor of A, in the category of R-modules. The character group of an abelian group A is the group $\mathrm{Ch}\, A = \mathrm{Hom}_Z(A, P)$, where P is the additive group of real numbers modulo 1. With the mapping function $\mathrm{Ch}\,\alpha = \alpha^*$ defined as in § 6, Ch is a contravariant functor on the category of abelian groups to itself — or, with the standard topology on $\mathrm{Ch}\, A$, a contravariant functor on discrete abelian groups to compact abelian groups. For any category $\mathscr{D}$ and its opposite $\mathscr{D}^{\mathrm{op}}$, the pair of functions P with $PD = D^*$ and $P\delta = \delta^*$ is a contravariant functor on $\mathscr{D}$ to $\mathscr{D}^{\mathrm{op}}$. Each contravariant functor T on $\mathscr{C}$ to $\mathscr{D}$ may be regarded as a covariant functor on $\mathscr{C}$ to $\mathscr{D}^{\mathrm{op}}$, namely, as the composite PT.

A *natural transformation* $h: S \to T$ between two contravariant functors on $\mathscr{C}$ to $\mathscr{D}$ is a function which assigns to every object $C \in \mathscr{C}$ a morphism $h(C): S(C) \to T(C)$ of $\mathscr{D}$ such that for each $\gamma: C \to C'$ in $\mathscr{C}$ the diagram

$$\begin{array}{ccc} S(C') & \xrightarrow{h(C')} & T(C') \\ \downarrow{\scriptstyle S(\gamma)} & & \downarrow{\scriptstyle T(\gamma)} \\ S(C) & \xrightarrow{h(C)} & T(C) \end{array} \tag{8.6}$$

is commutative. This diagram is just (8.3) upside down.

If T is a functor on $\mathscr{C}$ to $\mathscr{D}$ and S a functor on $\mathscr{D}$ to a third category $\mathscr{E}$, the composite functions $S \circ T$ yield a functor on $\mathscr{C}$ to $\mathscr{E}$ with variance the product of the variances of S and T (covariant $= +1$, contravariant, -1). For instance, let $\mathscr{M}_F$ be the category of vector spaces over a fixed field F, and let D be the functor on $\mathscr{M}_F$ to $\mathscr{M}_F$ which assigns to each vector space V its dual $D(V) = \mathrm{Hom}_F(V, F)$ and to each linear transformation ($=$ morphism of $\mathscr{M}_F$) $\alpha: V \to V'$ its induced map $\alpha^*: D(V') \to D(V)$, defined as in (6.3). Then D is contravariant, while the composite $D^2 = D \circ D$ is the covariant functor which assigns to each vector space V its double dual. There is a homomorphism

$$h = h(V): V \to D(DV)$$

which assigns to each vector v that function $hv: DV \to F$ with $(hv)f = f(v)$ for $f \in DV$. For finite dimensional V, $h(V)$ is the familiar isomorphism of V to its double dual. One verifies readily that h is a natural transformation $h: I \to D^2$ (where I denotes the identity functor).

There is a similar natural isomorphism of a finite abelian group to its double character group.

As an example of a *non*-natural isomorphism, recall that there is an isomorphism $k: V \cong D(V)$ for any finite dimensional vector space V. Specifically, for each such V choose a fixed basis $v_1, \ldots, v_n$, construct in $D(V)$ the dual basis $v^1, \ldots, v^n$, with v^i defined by the requirement that $v^i(v_j)$ is 0 or 1 according as $i \neq j$ or $i = j$, and set $k v_i = v^i$. This linear transformation $k = k(V): V \to D(V)$ is defined for each V; it maps the covariant identity functor I to the contravariant functor D. If we restrict attention to the category whose objects are finite dimensional vector spaces and whose morphisms are isomorphisms α of such, we may replace D by a covariant functor $\bar{D}$ with $\bar{D}(V) = D(V)$, $\bar{D}(\alpha) = D(\alpha^{-1})$. But $k(V): V \to \bar{D}(V)$ is not natural. For example, if V is 1-dimensional and $\alpha: V \to V$ is defined by $\alpha v_1 = c v_1$ for some scalar $c \in F$ with $0 \neq c^2 \neq 1$, then $\bar{D}(\alpha) k(V) v_1 = (1/c) v^1$; however $k(V) \alpha v_1 = c v^1$, so (8.3) is not commutative.

Functors in several variables may be covariant in some of their arguments and contravariant in others. Two arguments, contra and co, suffice to illustrate. Let $\mathscr{B}, \mathscr{C}$, and $\mathscr{D}$ be three categories. A *bifunctor* T on $\mathscr{B} \times \mathscr{C}$ to $\mathscr{D}$, contravariant in $\mathscr{B}$ and covariant in $\mathscr{C}$, is a pair of functions: An object function which assigns to $B \in \mathscr{B}$ and $C \in \mathscr{C}$ an object $T(B, C) \in \mathscr{D}$, and a mapping function which assigns to morphisms $\beta: B \to B'$ and $\gamma: C \to C'$ a morphism

$$T(\beta, \gamma): T(B', C) \to T(B, C') \tag{8.7}$$

of $\mathscr{D}$ (Note that the direction in B is reversed, that in C is preserved). These functions must satisfy the conditions

$$T(1_B, 1_C) = 1_{T(B,C)}, \tag{8.8}$$

$$T(\beta'\beta, \gamma'\gamma) = T(\beta, \gamma') \, T(\beta', \gamma), \tag{8.9}$$

the latter to hold whenever both composites $\beta'\beta$ and $\gamma'\gamma$ are defined. The composite on the right is then defined, for $\beta': B' \to B''$ and $\gamma': C' \to C''$ with (8.7) give

$$T(B'', C) \xrightarrow{T(\beta', \gamma)} T(B', C') \xrightarrow{T(\beta, \gamma')} T(B, C'').$$

It is convenient to set $T(\beta, 1_C) = T(\beta, C)$ and $T(1_B, \gamma) = T(B, \gamma)$. When B is fixed, $T(B, C)$ and $T(B, \gamma)$ are object and mapping functions of a covariant functor on $\mathscr{C}$ to $\mathscr{D}$, while, for fixed C, $T(B, C)$ and $T(\beta, C)$ provide a contravariant functor on $\mathscr{B}$ to $\mathscr{D}$. These mapping functions $T(B, \gamma)$ and $T(\beta, C)$ determine all $T(\beta, \gamma)$ for, by (8.9), $T(\beta, \gamma) = T(\beta 1_B, \gamma 1_C) = T(B, \gamma) \, T(\beta, C)$. We leave the reader to carry out the rest of the proof of

Proposition 8.1. *For given categories $\mathscr{B}, \mathscr{C}$, and $\mathscr{D}$ let T be a function assigning to each B and C an object $T(B, C) \in \mathscr{D}$. For each fixed $B \in \mathscr{B}$, let $T(B, C)$ be the object function of a covariant functor $\mathscr{C} \to \mathscr{D}$ with the mapping function $T(B, \gamma)$. For each fixed $C \in \mathscr{C}$, let $T(B, C)$ be the object function of a contravariant functor $\mathscr{B} \to \mathscr{D}$ with mapping function $T(\beta, C)$. Suppose that for each $\beta: B \to B'$ and each $\gamma: C \to C'$ the diagram*

$$\begin{array}{ccc} T(B', C) & \xrightarrow{T(B',\gamma)} & T(B', C') \\ {\scriptstyle T(\beta,C)}\downarrow & & \downarrow{\scriptstyle T(\beta,C')} \\ T(B, C) & \xrightarrow{T(B,\gamma)} & T(B, C') \end{array} \tag{8.10}$$

is commutative. Then the diagonal map

$$T(\beta, \gamma) = T(B, \gamma)\, T(\beta, C) = T(\beta, C')\, T(B', \gamma)$$

of this diagram makes T a bifunctor on $\mathscr{B}$ and $\mathscr{C}$ to $\mathscr{D}$, contravariant in $\mathscr{B}$ and covariant in $\mathscr{C}$. Every such functor can be so obtained.

If we write more simply β^* for $T(\beta, C)$ and γ_* for $T(B, \gamma)$, the commutativity condition of this proposition can be put with less accuracy and more vigor as $\beta^* \gamma_* = \gamma_* \beta^*$. This proposition usually provides the easiest verification that a given T is indeed a bifunctor. A typical example of such a bifunctor T is $\operatorname{Hom}_R(A, B)$, covariant in B and contravariant in A.

If S and T are two such bifunctors on $\mathscr{B} \times \mathscr{C}$ to $\mathscr{D}$, a *natural transformation* $f: S \to T$ is a function which assigns to each pair of objects B, C a morphism $f(B, C): S(B, C) \to T(B, C)$ in such wise that for all morphisms $\beta: B \to B'$ and $\gamma: C \to C'$ the diagram

$$\begin{array}{ccc} S(B', C) & \xrightarrow{f(B',C)} & T(B', C) \\ {\scriptstyle S(\beta,\gamma)}\downarrow & & \downarrow{\scriptstyle T(\beta,\gamma)} \\ S(B, C') & \xrightarrow{f(B,C')} & T(B, C') \end{array} \tag{8.11}$$

is commutative. In view of the decomposition of $T(\beta, \gamma)$ above, it suffices to require this condition only for β and 1_C, and for 1_B and γ. In other words, it suffices to require that $f(B, C)$ with either variable fixed be a natural transformation in the remaining variable.

Direct products provide an example of a bifunctor covariant in two arguments. Let $\mathscr{C}$ be a category in which each pair of objects has a direct product diagram, and choose such a diagram $\{\pi_i: A_1 \times A_2 \to A_i \mid i = 1, 2\}$ for each pair; this includes the choice of a direct product $A_1 \times A_2$ for each A_1 and A_2. Let $\alpha_i: A_i \to A_i'$ for $i = 1, 2$ be morphisms of $\mathscr{C}$, as in the diagram

$$\begin{array}{ccccc} A_1 & \xleftarrow{\pi_1} & A_1 \times A_2 & \xrightarrow{\pi_2} & A_2 \\ \downarrow{\scriptstyle \alpha_1} & & \vdots{\scriptstyle \beta} & & \downarrow{\scriptstyle \alpha_2} \\ A_1' & \xleftarrow{\pi_1'} & A_1' \times A_2' & \xrightarrow{\pi_2'} & A_2' \end{array}, \quad \beta = \alpha_1 \times \alpha_2 . \tag{8.12}$$

Then $\alpha_i \pi_i: A_1 \times A_2 \to A_i'$; by the couniversal property of the bottom row, there is a unique morphism $\beta: A_1 \times A_2 \to A_1' \times A_2'$ with $\pi_i' \beta = \alpha_i \pi_i$; that is, a unique β which makes the diagram commutative. For example, if $\mathscr{C}$ is the category of sets or of R-modules, and if each $A_1 \times A_2$ is chosen in the usual way as the set of all pairs (a_1, a_2), then $\beta(a_1, a_2) = (\alpha_1 a_1, \alpha_2 a_2)$. Call $\beta = \alpha_1 \times \alpha_2$ the *direct product* of the given morphisms. By the couniversal property, $1 \times 1 = 1$, and $(\gamma_1 \times \gamma_2)(\alpha_1 \times \alpha_2) = \gamma_1 \alpha_1 \times \gamma_2 \alpha_2$ wherever $\gamma_i \alpha_i$ is defined for $i = 1, 2$. Hence $P(A_1, A_2) = A_1 \times A_2$, $P(\alpha_1, \alpha_2) = \alpha_1 \times \alpha_2$ defines a covariant bifunctor P on $\mathscr{C}, \mathscr{C}$ to $\mathscr{C}$. For three objects A_1, A_2, A_3 the usual map $(A_1 \times A_2) \times A_3 \to A_1 \times (A_2 \times A_3)$ is a natural homomorphism of covariant trifunctors.

The notions of category and functor provide not profound theorems, but a convenient language. For example, consider the notion of diagrams of the "same form", say of diagram of modules of the form $D: A \to B \to C$. Any such diagram may be regarded as a functor. Indeed, introduce the finite category $\mathscr{H}$, which has three objects a, b, and c, the corresponding identity morphisms and the morphisms $\varkappa_0: a \to b$, $\lambda_0: b \to c$, and $\mu_0: a \to c$ with $\lambda_0 \varkappa_0 = \mu_0$. Then any diagram of modules as exhibited is a covariant functor on $\mathscr{H}$ to the category ${}_R\mathscr{M}$ of modules: Such a covariant functor D does provide three modules $D(a) = A$, $D(b) = B$, and $D(c) = C$ plus the homomorphisms $D(\varkappa_0)$, $D(\lambda_0)$, $D(\mu_0) = D(\lambda_0) D(\varkappa_0)$. Furthermore a map of a diagram D to a another diagram D' of the same form is exactly a natural transformation $D \to D'$ of functors. In this formulation we also can include the notions of diagrams with commutativity conditions; thus a commutative square diagram is a functor on the finite category

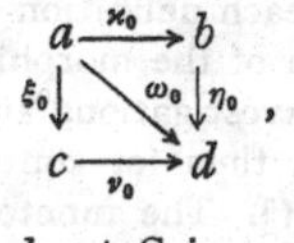

$$\eta_0 \varkappa_0 = \omega_0 = \nu_0 \xi_0. \qquad (8.13)$$

A *partly ordered set* S is a set with a binary relation $r \leqq s$ which is reflexive ($r \leqq r$), transitive ($r \leqq s$ and $s \leqq t$ imply $r \leqq t$) and such that $r \leqq s$ and $s \leqq r$ imply $r = s$. The partly ordered set S has a *zero* if there is an element $0 \in S$, necessarily unique, with $0 \leqq s$ for every s. An element $u \in S$ is a *least upper bound* (l. u. b.) of $s, t \in S$ if $s \leqq u, t \leqq u$ and $s \leqq v, t \leqq v$ imply $u \leqq v$. This l. u. b. is unique if it exists, and is written $u = s \cup t$. Similarly, $w = s \cap t$ is a g. l. b. of s and t if $w \leqq s$, $w \leqq t$ and $x \leqq s$, $x \leqq t$ imply $x \leqq w$. The partly ordered set S is a *lattice* if $s \cup t$ and $s \cap t$ exist for all s and t.

Each partly ordered set S may be regarded as a category $\mathscr{S}$, with objects the elements $s \in S$, morphisms the pairs $(s, r): r \to s$ with $r \leqq s$, and composition of morphisms defined by $(t, s)(s, r) = (t, r)$ when $r \leqq s \leqq t$. For example, the finite category (8.13) arises so from the partly ordered set with four elements a, b, c, d and partial order $a \leqq b \leqq d$,

$a \leqq c \leqq d$. If S is a lattice, any two objects s, t of $\mathcal{S}$ have a direct sum (given by $s \cup t$) and a direct product $s \cap t$; conversely, if $\mathcal{S}$ has direct sums and products, S is a lattice.

For any partly ordered S, a covariant functor $T: \mathcal{S} \to {}_R\mathcal{M}$ is a family $\{T_s \mid s \in S\}$ of R-modules together with homomorphisms $T(s, r): T_r \to T_s$ defined for each $r \leqq s$ and such that $T(t, s)\, T(s, r) = T(t, r)$ whenever $r \leqq s \leqq t$. The "direct limit" of such a family may be described conveniently in categorical terms [EILENBERG-MAC LANE 1945, Chap. IV; KAN 1958].

Exercises

1. If $\alpha_j: A_j \to A'_j$ are module homomorphisms, show that the map $\beta = \alpha_1 \times \alpha_2$: $A_1 \oplus A_2 \to A'_1 \oplus A'_2$ characterized by $\pi'_j \beta = \alpha_j \pi_j$, $j = 1, 2$, is also characterized by the conditions $\beta \iota_j = \iota'_j \alpha_j$, $j = 1, 2$.

2. Show that the associative law for the (external) direct sum of modules can be expressed as a *natural* isomorphism $(A \oplus B) \oplus C \cong A \oplus (B \oplus C)$.

3. Prove that the isomorphisms (6.5) are natural.

4. Let $\mathcal{C}$ be a small category in which each set hom(A, B) of morphisms has at most one element, and in which each equivalence is an identity. Prove that $\mathcal{C}$ may be obtained from a partly ordered set.

Notes. The idea of a module goes back at least to KRONECKER, who considered modules over polynomial rings; only in the last twenty years has this idea taken on its present central role in algebra. Projective modules were first used effectively in CARTAN-EILENBERG; now it is clear that they provide for linear algebra the appropriate generalization of a vector space (which is always a free module). EMMY NOETHER, in lectures at Göttingen, emphasized the importance of homomorphisms. The initial restriction to homomorphisms $\alpha: A \to B$ with $\alpha(A) = B$, as in VAN DER WAERDEN's influential *Moderne Algebra*, soon proved to be needlessly restrictive, and was dropped. By now it is expected that each definition of a type of Mathematical system be accompanied by a definition of the morphisms of this system. The arrow notation developed in topological investigations about 1940, probably starting with the use for correspondences and then for continous maps. Exact sequences were first noted in HUREWICZ [1941]. The functor "Hom" was long known, but apparently first appeared by this name in EILENBERG-MAC LANE [1942]. Categories and functors were introduced by the same authors in 1945. They have proved useful in the formulation of axiomatic homology (Chap. II below), in the cohomology of a sheaf over a topological space [GODEMENT 1958], in differential geometry [EHRESMANN 1957], and in algebraic geometry (GROTHENDIECK-DIEUDONNÉ [1960], cf. also the review by LANG [1961]). Foundational questions about the theory of categories, using sets and classes, are formulated in MAC LANE [1961].

Chapter two

Homology of Complexes

Here we first meet the basic notions of homology in simple geometric cases where the homology group arises from a boundary operator. In general, an abelian group with a boundary operator is called a "differential group" or, when provided with dimensions, a "chain complex".

This chapter considers the algebraic process of constructing homology and cohomology groups from chain complexes. Basic is the fact (§ 4) that a short exact sequence of complexes gives a long exact sequence of homology groups. As illustrative background, the last sections provide a brief description of the singular homology groups of a topological space.

1. Differential Groups

A *differential group* C is an abelian group C together with an endomorphism $d: C \to C$ such that $d^2=0$; call d the "differential" or "boundary operator" of C. Elements of C are often called *chains*, elements of $\operatorname{Ker} d$, *cycles*, and elements of $\operatorname{Im} d$, *boundaries*. The requirement that $d^2=0$ is equivalent to the inclusion $\operatorname{Im} d \subset \operatorname{Ker} d$. The homology group of the differential group C is defined to be the factor group of cycles modulo boundaries,

$$H(C) = \operatorname{Ker} d/\operatorname{Im} d = \operatorname{Ker} d/dC. \tag{1.1}$$

Its elements are the cosets $c + \operatorname{Im} d$ of cycles c; we call them *homology classes* and write them as

$$\operatorname{cls}(c) = c + dC \in H(C). \tag{1.2}$$

Two cycles c and c' in the same homology class are said to be *homologous*; in symbols $c \sim c'$.

As first examples we shall give a number of specific differential groups with their homology. Most of these examples will be found by dissecting a simple geometric figure into cells and taking d to be the operator which assigns to each cell the sum of its boundary cells, each affected with a suitable sign.

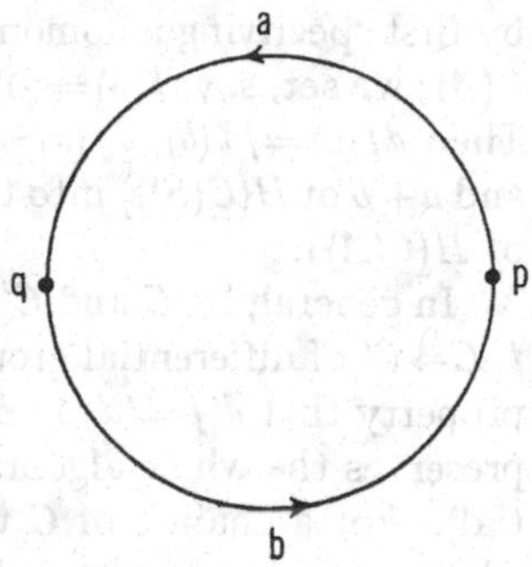

Fig. 1

Example 1. Take two points p and q on a circle S^1 which divide the circle into two semicircular arcs a and b. The "boundary" or "ends" of the arc a are the points q and p. Hence introduce the free abelian group $C(S^1)$ with the four free generators a, b, p, and q, and define an endomorphism d of $C(S^1)$ by setting

$$da = q - p, \qquad db = p - q, \qquad dp = 0 = dq. \tag{1.3}$$

Any element of $C(S^1)$ is represented uniquely as a linear combination $m_1 a + m_2 b + m_3 p + m_4 q$ with integral coefficients m_1, m_2, m_3, and m_4, while

$$d(m_1 a + m_2 b + m_3 p + m_4 q) = m_1(q-p) + m_2(p-q) = (m_1 - m_2)(q-p).$$

Thus $C(S^1)$ is a differential group. Its cycles are all the integral linear combinations of p, q, and $a+b$, while its boundaries are all the multiples of $q-p$. Hence there is a homology $p \sim q$, and the homology group is the direct sum

$$H(C(S^1)) = Z_\infty(\operatorname{cls}(p)) \oplus Z_\infty(\operatorname{cls}(a+b)), \tag{1.4}$$

where $Z_\infty(\operatorname{cls}(p))$ denotes the infinite cyclic group generated by the homology class $\operatorname{cls}(p)$. Thus the circle S^1 has two basic homology classes, the point p (dimension 0), and the circumference $a+b$ (dimension 1).

In this example the same circle could have been subdivided otherwise, say into more arcs. The homology groups turn out to be independent of the subdivision chosen. For example, isomorphic homology groups arise when the circle is cut into three arcs so as to form a triangle!

Example 2. Take a triangle Δ with vertices 0, 1, and 2, and edges 01, 12, and 02. The corresponding differential group $C(\Delta)$ is the free abelian group on six generators (0), (1), (2), (01), (12), (02), with the differential given by $d(0)=d(1)=d(2)=0$, $d(01)=(1)-(0)$, $d(02)=(2)-(0)$, $d(12)=(2)-(1)$; in other words, the boundary of each edge is the difference of its two end vertices. One finds

$$H(C(\Delta)) = Z_\infty(\operatorname{cls}(0)) \oplus Z_\infty(\operatorname{cls}[(12)-(02)+(01)]).$$

This group is indeed isomorphic to that found for the circle in Example 1; both are free abelian with two generators. An isomorphism may be given by first specifying a homomorphism f of the differential group $C(S^1)$ into $C(\Delta)$; we set, say, $f(p)=(0)$, $f(q)=(1)$, $f(a)=(01)$, and $f(b)=(12)-(02)$. Then $df(b)=fd(b)$, $df(a)=fd(a)$, and f carries the generating cycles p and $a+b$ of $H(C(S^1))$ into the generating cycles (0) and $(12)-(02)+(01)$ of $H(C(\Delta))$.

In general, let C and C' be two differential groups. A ***homomorphism*** $f: C \to C'$ of differential groups is a group homomorphism with the added property that $d'f = fd$; in other words, it is a function on C to C' which preserves the whole algebraic structure involved (addition *and* differential). For a chain c of C this implies that fc is a cycle or a boundary whenever c is a cycle or boundary, respectively. Hence the function $H(f)=f_*$, defined by $f_*(\operatorname{cls}(c)) = \operatorname{cls}(fc)$, is a group homomorphism

$$H(f): H(C) \to H(C') \qquad (\text{for } f: C \to C'). \tag{1.5}$$

We call $H(f)$ the homomorphism *induced* by f. Since $H(1_C)=1_{H(C)}$ and $H(f'f)=H(f')H(f)$, H is a covariant functor on differential groups to groups.

Example 3. The circular disc D is had by adding the inside c to the circle S^1; construct a corresponding differential group $C(D)$ by adjoining to $C(S^1)$ one new free generator c with boundary $dc=a+b$. Then $H(C(D))=Z_\infty(\operatorname{cls} p)$. The injection $j: C(S^1) \to C(D)$ thus induces a map $H(j): H(C(S^1)) \to H(C(D))$ which maps the second summand of (1.4) onto zero. In other words, the map $H(j)$ induced by an injection need not be a monomorphism; that is, the homology group of a subspace need not be a subgroup of the homology of the original space. This is why the injection j has a label different from the identity.

Example 4. For the sphere S^2 with the equator S^1, labelled as in Fig. 2, let u be the upper and l the lower hemisphere. Construct a differential group $C(S^2)$ by adjoining to $C(S^1)$ two new free generators u and l, with boundaries $du=a+b=-dl$. Then

$$H(C(S^2))=Z_\infty(\mathrm{cls}(p))\oplus Z_\infty(\mathrm{cls}(u+l))\,; \quad (1.6)$$

there is a cycle p in dimension 0 and one in dimension 2.

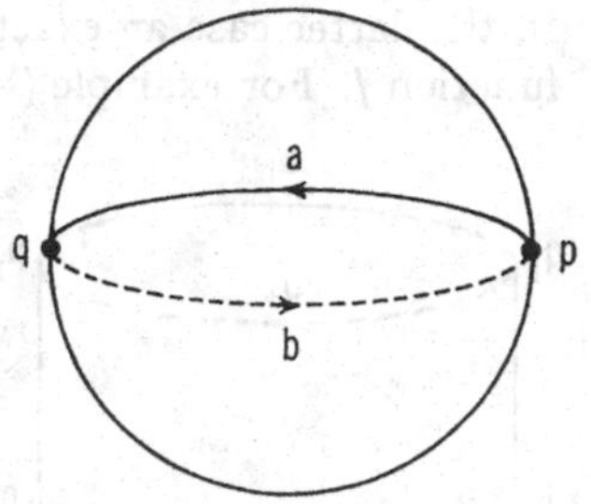

Fig. 2

Example 5. The real projective plane P^2, regarded as a topological space, may be obtained from the sphere S^2 by identifying each point of S^2 with the diametrically opposite point. In particular, each point in the upper hemisphere is identified with a point in the lower hemisphere. This suggests that we proceed algebraically by setting $u=-l$, $a=b$, and $p=q$ in the differential group $C(S^2)$ above. This will yield a new differential group $C(P^2)$, which is the free abelian group generated by u, a, and p with $du=2a$, $da=0$, $dp=0$. Then a is a cycle which is not a boundary, though $2a$ is a boundary. Hence

$$H(C(P^2))=Z_\infty(\mathrm{cls}(p))\oplus Z_2(\mathrm{cls}(a)),$$

where $Z_2(\mathrm{cls}(a))$ designates the cyclic group of order 2 with generator $\mathrm{cls}(a)$.

Example 6. Let $f(x, y)$ be a real valued function of class C^∞ (i.e., with continuous partial derivatives of all orders) defined in a connected open set D of points (x, y) in the Cartesian plane. For fixed D, the set A of all such functions is an abelian group under the operation of addition of function values. Take C to be the direct sum $A\oplus A\oplus A\oplus A$; an element of C is then a quadruple (f, g, h, k) of such functions, which we denote more suggestively as a formal "differential":

$$(f, g, h, k)=f+g\,dx+h\,dy+k\,dx\,dy.$$

Define $d\colon C\to C$ by setting

$$d(f, g, h, k)=\frac{\partial f}{\partial x}\,dx+\frac{\partial f}{\partial y}\,dy+\left(\frac{\partial h}{\partial x}-\frac{\partial g}{\partial y}\right)dx\,dy.$$

That $d^2=0$ is a consequence of the fact that $\dfrac{\partial^2 f}{\partial x\,\partial y}=\dfrac{\partial^2 f}{\partial y\,\partial x}$. Any cycle in C is a sum of the following three types: a constant $f=a$; an expression

$g\,dx+h\,dy$ with $\partial g/\partial y=\partial h/\partial x$ (in other words, an exact differential); an expression $k\,dx\,dy$. If the domain D of definition is, say, the interior of the square we can write the function k as $\partial h/\partial x$ for a suitable h, while any exact differential can be expressed (by suitable integration) as the differential of a function f. Hence, for this D the only homology classes are those yielded by the constant functions, and $H(C)$ is the additive group of real numbers. The same conclusion holds if D is the interior of a circle, but fails if D is, say, the interior of a circle with the origin deleted. In this latter case an exact differential need not be the differential of a function f. For example $(-y\,dx+x\,dy)/(x^2+y^2)$ is not such.

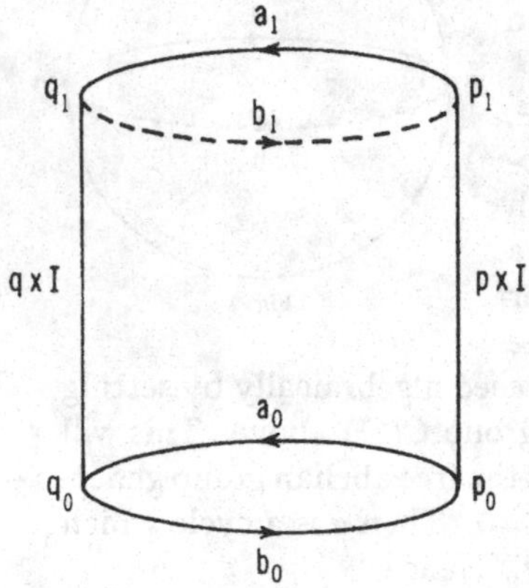

Fig. 3

Example 7. A circular cylinder may be regarded as the cartesian product $S^1\times I$ of a circle S^1 and a unit interval I. We subdivide this as shown, so that the circle S^1 at the base has vertices p_0, q_0 and edges a_0 and b_0, while those at the top are given by the same letters with subscript 1. The sides of the cylinder are the intervals $p\times I$ and $q\times I$ above p_0 and q_0, respectively, and the curved faces $a\times I$ and $b\times I$ above a_0 and b_0. Introduce the free abelian group $C(S^1\times I)$ with the twelve free generators $p\times I$, $q\times I$, $a\times I$, $b\times I$, and p_i, q_i, a_i, b_i, $(i=0, 1)$. Define $d\colon C\to C$ on the base and top exactly as for a circle $(da_i=q_i-p_i,\ db_i=p_i-q_i,\ dp_i=0=dq_i)$. Also set $d(p\times I)=p_1-p_0$ and $d(q\times I)=q_1-q_0$. Inspection of the geometric boundary of the curved surface $a\times I$ suggests that we set

$$d(a\times I)=a_1-(q\times I)-a_0+(p\times I) \text{ and } d(b\times I)=b_1-b_0+(q\times I)-(p\times I).$$

This defines d so that $d^2=0$. Inspection of the cycles and boundaries shows that

$$H(C(S^1\times I))=Z_\infty(\operatorname{cls}(p_0))\oplus Z_\infty(\operatorname{cls}(a_0+b_0)).$$

This homology group is isomorphic to the homology group $H(S^1)$ found for the circle in Example 1 above. The isomorphism can be written as $H(f_0)\colon H(S^1)\cong H(S^1\times I)$ if we take f_0 to be the homomorphism $f_0\colon C(S^1)\to C(S^1\times I)$ of differential groups defined by $f_0p=p_0$, $f_0q=q_0$, $f_0a=a_0$, $f_0b=b_0$. One could equally well give the isomorphism as $H(f_1)$, where the homomorphism $f_1=C(S^1)\to C(S^1\times I)$ is similarly defined. This equality $H(f_0)=H(f_1)$ holds because the cycles a_0+b_0 and a_1+b_1 on the cylinder are homologous, for their difference is the boundary

$$d(a\times I+b\times I)=(a_1+b_1)-(a_0+b_0).$$

To explicitly compare f_0 and f_1, let us define a function s by

$$sp=p\times I,\quad sq=q\times I,\quad sa=a\times I,\quad sb=b\times I.$$

This determines a homomorphism $s\colon C(S^1)\to C(S^1\times I)$ of abelian groups (not of differential groups) with the property that

$$ds\,c+s\,dc=f_1c-f_0c \tag{1.7}$$

for all c in $C(S^1)$. This equation may be read: The boundary $d(sc)$ of the cylinder sc over c consists of the top f_1c minus the bottom f_0c minus the cylinder $s(dc)$ over the boundary of c. This equation implies that the homomorphisms $H(f_1)$ and $H(f_0)$ are equal, for if c is any cycle ($dc=0$), then (1.7) gives $f_1c-f_0c=d(sc)$, whence $f_1c\sim f_0c$.

Maps with the property (1.7) will appear frequently under the name of "chain homotopies".

Exercises

1. Let C be a differential group. The definition $H(C)=\operatorname{Ker}d/\operatorname{Im}d$ can be written as $H(C)=\operatorname{Coker}(d'\colon C\to\operatorname{Ker}d)$, where d' is induced by d. Using $C/\operatorname{Ker}d\cong\operatorname{Im}d$, show that $H(C)$ has a dual description as $\operatorname{Ker}(d''\colon(\operatorname{Coker}d)\to C)$.

2. For a family C_t, $t\in T$, of differential groups, define the direct sum $\sum C_t$ and the direct product $\prod C_t$ and prove that $H(\sum C_t)\cong\sum H(C_t)$, $H(\prod C_t)\cong\prod H(C_t)$.

2. Complexes

In the usual differential groups C of § 1 we can assign integral dimensions to certain elements of C. The set C_n of all elements of dimension n is a group, C is the direct sum of the C_n, and $\partial C_n\subset C_{n-1}$. It is more effective to work directly with this collection of groups. The resulting object is called a "complex" of abelian groups.

For any ring R, a *chain complex* K of R-modules is a family $\{K_n,\partial_n\}$ of R-modules K_n and R-module homomorphisms $\partial_n\colon K_n\to K_{n-1}$, defined for all integers n, $-\infty<n<\infty$, and such that $\partial_n\partial_{n+1}=0$. This last condition is equivalent to the statement that $\operatorname{Ker}\partial_n\supset\operatorname{Im}\partial_{n+1}$. A complex K thus appears as a doubly infinite sequence

$$K\colon\cdots\leftarrow K_{-2}\leftarrow K_{-1}\leftarrow K_0\leftarrow K_1\leftarrow K_2\leftarrow\cdots$$

with each composite map zero. The *homology* $H(K)$ is the family of modules

$$H_n(K)=\operatorname{Ker}\partial_n/\operatorname{Im}\partial_{n+1}=(\operatorname{Ker}[K_n\to K_{n-1}])/\partial_{n+1}K_{n+1}. \tag{2.1}$$

Thus $H_n(K)=0$ means that the sequence K is exact at K_n. An *n-cycle* of K is an element of the submodule $C_n(K)=\operatorname{Ker}\partial_n$; an *n-boundary* is an element of $\partial_{n+1}K_{n+1}$. Then $H_n=C_n/\partial K_{n+1}$ (cycles mod boundaries). The coset of a cycle c in H_n is written as $\operatorname{cls}c=c+\partial K_{n+1}$, or as $\{c\}$, in much of the literature. Two n-cycles in the same homology class ($\operatorname{cls} c = \operatorname{cls} c'$) are said to be *homologous* ($c \sim c'$); this is the case if and only if $c - c' \in \partial K_{n+1}$.

If K and K' are complexes, a *chain transformation* $f\colon K \to K'$ is a family of module homomorphisms $f_n\colon K_n \to K'_n$, one for each n, such that $\partial'_n f_n = f_{n-1} \partial_n$ for all n. This last condition asserts the commutativity of the diagram (neglect the dotted arrows)

$$\begin{array}{lccccccc}
K\colon & \cdots \longleftarrow & K_{n-1} & \xleftarrow{\partial_n} & K_n & \xleftarrow{\partial_{n+1}} & K_{n+1} & \longleftarrow \cdots \\
 & & \downarrow f_{n-1} & \searrow & \downarrow f_n & \searrow & \downarrow f_{n+1} & \\
K'\colon & \cdots \longleftarrow & K'_{n-1} & \xleftarrow[\partial'_n]{} & K'_n & \xleftarrow[\partial'_{n+1}]{} & K'_{n+1} & \longleftarrow \cdots .
\end{array} \tag{2.2}$$

(Subsequently, we usually omit the subscript n on ∂_n and the prime on $\partial'\colon K'_n \to K'_{n-1}$.) The function $H_n(f) = f_*$ defined by $f_*(c + \partial K_{n+1}) = fc + \partial K'_{n+1}$ is a homomorphism $H_n(f)\colon H_n(K) \to H_n(K')$. With this definition, each H_n is a covariant functor on the category of chain complexes and chain transformations to the category of modules.

A *chain homotopy* s between two chain transformations $f, g\colon K \to K'$ is a family of module homomorphisms $s_n\colon K_n \to K'_{n+1}$, one for each dimension n, as in the dotted arrows of (2.2), such that

$$\partial'_{n+1} s_n + s_{n-1} \partial_n = f_n - g_n . \tag{2.3}$$

We write $s\colon f \simeq g$. The geometric background of this relation is sketched in Example 7 of § 1. Algebraically we have

Theorem 2.1. *If* $s\colon f \simeq g\colon K \to K'$, *then*

$$H_n(f) = H_n(g)\colon H_n(K) \to H_n(K'), \qquad -\infty < n < \infty . \tag{2.4}$$

Proof. If c is a cycle of K_n, then $\partial_n c = 0$; hence, by (2.3), $f_n c - g_n c = \partial s_n c$. This states that $f_n c$ and $g_n c$ are homologous, hence that $\operatorname{cls} f_n c = \operatorname{cls} g_n c$ in $H_n(K')$, as required.

A chain transformation $f\colon K \to K'$ is said to be a *chain equivalence* if there is another chain transformation $h\colon K' \to K$ (backwards!) and homotopies $s\colon hf \simeq 1_K$, $t\colon fh \simeq 1_{K'}$. Since $H_n(1_K) = 1$, the theorem yields

Corollary 2.2. *If $f\colon K \to K'$ is a chain equivalence, the induced map $H_n(f)\colon H_n(K) \cong H_n(K')$ is an isomorphism for each dimension n.*

Proposition 2.3. *Chain homotopies $s\colon f \simeq g\colon K \to K'$ and $s'\colon f' \simeq g'\colon K' \to K''$ yield a composite chain homotopy*

$$f' s + s' g\colon f' f \simeq g' g\colon K \to K'' .$$

Proof. Both $\partial s + s \partial = f - g$ and $\partial s' + s' \partial = f' - g'$ are given. Multiply the first by f', on the left, and the second by g on the right, and add.

Subcomplexes and quotient complexes have properties like those of submodules and quotient modules. A *subcomplex* S of K is a family of submodules $S_n \subset K_n$, one for each n, such that always $\partial S_n \subset S_{n-1}$. Hence S itself is a complex with boundary induced by $\partial = \partial_K$, and the injection $j: S \to K$ is a chain transformation. If $S \subset K$, the *quotient complex* K/S is the family $(K/S)_n = K_n/S_n$ of quotient modules with boundary $\partial': K_n/S_n \to K_{n-1}/S_{n-1}$ induced by ∂_K. The projection is a chain transformation $K \to K/S$, and the short sequence $S_n \rightarrowtail K_n \twoheadrightarrow (K/S)_n$ of modules is exact for each n.

If $f: K \to K'$ is a chain transformation, then $\operatorname{Ker} f = \{\operatorname{Ker} f_n\}$ is a subcomplex of K, $\operatorname{Im} f = \{f_n K_n\}$ a subcomplex of K', while $K'/\operatorname{Im} f$ is the cokernel of f and $K/\operatorname{Ker} f$ the coimage. A pair of chain transformations $K \xrightarrow{f} K' \xrightarrow{g} K''$ is *exact* at K' if $\operatorname{Im} f = \operatorname{Ker} g$; that is, if each sequence $K_n \to K'_n \to K''_n$ of modules is exact at K'_n. For any $f: K \to K'$,

$$0 \to \operatorname{Ker} f \to K \xrightarrow{f} K' \to \operatorname{Coker} f \to 0$$

is an exact sequence of complexes.

Instead of using lower indices, as in K_n, it is often notationally convenient to write K^n for K_{-n} and $\delta^n: K^n \to K^{n+1}$ in place of $\partial_{-n}: K_{-n} \to K_{-n-1}$. This is simply a different "upper index" notation for the same complex.

A complex K is *positive* (i. e., non-negative) if $K_n = 0$ for $n < 0$; its homology is then positive $(H_n(K) = 0$ for $n < 0)$. A complex K is *negative* if $K_n = 0$ for $n > 0$; equivalently, it is positive in the upper indices and has the form

$$0 \to K^0 \xrightarrow{\delta^0} K^1 \xrightarrow{\delta^1} K^2 \xrightarrow{\delta^2} \cdots, \qquad \delta\delta = 0,$$

with homology $H^n(K) = \operatorname{Ker} \delta^n / \delta K^{n-1}$ positive in the upper indices. In this form, it is often called a "right complex" or a "cochain complex". By a "cochain" homotopy $s: f \simeq g: K \to K'$ is meant a chain homotopy written with upper indices; that is, a family of maps $s^n: K^n \to K'^{\,n-1}$ with $\delta s + s\delta = f - g$. The complexes arising in practice are usually positive or negative; the general notion of a chain complex is useful to provide common proofs of formal properties like those expressed in Thm. 2.1.

Each module A may be regarded as a "trivial" positive complex, with $A_0 = A$, $A_n = 0$ for $n \neq 0$, and $\partial = 0$. A *complex over* A is a positive complex K together with a chain transformation $\varepsilon: K \to A$; such an ε is simply a module homomorphism $\varepsilon: K_0 \to A$ such that $\varepsilon\partial = 0: K_1 \to A$. A *contracting homotopy* for $\varepsilon: K \to A$ is a chain transformation $f: A \to K$ such that $\varepsilon f = 1_A$ together with a homotopy $s: 1 \simeq f\varepsilon$. In other words, a contracting homotopy consists of module homomorphisms $f: A \to K_0$ and $s_n: K_n \to K_{n+1}$, $n = 0, 1, \ldots$ such that

$$\varepsilon f = 1, \quad \partial_1 s_0 + f\varepsilon = 1_{K_0}, \quad \partial_{n+1} s_n + s_{n-1}\partial_n = 1 \quad (n > 0). \qquad (2.5)$$

Equivalently, extend the complex by setting $K_{-1}=A$, $\partial_0=\varepsilon\colon K_0\to K_{-1}$ and $s_{-1}=f$. Then (2.5) states simply that $s\colon 1\simeq 0$ for the maps 1, 0 of the extended complex to itself. If $\varepsilon\colon K\to A$ has a contracting homotopy, its homology groups are $\varepsilon_*\colon H_0(K)\cong A$ for $n=0$ and $H_n(K)=0$ for $n>0$.

Complexes K of free abelian groups arise in topology. If each K_n is finitely generated, then each $H_n(K)$ is a finitely generated abelian group. The structure theorem for such groups presents $H_n(K)$ as a direct sum $Z\oplus\cdots\oplus Z\oplus Z_{m_1}\oplus\cdots\oplus Z_{m_k}$, where the number b_n of infinite cyclic summands and the integers $m_1,\ldots,m_k$ (each a divisor of the next) depend only on $H_n(K)$. The integer b_n is called the n-th *Betti number* of K, and the $\{m_i\}$ the n-th *torsion coefficients*.

Exercises

1. Call a complex S *q-special* if $S_n=0$ for $n\neq q, q+1$ and $\partial\colon S_{q+1}\to S_q$ is a monomorphism. Prove that any complex K of free abelian groups K_n is a direct sum of q-special complexes (one for each q).

2. Call a q-special complex S of abelian groups *elementary* if either $S_q=S_{q+1}=Z$ or $S_q=Z$, $S_{q+1}=0$. Prove that each special S with S_q, S_{q+1} finitely generated free groups is a direct sum of elementary complexes. (Hint: use row and column operations on matrices of integers to choose new bases for S_q and S_{q+1}.)

3. Prove that any complex with each K_n a finitely generated free abelian group is a direct sum of elementary complexes.

3. Cohomology

Let C be a differential group and G an abelian group. Form the abelian group $C^*=\mathrm{Hom}_Z(C,G)$; its elements are the group homomorphisms $f\colon C\to G$, called *cochains* of C with "coefficients" in G. The differential $d\colon C\to C$ induces a map $d^*\colon C^*\to C^*$ defined by $d^*f=fd\colon C\to G$; call d^*f the *coboundary* of the cochain f; it is often written as $\delta f=d^*f$. Since $d^2=0$, $(d^*)^2=0$. Hence C^* with differential d^* is a differential group. Its homology is called the *cohomology* of C with coefficients G and is written $H^*(C,G)=H(\mathrm{Hom}(C,G))$.

Let K be a complex of R-modules and G an R-module. Form the abelian group $\mathrm{Hom}_R(K_n,G)$; its elements are the module homomorphisms $f\colon K_n\to G$, called *n-cochains* of K. The *coboundary* of f is the $(n+1)$-cochain

$$\delta^n f=(-1)^{n+1}f\,\partial_{n+1}\colon K_{n+1}\to G. \tag{3.1}$$

In other words, $\partial_{n+1}\colon K_{n+1}\to K_n$ induces $\partial^*_{n+1}\colon \mathrm{Hom}(K_n,G)\to \mathrm{Hom}(K_{n+1},G)$ and $\delta^n=(-1)^{n+1}\partial^*_{n+1}$ (the sign will be explained below). Since $\delta^n\delta^{n-1}=0$, the sequence

$$\cdots\longrightarrow \mathrm{Hom}_R(K_{n-1},G)\xrightarrow{\delta^{n-1}}\mathrm{Hom}_R(K_n,G)\xrightarrow{\delta^n}\mathrm{Hom}_R(K_{n+1},G)\longrightarrow\cdots \tag{3.2}$$

is a complex of abelian groups called $\mathrm{Hom}_R(K,G)$, usually written with upper indices as $\mathrm{Hom}^n(K,G)=\mathrm{Hom}(K_n,G)$. If K is positive in lower indices, $\mathrm{Hom}(K,G)$ is positive in upper indices.

The homology of this complex $\mathrm{Hom}(K,G)$ is called the *cohomology* of K with *coefficients* in G. With upper indices, it is the family of abelian groups

$$H^n(K,G)=H^n(\mathrm{Hom}(K,G))=\mathrm{Ker}\,\delta^n/\delta\,\mathrm{Hom}(K_{n-1},G). \qquad (3.3)$$

An element of $\delta\,\mathrm{Hom}(K_{n-1},G)$ is called an *n-coboundary* and an element of $\mathrm{Ker}\,\delta^n$ an *n-cocycle*. Thus a cocycle is a homomorphism $f\colon K_n\to G$ with $f\partial=0\colon K_{n+1}\to G$. Any chain transformation $h\colon K\to K'$ induces a chain transformation $h^*=\mathrm{Hom}(h,1)\colon \mathrm{Hom}(K',G)\to\mathrm{Hom}(K,G)$. Thus $\mathrm{Hom}(K,G)$ and $H^n(K,G)$ are bifunctors, covariant in G and contravariant in K. If $s\colon h\simeq g$ is a homotopy, then (2.3) implies that $s_n^*\partial_{n+1}^*+\partial_n^* s_{n-1}^*=h_n^*-g_n^*$. Hence $t^{n+1}=(-1)^{n+1}s_n^*$ is a homotopy $t\colon h^*\simeq g^*$.

More generally, we may define a complex $\mathrm{Hom}_R(K,L)$ from any pair of complexes K and L of R-modules. With lower indices, set

$$\mathrm{Hom}_n(K,L)=\prod_{p=-\infty}^{\infty}\mathrm{Hom}_R(K_p,L_{p+n}), \qquad (3.4)$$

so that an element f of Hom_n is a family of homomorphisms $f_p\colon K_p\to L_{p+n}$ for $-\infty<p<\infty$. The boundary $\partial_H f$ is the family $(\partial_H f)_p\colon K_p\to L_{p+n-1}$ defined by

$$(\partial_H f)_p k=\partial_L(f_p k)+(-1)^{n+1}f_{p-1}(\partial_K k),\qquad k\in K_p,\qquad f\in\mathrm{Hom}_n, \qquad (3.5)$$

where ∂_L and ∂_K denote the boundary operators in K and L, respectively. That this definition yields a complex is proved by the calculation:

$$(\partial_H\partial_H f)_p k=\partial_L\partial_L(f_p k)+(-1)^n\partial_L f_{p-1}\partial_K k$$
$$+(-1)^{n+1}\partial_L f_{p-1}\partial_K k+(-1)^1 f_{p-2}(\partial_K\partial_K k)=0,$$

since $\partial_L\partial_L=0=\partial_K\partial_K$. Clearly, $\mathrm{Hom}_R(K,L)$ is a bifunctor covariant in L and contravariant in K.

The signs in the definition (3.5) have been chosen so as to give the following two results.

Proposition 3.1. *When the ring R is regarded as a trivial complex, then* $\mathrm{Hom}(R,L)\cong L$ *under the natural homomorphism which assigns to each* $f_p\colon R\to L_p$ *its image* $f_p(1)\in L_p$.

Proof. This correspondence gives an isomorphism $\mathrm{Hom}(R,L_p)\cong L_p$ for each p. In this case the boundary formula (3.5) has no terms with ∂_K; the remaining term with $+\partial_L f$ shows that this isomorphism commutes with boundaries.

Proposition 3.2. *A 0-dimensional cycle of* Hom(K,L) *is a chain transformation* $f\colon K\to L$; *it is the boundary of an element s in* $\mathrm{Hom}_1(K,L)$ *exactly when s is a homotopy* $s\colon f\simeq 0$.

Proof. The formula (3.5) for the boundary (*with* signs) becomes

$$(\partial_H f)_p=\partial_L f_p - f_{p-1}\partial_K,\qquad n=0,$$
$$(\partial_H s)_p=\partial_L s_p + s_{p-1}\partial_K,\qquad n=1.$$

Thus $\partial_H f=0$ asserts that $f\colon K\to L$ is a chain transformation and $\partial_H s=f$ asserts that $f=\partial_L s+s\partial_K$, whence $s\colon f\simeq 0$, as asserted. These conclusions may be reformulated as

Corollary 3.3. *The homology group* $H_0(\mathrm{Hom}(K,L))$ *is the abelian group of homotopy classes of chain transformations* $f\colon K\to L$.

In particular, when $L=G$ is a trivial complex, the boundary ∂_L is zero, an element f of $\mathrm{Hom}_n(K,G)$ is a single homomorphism $f\colon K_{-n}\to G$, and $\partial_H f=(-1)^{n+1}f\partial\colon K_{-n+1}\to G$. With upper indices, this states that an element of $\mathrm{Hom}^n(K,G)$ is a homomorphism $f\colon K_n\to G$ with coboundary $\delta f=(-1)^{n+1}f\partial$. This agrees with the sign already used in (3.1), and explains the sign there. The reader should be warned, however, that most of the present literature on cohomology does not use this sign, and writes instead $\delta f=f\partial$.

4. The Exact Homology Sequence

Consider any short exact sequence

$$E\colon\ 0\longrightarrow K\xrightarrow{\varkappa}L\xrightarrow{\sigma}M\longrightarrow 0 \tag{4.1}$$

of chain complexes and chain transformations $\varkappa,\sigma$. The first transformation $\varkappa$ has kernel zero, but the induced map $H_n(\varkappa)\colon H_n(K)\to H_n(L)$ on homology may have a non-trivial kernel, as in Example 1.3. To study when this can happen, identify K with the subcomplex $\varkappa K$ of L and consider a cycle c of K_n whose homology class becomes zero in L. This means that $c=\partial l$ for some $(n+1)$-chain $l\in L$, and hence that the coset $l+K_{n+1}$ is a cycle of the quotient complex $L/K\cong M$. Conversely, any homology class of $H_{n+1}(L/K)$ consists of cycles $l+K_{n+1}$ with $\partial l=c\in K_n$, hence yields a homology class $\mathrm{cls}\,c$ in $H_n(K)$ which is in the kernel of $H_n(\varkappa)$. This correspondence of $l+K_{n+1}$ to c is a homomorphism $H_{n+1}(L/K)\to H_n(K)$ which we now describe systematically.

In (4.1), let m be a cycle in M_{n+1}. Since σ is an epimorphism, one can choose $l\in L_{n+1}$ with $\sigma l=m$. Since $\partial m=0$, one has $\sigma\partial l=0$; since E is exact, there is a unique cycle $c\in K_n$ with $\varkappa c=\partial l$, as in

$$\begin{array}{ccccccccc} & l & \longrightarrow & m & & & L_{n+1} & \xrightarrow{\sigma} & M_{n+1}\longrightarrow 0\\ & \downarrow & & \downarrow & & & \downarrow{\scriptstyle\partial} & & \downarrow\\ c\longrightarrow & \partial l & \longrightarrow & 0 & \text{in} & K_n\xrightarrow{\varkappa} & L_n & \longrightarrow & M_n. \end{array}$$

The homology class $\operatorname{cls}(c) \in H_n(K)$ is independent of the choice of l with $\sigma l = m$, depends only on the homology class of m, and is additive in m. Hence $\partial_E(\operatorname{cls} m) = \operatorname{cls} c$ defines a homomorphism

$$\partial_E\colon H_{n+1}(M) \to H_n(K) \tag{4.2}$$

called the *invariant boundary* or the *connecting homomorphism* for E. Specifically

$$\partial_E \operatorname{cls} m = \operatorname{cls} c \quad \text{when} \quad \varkappa c = \partial l, \quad \sigma l = m \quad \text{for some } l. \tag{4.3}$$

This suggests the notation $c = \varkappa^{-1}\partial\sigma^{-1} m$; or regard cls as a homomorphism $\operatorname{cls}_K\colon C_n(K) \to H_n(K)$; then ∂_E is defined by a "switchback" formula $\partial_E = (\operatorname{cls}_K)\,\varkappa^{-1}\partial\,\sigma^{-1}(\operatorname{cls}_M)^{-1}$ — even though the inverses cls^{-1}, $\varkappa^{-1}$, σ^{-1} are not strictly defined (but see §6 below).

Theorem 4.1. *(Exact homology sequence.) For each short exact sequence* (4.1) *of chain complexes the corresponding long sequence*

$$\cdots \to H_{n+1}(M) \xrightarrow{\partial_E} H_n(K) \xrightarrow{\varkappa_*} H_n(L) \xrightarrow{\sigma_*} H_n(M) \xrightarrow{\partial_E} H_{n-1}(K) \to \cdots \tag{4.4}$$

of homology groups, with maps the connecting homomorphism ∂_E, $\varkappa_* = H_n(\varkappa)$, *and* $\sigma_* = H_n(\sigma)$, *is exact.*

This sequence (4.4) is infinite in both directions, but is zero for $n<0$ when the complexes are positive. It gives the desired description of the kernel and cokernel of $H_n(\varkappa)\colon H_n(K) \to H_n(L)$ when $\varkappa$ is a monomorphism; namely the kernel is $\partial_E H_{n+1}(M)$, and the cokernel is isomorphic to $\sigma_* H_n(L)$.

Proof. By the definitions, the composite of any two successive homomorphisms in the sequence (4.4) is the zero homomorphism. It remains to show for each dimension n that (i) $\operatorname{Ker}\varkappa_* \subset \partial_E H_{n+1}(M)$; (ii) $\operatorname{Ker}\sigma_* \subset \varkappa_* H_n(K)$; (iii) $\operatorname{Ker}\partial_E \subset \sigma_* H_n(L)$. Our preparatory discussion showed the first true.

To prove the second inclusion, suppose that $\operatorname{cls}(c)$ is the homology class of a cycle c of L_n such that $\sigma_* \operatorname{cls}(c) = 0$. This means that $\sigma c = \partial m$ for some $m \in M_{n+1}$. Since σ is an epimorphism, there is a $l \in L_{n+1}$ with $\sigma l = m$. Hence $\sigma(c - \partial l) = 0$, so that $c - \partial l = \varkappa k$ for some $k \in K_n$ with $\partial k = 0$. This asserts that $\operatorname{cls}(c) = \operatorname{cls}(c - \partial l) = \varkappa_* \operatorname{cls}(k)$ is in the image of $\varkappa_*$.

To prove the third inclusion, recall that $\partial_E \operatorname{cls}(m) = \operatorname{cls} c$, where $c \in K_{n-1}$ and $l \in L_n$ have $\varkappa c = \partial l$, $\sigma l = m$, as in (4.3). If $\operatorname{cls} c = 0$, there is a k' in K_n with $\partial k' = c$. Then $\varkappa\partial k' = \partial l$, hence $\partial(l - \varkappa k') = 0$. Thus $l - \varkappa k'$ is a cycle of L, and $\sigma(l - \varkappa k') = \sigma l = m$, so that $\operatorname{cls}(m) \in \operatorname{Im}\sigma_*$, as asserted. This completes the proof.

Consider the category $\mathscr{E}$ of short exact sequences of chain complexes. A morphism $E \to E'$ in this category is a triple (f, g, h) of chain transformations which render the diagram

$$\begin{array}{ccccccccc} E\colon & 0 \to & K & \to & L & \xrightarrow{\sigma} & M & \to 0 \\ & & \downarrow f & & \downarrow g & & \downarrow h & \\ E'\colon & 0 \to & K' & \to & L' & \to & M' & \to 0 \end{array} \tag{4.5}$$

commutative. For each n, $H_n(K)$, $H_n(L)$, and $H_n(M)$ are functors of E.

Proposition 4.2. *For each $E \in \mathscr{E}$, the connecting homomorphism $\partial_E\colon H_{n+1}(M) \to H_n(K)$ is natural.*

The statement that ∂_E is natural means exactly that the diagram

$$\begin{array}{ccc} H_{n+1}(M) & \xrightarrow{\partial_E} & H_n(K) \\ \downarrow H_{n+1}(h) & & \downarrow H_n(f) \\ H_{n+1}(M') & \xrightarrow{\partial_{E'}} & H_n(K') \end{array} \tag{4.6}$$

is commutative. The proof is an easy diagram chase in (4.5) with the definition of ∂_E. The conclusion can be expressed in a bigger diagram:

$$\begin{array}{ccccccccc} \cdots \to & H_{n+1}(L) & \xrightarrow{\sigma_*} & H_{n+1}(M) & \xrightarrow{\partial_E} & H_n(K) & \xrightarrow{\varkappa_*} & H_n(L) & \to \cdots \\ & \downarrow g_* & & \downarrow h_* & & \downarrow f_* & & \downarrow g_* & \\ \cdots \to & H_{n+1}(L') & \xrightarrow{\sigma'_*} & H_{n+1}(M') & \xrightarrow{\partial_{E'}} & H_n(K') & \xrightarrow{\varkappa'_*} & H_n(L') & \to \cdots. \end{array} \tag{4.7}$$

Here the rows are the exact homology sequences of Thm. 4.1 for E and E' and the whole diagram is commutative; for example, the left hand square because $\sigma'_* g_* = (\sigma' g)_*$, $h_* \sigma_* = (h\sigma)_*$ and $\sigma' g = h\sigma$ by the commutativity of (4.5). The conclusion may be formulated thus: A morphism of E to E' induces a morphism of the exact homology sequence of E to that of E'.

The mapping cone of a chain transformation $f\colon K \to K'$ gives an example of this exact sequence. The problem is that of fitting the induced maps $f_*\colon H_n(K) \to H_n(K')$ on homology into an exact sequence. For this purpose, construct a complex $M = M(f)$, called the *mapping cone* of f (or sometimes, with less accuracy, the *mapping cylinder* of f), with

$$M_n = K_{n-1} \oplus K'_n, \qquad \partial(k, k') = (-\partial k, \partial k' + fk).$$

Then $\partial\colon M_n \to M_{n-1}$ satisfies $\partial^2 = 0$, so M is a complex, and the injection $\iota\colon K' \to M$ is a chain transformation. The projection $\pi\colon M \to K^+$ with $\pi(k, k') = k$ is also a chain transformation, if by K^+ we mean the complex K with the dimensions all raised by one and the sign of the boundary changed (i.e., $(K^+)_n = K_{n-1}$). Moreover $E_f\colon K' \rightarrowtail M \twoheadrightarrow K^+$ is a short exact sequence of complexes. Hence

Proposition 4.3. *A chain transformation $f\colon K \to K'$ with mapping cone $M(f)$ determines an exact sequence*

$$\cdots \to H_n(K') \xrightarrow{i_*} H_n(M(f)) \xrightarrow{\pi_*} H_{n-1}(K) \xrightarrow{f_*} H_{n-1}(K') \to \cdots.$$

Proof. This is the exact sequence of E_f, with $H_n(K^+) \cong H_{n-1}(K)$; moreover the connecting homomorphism $\partial_{E_f}\colon H_n(K^+) \to H_{n-1}(K')$ can be seen to be identical with the homomorphism induced by f.

The mapping cone is the algebraic analogue of the following geometric construction. Let $f\colon X \to X'$ be a continuous map of topological spaces. Form the cone over X by taking the cartesian product $X \times I$ with the unit interval I and identifying all points $(x,0)$ for $x \in X$. Attach this cone to X' by identifying each point $(x,1)$ of $X \times I$ with $f(x) \in X'$; the resulting space is the mapping cone of f, and suggests our boundary formula. DOLD [1960] gives a further development of these ideas.

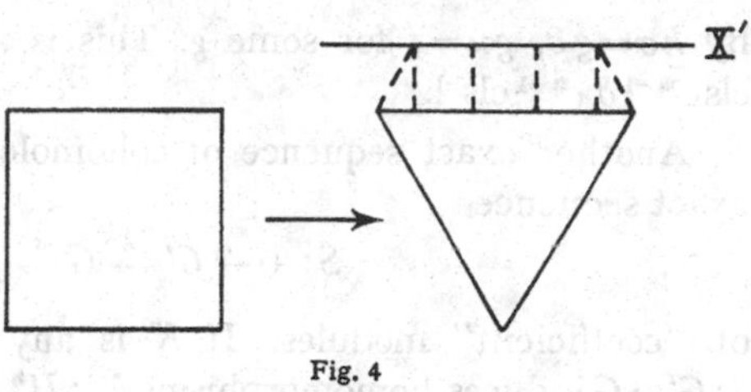

Fig. 4

We now consider exact cohomology sequences. A short exact sequence E of complexes of R-modules is said to split as a sequence of modules if for each n the sequence $K_n \rightarrowtail L_n \twoheadrightarrow M_n$ splits; that is, if for each n, K_n is a direct summand of L_n. For example, if each M_n is a projective module, then E of (4.1) splits as a sequence of modules, by Thm. I.6.3.

Theorem 4.4. *If G is an R-module and E a short exact sequence* (4.1) *of complexes of R-modules which splits as a sequence of modules, then there is for each dimension n a natural connecting homomorphism $\delta_E\colon H^n(K,G) \to H^{n+1}(M,G)$ such that the sequence of cohomology groups*

$$\cdots \to H^n(M,G) \xrightarrow{\sigma^*} H^n(L,G) \xrightarrow{\varkappa^*} H^n(K,G) \xrightarrow{\delta_E} H^{n+1}(M,G) \to \cdots \qquad (4.8)$$

is exact.

Proof. To construct the cohomology of E, first apply the contravariant functor $\mathrm{Hom}_R(-,G)$ to E to get the reversed sequence of complexes

$$E^*\colon\ 0 \to \mathrm{Hom}(M,G) \to \mathrm{Hom}(L,G) \to \mathrm{Hom}(K,G) \to 0.$$

Since the given sequence splits as a sequence of modules, E^* is exact. The connecting homomorphism $\partial_{E^*}\colon H_{-n+1}(\mathrm{Hom}(K,G)) \to H_{-n}(\mathrm{Hom}(M,G))$ for E^*, when written with upper indices $H^{n-1} = H_{-n+1}$, is the desired connecting homomorphism δ_E. By Prop. 4.2 it follows that δ_E is natural when the arguments $H^n(K,G)$ and $H^{n+1}(M,G)$ are regarded as contravariant functors on the category of those short exact sequences of complexes which split as modules. For that matter, δ_E is also natural when its arguments are regarded as covariant functors of the R-module G. Finally, the exact homology sequence for E^*, with indices shifted up, becomes the desired exact cohomology sequence (4.8).

For reference we describe the action of δ_E in terms of cochains. Since E^* is exact, each n-cocycle of K, regarded as a homomorphism $f: K_n \to G$, can be written as $f = g\varkappa$ where $g: L_n \to G$ is an n-cochain of L. Then $g\partial\varkappa = g\varkappa\partial = f\partial = 0$, so $g\partial$ factors through σ as $g\partial = h\sigma$ for some $h: M_{n+1} \to G$. Since $h\partial\sigma = h\sigma\partial = g\partial\partial = 0$, and σ is an epimorphism, it follows that $h\partial = 0$: h is a cocycle of M. Then

$$\delta_E \operatorname{cls} f = \operatorname{cls} h \quad \text{defines} \quad \delta_E: H^n(K,G) \to H^{n+1}(M,G) \tag{4.9}$$

by $h\sigma = g\partial$, $g\varkappa = f$ for some g. This is again a switchback rule: $\delta_E = \operatorname{cls} \sigma^{*-1} \delta \varkappa^{*-1} \operatorname{cls}^{-1}$.

Another exact sequence of cohomology groups arises from a short exact sequence

$$S:\ 0 \to G' \xrightarrow{\lambda} G \xrightarrow{\tau} G'' \to 0 \tag{4.10}$$

of "coefficient" modules. If K is any complex, the monomorphism $\lambda: G' \to G$ induces homomorphisms $\lambda_*: H^n(K,G') \to H^n(K,G)$. The inquiry as to the kernel and the cokernel of λ_* is met by the following exact sequence (which is *not* a dual to that of Thm. 4.4):

Theorem 4.5. *If K is a complex of R-modules with each module K_n projective and if S is a short exact sequence of R-modules, as in* (4.10), *there is for each dimension a connecting homomorphism $\delta_S: H^n(K,G'') \to H^{n+1}(K,G')$ which is natural when its arguments are regarded as covariant functors of the exact sequence S or as contravariant functors of K and which yields the long exact sequence:*

$$\cdots \to H^n(K,G') \xrightarrow{\lambda_*} H^n(K,G) \xrightarrow{\tau_*} H^n(K,G'') \xrightarrow{\delta_S} H^{n+1}(K,G') \to \cdots. \tag{4.11}$$

Proof. Since each K_n is projective,

$$S_*:\ 0 \to \operatorname{Hom}(K,G') \to \operatorname{Hom}(K,G) \to \operatorname{Hom}(K,G'') \to 0$$

is exact, and yields δ_S as ∂_{S_*}, with the usual shift to upper indices, and with (4.11) as a consequence of Thm. 4.1.

Note the explicit rule for constructing δ_S. Let $f: K_n \to G''$ be a cocycle. Since S_* is exact, we may write $f = \tau g$ for $g: K_n \to G$ a cochain; since f is a cocycle, $g\partial = \lambda h$, where $h: K_{n+1} \to G'$ is a cocycle. Then

$$\delta_S \operatorname{cls} f = \operatorname{cls} h, \quad \lambda h = g\partial, \quad \tau g = f. \tag{4.12}$$

This is again a switchback rule: $\delta_S = \operatorname{cls} \lambda^{-1} \delta\, \tau^{-1} \operatorname{cls}^{-1}$.

Exercises

1. If $f, g: K \to K'$ are chain homotopic, show that the associated exact sequences for the mapping cones $M(f)$ and $M(g)$ are isomorphic.

2. (The BOCKSTEIN Operator.) Let K be a complex of free abelian groups, Z_p the additive group of integers modulo the prime p, and $S=(\lambda, \tau): Z \rightarrowtail Z \twoheadrightarrow Z_p$ the short exact sequence with λ multiplication by p. Construct the corresponding exact sequence (4.11) and show that $\beta = \tau_* \delta_S: H^n(K, Z_p) \to H^{n+1}(K, Z_p)$ can be described as follows. Lift each n-cocycle $c: K_n \to Z_p$ to an n-cochain $a: K_n \to Z$; then $\delta a = pb$ for some $b: K_{n+1} \to Z$, and $\beta(\operatorname{cls} c) = \operatorname{cls}(\tau b)$. This β is known as the BOCKSTEIN cohomology operator [cf. BROWDER 1961].

3. Let $f: K \to K'$ have mapping cone M, kernel L, and cokernel N, so that $F: L \rightarrowtail K \twoheadrightarrow fK$, $G: fK \rightarrowtail K' \twoheadrightarrow N$ are short exact sequences of complexes. Construct chain transformations $g: L^+ \to M$ and $h: M \to N$ by $g(l) = (l, 0)$, $h(k, k') = k' + fK$, and show the sequence

$$\cdots \to H_{n-1}(L) \xrightarrow{g_*} H_n(M) \xrightarrow{h_*} H_n(N) \xrightarrow{\eta} H_{n-2}(L) \to \cdots$$

exact, where $\eta = \partial_F \partial_G$ is the composite of the connecting homomorphisms for F, G.

4. Show that the exact sequence of Ex. 3, that of Prop. 4.3, and those for F and G all appear in a "braid" diagram

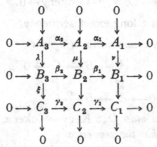

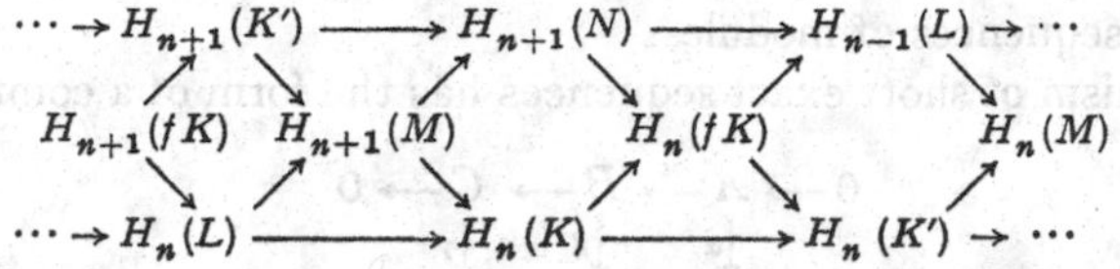

which is commutative except for a sign (-1) in the middle diamond [MAC LANE 1960b].

5. Some Diagram Lemmas

An application of the exact homology sequence is

Lemma 5.1. *(The 3×3 Lemma.) In the following commutative diagram of modules*

$$\begin{array}{ccccccccc}
 & & 0 & & 0 & & 0 & & \\
 & & \downarrow & & \downarrow & & \downarrow & & \\
0 & \to & A_3 & \xrightarrow{\alpha_2} & A_2 & \xrightarrow{\alpha_1} & A_1 & \to & 0 \\
 & & \downarrow{\scriptstyle\lambda} & & \downarrow{\scriptstyle\mu} & & \downarrow{\scriptstyle\nu} & & \\
0 & \to & B_3 & \xrightarrow{\beta_2} & B_2 & \xrightarrow{\beta_1} & B_1 & \to & 0 \\
 & & \downarrow{\scriptstyle\xi} & & \downarrow & & \downarrow & & \\
0 & \to & C_3 & \xrightarrow{\gamma_2} & C_2 & \xrightarrow{\gamma_1} & C_1 & \to & 0 \\
 & & \downarrow & & \downarrow & & \downarrow & & \\
 & & 0 & & 0 & & 0 & &
\end{array}$$

suppose that all three columns and the first two rows (or the last two rows) are short exact sequences. Then the remaining row is exact.

Proof. Any sequence $A_3 \to A_2 \to A_1$ with maps α_2, α_1 such that $\alpha_1\alpha_2 = 0$ may be regarded as a chain complex A with α_2, α_1 as the boundary homomorphisms and with non-zero chains only in dimensions 1, 2, and 3. The homology of this complex will vanish (in dimensions 1, 2, and 3) precisely when it is a short exact sequence.

Suppose now that the last two rows are exact. Then, for $a \in A_3$, $\nu\alpha_1\alpha_2 a = \beta_1\beta_2\lambda a = 0$; since ν is a monomorphism, $\alpha_1\alpha_2 a = 0$. Thus the first row is indeed a complex. Since the columns are exact we may now regard the whole 3×3 diagram as a short exact sequence $0 \to A \to B \to C \to 0$ of three complexes. The relative homology sequence now reads

$$\cdots \to H_{n+1}(C) \to H_n(A) \to H_n(B) \to \cdots.$$

But the exactness of rows B and C give $H_{n+1}(C) = 0 = H_n(B)$, so the exactness of the relative homology sequence makes $H_n(A) = 0$ for $n = 1$, 2, and 3.

The argument is similar, given that the first two rows are exact.

The chief result of this chapter — the exactness of the homology sequence (4.4) — can be proved in a different way from a lemma on short exact sequences of modules.

A morphism of short exact sequences has the form of a commutative diagram

$$\begin{array}{ccccccccc} 0 & \to & A & \xrightarrow{\varkappa} & B & \to & C & \to & 0 \\ & & \downarrow{\scriptstyle\alpha} & & \downarrow{\scriptstyle\beta} & & \downarrow{\scriptstyle\gamma} & & \\ 0 & \to & A' & \xrightarrow{\varkappa'} & B' & \to & C' & \to & 0 \end{array} \tag{5.1}$$

with exact rows; the kernel and the cokernel of this morphism are short sequences, but need not be exact (example: map $0 \rightarrowtail A = A$ to $A = A \twoheadrightarrow 0$ with $\beta = 1_A$). The horizontal maps of the diagram do induce maps which give exact sequences

$$0 \to \operatorname{Ker}\alpha \to \operatorname{Ker}\beta \to \operatorname{Ker}\gamma$$

and

$$\operatorname{Coker}\alpha \to \operatorname{Coker}\beta \to \operatorname{Coker}\gamma \to 0.$$

They can be combined in a long exact sequence:

Lemma 5.2. *For any commutative diagram*

$$(D) \qquad \begin{array}{ccccccccc} & & A & \to & B & \xrightarrow{\sigma} & C & \to & 0 \\ & & \downarrow{\scriptstyle\alpha} & & \downarrow{\scriptstyle\beta} & & \downarrow{\scriptstyle\gamma} & & \\ 0 & \to & A' & \xrightarrow{\varkappa'} & B' & \xrightarrow{\sigma'} & C' & & \end{array}$$

with exact rows there is a map D_*: $\operatorname{Ker}\gamma \to \operatorname{Coker}\alpha$, *natural for functors of the diagram* D, *such that the sequence*

$$\operatorname{Ker}\alpha \to \operatorname{Ker}\beta \to \operatorname{Ker}\gamma \xrightarrow{D_*} \operatorname{Coker}\alpha \to \operatorname{Coker}\beta \to \operatorname{Coker}\gamma \tag{5.2}$$

is exact. We call (5.2) *the Ker-Coker sequence.*

Proof. Let ι: $\operatorname{Ker}\gamma \to C$ be the injection, η: $A' \to A'/\alpha A$, the projection. The switchback formula $D_* = \eta\, \varkappa'^{-1} \beta\, \sigma^{-1} \iota$ then defines D_* without ambiguity. To prove the exactness of (5.2), say at $\operatorname{Coker}\alpha$, suppose $\varkappa'_*(a' + \alpha A) = 0$ for some a'. Then $\varkappa' a' = \beta b$ for some b and $\sigma' \varkappa' a = \gamma \sigma b = 0$, so $\sigma b \in \operatorname{Ker}\gamma$ has $D_* \sigma b = a' + \alpha A$, which is the required exactness. The rest of the proof is similar.

We call D_* the *connecting homomorphism* of the diagram D.

Now we prove Thm. 4.1 for the short exact sequence E of complexes $K \rightarrowtail L \twoheadrightarrow M$. Let $C_n(K)$ denote the module of n-cycles of K and form the diagram

$$D(E):\quad \begin{array}{ccccccc} K_n/\partial K_{n+1} & \to & L_n/\partial L_{n+1} & \to & M_n/\partial M_{n+1} & \to & 0 \\ \downarrow \partial_* & & \downarrow \partial_* & & \downarrow \partial_* & & \\ 0 \to C_{n-1}(K) & \to & C_{n-1}(L) & \to & C_{n-1}(M) & & \end{array}$$

with exact rows and vertical maps induced by ∂. The first kernel is $C_n(K)/\partial K_{n+1} = H_n(K)$, and the first cokernel is $C_{n-1}(K)/\partial K_n = H_{n-1}(K)$, so the Ker-Coker sequence (5.2) is

$$H_n(K) \to H_n(L) \to H_n(M) \xrightarrow{D(E)_*} H_{n-1}(K) \to H_{n-1}(L) \to H_{n-1}(M).$$

The middle map $D(E)_*$, as defined by switchback, is identical with the connecting homomorphism ∂_E of Thm. 4.1.

Exercises

1. Prove the 3×3 lemma by diagram chasing, without using the exact homology sequence.

2. If in the hypotheses of the 3×3 lemma one assumes only the first and third rows exact, show that the second row need not be exact, but will be exact if $\beta_1 \beta_2 = 0$.

3. Under the hypotheses of the 3×3 lemma, establish exact sequences

$$0 \to A_3 \to B_3 \oplus A_2 \to B_2 \to C_1 \to 0$$
$$0 \to A_3 \to B_2 \to C_2 \oplus B_1 \to C_1 \to 0.$$

4. In a commutative 3×3 diagram assume only that all three columns are "left exact" (i.e., exact at A and B) and that the last two rows are left exact. Prove that the first row is left exact. If, in addition β_1 and ξ are epimorphisms, prove that the first row is exact.

5. Prove the Ker-Coker sequence from the exact homology sequence. [Hint: Replace A by $\operatorname{Coim}(A \to B)$ and dually for C'.]

6. For any homomorphisms α: $A \to B$, β: $B \to C$ establish an exact sequence

$$0 \to \operatorname{Ker}\alpha \to \operatorname{Ker}\beta\alpha \to \operatorname{Ker}\beta \to \operatorname{Coker}\alpha \to \operatorname{Coker}\beta\alpha \to \operatorname{Coker}\beta \to 0.$$

6. Additive Relations

The "switchback" formulas can be justified in terms of "additive relations". They will appear later in the treatment of spectral sequences.

An *additive relation* r: $A \to B$ is defined to be a submodule of the direct sum $A \oplus B$; in other words, r is a set of pairs (a, b) closed under addition and R-multiples and not empty. The *converse* is the relation $r^{-1}: B \rightharpoonup A$ consisting of all pairs (b,a) with $(a,b) \in r$. If $s: B \rightharpoonup C$ is another additive relation, the *composite* $sr: A \rightharpoonup C$ is the set of all those pairs (a,c) such that there is a $b \in B$ with $(a,b) \in r$ and $(b,c) \in s$. This composition is associative, when defined. The *graph* of a homomorphism $\alpha: A \to B$ is the additive relation consisting of all pairs $(a, \alpha a)$ for $a \in A$; since the composite of two graphs is the graph of the composite homomorphism, we may identify each homomorphism with its graph. The class with objects all modules and morphisms all additive relations $r: A \rightharpoonup B$ is a category — but note that rr^{-1} need not be the identity relation.

For each additive relation $r: A \rightharpoonup B$ introduce the submodules

$$\left.\begin{aligned} \operatorname{Def} r &= [a \mid (\exists b), (a,b) \in r] \qquad & \operatorname{Im} r &= \operatorname{Def} r^{-1}, \\ \operatorname{Ker} r &= [a \mid (a,0) \in r] \qquad & \operatorname{Ind} r &= \operatorname{Ker} r^{-1}. \end{aligned}\right\} \tag{6.1}$$

Here $\operatorname{Ker} r \subset \operatorname{Def} r \subset A$ and $\operatorname{Ind} r \subset \operatorname{Im} r \subset B$. $\operatorname{Def} r$ is the *domain of definition* of r, while Ind is the *"indeterminancy"* of r, and consists of all b with $(0,b) \in r$. Moreover, r is the graph of a homomorphism if and only if $\operatorname{Def} r = A$ and $\operatorname{Ind} r = 0$.

For example, the converse of a homomorphism $\beta: B \to A$ is an additive relation β^{-1} with $\operatorname{Def} \beta^{-1} = \operatorname{Im} \beta$, $\operatorname{Ind} \beta^{-1} = \operatorname{Ker} \beta$. In a complex K the set of pairs $(c, \operatorname{cls} c)$ for $c \in C_n(K)$ is an additive relation cls: $K_n \rightharpoonup H_n(K)$ with $\operatorname{Def}(\operatorname{cls}) = C_n(K)$. With these observations, the "switchback" formula for the connecting homomorphism appears as the composite of additive relations.

Any additive relation can be regarded as a "many-valued" homomorphism; more exactly, as a homomorphism of a submodule to a quotient module:

Proposition 6.1. *Each additive relation* $r: A \rightharpoonup B$ *determines a homomorphism* $r^0: \operatorname{Def} r \to B/(\operatorname{Ind} r)$ *such that*

$$r = \pi^{-1} r^0 j^{-1}, \qquad j: \operatorname{Def} r \to A, \qquad \pi: B \to B/\operatorname{Ind} r, \tag{6.2}$$

where j is the injection and π the projection. Conversely, given a submodule $S \subset A$, a quotient module B/L of B, and a homomorphism $\beta: S \to B/L$ there is a unique additive relation $r: A \rightharpoonup B$ with $r^0 = \beta$.

Proof. Given $a \in \operatorname{Def} r$, $(a,b) \in r$ and $(a,b') \in r$ imply $(0, b-b') \in r$, hence $b - b' \in \operatorname{Ind} r$. Then $r^0(a) = b + \operatorname{Ind} r$ defines a homomorphism r^0 with the desired form (6.2). Conversely, given β, r is the set of all pairs (s,b) with $b \in \beta(s)$.

A similar argument shows that each additive relation r induces an isomorphism θ_r: $(\operatorname{Def} r)/(\operatorname{Ker} r) \cong (\operatorname{Im} r)/(\operatorname{Ind} r)$; conversely, each isomorphism of a subquotient of A to a subquotient of B arises in this way from a unique additive relation r.

Given subquotients S/K of A and S'/K' of A', each homomorphism $\alpha: A \to A'$ *induces* an additive relation

$$\alpha_{\#} = \alpha(S/K, S'/K'): S/K \rightharpoondown S'/K', \tag{6.3}$$

defined to be the set of all pairs $(s+K, s'+K')$ of cosets with $s \in S$, $s' \in S'$, and $s' = \alpha s$. This includes the previous notion of induced homomorphisms.

For an equivalence one can determine the inverse of an induced relation.

Proposition 6.2. *(Equivalence principle.) If $\theta: A \to A'$ is an equivalence, then*

$$(\theta_{\#})^{-1} = (\theta^{-1})_{\#}: S'/K' \rightharpoondown S/K.$$

Indeed, each of $(\theta_{\#})^{-1}$ and $(\theta^{-1})_{\#}$ consists of the same pairs.

In Chap. XI we will use the composite of two induced relations. This is not always the relation induced by the composite homomorphism. For example, in the direct sum $A = B \oplus B$ let B_1 be the submodule of all $(b, 0)$, B_2 the submodule of all $(0, b)$ and Δ the submodule of all (b, b) (the "diagonal" submodule). Then 1_A induces isomorphisms $B_1 \cong A/B_2 \cong \Delta$, but the relation $B_1 \rightharpoondown \Delta$ induced by 1_A consists of $(0, 0)$ alone. Composition works reasonably well only under a restrictive hypothesis, as follows:

Proposition 6.3. *(Composition principle.) If homomorphisms $\alpha: A \to A'$ and $\beta: A' \to A''$ induce the additive relations $\alpha_{\#}: S/K \rightharpoondown S'/K'$ and $\beta_{\#}: S'/K' \rightharpoondown S''/K''$ on given subquotients, then*

$$\beta_{\#}\alpha_{\#} = (\beta\alpha)_{\#}: S/K \rightharpoondown S''/K'',$$

provided (i) *either $\alpha K \supset K'$ or $\beta K' \subset K''$ and* (ii) *either $\alpha S \subset S'$ or $\beta^{-1}S'' \subset S'$.*

Proof. Suppose first that $(s+K, s''+K'') \in \beta_{\#}\alpha_{\#}$. By definition of the composite of two relations, there are s_1' and s_2' in S' with $s_1' + K' = s_2' + K'$ and $\alpha s = s_1'$, $\beta s_2' = s''$. Thus $s_1' - s_2' = k' \in K'$, and $\beta\alpha s = s'' + \beta k'$. In case either $\beta K' \subset K''$ or $K' \subset \alpha K$ this gives $(s+K, s''+K'') \in (\beta\alpha)_{\#}$, so hypothesis (i) gives $\beta_{\#}\alpha_{\#} \subset (\beta\alpha)_{\#}$. Similarly, (ii) gives the opposite inclusion.

Exercises

1. For each additive relation $r: A \rightharpoondown B$, prove $rr^{-1}r = r$.
2. For additive relations r and s, prove $(rs)^{-1} = s^{-1}r^{-1}$.
3. If $u = A \rightharpoondown A$ is an additive relation with $u^{-1} = u = u^2$, prove that there are submodules $K \subset S \subset A$ with $u = [(s, s+k) \mid s \in S, k \in K]$. Establish the converse.
4. For each additive relation $r: A \rightharpoondown B$, describe rr^{-1} and $r^{-1}r$.
5. Under the hypotheses of the strong Four Lemma (Lemma I.3.2), prove $\xi\alpha^{-1} = \beta^{-1}\eta$.

7. Singular Homology

The use of complexes may be illustrated by a brief description of the singular homology groups of a topological space. We first introduce affine simplices.

Let E be an n-dimensional Euclidean space; that is, an n-dimensional vector space over the field of real numbers in which there is given a symmetric, bilinear, and positive definite inner product (u, v) for each pair of vectors $u, v \in E$. The usual distance function $\varrho(u, v) = (u-v, u-v)^{\frac{1}{2}}$ makes E a metric space and hence a topological space. In particular, E may be the space E^n of all n-tuples $u = (a_1, \ldots, a_n)$ of real numbers a_i, with termwise addition and with the standard inner product $(a_1, \ldots, a_n)$ $(b_1, \ldots, b_n) = \sum a_i b_i$.

The *line segment* joining two points $u, v \in E$ is the set of all points $tu + (1-t)v$, for t real and $0 \leqq t \leqq 1$; that is, of all points $x_0 u + x_1 v$, where x_0, x_1 are real numbers with $x_0 + x_1 = 1$, $x_0 \geqq 0$, $x_1 \geqq 0$. A subset C of E is *convex* if it contains the line segment joining any two of its points. If $u_0, \ldots, u_m$ are $m+1$ points of E, the set of all points

$$u = x_0 u_0 + \cdots + x_m u_m, \qquad x_0 + x_1 + \cdots + x_m = 1, \qquad x_i \geqq 0 \qquad (7.1)$$

is a convex set containing $u_0, \ldots, u_m$ and in fact the smallest convex set containing these points; it is called the *convex hull* of $u_0, \ldots, u_m$. The points $u_0, \ldots, u_m$ are said to be *affine independent* if every point of this convex hull has a unique representation in the form (7.1); the real numbers x_i are then the *barycentric coordinates* of u relative to $u_0, \ldots, u_m$. It can be shown that the points $u_0, \ldots, u_m$ are affine independent if and only if the vectors $u_1 - u_0, \ldots, u_m - u_0$ are linearly independent.

An *affine m-simplex* is by definition the convex hull of $m+1$ affine independent points. These points are the *vertices* of the simplex. Thus a 1-simplex is a line segment, a 2-simplex is a triangle (with interior), a 3-simplex is a tetrahedron, etc. For each dimension n we will take a *standard* affine n-simplex Δ^n in the space E^n, and we will label the vertices of Δ^n as $(0, 1, \ldots, n)$. (For example, take 0 to be the origin and $1, \ldots, n$ a basis of n orthogonal vectors in E^n.)

For any topological space X, a *singular n-simplex* T in X is a continuous map $T: \Delta^n \to X$. Thus a singular 0-simplex of X is just a point of X, or, more accurately, a map of the standard point Δ^0 into (a point of) X. We first construct certain singular simplices in convex subsets of E.

Let E and E' be Euclidean spaces, $L: E \to E'$ a linear transformation and u_0' a fixed vector of E'. The function $f(u) = u_0' + L(u)$ on E to E' is called an *affine transformation* $f: E \to E'$. As the composite of the linear transformation L with the translation by u_0', f is continuous.

Proposition 7.1. *If $u_0, \ldots, u_n$ are $n+1$ affine independent points in E^n, while $v_0, \ldots, v_n \in E'$, there is a unique affine transformation $f: E^n \to E'$ with $f(u_i) = v_i$, $i = 0, \ldots, n$.*

Proof. The vectors $u_i - u_0$, $i = 1, \ldots, n$, are a basis of E^n. Let L be the unique linear transformation with $L(u_i - u_0) = v_i - v_0$; $f(u) = v_0 - L(u_0) + L(u)$ is the required affine transformation: It may also be written in barycentric coordinates as

$$f(x_0 u_0 + \cdots + x_n u_n) = x_0 v_0 + \cdots + x_n v_n, \qquad \sum x_i = 1.$$

In particular, let $v_0, \ldots, v_n$ be an ordered set of points in a convex subset C of E'. The unique affine transformation $f: E^n \to E'$ which carries the vertices $0, 1, \ldots, n$ of the standard simplex Δ^n in order into $v_0, \ldots, v_n$ thus gives a continuous map $\Delta^n \to C$ which we write as

$$(v_0, \ldots, v_n)_C : \Delta^n \to C. \tag{7.2}$$

This we call the *affine singular n-simplex* (with standard vertices $0, \ldots, n$ mapped to $v_0, \ldots, v_n$). For example, if the $v_0, \ldots, v_n$ are affine independent, it is a homeomorphism of the standard simplex Δ^n to the affine simplex spanned by the v's. In particular $J_n = (0, 1, \ldots, n)_{\Delta^n}$ is the identity map of Δ^n onto itself. If the $v_0, \ldots, v_n$ are dependent, the corresponding map $(v_0, \ldots, v_n)_C$ collapses the standard Δ^n onto a simplex of lower dimension.

We may now describe the "boundary" of Δ^n to consist of certain $(n-1)$-dimensional singular simplices which are the "faces" of Δ^n. For example, the faces of the triangle $\Delta^2 = (0, 1, 2)$ are the three edges represented by the segments $(1\,2)$, $(0\,2)$, and $(0\,1)$; in the notation (7.2) they are the three continuous maps $(1, 2)_{\Delta^2}$, $(0, 2)_{\Delta^2}$, and $(0, 1)_{\Delta^2}$ of Δ^1 into Δ^2. In general Δ^n has $n+1$ faces; its i-th face is the affine singular $(n-1)$-simplex

$$\varepsilon^i = \varepsilon^i_n : (0, 1, \ldots, \hat{i}, \ldots, n)_{\Delta^n} : \Delta^{n-1} \to \Delta^n, \qquad i = 0, \ldots, n, \tag{7.3}$$

where the notation $\hat{i}$ indicates that the vertex i is to be omitted. Any singular n-simplex $T: \Delta^n \to X$ has $n+1$ faces $d_i T$ defined by

$$d_i T = T \varepsilon^i_n : \Delta^{n-1} \to X, \qquad i = 0, \ldots, n, \qquad n > 0. \tag{7.4}$$

In other words, $d_i T$ is the map obtained by restricting T to the i-th face of Δ^n and regarding this restriction (via ε^i) as a map defined on Δ^{n-1}. Any singular simplex T can be written as the composite $T = T J_n$, where $J_n: \Delta^n \to \Delta^n$ is the identity map of Δ^n, and hence a singular n-simplex of Δ^n. The faces of T are then given by the formula

$$d_i T = T(d_i J_n), \qquad i = 0, \ldots, n, \qquad n > 0. \tag{7.5}$$

For an affine singular simplex (7.2), the i-th face omits the i-th vertex:

$$d_i(v_0, \ldots, v_n)_C = (v_0, \ldots, \hat{v}_i, \ldots, v_n)_C. \tag{7.6}$$

The process of forming iterated faces satisfies the identity

$$d_i d_j T = d_{j-1} d_i T, \quad i < j. \tag{7.7}$$

By (7.5) it suffices to prove this in the case when $T = J_n$; here it is clear, since the process of first omitting vertex j and then vertex i amounts to the same as first omitting vertex i and then (in the new numbering of vertices of $d_i J_n$) vertex $j-1$. An alternative proof may be given by replacing each point of the standard n-simplex Δ^n by its barycentric coordinates $x_0, \ldots, x_n$. A singular n-simplex T in the space X is then a continuous function with values $T(x_0, \ldots, x_n) \in X$, defined for all real x_i with $x_i \geqq 0$ and $x_0 + \cdots + x_n = 1$. The i-th face is the function defined by

$$(d_i T)(x_0, \ldots, x_{n-1}) = T(x_0, \ldots, x_{i-1}, 0, x_i, \ldots, x_{n-1});$$

i.e., by letting the i-th variable in T be 0. Hence (7.7) follows, because first setting $x_j = 0$ and then $x_i = 0$ for $i < j$ in $T(x_0, \ldots, x_n)$ amounts to first setting $x_i = 0$ and then setting equal to 0 the variable with the new number $j-1$.

To each space X we now construct a complex $S(X)$ of abelian groups, called the *singular complex* of X. Take $S_n(X)$ to be the free abelian group with generators all singular n-simplices T of X. Then the i-th face operation defines a homomorphism $d_i \colon S_n(X) \to S_{n-1}(X)$ for $i = 0, \ldots, n$ and $n > 0$. Define the *boundary* homomorphism

$$\partial \colon S_n(X) \to S_{n-1}(X)$$

as the sum of the face homomorphisms with alternating signs; that is

$$\partial T = d_0 T - d_1 T + \cdots + (-1)^n d_n T = \sum_{i=0}^{n} (-1)^i d_i T, \quad n > 0. \tag{7.8}$$

An n-chain $c \in S_n(X)$ has a unique representation as a sum $c = \sum_T c(T)\, T$ where the coefficients $c(T)$ are integers, zero except for a finite number of T; its boundary is $\partial c = \sum c(T)\, \partial T$. To show that $S(X)$ is a complex, we must prove that the composite $\partial\partial \colon S_n \to S_{n-2}$ is the zero homomorphism for $n > 1$. It suffices to prove $\partial\, \partial T = 0$. But

$$\partial\, \partial T = \sum_{i<j} (-1)^{i+j} d_i d_j T + \sum_{i \geqq j} (-1)^{i+j} d_i d_j T.$$

Using (7.7) and switching the labels i and j in the second sum, this is

$$\partial\, \partial T = \sum_{j-1 \geqq i} (-1)^{i+j} d_{j-1} d_i T + \sum_{k \geqq i} (-1)^{i+k} d_k d_i T.$$

The two sums are equal except for sign, hence cancel to give $\partial\partial = 0$.

The n-dimensional *singular homology* group $H_n(X)$ of the space X is now defined to be the n-th homology group $H_n(S(X))$ of the complex $S(X)$.

Theorem 7.2. *The homology group $H_n(X)$ is a covariant functor of X.*

Proof. If Y is a second topological space and $f: X \to Y$ any continuous map, each singular simplex $T: \Delta^n \to X$ of X yields by composition a singular simplex $fT: \Delta^n \to Y$ of Y. The correspondence $T \to fT$ on the free generators T of $S_n(X)$ yields a homomorphism $S_n(f): S_n(X) \to S_n(Y)$. Moreover, $d_i(fT) = f(d_i T)$; hence $\partial S(f) = S(f)\partial$, so $S(f)$ is a chain transformation which induces homomorphisms $H_n(S(f)): H_n(X) \to H_n(Y)$ on the homology groups in each dimension. With these homomorphisms, H_n is a covariant functor on the category with objects all topological spaces, morphisms all continuous maps.

If G is any abelian group, the cohomology groups $H^n(S(X), G)$ are the *singular cohomology groups* of X with coefficients G. They are bifunctors, contravariant in X and covariant in G.

The homomorphism $\varepsilon: S_0(X) \to Z$ which carries each singular 0-simplex into $1 \in Z$ is called the *augmentation* of $S(X)$. Since $\varepsilon\partial = 0$: $S_1(X) \to Z$, $\varepsilon: S(X) \to Z$ is a complex over Z. Moreover, ε induces an epimorphism $\varepsilon_*: H_0(X) \twoheadrightarrow Z$. A space X is called *acyclic* if $H_n(X) = 0$ for $n > 0$ and ε_* is an isomorphism $H_0(X) \cong Z$.

Proposition 7.3. *A topological space with only one point is acyclic.*

Proof. Let $X = \{x\}$ be the space. In each dimension n, X has only one singular simplex, namely the map $T_n: \Delta^n \to \{x\}$ which collapses Δ^n to the point x. Hence each face $d_i T_n$ is T_{n-1}, for $i = 0, \dots, n$. Since ∂T is the alternating sum of faces, $\partial T_{2m} = T_{2m-1}$ and $\partial T_{2m-1} = 0$. Thus in even dimensions $S(X)$ has no cycles except 0, while in odd dimensions all elements of $S_{2m-1}(X)$ are cycles and also boundaries. Therefore $H_n(X) = 0$ for all $n > 0$; clearly $H_0(X) \cong Z$.

Exercises

1. Let the affine simplex Γ be the convex hull of the affine independent points $u_0, \dots, u_m$. Show that $u \in \Gamma$ is one of the points u_i if and only if $v, w \in \Gamma$ and u on the segment from v to w imply $u = v$ or $u = w$. Conclude that the simplex Γ, as a convex set, determines its vertices.

2. If X is pathwise connected, prove that $\varepsilon_*: H_0(X) \cong Z$. (Definition: Let I be the unit interval. X is *pathwise connected* if to each pair of points $x, y \in X$ there exists a continuous map $f: I \to X$ with $f(0) = x$, $f(1) = y$.)

8. Homotopy

Two continuous maps of a space X into a space Y are said to be "*homotopic*" if it is possible to continuously deform the first map into the second. Consider the deformation as taking place in a unit interval of time; then it can be regarded as a continuous map defined on the cartesian product $X \times I$ of the space X and the unit interval I, $0 \leq t \leq 1$ on the real t-axis. Hence we make the

Definition. Two continuous maps $f_0, f_1: X \to Y$ are homotopic if and only if there is a continuous map $F: X \times I \to Y$ such that

$$F(x,0)=f_0(x), \quad F(x,1)=f_1(x). \tag{8.1}$$

When this holds, we write $F: f_0 \simeq f_1: X \to Y$.

The condition (8.1) states that the homotopy starts, for $t=0$, with the initial map f_0 and that it ends, for $t=1$, with the final map f_1. For example, a space X is called *contractible* if the identity map $1: X \to X$ is homotopic to a map which sends X into a single point. Any convex set C in a Euclidean space is contractible to any one of its points w, via the homotopy D defined by

$$D(u,t)=tw+(1-t)u, \quad 0 \leq t \leq 1, \quad u \in C. \tag{8.2}$$

This function is clearly continuous and takes values in C, because C is convex.

This geometric notion is closely related to the algebraic notion of a chain homotopy. As a first example, we prove

Proposition 8.1. *Any convex set C in a Euclidean space is acyclic.*

The proof uses a chain homotopy $s: S_n(C) \to S_{n+1}(C)$. Since $S_n(C)$ is the free abelian group generated by the singular n-simplices T of C, it will suffice to define a singular $(n+1)$-simplex $sT: \Delta^{n+1} \to C$ for each T. In terms of the barycentric coordinates $(x_0, \ldots, x_{n+1})$ of a point of Δ^{n+1}, set

$$\left.\begin{aligned}(sT)(x_0, \ldots, x_{n+1}) &= x_0 w + (1-x_0)\, T\left(\frac{x_1}{1-x_0}, \ldots, \frac{x_{n+1}}{1-x_0}\right), \quad && x_0 \neq 1,\\ &= w, && x_0 = 1,\end{aligned}\right\} \tag{8.3}$$

where w is a fixed point of C. To see that sT is continuous at $x_0=1$, we rewrite the definition so that it resembles the geometric homotopy D of (8.2). Let $v_0=0$ be the initial vertex of Δ^{n+1}; then

$$\left(0,\, x_1/(1-x_0), \ldots, x_{n+1}/(1-x_0)\right)$$

can be viewed as the barycentric coordinates of some point u' on the opposite face. Each point of Δ^{n+1} can be written as a weighted average $x_0 v_0 + (1-x_0)u'$ for some u', unique except when $x_0=1$. The point u' on the opposite face determines $u \in \Delta^n$ with $\varepsilon^0 u = u'$. The definition (8.3) now reads, in all cases

$$(sT)\left(x_0 v_0 + (1-x_0)u'\right) = x_0 w + (1-x_0)\, T(u), \quad \varepsilon^0 u = u'.$$

In other words, the segment in Δ^{n+1} joining v_0 to each point u' of the opposite face is mapped by sT linearly onto the segment joining $w \in C$ to $T(u) \in C$. In particular, since Δ^n is compact, $T\Delta^n$ is compact and hence bounded, so $sT\colon \Delta^{n+1} \to C$ is continuous at $x_0 = 0$.

This $s\colon S_n(C) \to S_{n+1}(C)$ provides a contracting homotopy for the augmented complex $\varepsilon\colon S(X) \to Z$. In the notation of (2.5), let $f\colon Z \to S(X)$ be the chain transformation which carries $1 \in Z$ to the singular 0-simplex T_0 at the chosen point $w \in C$. The i-th face $d_i(sT)$ is the singular n-simplex obtained from (8.3) by setting $x_i = 0$. Hence $d_0(sT) = T$, while $d_{i+1}(sT) = s d_i T$ if $n > 0$ and $d_1 sT = T_0$ if $n = 0$. This gives $\partial(sT) = T - s(\partial T)$ for $n > 0$, $\partial sT = T - f\varepsilon T$ for $n = 0$, and $\varepsilon f = 1$, all as in (2.5). Hence $S(X)$ is acyclic, as required.

More generally, consider any homotopy $F\colon X \times I \to Y$. Regard $X \times I$ as a cylinder on the base X; the boundary of this cylinder is the top (where $F = f_1$) minus the bottom (where $F = f_0$) minus the sides (i.e., minus F on $(\partial X) \times I$). The resulting schematic formula $\partial F = f_1 - f_0 - F\partial$ suggests the definition $\partial s = f_1 - f_0 - s\partial$ of a chain homotopy. These indications can be made precise, as follows:

Theorem 8.2. *If $f_0 \simeq f_1\colon X \to Y$ are homotopic continuous maps, the induced chain transformations $S(f_0), S(f_1)\colon S(X) \to S(Y)$ are chain homotopic.*

We reduce this theorem to the special case of the cylinder $X \times I$. By the base b and the top t of this cylinder we mean the continuous maps $b, t\colon X \to X \times I$ defined by $b(x) = (x, 0)$ and $t(x) = (x, 1)$; they are clearly homotopic.

Lemma 8.3. *For any cylinder there is a chain homotopy $u\colon S(t) \simeq S(b)$.*

The lemma implies the theorem. For let $F\colon X \times I \to Y$ be any homotopy $F\colon f_0 \simeq f_1$. Then $Fb = f_0$, $Ft = f_1$, and $S(F)$ is a chain transformation. Define s as the composite

$$s = S_{n+1}(F)\, u\colon S_n(X) \xrightarrow{u} S_{n+1}(X \times I) \to S_{n+1}(Y).$$

Then $\partial s + s\partial = S(F)(\partial u + u\partial) = S(F)(S(t) - S(b)) = S(f_1) - S(f_0)$.

To prove the lemma, we prove more: That $u = u_X\colon S(X) \to S(X \times I)$ can be chosen simultaneously for all topological spaces X so as to be natural. For each continuous map $g\colon X \to X'$ of spaces, naturality requires that the diagram

$$\begin{array}{ccc} S_n(X) & \xrightarrow{u_X} & S_{n+1}(X \times I) \\ \downarrow{\scriptstyle S(g)} & & \downarrow{\scriptstyle S(g \times 1)} \\ S_n(X') & \xrightarrow{u_{X'}} & S_{n+1}(X' \times I) \end{array} \qquad (8.4)$$

be commutative. Observe that $b, t\colon X \to X \times I$ are already natural. We construct such a u by induction on n. For $n = 0$, a singular 0-simplex is just a point $T(0)$ of X. Take u_0T to be that singular 1-simplex of $X \times I$ defined by $(u_0T)(x_0, x_1) = (T(0), x_1)$, so that u_0T is the segment vertically above $T(0)$ in the cylinder $X \times I$. Then $d_0(u_0T) = t(T(0))$, $d_1(u_0T) = b(T(0))$, so $\partial(u_0T)$ is indeed $S(t)T - S(b)T$. Moreover u_0 is clearly natural.

For $n>0$, suppose u_m has been defined for all $m<n$, in particular, $\partial u_{n-1} + u_{n-2}\partial = S(t) - S(b)$; if $n=1$, u_{n-2} is zero here. Let J_n: $\Delta^n \to \Delta^n$ be the identity map of the standard simplex. We first define $uJ_n \in S_{n+1}(\Delta^n \times I)$; its boundary ought to be

$$\partial uJ_n = S(t)J_n - S(b)J_n - u_{n-1}\partial J_n. \tag{8.5}$$

Now the expression c on the right is a chain of $S_n(\Delta^n \times I)$; its boundary is

$$\partial c = \partial S(t)J_n - S(b)\,\partial J_n - \partial u_{n-1}\partial J_n = (S(t) - S(b) - \partial u_{n-1})\,\partial J_n,$$

which is zero by the induction assumption. Hence c is an n-cycle of $\Delta^n \times I$. But $\Delta^n \times I$ is a convex subset of a Euclidean space, hence is acyclic by Prop. 8.1. Therefore c is a boundary, say $c = \partial a$ for some $a \in S_{n+1}(\Delta^n \times I)$. We set $uJ_n = a$; then (8.5) holds.

If T: $\Delta^n \to X$ is now a singular simplex of any space X, $T = TJ_n = S(T)J_n$ and $T \times 1$: $\Delta^n \times I \to X \times I$. Define $uT = S(T\times 1)uJ_n = S(T \times 1)a$. This definition immediately satisfies the naturality requirement. To show that it gives the required homotopy, calculate

$$\partial uT = S(T\times 1)\,\partial a = S(T\times 1)\,[S(t)J_n - S(b)J_n - u_{n-1}\partial J_n],$$

where t and b are top and base for $\Delta^n \times I$. But t, b, and u_{n-1} are all natural, hence (8.5) gives $\partial uT = S(t)T - S(b)T - u_{n-1}\partial T$, as desired.

This type of proof consists in first constructing the desired object (here, the desired chain homotopy) on a model chain such as J_n by observing that the space $\Delta^n \times I$ in which the model lies is acyclic, and in then carrying the object around to other spaces by the maps T. It is an old method in topology; it will reappear later (Chap. VIII) as the method of acyclic models. Here it has the merit of avoiding an explicit formula for the homotopy u.

Corollary 8.4. *If the continuous maps f_0, f_1: $X \to Y$ are homotopic, the induced homomorphisms $H(f_0)$, $H(f_1)$: $H_n(X) \to H_n(Y)$ are equal.*

Exercises

1. Show that any contractible space is acyclic.
2. In the *prism* $\Delta^n \times I$ let $0, 1, \ldots, n$ denote the vertices of the base, $0', 1', \ldots, n'$, those of the top. Show that an explicit homotopy u for $X = \Delta^n$ in Lemma 8.3 is given, in the notation for affine singular simplices, by

$$uJ_n = \sum_{i=0}^{n} (-1)^i\,(0, 1, \ldots, i, i', (i+1)', \ldots, n')_{\Delta^n \times I}.$$

3. For $n=1, 2$ as in Ex. 2, show that the terms of uJ_n correspond to a "triangulation" of the prism $\Delta^n \times I$ (Draw a figure).

4. Show that $\Delta^n \times I$ can be "triangulated" as follows. Partly order the vertices of $\Delta^n \times \{0\}$ and $\Delta^n \times \{1\}$ by the rule that $(i, \varepsilon) \leqq (j, \eta)$ for $\varepsilon, \eta = 0, 1$, if $i \leqq j$ and $\varepsilon \leqq \eta$. Take as simplices of the triangulation all those formed by a linearly ordered subset of the whole set of vertices, and show that the resulting n-simplices are those appearing in uJ_n in Ex. 2.

9. Axioms for Homology

Let A be a subspace of X. Identify each singular simplex $T: \Delta^n \to A$ of A with the composite map $\Delta^n \to A \to X$; then T becomes a singular simplex of X and the singular complex $S(A)$ a subcomplex of $S(X)$. The homology groups of the quotient complex,

$$H_n(X,A) = H_n(S(X)/S(A)), \tag{9.1}$$

are called the ***relative homology*** groups of the ***pair*** (X,A) of spaces. They are subquotient groups of the quotient $S(X)/S(A)$, hence can be rewritten as subquotients

$$H_n(X,A) = C_n(X,A)/B_n(X,A) \tag{9.2}$$

of $S(X)$. Specifically $C_n(X,A)$ consists of those elements $c \in S_n(X)$ with $\partial c \in S_{n-1}(A)$, while $B_n(X,A) = S_n(A) \cup \partial S_{n+1}(X)$; the elements c of $C_n(X,A)$ are known as ***relative cycles***; those of $B_n(X,A)$ as ***relative boundaries***. A single space X may be regarded as a pair of spaces $(X, \emptyset)$, with $\emptyset$ the empty set; then $H_n(X, \emptyset) = H_n(X)$.

A map $f: (X,A) \to (Y,B)$ of pairs of spaces is by definition a continuous map $f: X \to Y$ with $f(A) \subset B$. With these maps as morphisms, the pairs of spaces constitute a category, and $H_n(X,A)$ is a covariant functor on this category to abelian groups.

Each pair (X,A) gives a short exact sequence of complexes $S(A) \rightarrowtail S(X) \twoheadrightarrow S(X)/S(A)$. The connecting homomorphism ∂_* for this sequence is called the ***invariant boundary*** of the pair (X,A); the exact homology sequence (Thm. 4.1) gives

Theorem 9.1. *If (X,A) is a pair of spaces, the long sequence*

$$\cdots \to H_n(A) \xrightarrow{i_*} H_n(X) \xrightarrow{j_*} H_n(X,A) \xrightarrow{\partial_*} H_{n-1}(A) \to \cdots, \tag{9.3}$$

ending in $\to H_0(X) \to H_0(X,A) \to 0$, is exact.

Specifically, i: $(A, \emptyset) \to (X, \emptyset)$ and j: $(X, \emptyset) \to (X, A)$ are maps of pairs induced by the identity function, while ∂_* is given for each relative cycle c as $\partial_*(\operatorname{cls} c) = \operatorname{cls}(\partial c)$. We have already noted (Example (1.3)) that $i_*: H_n(A) \to H_n(X)$ need not be a monomorphism; this exact sequence describes the kernel and the image of i_*.

Two maps f_0, f_1: $(X,A)\to(Y,B)$ of pairs are homotopic if there is a homotopy F: $f_0\simeq f_1$: $X\to Y$ with $F(A\times I)\subset B$; this last condition means that F cut down to $A\times I$ is a homotopy between f_0 and f_1 cut down to maps of A into B. An extension of the argument for Thm. 8.2 shows that homotopic maps f_0, f_1 of pairs have $H_n(f_0)=H_n(f_1)$: $H_n(X,A)\to H_n(Y,B)$.

The singular homology theory for pairs of spaces thus gives:

1. Functors $H_n(X,A)$ of pairs of spaces to abelian groups, $n=0, 1, \ldots$.

2. Natural homomorphisms ∂_*: $H_n(X,A)\to H_{n-1}(A)$, $n=1, 2, \ldots$.

These data satisfy the following additional conditions:

3. If X consists of a single point, $H_0(X)\cong Z$ and $H_n(X)=0$ for $n>0$.

4. For any pair (X,A) the relative homology sequence (9.3) is exact.

5. Homotopic maps of pairs induce equal homomorphisms on each H_n.

6. (Excision.) If $X\supset A\supset M$ are spaces such that the closure of M is contained in the interior of A, let $X-M\supset A-M$ denote the subspaces obtained by removing all points of M from X and from A, respectively. Then the injection k of $X-M$ into X induces isomorphisms on the relative homology groups

$$H_n(k)\colon H_n(X-M, A-M)\cong H_n(X,A). \tag{9.4}$$

Our discussion has indicated the proofs of all except the sixth property; a proof of this uses "barycentric subdivisions"; it may be found in EILENBERG-STEENROD [1952], WALLACE [1957], or HILTON-WYLIE [1960].

These six properties may be taken as axioms for homology. It can be proved that when the pair (X,A) can be "triangulated" by a finite number of affine simplices, any relative homology groups satisfying the axioms must agree with the singular homology groups. Moreover, from the axioms alone one can calculate the singular homology groups of elementary spaces to agree with those calculated from "naive" subdivisions in §1. In particular, if S^n is the n-sphere, one deduces that $H_n(S^n)\cong Z$, $H_0(S^n)\cong Z$ and $H_i(S^n)=0$ for $0\neq i\neq n$. This, and other striking geometric properties (Brouwer fixed point theorem, etc.) are presented in EILENBERG-STEENROD [1952], Chap. XI.

We have now completed our too brief indication of the use of homology theory in topology.

Notes. "Complex" originally meant simplicial complex; in topology "complex" has various geometric meanings, such as "cell complex" or "CW-complex". The chain complex in our purely algebraic sense was introduced by MAYER [1929,

1938]. The formulation of exact homology sequences, as codified in KELLEY-PITCHER [1947], allowed a systematic treatment of simple facts which previously were done "by hand" in each case. POINCARÉ introduced homology, via Betti numbers; it was Emmy NOETHER who emphasized that the homology of a space deals with a homology *group* rather than just with Betti numbers and torsion coefficients. Singular homology in its present form is due to EILENBERG; the axioms for homology theory, with application to other homology theories (ČECH theory) appear in the influential book by EILENBERG-STEENROD. Additive relations have been explicitly recognized only recently [LUBKIN 1960, MAC LANE 1961, PUPPE 1962]. The corresponding notion for multiplicative groups occurs in WEDDERBURN [1941], ZASSENHAUS [1958], and for general algebraic structures in LORENZEN [1954] and LAMBEK [1958].

Chapter three

Extensions and Resolutions

A long exact sequence of R-modules

$$0 \to A \to B_{n-1} \to \cdots \to B_1 \to B_0 \to C \to 0$$

running from A to C through n intermediate modules is called an "n-fold extension" of A by C. These extensions, suitably classified by a congruence relation, are the elements of a group $\operatorname{Ext}^n(C,A)$. To calculate this group, we present C as the quotient $C = F_0/S_0$ of a free module F_0; this process can be iterated as $S_0 = F_1/S_1$, $S_1 = F_2/S_2, \ldots$ to give an exact sequence

$$\cdots \to F_n \to F_{n-1} \to \cdots \to F_1 \to F_0 \to C \to 0$$

called a "free resolution" of C. The complex $\operatorname{Hom}(F_n, A)$ has cohomology $\operatorname{Ext}^n(C,A)$. Alternatively, one may imbed A in an injective module J_0 (§7) and then J_0/A in an injective module J_1; this process iterates to give an exact sequence

$$0 \to A \to J_0 \to J_1 \to \cdots \to J_{n-1} \to J_n \to \cdots$$

called an "injective coresolution" of A. The complex $\operatorname{Hom}(C, J_n)$ has cohomology $\operatorname{Ext}^n(C,A)$. In particular, $\operatorname{Ext}^1(C,A)$ is often called $\operatorname{Ext}(C,A)$.

The chapter starts with the definition of Ext^1, which is at once applied (§4) to calculate the cohomology of a complex of free abelian groups from its homology. The chapter ends with a canonical process for imbedding any module in a "minimal" injective.

1. Extensions of Modules

Let A and C be modules over a fixed ring R. An *extension* of A by C is a short exact sequence $E = (\varkappa, \sigma): A \rightarrowtail B \twoheadrightarrow C$ of R-modules and R-module homomorphisms. A *morphism* $\Gamma\colon E \to E'$ of extensions is a triple $\Gamma=(\alpha,\beta,\gamma)$ of module homomorphisms such that the diagram

$$\begin{array}{ccccccccc} E\colon & 0 \longrightarrow & A & \xrightarrow{\varkappa} & B & \xrightarrow{\sigma} & C & \longrightarrow 0 \\ \downarrow{\scriptstyle\Gamma} & & \downarrow{\scriptstyle\alpha} & & \downarrow{\scriptstyle\beta} & & \downarrow{\scriptstyle\gamma} & \\ E'\colon & 0 \longrightarrow & A' & \xrightarrow{\varkappa'} & B' & \xrightarrow{\sigma'} & C' & \longrightarrow 0 \end{array} \tag{1.1}$$

is commutative. In particular, take $A'=A$ and $C'=C$; two extensions E and E' of A by C are *congruent* ($E\equiv E'$) if there is a morphism $(1_A,\beta,1_C)\colon E\to E'$. When this is the case, the short Five Lemma shows that the middle homomorphism β is an isomorphism; hence congruence of extensions is a reflexive, symmetric, and transitive relation. Let $\operatorname{Ext}_R(C,A)$ denote the set of all congruence classes of extensions of A by C.

An extension of A by C is sometimes described as a pair (B,θ) where A is a submodule of B, and θ is an isomorphism $B/A\cong C$. Each such pair determines a short exact sequence $A\rightarrowtail B\twoheadrightarrow B/A$ and every extension of A by C is congruent to one so obtained.

One extension of A by C is the direct sum $A\rightarrowtail A\oplus C\twoheadrightarrow C$. An extension $E=(\varkappa,\sigma)$ is said to be *split* if it is congruent to this direct sum extension; as in Prop. I.4.3, this is the case if and only if σ has a right inverse $\mu\colon C\to B$ (or, equivalently, $\varkappa$ has a left inverse). Any extension by a projective module P is split, so $\operatorname{Ext}_R(P,A)$ has but one element. To illustrate a non-trivial case, take $R=Z$. Then, for example, the additive group $2Z$ of even integers has two extensions by a cyclic group Z_2 of order 2: The direct sum $2Z\oplus Z_2$ and the group $Z\supset 2Z$. This is a special case of the following fact:

Proposition 1.1. *For any abelian group A and for $Z_m(c)$ the cyclic group of order m and generator c there is a $1-1$ correspondence*

$$\eta\colon \operatorname{Ext}_Z(Z_m(c),A)\cong A/mA,$$

where mA is the subgroup of A consisting of all ma for $a\in A$.

Proof. Take any extension E of A by Z_m; in the middle group B choose an element u with $\sigma u=c$ to serve as a sort of "representative" of the generator c. Each element of B can be written uniquely as $b=\varkappa a+hu$ for some $a\in A$ and some integer h, $h=0,\ldots,m-1$. Since $mc=0$, $\sigma(mu)=0$, so $mu=\varkappa g$ for a unique $g\in A$. This g determines the "addition table" for B, because

$$\begin{aligned}(\varkappa a+hu)+(\varkappa a'+h'u)&=\varkappa(a+a')+(h+h')\,u, && h+h'<m,\\ &=\varkappa(a+a'+g)+(h+h'-m)\,u, && h+h'\geqq m.\end{aligned}$$

The element g is not invariant; the representative u may be replaced by any $u'=u+\varkappa f$ for $f\in A$, thus replacing g by $g'=g+mf$. The coset $g+mA$ in A/mA is uniquely determined by the extension E. Set $\eta(E)=$

$g+mA$. If $E \equiv E'$, $\eta(E)=\eta(E')$. If g is any element of A, take for B the set of all pairs (a,h) with $a \in A$, $h=0, \ldots, m-1$ and define addition of pairs, using g, as in the table above. This addition is associative, makes B a group, and gives an extension E with $\eta(E)=g+mA$. Hence η is $1-1$ onto A/mA.

Now η is a correspondence between a set Ext_Z and an abelian group A/mA; this suggests that $\mathrm{Ext}_R(C,A)$ is always an abelian group. We shall shortly show this to be so. First we show that Ext is a functor, on the category of modules to that of sets.

Let A be fixed. To show $\mathrm{Ext}_R(C,A)$ a contravariant functor of C requires for each $E \in \mathrm{Ext}_R(C,A)$ and each $\gamma\colon C' \to C$ a suitable extension $E'=\gamma^* E \in \mathrm{Ext}_R(C',A)$. This E' may be denoted by $E\gamma$, and is described by the following lemma, which shows E' unique and which hence easily implies the congruences

$$E\,1_C \equiv E, \qquad E(\gamma\gamma') \equiv (E\gamma)\,\gamma'. \tag{1.2}$$

They state that E depends contravariantly upon C; note, in particular, that the notation $E\gamma$, with γ behind, gives the good order for multiplication of γ's in the second equation (1.2).

Lemma 1.2. *If E is an extension of an R-module A by an R-module C and if $\gamma\colon C' \to C$ is a module homomorphism, there exists an extension E' of A by C' and a morphism $\Gamma=(1_A, \beta, \gamma)\colon E' \to E$. The pair (Γ, E') is unique up to a congruence of E'.*

Existence proof: In the diagram

$$\begin{array}{lccccccccc} E'\colon & 0 \longrightarrow & A & \overset{\varkappa'}{\cdots\rightarrow} & ? & \overset{\sigma'}{\cdots\rightarrow} & C' & \longrightarrow 0 \\ & & \| & & \vdots\downarrow \beta & & \downarrow \gamma & \\ E\colon & 0 \longrightarrow & A & \xrightarrow{\varkappa} & B & \xrightarrow{\sigma} & C & \longrightarrow 0 \end{array} \tag{1.3}$$

the sides and the bottom are given; we wish to fill in the module at "?" and the dotted arrows so as to make the diagram commutative and the top row exact. To do so, put at ? that subgroup $B' \subset B \oplus C'$ which consists of the pairs (b, c') with $\sigma b=\gamma c'$; define σ' and β as $\sigma'(b, c')=c'$, $\beta(b, c')=b$. This choice insures commutativity in the right-hand square of (1.3). With the definition $\varkappa' a=(\varkappa a, 0)$ the diagram is completed; the remaining conditions may be verified.

Uniqueness proof: Take any other such E'' with a morphism $\Gamma''=(1_A, \beta'', \gamma)\colon E'' \to E$. If B'' is the middle module in E'', define $\beta'\colon B'' \to B'$ by $\beta' b''=(\beta'' b'', \sigma'' b'')$; then $\Gamma_0=(1_A, \beta', 1_{C'})\colon E'' \to E'$ is a congruence and the composite $E'' \to E' \to E$ is Γ'', so that the diagram $\Gamma\colon E' \to E$ is unique up to a congruence Γ_0 of E', as asserted.

We call $E'=E\gamma$ the *composite* of the extension E and the homomorphism γ; the type of construction involved occurs repeatedly, for instance, in the study of induced fiber spaces (where γ is a fiber map!). Algebraically, E' has the following "couniversal" property:

Lemma 1.3. *Under the hypotheses of Lemma 1.2 each morphism $\Gamma_1=(\alpha_1,\beta_1,\gamma_1)\colon E_1\to E$ of extensions with $\gamma_1=\gamma$ can be written uniquely as a composite*

$$E_1\xrightarrow{(\alpha_1,\beta',1)}E\gamma\xrightarrow{(1,\beta,\gamma)}E. \tag{1.4}$$

More briefly, Γ_1 can be "factored through" $\Gamma\colon E\gamma\to E$.

Proof. Here $E_1=(\varkappa_1,\sigma_1)$ has the form $A_1\rightarrowtail B_1\twoheadrightarrow C'$. (Draw the diagram!) Define $\beta'\colon B_1\to B'$ as $\beta' b_1=(\beta_1 b_1,\sigma_1 b_1)$. This is the only way of defining β' so that $\beta_1=\beta\beta'$ and so that the diagram $(\alpha_1,\beta',1)\colon E_1\to E'$ will be commutative. The verification that this β' yields the desired factorization (1.4) is routine.

Incidentally, this factorization includes the uniqueness assertion of Lemma 1.2, in as much as $\Gamma''=(1_A,\beta'',\gamma)\colon E''\to E$ has by (1.4) the factorization $(1,\beta'',\gamma)=(1,\beta,\gamma)(1,\beta',1)$ with the factor $(1,\beta',1)\colon E''\to E'$ a congruence.

Next we show $\operatorname{Ext}(C,A)$ to be a covariant functor of A, for fixed C, by constructing for each E and for each $\alpha\colon A\to A'$ a "composite" extension $E'=\alpha E$, characterized as follows:

Lemma 1.4. *For $E\in\operatorname{Ext}(C,A)$ and $\alpha\colon A\to A'$ there is an extension E' of A' by C and a morphism $\Gamma=(\alpha,\beta,1_C)\colon E\to E'$. The pair (Γ,E') is unique up to a congruence of E'.*

Proof. We are required to fill in the diagram

$$\begin{array}{rccccccccc} E\colon & 0 & \longrightarrow & A & \xrightarrow{\varkappa} & B & \xrightarrow{\sigma} & C & \longrightarrow & 0\\ & & & \downarrow{\scriptstyle\alpha} & & \vdots{\scriptstyle\beta} & & \| & & \\ E'\colon & 0 & \longrightarrow & A' & \xrightarrow{\varkappa'} & ? & \xrightarrow{\sigma'} & C & \longrightarrow & 0 \end{array} \tag{1.5}$$

at the question mark and the dotted arrows so as to make the diagram commutative and the bottom row exact. To do so, take in $A'\oplus B$ the subgroup N of all elements $(-\alpha a,\varkappa a)$ for $a\in A$. At "?" in the diagram put the quotient group $(A'\oplus B)/N$, and write elements of this quotient group as cosets $(a',b)+N$. Then the equations $\varkappa' a'=(a',0)+N$, $\sigma'[(a',b)+N]=\sigma b$ and $\beta b=(0,b)+N$ define maps which satisfy the required conditions. That the E' so constructed is unique may be proved directly or deduced from the following "universal" property of E'.

Lemma 1.5. *Under the hypotheses of Lemma 1.4, any morphism $\Gamma_1=(\alpha_1,\beta_1,\gamma_1)\colon E\to E_1$ of extensions with $\alpha_1=\alpha$ can be written uniquely as a composite*

$$E\xrightarrow{(\alpha,\beta,1)}\alpha E\xrightarrow{(1,\beta',\gamma_1)}E_1.$$

More briefly, Γ_1 can be "factored through" $E\to\alpha E$.

Proof. If $E_1=(\varkappa_1, \sigma_1)$ with middle module B_1, a homomorphism $\beta'\colon (A'\oplus B)/N \to B_1$ may be defined by $\beta'[(a', b)+N]=\varkappa_1 a' + \beta_1 b$. One then verifies that $\beta_1=\beta'\beta$, that $(1_{A'}, \beta', \gamma_1)$ is a morphism of extensions, and that this β' is uniquely determined, completing the proof.

The uniqueness properties of αE yield congruences

$$1_A E \equiv E, \qquad (\alpha\alpha')E \equiv \alpha(\alpha' E).$$

Hence $\operatorname{Ext}(C,A)$ is a covariant functor of A. The fact that it is a bifunctor (of A and C) is demonstrated by the following result:

Lemma 1.6. *For α, γ, and E as in Lemmas* 1.2 *and* 1.4 *there is a congruence of extensions $\alpha(E\gamma)\equiv(\alpha E)\gamma$.*

Proof. By the definitions of $E\gamma$ and αE there are morphisms

$$E\gamma \xrightarrow{(1,\beta_1,\gamma)} E \xrightarrow{(\alpha,\beta_2,1)} \alpha E$$

with composite $(\alpha, \beta_2\beta_1, \gamma)\colon E\gamma \to \alpha E$. By Lemma 1.3, the extension $(\alpha E)\gamma$ is couniversal for such maps; that is, $(\alpha, \beta_2\beta_1, \gamma)$ has a factorization

$$E\gamma \xrightarrow{(\alpha,\beta',1)} (\alpha E)\gamma \xrightarrow{(1,\beta,\gamma)} \alpha E.$$

Now the left hand map is exactly the sort of morphism of extensions used in Lemma 1.4 to define $\alpha(E\gamma)$ from $E\gamma$. Hence, by the uniqueness assertion of that lemma, $\alpha(E\gamma)\equiv(\alpha E)\gamma$, q.e.d.

To illustrate one use of these lemmas, we prove:

Proposition 1.7. *For any extension $E=(\varkappa,\sigma)$ the composite extensions $\varkappa E$ and $E\sigma$ are split.*

Proof. The diagram

$$\begin{array}{lccccccccc}
E & 0 & \longrightarrow & A & \xrightarrow{\varkappa} & B & \xrightarrow{\sigma} & C & \longrightarrow & 0\\
 & & & \downarrow{\scriptstyle\varkappa} & & \downarrow{\scriptstyle\nu} & & \| & & \\
E'\colon & 0 & \longrightarrow & B & \longrightarrow & B\oplus C & \longrightarrow & C & \longrightarrow & 0,
\end{array}$$

with ν defined by $\nu b=(b, \sigma b)$, is commutative. Hence the definition of $\varkappa E$ in Lemma 1.4 shows that $\varkappa E$ is given by the bottom row, hence is split. Let the reader display the dual diagram which splits $E\sigma$.

Proposition 1.8. *Any morphism $\Gamma_1=(\alpha,\beta,\gamma)\colon E\to E'$ of extensions implies a congruence $\alpha E\equiv E'\gamma$.*

Proof. By the universal property of αE (Lemma 1.5), the map Γ_1 can be factored through $\Gamma\colon E\to\alpha E$ as $\Gamma_1=\Gamma_2\Gamma$, where $\Gamma_2=(1_{A'}, \beta', \gamma)\colon \alpha E\to E'$. This last map characterizes αE as $E'\gamma$, by Lemma 1.2.

2. Addition of Extensions

The direct sum $A\oplus C$ of two modules may be regarded as a covariant bifunctor of A and C, since there is for any two homomorphisms

$\alpha\colon A\to A'$ and $\gamma\colon C\to C'$ a homomorphism

$$\alpha\oplus\gamma\colon A\oplus C\to A'\oplus C'$$

with the usual properties $(\alpha\oplus\gamma)(\alpha'\oplus\gamma')=\alpha\alpha'\oplus\gamma\gamma'$ and $1_A\oplus 1_C=1_{A\oplus C}$. This homomorphism may be defined by setting $(\alpha\oplus\gamma)(a,c)=(\alpha a,\gamma c)$ or as the unique homomorphism in the middle which renders the diagram

$$\begin{array}{ccccc} A & \leftarrow & A\oplus C & \rightarrow & C \\ \downarrow{\scriptstyle\alpha} & & \vdots{\scriptstyle\alpha\oplus\gamma} & & \downarrow{\scriptstyle\gamma} \\ A' & \leftarrow & A'\oplus C' & \rightarrow & C' \end{array}$$

commutative. Here each row consists of projections of the direct sum, as in (I.8.12).

The *diagonal* homomorphism for a module C is

$$\Delta=\Delta_C\colon C\to C\oplus C,\qquad \Delta(c)=(c,c). \tag{2.1}$$

It may also be described as that map which renders

$$\begin{array}{ccccc} C & = & C & = & C \\ \| & & \vdots{\scriptstyle\Delta} & & \| \\ C & \xleftarrow{\pi_1} & C\oplus C & \xrightarrow{\pi_2} & C \end{array}$$

commutative. The codiagonal map for a module A is

$$\nabla=\nabla_A\colon A\oplus A\to A,\qquad \nabla(a_1,a_2)=a_1+a_2; \tag{2.1'}$$

it has a dual diagrammatic description by $\nabla\iota_1=1_A=\nabla\iota_2\colon A\to A$. The maps Δ and ∇ may be used to rewrite the usual definition of the sum $f+g$ of two homomorphisms $f,g\colon C\to A$ as

$$f+g=\nabla_A(f\oplus g)\Delta_C; \tag{2.2}$$

the reader should verify that $(f+g)c$ is still $fc+gc$ under this formula.

Given two extensions $E_i=(\varkappa_i,\sigma_i)\colon A_i\rightarrowtail B_i\twoheadrightarrow C_i$ for $i=1,2$, we define their direct sum to be the extension

$$E_1\oplus E_2\colon 0\to A_1\oplus A_2\to B_1\oplus B_2\to C_1\oplus C_2\to 0. \tag{2.3}$$

We now make $\operatorname{Ext}(C,A)$ a group under an addition which utilizes (2.3).

Theorem 2.1. *For given R-modules A and C the set $\operatorname{Ext}_R(C,A)$ of all congruence classes of extensions of A by C is an abelian group under the binary operation which assigns to the congruence classes of extensions E_1 and E_2 the congruence class of the extension*

$$E_1+E_2=\nabla_A(E_1\oplus E_2)\Delta_C. \tag{2.4}$$

The class of the split extension $A \rightarrowtail A \oplus C \twoheadrightarrow C$ is the zero element of this group, while the inverse of any E is the extension $(-1_A)E$. For homomorphisms $\alpha\colon A \to A'$ and $\gamma\colon C' \to C$ one has

$$\alpha(E_1+E_2) \equiv \alpha E_1 + \alpha E_2, \qquad (E_1+E_2)\gamma \equiv E_1\gamma + E_2\gamma, \tag{2.5}$$

$$(\alpha_1+\alpha_2)E \equiv \alpha_1 E + \alpha_2 E, \qquad E(\gamma_1 \mid \gamma_2) \equiv E\gamma_1 + E\gamma_2. \tag{2.6}$$

The composition (2.4) is known as the *Baer sum*; the rules (2.5) state that the maps $\alpha_*\colon \operatorname{Ext}(C,A) \to \operatorname{Ext}(C,A')$ and $\gamma^*\colon \operatorname{Ext}(C,A) \to \operatorname{Ext}(C',A)$ are group homomorphisms.

We give two different proofs. The first is "computational"; it is like the calculation made in §1 to show that $\operatorname{Ext}_Z(Z_m, A)$ is the group A/mA.

Take any extension $E=(\varkappa,\sigma)$ of A by C, with $\sigma\colon B \twoheadrightarrow C$. To each c in C choose a representative $u(c)$; that is, an element $u(c) \in B$ with $\sigma u(c) = c$. For each $r \in R$, the exactness of E gives $ru(c) - u(rc) \in \varkappa A$; similarly, c, $d \in C$ have $u(c+d) - u(c) - u(d) \in \varkappa A$. Hence there are elements $f(c,d)$ and $g(r,c) \in A$ with

$$u(c) + u(d) = \varkappa f(c,d) + u(c+d), \qquad c, d \in C, \tag{2.7a}$$

$$ru(c) = \varkappa g(r,c) + u(rc), \quad r \in R, \; c \in C. \tag{2.7b}$$

Call the pair of functions (f,g) a *factor system* for E. Let $F_R(C,A)$ denote, during this proof, the set of all pairs (f,g) of functions f on $C \times C$ to A and g on $R \times C$ to A. Each factor system is an element of $F_R(C,A)$, and F_R is a group under termwise addition; that is, with $(f_1+f_2)(c,d) = f_1(c,d) + f_2(c,d)$.

The factor system for E is not unique. For any different choice of representatives $u'(c)$ we must have $u'(c) = \varkappa h(c) + u(c)$ for some function h on C to A. One calculates that

$$u'(c) + u'(d) = \varkappa[h(c) + h(d) - h(c+d) + f(c,d)] + u'(c+d),$$

$$ru'(c) = \varkappa[rh(c) - h(rc) + g(r,c)] + u'(rc).$$

The new factor system $f'(c,d)$, $g'(r,c)$ for the representatives u' is then given by the expressions in brackets in these equations. We may express this fact differently: To each function h on C to A there is an element $(\delta_C h, \delta_R h) \in F_R(C,A)$ defined by

$$(\delta_C h)(c,d) = h(c) + h(d) - h(c+d), \qquad (\delta_R h)(r,c) = rh(c) - h(rc).$$

The factor system f', g' for representatives u' then has the form $(f', g') = (f,g) + (\delta_C h, \delta_R h)$. Conversely, any such function h can be used to change representatives in an extension. Thus, if we denote by $S_R(C,A)$ the subgroup of all those pairs of functions in $F_R(C,A)$ of the form $(\delta_C h, \delta_R h)$, the factor system (f,g) of E is uniquely defined modulo $S_R(C,A)$.

Use the factor group $F_R(C,A)/S_R(C,A)$; to each extension E assign the coset $\omega(E)$ of any one of its factor systems (f,g) in this group F_R/S_R. Then $\omega(E)$ is uniquely determined by E.

A congruence of extensions maps representatives to representatives; hence congruent extensions have the same factor systems. It follows that ω is a $1-1$ mapping of the congruence classes of extensions to a subset of the abelian group $F_R(C,A)/S_R(C,A)$. To show $\operatorname{Ext}(C,A)$ an abelian group under the Baer sum it now suffices to show that

$$\omega(E_1+E_2)=\omega(E_1)+\omega(E_2), \quad \omega[(-1_A)E]=-\omega(E).$$

The first follows by calculating a factor system for $E_1\oplus E_2$, and thence for E_1+E_2. The second follows by observing (draw the diagram!) that $(-1_A)E$ is obtained from E just by changing the sign of the map $\varkappa: A\rightarrowtail B$ and hence by changing the signs of f and g in the factor system. Finally, the split extension E_0 has $(0,0)$ as one of its factor systems, hence is the zero of this addition.

It is also possible (see the exercises) to characterize directly those pairs of functions (f,g) which can occur as factor systems for an extension, and hence to show that $\operatorname{Ext}_R(C,A)$ is an abelian group without using the Baer sum at all.

The proof of (2.5) is easy; $F_R(C,A)/S_R(C,A)$ is a bifunctor, and ω is a natural homomorphism. The proof of (2.6) is similar.

We now turn to the second (conceptual) proof of the theorem. For the direct sum (2.3) of two extensions E_i the congruences

$$(\alpha_1\oplus\alpha_2)(E_1\oplus E_2)\equiv\alpha_1E_1\oplus\alpha_2E_2, \tag{2.8}$$

$$(E_1\oplus E_2)(\gamma_1\oplus\gamma_2)\equiv E_1\gamma_1\oplus E_2\gamma_2, \tag{2.9}$$

may be proved by the lemmas of §1 which characterize the composite extensions $E_i\gamma_i$ and α_iE_i. For $\alpha: A\to A'$ one calculates easily that

$$\alpha\nabla=\nabla(\alpha\oplus\alpha): A\oplus A\to A', \tag{2.10}$$

and similarly for $\gamma: C'\to C$ that

$$\Delta\gamma=(\gamma\oplus\gamma)\Delta: C'\to C\oplus C. \tag{2.10'}$$

Now we can prove the assertion (2.5) of the theorem by the string of congruences

$$\begin{aligned}\alpha(E_1+E_2)&\equiv\alpha\nabla(E_1\oplus E_2)\Delta\equiv\nabla(\alpha\oplus\alpha)(E_1\oplus E_2)\Delta\\ &\equiv\nabla(\alpha E_1\oplus\alpha E_2)\Delta\equiv\alpha E_1+\alpha E_2;\end{aligned}$$

the second half is similar. The proof of (2.6) is parallel to this once we know that

$$\Delta E\equiv(E\oplus E)\Delta, \quad E\nabla\equiv\nabla(E\oplus E). \tag{2.11}$$

Since $(\Delta_A, \Delta_B, \Delta_C): E \to E \oplus E$ is a morphism of extensions, the first of these identities follows from Prop. 1.8. Similarly, $(\nabla, \nabla, \nabla): E \oplus E \to E$ gives the second.

Now let us show that the Baer sum (2.4) makes Ext a group. The associative law follows from the definition (2.4) once we know that the diagonal and codiagonal satisfy the identities

$$(\Delta \oplus 1_C)\Delta = (1_C \oplus \Delta)\Delta\colon\ C \to C \oplus C \oplus C, \tag{2.12}$$

$$\nabla(\nabla \oplus 1_A) = \nabla(1_A \oplus \nabla)\colon\ A \oplus A \oplus A \to A. \tag{2.12'}$$

These follow directly from the definition of Δ or ∇, provided we identify $(C \oplus C) \oplus C$ with $C \oplus (C \oplus C)$ by the obvious isomorphism. To prove the commutative law for the Baer sum, use the isomorphism $\tau_A: A_1 \oplus A_2 \to A_2 \oplus A_1$ given by $\tau_A(a_1, a_2) = (a_2, a_1)$ (or, if you wish, by universality and a suitable diagram!). The morphism $(\tau_A, \tau_B, \tau_C): (E_1 \oplus E_2) \to E_2 \oplus E_1$ shows that $\tau_A(E_1 \oplus E_2) \equiv (E_2 \oplus E_1)\tau_C$; a calculation or a diagram proves that $\nabla_A \tau_A = \nabla_A$ and that $\Delta_C = \tau_C \Delta_C$. Hence we get the commutative law by

$$E_1 + E_2 = \nabla(E_1 \oplus E_2)\Delta = \nabla\tau(E_1 \oplus E_2)\Delta$$
$$\equiv \nabla(E_2 \oplus E_1)\tau\Delta \equiv \nabla(E_2 \oplus E_1)\Delta = E_2 + E_1.$$

To show that the split extension E_0 acts as the zero for the Baer sum, first observe that for any $E \in \operatorname{Ext}(C,A)$ there is a commutative diagram

$$\begin{array}{cccccccccc} E\colon & 0 \to & A & \longrightarrow & B & \xrightarrow{\sigma} & C & \to 0 \\ & & \downarrow{\scriptstyle 0} & & \downarrow{\scriptstyle \nu} & & \| & \\ E_0\colon & 0 \to & A & \to & A \oplus C & \to & C & \to 0, \end{array}$$

where ν is the map $\nu b = (0, \sigma b) = \iota_2 \sigma b$. This diagram asserts that the split extension E_0 can be written as the composite $E_0 = 0_A E$, with $0_A: A \to A$ the zero homomorphism. Now the distributive law gives $E + E_0 \equiv 1_A E + 0_A E \equiv (1_A + 0_A) E \equiv 1_A E \equiv E$. A similar argument shows that $(-1_A) E$ acts as the additive inverse of E under the Baer sum. Our second proof of the theorem is complete.

The second distributive law (2.6) contained in this theorem may be expressed as follows. For each $\alpha: A \to A'$ let $\alpha_*: \operatorname{Ext}(C,A) \to \operatorname{Ext}(C,A')$ be the induced homomorphism, and similarly set $\gamma^* E = E\gamma$. Then $(\gamma_1^* + \gamma_2^*) E = \gamma_1^* E + \gamma_2^* E$, so (2.6) may now be written

$$(\alpha_1 + \alpha_2)_* = (\alpha_1)_* + (\alpha_2)_*, \qquad (\gamma_1 + \gamma_2)^* = (\gamma_1)^* + (\gamma_2)^*.$$

A bifunctor with this property is said to be *additive*. Exactly as in (I.6.5), this property gives natural isomorphisms

$$\operatorname{Ext}(C, A_1 \oplus A_2) \cong \operatorname{Ext}(C, A_1) \oplus \operatorname{Ext}(C, A_2),$$
$$\operatorname{Ext}(C_1 \oplus C_2, A) \cong \operatorname{Ext}(C_1, A) \oplus \operatorname{Ext}(C_2, A).$$

For $R=Z$ and C a finitely generated abelian group, these formulas, with Prop. 1.1 and $\mathrm{Ext}_Z(Z,A)=0$, allow us to calculate $\mathrm{Ext}_Z(C,A)$.

Corollary 2.2. *If the finite abelian groups A and C have relatively prime orders, then every extension of A by C splits.*

Proof. Let m and n be the orders of A and C, and let $\mu_m\colon C\to C$ be the homomorphism $\mu_m c=mc$ given by multiplication by m in C. Since m and n are relatively prime, there is an m' with $m'm\equiv 1 \pmod n$; hence μ_m is an automorphism, and every element of $\mathrm{Ext}(C,A)$ has the form $E\mu_m$ for some E. But $\mu_m=1_C+\cdots+1_C$, with m summands, so

$$E\,\mu_m=E(1_C+\cdots+1_C)\equiv(1_A+\cdots+1_A)E=\nu_m E=0,$$

where $\nu_m\colon A\to A$ is $\nu_m(a)=ma=0$, q.e.d.

Exercises

In the following exercises it is convenient to assume that all factor systems (f,g) satisfy the "normalization conditions"

$$f(c,0)=0=f(0,d),\qquad g(r,0)=0.$$

This can always be accomplished by using representatives u with $u(0)=0$.

1. For abelian groups (i.e., with $R=Z$) show that a "normalized" function on $C\times C$ to A is a factor system for extensions of abelian groups if and only if it satisfies the identities

$$f(c,d)+f(c+d,e)=f(c,d+e)+f(d,e),\qquad f(c,d)=f(d,c),$$

which correspond respectively to the associative and commutative laws.

2. If $G_Z(C,A)$ is the set of normalized functions f satisfying the identities of Ex. 1, show that $\mathrm{Ext}_Z(C,A)\cong G_Z(C,A)/S_Z(C,A)$.

3. Do the analogue of Ex. 1 for any ring (identities on factor systems consisting of two functions f and g).

3. Obstructions to the Extension of Homomorphisms

We have already observed that the functor Hom does not preserve exact sequences, because a homomorphism $\alpha\colon A\to G$ on a submodule $A\subset B$ cannot always be extended to a homomorphism of B into G We can now describe a certain element αE of $\mathrm{Ext}(B/A,G)$ which presents the "obstruction" to this extension.

Lemma 3.1. *Let A be a submodule of B, and $E\colon A\rightarrowtail B\twoheadrightarrow C$ the corresponding exact sequence, with $C=B/A$. A homomorphism $\alpha\colon A\to G$ can be extended to a homomorphism $B\to G$ if and only if the extension αE splits.*

Proof. Suppose first that α is extendable to $\hat{\alpha}\colon B \to G$. Form the diagram

$$\begin{array}{rccccccccc} E\colon & 0 & \to & A & \longrightarrow & B & \xrightarrow{\sigma} & C & \to & 0 \\ & & & \downarrow{\scriptstyle\alpha} & & \vdots & & \| & & \\ E'\colon & 0 & \to & G & \xrightarrow{\iota_1} & G \oplus C & \xrightarrow{\pi_2} & C & \to & 0, \end{array}$$

where E' is the external direct sum with injection ι_1, projection π_2. Fill in the dotted arrow with the map $b \to (\hat{\alpha} b, \sigma b) = \iota_1 \hat{\alpha} b + \iota_2 \sigma b$. The resulting diagram is commutative, hence yields a morphism $E \to E'$. According to Lemma 1.4, $E' \equiv \alpha E$. Since E' splits, so does αE.

Conversely, assume that αE splits. The diagram

$$\begin{array}{rccccccccc} E\colon & 0 & \to & A & \longrightarrow & B & \longrightarrow & C & \to & 0 \\ & & & \downarrow{\scriptstyle\alpha} & & \downarrow{\scriptstyle\beta} & & \| & & \\ \alpha E\colon & 0 & \to & G & \underset{\pi_1}{\overset{\iota_1}{\rightleftarrows}} & B' & \xrightarrow{\pi_2} & C & \to & 0, \end{array}$$

used to define αE yields a map $\pi_1 \beta\colon B \to G$ which extends α. The lemma is proved.

The assignment to each $\alpha\colon A \to G$ of its obstruction αE is, by (2.6), a group homomorphism

$$E^*\colon \operatorname{Hom}_R(A, G) \to \operatorname{Ext}_R(C, G).$$

Call this the connecting homomorphism for the exact sequence E.

Theorem 3.2. *If $E\colon A \overset{\varkappa}{\rightarrowtail} B \overset{\sigma}{\twoheadrightarrow} C$ is a short exact sequence of R-modules, then the sequence*

$$\left.\begin{array}{l} 0 \to \operatorname{Hom}_R(C, G) \to \operatorname{Hom}_R(B, G) \to \operatorname{Hom}_R(A, G) \\ \quad \xrightarrow{E^*} \operatorname{Ext}_R(C, G) \xrightarrow{\sigma^*} \operatorname{Ext}_R(B, G) \xrightarrow{\varkappa^*} \operatorname{Ext}_R(A, G) \end{array}\right\} \tag{3.1}$$

of abelian groups is exact for any R-module G.

Proof. We already know exactness at $\operatorname{Hom}(C, G)$ and $\operatorname{Hom}(B, G)$ by (I.6.7). Lemma 3.1 gives exactness at $\operatorname{Hom}(A, G)$. By Prop. 1.7, $\sigma^* E^* = (E\sigma)^* = 0$. Conversely, to show kernel contained in image at $\operatorname{Ext}(C, G)$ we must take an $E_1 \in \operatorname{Ext}(C, G)$ such that $E_1\sigma$ splits and show E_1 the obstruction for some map $A \to G$. The fact that $E_1\sigma$ splits gives a commutative diagram

$$\begin{array}{rccccccccc} & & & & & & & A & & \\ & & & & & & & \downarrow{\scriptstyle\varkappa} & & \\ E_1\sigma\colon & 0 & \to & G & \to & G \oplus B & \underset{\mu}{\rightleftarrows} & B & \to & 0 \\ & & & \downarrow & & \downarrow{\scriptstyle\beta} & \swarrow & \downarrow{\scriptstyle\sigma} & & \\ E_1\colon & 0 & \to & G & \xrightarrow{\varkappa_1} & B_1 & \xrightarrow{\sigma_1} & C & \to & 0. \end{array}$$

The splitting map μ followed by β yields $\beta_1 = \beta\mu\colon B \to B_1$ (dotted arrow above) which makes the right hand lower triangle commutative.

Therefore $\sigma_1\beta_1\varkappa=\sigma\varkappa=0$. But E_1 is exact, so $\beta_1\varkappa$ factors as $\varkappa_1\alpha_1$ for some $\alpha_1: A\to G$. Then $(\alpha_1, \beta_1, 1): E\to E_1$ is a morphism of exact sequences which states $E_1\equiv\alpha_1 E$.

An analogous argument yields exactness of the sequence at $\mathrm{Ext}_R(B, G)$ and hence completes the proof of the theorem.

This theorem asserts that the functor Ext repairs the inexactitude of Hom on the right. At the same time Ext presents a new inexactitude: On the right in (3.1), $\mathrm{Ext}_R(B, G)\to\mathrm{Ext}_R(A, G)$ is not always an epimorphism (see exercise). To describe the cokernel we need a new functor Ext^2.

Turn now to the problem: When can a homomorphism $\gamma: G\to B/A$ be "lifted" to B; that is, when is there a $\hat{\gamma}: G\to B$ such that γ is the composite $G\to B\to B/A$? This yields a dual to the previous lemma.

Lemma 3.3. *Let $C=B/A$ be a quotient module, E the corresponding exact sequence. A homomorphism $\gamma: G\to B/A$ can be lifted to a homomorphism $\hat{\gamma}: G\to B$ if and only if the extension $E\gamma$ splits.*

The proof is exactly dual to that of Lemma 3.1, in the sense that all arrows are reversed and that direct sums are replaced by direct products. Again, call $E\gamma\in\mathrm{Ext}(G, A)$ the *obstruction* to lifting γ. The assignment to each $\gamma: G\to C$ of its obstruction $E\gamma$ is a group homomorphism

$$E_*: \mathrm{Hom}(G, C)\to\mathrm{Ext}(G, A)$$

called the connecting homomorphism for E.

Theorem 3.4. *If $E: A\rightarrowtail B\twoheadrightarrow C$ is a short exact sequence of R-modules, then the sequence*

$$\left.\begin{aligned} 0\to\mathrm{Hom}_R(G, A)\to\mathrm{Hom}_R(G, B)\to\mathrm{Hom}_R(G, C)\\ \xrightarrow{E_*}\mathrm{Ext}_R(G, A)\to\mathrm{Ext}_R(G, B)\to\mathrm{Ext}_R(G, C)\end{aligned}\right\} \tag{3.2}$$

is exact for any R-module G.

The proof is dual to that of Thm. 3.2.

Theorem 3.5. *An R-module P is projective if and only if $\mathrm{Ext}_R(P, G)=0$ for every R-module G.*

By Thm. I.6.3, P is projective if and only if every extension by P splits. Thm. 3.2 provides the following way to calculate the group Ext.

Theorem 3.6. *If C and G are given modules and if $F: K\overset{\varkappa}{\rightarrowtail} P\twoheadrightarrow C$ is an exact sequence with P projective, then*

$$\mathrm{Ext}_R(C, G)\cong\mathrm{Hom}_R(K, G)/\varkappa^*\mathrm{Hom}_R(P, G). \tag{3.3}$$

In particular, the group on the right is independent (up to isomorphism) of the choice of the short exact sequence F.

Proof. In (3.1) replace E by F. Since P is projective, $\operatorname{Ext}_R(P,G)=0$, and the exactness of (3.1) gives the formula (3.3) for $\operatorname{Ext}_R(C,G)$.

Since any module C can be represented as a quotient of a free module, one may always calculate $\operatorname{Ext}_R(C,G)$ by (3.3) with P free. For example, the exact sequence $Z \overset{\varkappa}{\rightarrowtail} Z \twoheadrightarrow Z/mZ$, with $\varkappa$ multiplication by the integer m, provides a representation of the cyclic group Z_m as a quotient of Z. Since $\operatorname{Hom}(Z,A)\cong A$ under the correspondence which maps each $f\colon Z\to A$ into $f(1)$, we obtain an isomorphism $\operatorname{Ext}_Z(Z_m,A)\cong A/mA$. The correspondence is that already used in Prop. 1.1.

Proposition 3.7. *For abelian groups the sequences of Thms.* 3.2 *and* 3.4 *remain exact when a zero is added on the right.*

Proof. In the case of Thm. 3.2 we must show that $\varkappa\colon A\rightarrowtail B$ a monomorphism implies $\varkappa^*\colon \operatorname{Ext}(B,G)\to\operatorname{Ext}(A,G)$ an epimorphism. To this end, take a free abelian group F, an epimorphism $\varphi\colon F\twoheadrightarrow B$ with kernel K, and let L be $\varphi^{-1}(\varkappa A)$. Then φ maps L onto $\varkappa A$ with the same kernel K, giving a commutative diagram

$$\begin{array}{rccccccccc}
E_1\colon & 0 & \to & K & \to & L & \to & A & \to & 0\\
 & & & \| & & \downarrow & & \downarrow{\scriptstyle\varkappa} & & \\
E_2\colon & 0 & \to & K & \to & F & \xrightarrow{\varphi} & B & \to & 0
\end{array}$$

with exact rows E_1, E_2 and hence $E_1\equiv E_2\varkappa$. This yields a commutative diagram

$$\begin{array}{ccccc}
\operatorname{Hom}(K,G) & \xrightarrow{E_2^*} & \operatorname{Ext}(B,G) & & \\
\| & & \downarrow{\scriptstyle\varkappa^*} & & \\
\operatorname{Hom}(K,G) & \xrightarrow{E_1^*} & \operatorname{Ext}(A\ G) & \to & \operatorname{Ext}(L,G);
\end{array}$$

the bottom row is exact by Thm. 3.2. But L, as a subgroup of the free abelian group F, is itself free. By Thm. 3.5, $\operatorname{Ext}(L,G)=0$, hence E_1^* is an epimorphism in the diagram, and so is $\varkappa^*$, q.e.d.

In the case of Thm. 3.4 we are given $E\colon A\rightarrowtail B\twoheadrightarrow C$ exact and we must show $\operatorname{Ext}(G,B)\to\operatorname{Ext}(G,C)$ an epimorphism. Represent any element of $\operatorname{Ext}(G,C)$ by an exact sequence $S\colon C\rightarrowtail D\twoheadrightarrow G$. Since $\mu\colon C\to D$ is a monomorphism, the case just treated shows that there is an exact sequence $E'\colon A\rightarrowtail M\twoheadrightarrow D$ with $\mu^*E'=E$. This states that we can fill out the following commutative diagram so that the first two rows and the last column will be exact

$$\begin{array}{rccccccccc}
E\colon & 0 & \to & A & \to & B & \to & C & \to & 0\\
 & & & \| & & \vdots & & \downarrow{\scriptstyle\mu} & & \\
E'\colon & 0 & \to & A & \dashrightarrow & ? & \dashrightarrow & D & \to & 0\\
 & & & & & \vdots & & \downarrow & & \\
 & & & & & G & = & G & &
\end{array}$$

A diagram chase then shows the middle column exact. This middle column provides an element of $\operatorname{Ext}(G, B)$ mapping on the last column $S \in \operatorname{Ext}(G, C)$, as desired.

Note that the diagram above is symmetric: Given exactness of the top row and the right column, exactness of the middle row is equivalent to exactness of the middle column. The case of Thm. 3.2 asserts that the diagram can be filled out so that the middle row is exact, while the case of Thm. 3.4 asserts that the diagram can be filled out so the middle column is exact. The same fact can be stated in subgroup language as follows:

Corollary 3.8. *Given abelian groups D and $A \subset B$ and a monomorphism $\mu: B/A \rightarrowtail D$, there exists an abelian group $M \supset B$ and an extension of μ to an isomorphism $M/A \cong D$.*

This amounts to the construction of a group M from a given subgroup B and an "overlapping" quotient group D.

Exercises

1. (Inexactitude of Ext on the right.) Let $R = \mathsf{K}[x,y]$ be the polynomial ring in two indeterminates x and y with coefficients in a field K and (x,y) the ideal of all polynomials with constant term 0. The quotient module $R/(x,y)$ is isomorphic to K, where K is regarded as an R-module with $xk = 0 = yk$, for all $k \in \mathsf{K}$, and $E: (x,y) \rightarrowtail R \twoheadrightarrow \mathsf{K}$ is an exact sequence of R-modules. Show that $\operatorname{Ext}_R(R, G) \to \operatorname{Ext}_R((x,y), G)$ is not an epimorphism for all G, by choosing an extension on the right in which (x,y) is represented as the quotient of a free module in two generators.

2. Show similarly that the sequence of Thm. 3.4 cannot be completed with a zero on the right.

3. Show that Cor. 3.8 amounts to the following (self-dual) assertion: Any homomorphism $\alpha: B \to D$ of abelian groups can be written as a composite $\alpha = \tau\nu$ with ν a monomorphism, τ an epimorphism, and $\operatorname{Ker}\tau = \nu(\operatorname{Ker}\alpha)$.

4. Give a direct proof of the second half of Prop. 3.7. (Write G as quotient of a free group.)

5. Prove Prop. 3.7 for modules over a principal ideal domain.

6. For p a prime number and C an abelian group with $pC = 0$, prove

$$\operatorname{Ext}_Z(C, G) \cong \operatorname{Hom}_Z(C, G/pG)$$ [Eilenberg-Mac Lane 1954, Thm. 26.5].

7. For p a prime, P the additive group of all rational numbers of the form m/p^e, $m, e \in Z$, and $Z^{(p)}$ the additive group of p-adic integers, prove

$$\operatorname{Ext}_Z(P, Z) \cong Z^{(p)}/Z$$ [Eilenberg-Mac Lane 1942, Appendix B].

4. The Universal Coefficient Theorem for Cohomology

As a first application of the functor Ext we give a method of "calculating" the cohomology groups of a complex for any coefficient group from the homology of that complex — provided we are dealing with complexes of free abelian groups or of free modules over a principal ideal domain.

Theorem 4.1. *(Universal Coefficients.) Let K be a complex of free abelian groups K_n and let G be any abelian group. Then for each dimension n there is an exact sequence*

$$0 \to \operatorname{Ext}(H_{n-1}(K), G) \xrightarrow{\beta} H^n(K,G) \xrightarrow{\alpha} \operatorname{Hom}(H_n(K), G) \to 0 \qquad (4.1)$$

with homomorphisms β and α natural in K and G. This sequence splits, by a homomorphism which is natural in G but not in K.

The second map α is defined on a cohomology class, $\operatorname{cls} f$, *as follows. Each n-cocycle of* $\operatorname{Hom}(K,G)$ *is a homomorphism $f: K_n \to G$ which vanishes on ∂K_{n+1}, so induces $f_*: H_n(K) \to G$. If $f = \delta g$ is a coboundary, it vanishes on cycles, so $(\delta g)_* = 0$. Define $\alpha(\operatorname{cls} f) = f_*$.*

Proof. Write C_n for the group of n-cycles of K; then $D_n = K_n/C_n$ is isomorphic to the group B_{n-1} of $(n-1)$-boundaries of K. The boundary homomorphism $\partial: K_n \to K_{n-1}$ factors as

$$K_n \xrightarrow{j} D_n \xrightarrow{\partial'} C_{n-1} \xrightarrow{i} K_{n-1}, \qquad (4.2)$$

with j the projection, i the injection. The short sequences

$$T_n\colon C_n \rightarrowtail K_n \twoheadrightarrow D_n, \qquad S_n\colon D_{n+1} \overset{\partial'}{\rightarrowtail} C_n \twoheadrightarrow H_n(K) \qquad (4.3)$$

are exact, the second by the definition of H_n as $C_n/\partial K_n$. The coboundary in the complex $\operatorname{Hom}(K,G)$ is $\delta = \pm\partial^*$, where $\partial^*: \operatorname{Hom}(K_{n-1},G) \to \operatorname{Hom}(K_n,G)$ is induced by ∂. This complex appears as the middle row in the diagram

$$\begin{array}{ccccccc}
 & & & & 0 & & 0 \\
 & & & & \uparrow & & \downarrow \\
0 \to & \operatorname{Hom}(H_n,G) & \to & \operatorname{Hom}(C_n,G) & \xrightarrow{\partial'^*} & \operatorname{Hom}(D_{n+1},G) & \\
 & & & \uparrow{\scriptstyle i^*} & & \downarrow{\scriptstyle j^*} & \\
\cdots \to & \operatorname{Hom}(K_{n-1},G) & \xrightarrow{\delta} & \operatorname{Hom}(K_n,G) & \xrightarrow{\delta} & \operatorname{Hom}(K_{n+1},G) & \to \cdots \\
 & \downarrow{\scriptstyle i^*} & & \uparrow{\scriptstyle j^*} & & & \\
 & \operatorname{Hom}(C_{n-1},G) & \xrightarrow{\partial'^*} & \operatorname{Hom}(D_n,G) & \xrightarrow{S_{n-1}^*} & \operatorname{Ext}(H_{n-1},G) & \to 0. \\
 & \downarrow & & \uparrow & & & \\
 & 0 & & 0 & & &
\end{array}$$

This diagram is commutative up to a sign (that involved in the definition $\delta = \pm\partial^*$). In the diagram the fundamental exact sequence (Thm. 3.2) for Hom and Ext appears several times. The top row is the exact sequence for S_n, the bottom that for S_{n-1}, with the right-hand zero standing for $\operatorname{Ext}(C_{n-1},G)$ which vanishes because $C_{n-1} \subset K_{n-1}$ is free. The columns are (parts of) the exact sequences for T_{n-1}, T_n, and T_{n+1}; the zero at the middle top is $\operatorname{Ext}(D_n,G)$, zero because D_n is free.

The cohomology of the middle row is $\operatorname{Ker}\delta/\operatorname{Im}\delta$. Since j^* is a monomorphism and i^* an epimorphism, it is $\operatorname{Ker}(\partial'^* i^*)/\operatorname{Im}(j^*\partial'^*)$, and is mapped by i^* onto $\operatorname{Ker}\partial'^*$, isomorphic to $\operatorname{Hom}(H_n, G)$ by exactness of the top row. The combined map is α. Its kernel is $\operatorname{Im} j^*/\operatorname{Im}(j^*\partial'^*)$; as j^* is a monomorphism, this is $\operatorname{Ext}(H_{n-1}, G)$, by exactness of the bottom row. This proves (4.1) exact, with β described in "switchback" notation as $\operatorname{cls} j^*(S_{n-1}^*)^{-1}$, hence natural.

To split the sequence (4.1), observe that $D_n \cong B_{n-1} \subset K_{n-1}$ is free, so the sequence T_n of (4.3) splits by a homomorphism $\varphi\colon D_n \to K_n$ with $j\varphi = 1_D$. Then $\varphi^* j^* = 1$, so $S_{n-1}^* \varphi^* \operatorname{cls}^{-1}$ is defined and is a left inverse for $\beta = j^*(S_{n-1}^*)^{-1}$. This left inverse depends on the choice of the maps φ splitting T_n. Such a choice cannot be made uniformly for all free complexes K, hence φ^* is not natural in K (but is natural in G for K fixed).

This proof uses several times the fact that subgroups of free abelian groups are free. The analogous statement holds for free modules over a principal ideal domain; hence the theorem holds for K a complex of free modules over such a domain D (and G a D-module). The most useful case is that for vector spaces over a field. Here Thm. 4.1 gives

Corollary 4.2. *If K is a chain complex composed of vector spaces K_n over a field F, and if V is any vector space over that field, there is a natural isomorphism $H^n(K, V) \cong \operatorname{Hom}(H_n(K), V)$.*

In particular, when $V = F$, $H^n(K, F)$ is the vector space dual of $H_n(K)$.

Thm. 4.1 is a special case of a more general result which "calculates" the homology of the complex $\operatorname{Hom}(K, L)$ formed from two complexes K and L. Recall (II.3.4) that $\operatorname{Hom}(K, L)$ is a complex with $\operatorname{Hom}_n(K, L) = \prod_p \operatorname{Hom}(K_p, L_{p+n})$ and with the boundary $\partial = \partial_H$ of any n-chain $f = \{f_p\colon K_p \to L_{p+n}\}$ given by

$$(\partial_H f)_p k = \partial_L(f_p k) + (-1)^{n+1} f_{p-1}(\partial_K k), \quad k \in K_p. \tag{4.4}$$

The general theorem reads

Theorem 4.3. *(Homotopy Classification Theorem.) For K and L complexes of abelian groups with each K_n free as an abelian group, there is for each n a short exact sequence*

$$\left.\begin{aligned} &\prod_{p=-\infty}^{\infty} \operatorname{Ext}(H_p(K), H_{p+n+1}(L)) \overset{\beta}{\rightarrowtail} H_n(\operatorname{Hom}(K, L)) \\ &\qquad \overset{\alpha}{\twoheadrightarrow} \prod_{p=-\infty}^{\infty} \operatorname{Hom}(H_p(K), H_{p+n}(L)), \end{aligned}\right\} \tag{4.5}$$

with homomorphisms β and α which are natural in K and L. This sequence splits by a homomorphism which is not natural.

Change lower indices here to upper indices by the usual convention $H_{-n} = H^n$ and assume $L = L_0 = G$ with boundary zero; then each of the products has at most one non-zero term, and (4.5) becomes (4.1). In general, if we shift the indices of L by n (and change the sign of the boundaries in L by $(-1)^n$) we shift $H_n(\mathrm{Hom}(K,L))$ to $H_0(\mathrm{Hom}(K,L))$; hence it suffices to prove the theorem when $n=0$. Now a 0-cycle of $\mathrm{Hom}(K,L)$ is by (4.4) just a chain transformation $f: K\to L$; as such it induces for each dimension p a homomorphism $(f_p)_*: H_p(K)\to H_p(L)$. The family of these homomorphisms is an element $f_*=\{(f_p)_*\}\in \prod_p \mathrm{Hom}(H_p(K),H_p(L))$. Any f' homotopic to f induces the same homomorphism f_*. Since an element of $H_0(\mathrm{Hom}(K,L))$ is just a homotopy class, cls f, of such chain transformations (Prop. II.3.2), the assignment $\alpha(\mathrm{cls} f)=f_*$ determines the natural homomorphism α for the theorem. The definition of the homomorphism β is more subtle and will be given below. We first treat a special case of the theorem.

Lemma 4.4. *If the boundary in K is identically zero, then $\alpha=\alpha_0$ is an isomorphism*

$$\alpha_0: H_0(\mathrm{Hom}(K,L))\cong \prod_{p=-\infty}^{\infty} \mathrm{Hom}(K_p, H_p(L)).$$

Proof. Since $\partial_K=0$, $H_p(K)=K_p$. Let $C_p(L)$ denote the group of p-cycles of L, while $B_p(L)$ is that of p-boundaries. Any $g=\{g_p\}\in \prod \mathrm{Hom}(K_p, H_p(L))$ consists of homomorphisms $g_p: K_p\to H_p(L)$; since K is free and $C_p(L)\twoheadrightarrow H_p(L)$ is an epimorphism, each g_p can be lifted to $g'_p: K_p\to C_p(L)$. These g'_p with range extended to $L_p\supset C_p(L)$ constitute a chain transformation $f: K\to L$ with $\alpha_0(\mathrm{cls} f)=g$. Thus α_0 is an epimorphism.

To show α_0 a monomorphism, suppose $\alpha_0(\mathrm{cls} f)=0$ for some f. For each p this means that $f_p(K_p)\subset B_p(L)$. Since $\partial: L_{p+1}\twoheadrightarrow B_p(L)$ and K_p is free, the map f_p can be lifted to $s_p: K_p\to L_{p+1}$ with $\partial s_p=f_p$. Since $s_{p-1}\partial= s_{p-1}\partial_K=0$, this equation may be written $f_p=\partial s_p+s_{p-1}\partial$. This states that f is chain homotopic to zero, hence $\mathrm{cls} f=0$ in $H_0(\mathrm{Hom}(K,L))$. Thus $\mathrm{Ker}\,\alpha_0=0$, and the lemma is proved.

Now consider the general case of Thm. 4.3, using the notation (4.2) and (4.3) in K. The family of groups $C_n\subset K_n$ can be regarded as a complex with boundary zero. A similar convention for D gives an exact sequence

$$0\to C\xrightarrow{i} K\xrightarrow{j} D\to 0 \qquad (4.6)$$

of complexes. Apply the functor $\mathrm{Hom}(-,L)$ to get another exact sequences of complexes

$$E:\ 0\to \mathrm{Hom}(D,L)\xrightarrow{j^*} \mathrm{Hom}(K,L)\xrightarrow{i^*} \mathrm{Hom}(C,L)\to 0,$$

where the zero on the right stands for $\mathrm{Ext}(D,L)$, which vanishes because $D_n\cong B_{n-1}\subset K_{n-1}$ is a subgroup of a free group, hence free. The exact homology sequence of E reads

$$\cdots \xrightarrow{\partial_E} H_0(\operatorname{Hom}(D,L)) \xrightarrow{j^*} H_0(\operatorname{Hom}(K,L)) \xrightarrow{i^*} H_0(\operatorname{Hom}(C,L)) \xrightarrow{\partial_E} \cdots,$$

with the connecting homomorphisms, for $n=1$ and $n=0$,

$$\partial_{E,n}\colon\ H_n(\operatorname{Hom}(C,L)) \to H_{n-1}(\operatorname{Hom}(D,L)).$$

The middle portion of this sequence can be expressed in terms of ∂_E as a short exact sequence

$$0 \to \operatorname{Coker} \partial_{E,1} \to H_0(\operatorname{Hom}(K,L)) \to \operatorname{Ker} \partial_{E,0} \to 0. \tag{4.7}$$

This is a short exact sequence with middle term $H_0(\operatorname{Hom}(K,L))$, exactly as in our theorem; it remains only to identify the end terms by analysing ∂_E.

Now $\partial'\colon D \to C$ induces maps $\partial'^*\colon \operatorname{Hom}_n(C,L) \to \operatorname{Hom}_{n-1}(D,L)$ anticommuting with ∂_L and hence also induces maps on homology. These maps (up to sign) are the connecting homomorphisms ∂_E. Indeed, ∂_E was defined on cycles by the "switchback" $j^{*-1}\partial_H i^{*-1}$. A cycle g of $\operatorname{Hom}_n(C,L)$ is a family of maps $\{g_p\colon C_p \to L_{p+n}\}$ with $\partial_L g_p = 0$; since D_p is free, $K_p \cong C_p \oplus D_p$, so each g_p can be extended to $f_p\colon K_p \to L_{p+n}$ with $\partial_L f_p = 0$. Since $i^* f = f i = g$, take $i^{*-1} g$ to be f. Since $\partial_L f = 0$, the formula (4.4) for the boundary ∂_H in $\operatorname{Hom}(K,L)$ reduces to $\partial_H f = \pm \partial_K^* f$. Now $\partial_K = i\,\partial' j$ by (4.2), so $\partial_H f = \pm j^* \partial'^* i^* f$ and we may take $j^{*-1}\partial_H i^{*-1} g$ to be $\pm \partial'^* g$: Thus ∂_E is indeed induced by $\pm \partial'^*$. But the isomorphism α_0 of Lemma 4.4 is natural, so we have commutativity up to a sign in the diagrams

$$\begin{array}{ccc} H_n(\operatorname{Hom}(C,L)) & \xrightarrow{\partial_E = \pm\partial'^*} & H_{n-1}(\operatorname{Hom}(D,L)) \\ \downarrow{\scriptstyle \alpha_0} & & \downarrow{\scriptstyle \alpha_0} \\ \prod_p \operatorname{Hom}(C_p, H_{p+n}(L)) & \xrightarrow{\partial'^*} & \prod_p \operatorname{Hom}(D_{p+1}, H_{p+n}(L)). \end{array}$$

We may thus read off the kernel of ∂_E as isomorphic to that of ∂'^* (lower line).

Now apply $\operatorname{Hom}(-, H_{p+n}(L))$ to the exact sequence S_p of (4.3). According to the fundamental exact sequence (Thm. 3.2) for Hom and Ext, we get an exact sequence

$$\left.\begin{array}{l} 0 \to \operatorname{Hom}(H_p(K), H_{p+n}(L)) \to \operatorname{Hom}(C_p, H_{p+n}(L)) \\ \xrightarrow{\partial'^*} \operatorname{Hom}(D_{p+1}, H_{p+n}(L)) \xrightarrow{S^*} \operatorname{Ext}(H_p(K), H_{p+n}(L)) \to 0, \end{array}\right\} \tag{4.8}$$

where the last zero stands for $\operatorname{Ext}(C_p, H_{p+n}(L))$, which vanishes because $C_p \subset K_p$ is free. The direct product of these sequences over all p is still exact, and gives the kernels and cokernels of ∂'^* as

$$\operatorname{Ker} \partial_{E,0} \cong \operatorname{Ker} \partial'^* = \prod_p \operatorname{Hom}(H_p(K), H_p(L)),$$

$$\operatorname{Coker} \partial_{E,1} \cong \operatorname{Coker} \partial'^* = \prod_p \operatorname{Ext}(H_p(K), H_{p+1}(L)).$$

Substituting these values in (4.7) gives the desired exact sequence (4.5) of Thm. 4.3. The homomorphism α thereby is the composite

$$H_0(\operatorname{Hom}(K,L)) \xrightarrow{i^*} H_0(\operatorname{Hom}(C,L)) \xrightarrow{\alpha_0} \prod_p \operatorname{Hom}(C_p, H_p(L))$$
$$\to \prod_p \operatorname{Hom}(H_p(K), H_p(L));$$

here the last arrow stands for the additive relation which is the converse (or, the "inverse") of the first monomorphism of (4.8). This composite α assigns to each $f\colon K \to L$ the family of induced maps of homology classes, so is the map already described. The homomorphism β is the composite $H(j^*)\alpha_0^{-1}S^{*-1}$ of natural maps, hence is natural.

To split (4.5), first construct L' and a chain transformation $h\colon L' \to L$ so that each $H_p(h)$ is an isomorphism and each L'_n is free. By naturality, h maps the sequence (4.5) for L' to that for L, isomorphically by the short five lemma; hence it suffices to split (4.5) for L' not L. Now regard $H(L')$ as a complex with zero boundary; because L' is free, there is a chain transformation $g\colon L' \to H(L')$ which is the identity on homology. By naturality, it now suffices to split (4.5) with L' replaced by $H(L')$. Finally, use $f\colon K \to H(K)$ the identity on homology. The map α for $H(K)$ and $H(L')$ is the identity map, while by naturality f maps this to α for K, $H(L')$ and so is a right inverse, splitting that α.

Corollary 4.5. *If K and L are complexes of abelian groups with each K_n and each $H_n(K)$ free, then two chain transformations f, $f'\colon K \to L$ are chain homotopic if and only if $f_* = f'_*\colon H_n(K) \to H_n(L)$ for every dimension n.*

The proof depends on observing that $f \simeq f'$ means exactly that $\operatorname{cls} f = \operatorname{cls} f'$ in $H_0(\operatorname{Hom}(K, L))$. On the other hand, when some $\operatorname{Ext}(H_p(K), H_{p+1}(L)) \neq 0$, the condition $f_* = f'_*$ for all n is not sufficient to make f chain homotopic to f'.

Corollary 4.6. *If $f\colon K \to K'$ is a chain transformation between complexes K and K' of free abelian groups with $f_*\colon H_n(K) \cong H_n(K')$ for all n, then, for any coefficient group G, $f^*\colon H^n(K', G) \to H^n(K, G)$ is an isomorphism.*

Proof. Since the maps α and β are natural in K, the diagram

$$\begin{array}{ccccccccc}
0 \to & \operatorname{Ext}(H_{n-1}(K'), G) & \to & H^n(K', G) & \to & \operatorname{Hom}(H_n(K'), G) & \to 0 \\
& \downarrow f^* & & \downarrow f^* & & \downarrow f^* & \\
0 \to & \operatorname{Ext}(H_{n-1}(K), G) & \to & H^n(K, G) & \to & \operatorname{Hom}(H_n(K), G) & \to 0
\end{array}$$

is commutative. Since the maps $f_n\colon H_n(K) \to H_n(K')$ are isomorphisms, so are the outside vertical maps $\operatorname{Ext}(f_{n-1}, 1_G)$ and $\operatorname{Hom}(f_n, 1_G)$. By the short five lemma, the middle map is an isomorphism, q.e.d.

Exercises

1. Give a direct proof of Cor. 4.2.

2. Show that Thms. 4.1 and 4.3 hold for complexes of R-modules if the hypothesis that K_n is free is replaced by the assumption that $C_n(K)$ and $K_n/C_n(K)$ are projective modules for every n.

3. If K and L are complexes of abelian groups with each K_n free, then to any family $\gamma_n: H_n(K) \to H_n(L)$ of homomorphisms, one for each n, there is a chain transformation $f: K \to L$ with $\gamma_n = H_n(f)$.

5. Composition of Extensions

Return now to the study of the extensions of modules. Two short exact sequences

$$E: 0 \to A \to B_1 \xrightarrow{\sigma} K \to 0, \qquad E': 0 \to K \xrightarrow{\lambda} B_0 \to C \to 0,$$

the first ending at the module K where the second starts, may be *spliced* together by the composite map $B_1 \to K \to B_0$ to give a longer exact sequence

$$E \circ E': 0 \to A \to B_1 \xrightarrow{\lambda\sigma} B_0 \to C \to 0 \tag{5.1}$$

called the Yoneda *composite* of E and E'. Conversely, any exact sequence $A \rightarrowtail B_1 \to B_0 \twoheadrightarrow C$ has such a factorization, with $K = \operatorname{Ker}(B_0 \to C) = \operatorname{Im}(B_1 \to B_0)$.

Longer exact sequences work similarly. Consider

$$S: 0 \to A \to B_{n-1} \to B_{n-2} \to \cdots \to B_0 \to C \to 0$$

an n-fold exact sequence *starting* at A and *ending* at C. If T is any m-fold exact sequence, starting at the module C where S ends, a splice at C gives the Yoneda composite $S \circ T$, which is an $(n+m)$-fold exact sequence starting where S starts and ending where T ends. This composition of sequences is clearly associative, but it need not be associative under the composition with homomorphisms. For example, for E and E' as in (5.1), let M be any module and $\pi: K \oplus M \to K$ the projection of the direct sum. The commutative diagrams

$$\begin{array}{ccccc} E_1: A & \rightarrowtail & B_1 \oplus M & \twoheadrightarrow & K \oplus M \\ \| & & \downarrow & & \downarrow \pi \\ E: A & \rightarrowtail & B_1 & \twoheadrightarrow & K, \end{array} \qquad \begin{array}{ccccc} E_1': K \oplus M & \rightarrowtail & B_0 \oplus M & \twoheadrightarrow & C \\ \downarrow \pi & & \downarrow & & \| \\ E': K & \rightarrowtail & B_0 & \twoheadrightarrow & C \end{array}$$

and the definitions of composites show that $E_1 \equiv E\pi$ and $E' \equiv \pi E_1'$; in the top row, the composite

$$E_1 \circ E_1': 0 \to A \to B_1 \oplus M \xrightarrow{\lambda\sigma \oplus 1} B_0 \oplus M \to C \to 0 \tag{5.1'}$$

is not the same as (5.1); in other words $(E\pi) \circ E_1' \neq E \circ (\pi E_1')$, and the associative law fails.

For short exact sequences we have already defined congruence as isomorphism with end maps the identity. For long sequences we need a wider congruence relation "$\equiv$" with the property that

$$(E''\beta)\circ E' \equiv E''\circ(\beta E') \tag{5.2}$$

whenever the composites involved are all defined; that is, for E'' ending at some module K, for $\beta\colon L\to K$ for some L, and E' starting at L. Let us then define congruence as the weakest reflexive, symmetric, and transitive relation including (5.2) and the previous congruence for short exact sequences. This definition can be restated as follows. Write any n-fold exact sequence S as the composite of n exact sequences E_i in the form

$$S = E_n\circ E_{n-1}\circ\cdots\circ E_1; \tag{5.3}$$

the E_i are unique up to isomorphism. A second n-fold sequence S' with the same start and end as S is *congruent* to S if S' can be obtained from S by a finite sequence of replacements of the following three types

(i) Replace any one factor E_i by a congruent short exact sequence;

(ii) If two successive factors have the form $E''\beta\circ E'$ for some E'', β, and E', as in (5.2), replace them by $E''\circ\beta E'$;

(iii) If two successive factors have the form $E''\circ\beta E'$, replace by $E''\beta\circ E'$.

For example, the 2-fold sequences (5.1) and (5.1′) are congruent.

We also define the composite of a long exact sequence or its congruence class with a "matching" homomorphism. Specifically, if S is an n-fold exact sequence starting at A and ending at C, then we define αS whenever α is a homomorphism with domain A and $S\gamma$ whenever γ has range C by the formulas (for S as in (5.3)):

$$\alpha(E_n\circ\cdots\circ E_1) = (\alpha E_n)\circ E_{n-1}\circ\cdots\circ E_1,$$
$$(E_n\circ\cdots\circ E_2\circ E_1)\,\gamma = E_n\circ\cdots\circ E_2\circ(E_1\gamma).$$

If S and S' are n-fold exact sequences, a *morphism* $\Gamma\colon S\to S'$ is a family of homomorphisms $(\alpha,\ldots,\gamma)$ forming a commutative diagram

$$\begin{array}{ccccccccccccc} S\colon & 0 & \to & A & \to & B_{n-1} & \to & \cdots & \to & B_0 & \to & C & \to 0 \\ \downarrow{\scriptstyle\Gamma} & & & \downarrow{\scriptstyle\alpha} & & \downarrow & & & & \downarrow & & \downarrow{\scriptstyle\gamma} & \\ S'\colon & 0 & \to & A' & \to & B'_{n-1} & \to & \cdots & \to & B'_0 & \to & C' & \to 0\,. \end{array}$$

We say that Γ *starts* with the homomorphism α and *ends* with γ. Now αE was defined by just such a diagram $E\to\alpha E$, so our definition of αS above yields a morphism $S\to\alpha S$ starting with α and ending with 1, as well as $S\gamma\to S$ starting with 1 and ending with γ. More generally we have, as in Prop. 1.8,

Proposition 5.1. *Each morphism $\Gamma: S \to S'$ of n-fold exact sequences S and S′ starting with α and ending with γ yields a congruence* $\alpha S \equiv S'\gamma$.

Proof. For notational symmetry, set $B_n = A$ and $B_{-1} = C$. Write $K_i = \operatorname{Im}(B_i \to B_{i-1}) = \operatorname{Ker}(B_{i-1} \to B_{i-2})$ for $i = n-1, \ldots, 1$; thus S factors as $E_n \circ \cdots \circ E_1$, where $E_i: K_i \rightarrowtail B_{i-1} \twoheadrightarrow K_{i-1}$ and $K_n = A$, $K_0 = C$. Factor S' similarly. The given morphism $\Gamma: S \to S'$ induces homomorphisms $\beta_i: K_i \to K_i'$ which form a commutative diagram

$$\begin{array}{rl} E_i\colon & 0 \to K_i \to B_{i-1} \to K_{i-1} \to 0 \\ & \quad\;\; \downarrow \beta_i \quad \downarrow \quad\;\; \downarrow \beta_{i-1} \\ E_i'\colon & 0 \to K_i' \to B_{i-1}' \to K_{i-1}' \to 0 . \end{array}$$

By Prop. 1.8 this diagram implies that $\beta_i E_i \equiv E_i' \beta_{i-1}$; at the ends, $\beta_n = \alpha$ and $\beta_0 = \gamma$. Hence, by our definition of congruence,

$$\begin{aligned} \alpha S = (\alpha E_n) \circ E_{n-1} \cdots &\equiv (E_n' \beta_{n-1}) \circ E_{n-1} \circ \cdots \equiv E_n' \circ (\beta_{n-1} E_{n-1}) \circ \cdots \\ &\equiv E_n' \circ (E_{n-1}' \beta_{n-2}) \circ \cdots \equiv \cdots \equiv S' \beta_0 = S' \gamma . \end{aligned}$$

This result also gives an alternative definition of congruence, as follows.

Proposition 5.2. *A congruence $S \equiv S'$ holds between two n-fold exact sequences starting at A and ending at C if and only if there is an integer k and 2k morphisms of n-fold exact sequences*

$$S = S_0 \to S_1 \leftarrow S_2 \to \cdots \leftarrow S_{2k-2} \to S_{2k-1} \leftarrow S_{2k} = S',$$

running alternately to the left and to the right, all starting with 1_A and ending with 1_C.

This proposition states that $S \equiv S'$ is the weakest reflexive, symmetric, and transitive relation such that $\Gamma: S \to S'$ with ends 1 implies $S \equiv S'$.

Proof. First suppose $S \equiv S'$. In the elementary congruence (5.2), the definition of $E''\beta$ yields a morphism $E''\beta \to E''$, while the definition of $\beta E'$ yields a morphism $E' \to \beta E'$ of exact sequences. Placing these morphisms side by side yields a diagram

$$\begin{array}{ll} E''\beta\colon A \rightarrowtail B_1 \twoheadrightarrow L & L \rightarrowtail B_0' \twoheadrightarrow C\colon E' \\ \quad\;\;\; \| \qquad \downarrow \quad\; \downarrow \beta & \downarrow \beta \quad\; \downarrow \qquad \| \\ E''\colon A \rightarrowtail B_1'' \twoheadrightarrow K & K \rightarrowtail B_0 \twoheadrightarrow C\colon \beta E' . \end{array}$$

Splicing these two diagrams together on the common map β yields a morphism $(E''\beta) \circ E' \to E'' \circ (\beta E')$. Hence a string of congruences (5.2) yields a string of morphisms, as displayed. The converse is immediate, by Prop. 5.1.

Let $\operatorname{Ext}_R^n(C, A)$, for fixed R-modules C and A, stand for the set of all congruence classes $\sigma = \operatorname{cls} S$ of n-fold exact sequences S starting at A and ending at C. Write $S \in \in \mathrm{Ext}^n(C,A)$ for $S \in \sigma \in \mathrm{Ext}^n(C,A)$. If $T \in \tau \in \mathrm{Ext}^m(D, C')$ the composite $S \circ T$ is defined when $C = C'$; the class of $S \circ T$ is determined by σ and τ, and is an element of $\mathrm{Ext}^{n+m}(D,A)$ which we denote as $\sigma\tau$ (without the circle notation for composition). The "matching" condition needed to define $\sigma\tau$ can be remembered if one regards $\sigma \in \mathrm{Ext}^n(C,A)$ as a "morphism" from the end module C to the starting module A; then $\sigma\tau$ is defined when the range of the morphism τ equals the domain of σ. This rule will include the matching conditions for the composition of a homomorphism $\alpha\colon A \to A'$ with an extension $\sigma \in \mathrm{Ext}^n(C,A)$. This rule will also include the composition of two ordinary homomorphisms if we interpret $\mathrm{Ext}^0(C,A)$ to be $\mathrm{Hom}(C,A)$. This we do.

Each $\mathrm{Ext}^n_R(C,A)$ is a bifunctor on R-modules to sets, contravariant in C and covariant in A. It is also an abelian group, under addition by a Baer sum. Indeed, two n-fold exact sequences $S \in \sigma \in \mathrm{Ext}^n(C,A)$ and $S' \in \sigma' \in \mathrm{Ext}^n(C',A')$ have a direct sum $S \oplus S' \in \in \mathrm{Ext}^n(C \oplus C', A \oplus A')$ found by taking direct sums of corresponding modules and maps in S and S'. The congruence class of $S \oplus S'$ depends only on the classes σ and σ', and hence may be denoted as $\sigma \oplus \sigma'$; to see this, note that the congruence $(E''\beta) \circ E' \equiv E'' \circ (\beta E')$ of (5.2) will carry over to a congruence on the direct sum as in

$$\begin{aligned}(E''\beta \oplus F'') \circ (E' \oplus F') &\equiv (E'' \oplus F'')(\beta \oplus 1) \circ (E' \oplus F')\\ &\equiv (E'' \oplus F'') \circ (\beta \oplus 1)(E' \oplus F')\\ &\equiv (E'' \oplus F'') \circ (\beta E' \oplus F').\end{aligned}$$

Finally, the Baer sum is defined for $\sigma_1, \sigma_2 \in \mathrm{Ext}^n(C,A)$, $i = 1, 2$, by the familiar formula

$$\sigma_1 + \sigma_2 = \nabla_A (\sigma_1 \oplus \sigma_2) \Delta_C. \tag{5.4}$$

Theorem 5.3. *Let* Ext_R *be the collection of all congruence classes* $\sigma, \tau, \ldots$ *of multiple exact sequences of* R*-modules. Each* σ *has a degree* n ($n = 0, 1, 2, \ldots$), *an* R*-module* C *as domain, and a module* A *as range; we then write* $\sigma \in \mathrm{Ext}^n(C,A)$, *and* $\mathrm{Ext}^0(C,A) = \mathrm{Hom}(C,A)$. *The composite* $\sigma\tau$ *is defined when* $\mathrm{range}\,\tau = \mathrm{domain}\,\sigma$, *and*

$$\mathrm{degree}(\sigma\tau) = \deg\sigma + \deg\tau, \quad \mathrm{range}\,\sigma\tau = \mathrm{range}\,\sigma, \quad \mathrm{domain}\,\sigma\tau = \mathrm{domain}\,\tau.$$

The sum $\sigma_1 + \sigma_2$ *is defined for* σ_1, σ_2 *in the same* $\mathrm{Ext}^n(C,A)$ *and makes* $\mathrm{Ext}^n(C,A)$ *an abelian group. The distributive laws*

$$(\sigma_1 + \sigma_2)\tau = \sigma_1\tau + \sigma_2\tau, \qquad \sigma(\tau_1 + \tau_2) = \sigma\tau_1 + \sigma\tau_2 \tag{5.5}$$

and the associative law $\varrho(\sigma\tau) = (\varrho\sigma)\tau$ *all hold when the addition and composition involved are defined.*

In brief, Ext_R is like a ring, except that the sum $\sigma+\tau$ and the product $\sigma\tau$ are not always defined.

This theorem clearly includes the previous Thm. 2.1 on $\operatorname{Ext}^1=\operatorname{Ext}$, and the proof is exactly like the "conceptual" proof of that theorem. That proof rested on certain rules for "direct sums". In the present case these rules (and their prior counterparts) are

$$(\sigma\oplus\sigma')(\tau\oplus\tau')=\sigma\tau\oplus\sigma'\tau', \qquad (2.8) \text{ and } (2.9), \qquad (5.6)$$

$$\sigma\nabla=\nabla(\sigma\oplus\sigma), \qquad (2.10), (2.11), \qquad (5.7)$$

$$\Delta\tau=(\tau\oplus\tau)\Delta, \qquad (2.10'), (2.11'), \qquad (5.8)$$

$$\omega(\sigma\oplus\sigma')=(\sigma'\oplus\sigma)\omega, \qquad (5.9)$$

where ω is the natural isomorphism $\omega\colon A\oplus A'\to A'\oplus A$. It remains to prove these rules.

First take (5.6). If σ and τ both have degree zero, they are ordinary homomorphisms, and (5.6) is the usual (functorial) rule for computing direct sums of homomorphisms. If σ and τ both have positive degree, (5.6) is an obvious rule about the composition of direct sums of exact sequences. If σ has degree zero and τ has positive degree, then σ and σ' actually operate just on the leftmost factor of τ and τ', hence (5.6) is reduced to the case where τ and τ' are short exact sequences; this case is (2.8). Similarly, when σ has positive degree and τ degree zero, (5.6) reduces to (2.9).

Next take (5.7). When σ has degree zero, (5.7) becomes (2.10); when σ has degree 1 and is a short exact sequence, it is the second of (2.11). When σ has degree 2, (2.11) gives the congruences

$$\begin{aligned}(E_2\circ E_1)\nabla &= E_2\circ(E_1\nabla)\equiv E_2\circ\nabla(E_1\oplus E_1)\\ &\equiv E_2\nabla\circ(E_1\oplus E_1)\equiv\nabla(E_2\oplus E_2)\circ(E_1\oplus E_1),\end{aligned}$$

which is (5.7). Longer cases are similar.

The proof of (5.8) is analogous, and (5.9) comes from the rule $\omega(E_1\oplus E_2)\cong(E_2\oplus E_1)\omega$, obtained by applying Prop. 1.8 to the morphism $(\omega,\omega,\omega)\colon E_1\oplus E_2\to E_2\oplus E_1$.

It remains only to exhibit the zero and the inverse for the abelian group $\operatorname{Ext}^n(C,A)$. The inverse of $\operatorname{cls} S$ will be $\operatorname{cls}((-1_A)S)$. The zero element of Ext^n is for $n=0$ the zero homomorphism, for $n=1$ the direct sum extension, and for $n>1$ the congruence class of the n-fold exact sequence

$$S_0\colon 0\to A\xrightarrow{1}A\to 0\to\cdots\to 0\to C\xrightarrow{1}C\to 0.$$

Indeed, for each $S\in\in\operatorname{Ext}^n(C,A)$ there is a morphism $(0,\ldots,1)\colon S\to S_0$, so, by Prop. 5.1, $S_0\equiv 0_A S$ and $\operatorname{cls} S+\operatorname{cls} S_0=\operatorname{cls} S$.

The rules (5.5) show that Ext^n is additive, so we obtain, as for Ext^1, the isomorphisms $\operatorname{Ext}^n(A\oplus B,G)\cong\operatorname{Ext}^n(A,G)\oplus\operatorname{Ext}^n(B,G)$, $\operatorname{Ext}^n(A,G\oplus H)\cong\operatorname{Ext}^n(A,G)\oplus\operatorname{Ext}^n(A,H)$. Furthermore, any short extension by a projective module splits, hence

$$\operatorname{Ext}^n(P,G)=0, \quad n>0, \quad P \text{ projective}. \tag{5.10}$$

Our construction of an element $\sigma\in\operatorname{Ext}^n(C,A)$ as a class of all (possible) n-fold sequences congruent to one given sequence S yields a "big" class, and the class $\operatorname{Ext}^n(C,A)$ of such classes is then not well defined in the usual axiomatics of set theory. This "wild" use of set theory can be repaired: It is intuitively clear that it suffices to limit the cardinal numbers of the sets used in constructing sequences S for given modules A and C.

We turn now to find means of computing the groups Ext^n.

6. Resolutions

Any module C is a quotient $C=F_0/R_0$ of some free module F_0. The submodule R_0 is again a quotient $R_0=F_1/R_1$ of a suitable free module F_1. Continuation of this process yields an exact sequence $\cdots\to F_1\to F_0\to C\to 0$ which will be called a "free resolution" of C. We aim to compare any two such.

In more detail, a *complex* (X,ε) *over* the R-module C is a sequence of R-modules X and homomorphisms

$$\to X_n\xrightarrow{\partial}X_{n-1}\xrightarrow{\partial}\cdots\to X_1\xrightarrow{\partial}X_0\xrightarrow{\varepsilon}C\to 0, \tag{6.1}$$

such that the composite of any two successive homomorphisms is zero. In other words, X is a positive complex of R-modules, C is a trivial chain complex ($C=C_0$, $\partial=0$), and $\varepsilon\colon X\to C$ is a chain transformation of the complex X to the complex C. A *resolution* of C is an *exact* sequence (6.1); that is, a complex (X,ε) over C with the homology $H_n(X)=0$, for $n>0$, and $\varepsilon\colon H_0(X)\cong C$. The complex X is *free* if each X_n is a free module and *projective* if each X_n is projective. We compare any projective complex with any resolution.

Theorem 6.1. *(Comparison Theorem.) If $\gamma\colon C\to C'$ is a homomorphism of modules, while $\varepsilon\colon X\to C$ is a projective complex over C and $\varepsilon'\colon X'\to C'$ is a resolution of C', then there is a chain transformation $f\colon X\to X'$ with $\varepsilon' f=\gamma\varepsilon$ and any two such chain transformations are chain homotopic.*

We say that such an f *lifts* γ.

The proof uses only categorical properties of projectives and of exactness. Since X_0 is projective and ε' an epimorphism, $\gamma\varepsilon\colon X_0\to C'$ can be lifted to $f_0\colon X_0\to X'_0$ with $\varepsilon' f_0=\gamma\varepsilon$. By induction it then suffices to construct f_n, given $f_{n-1}, \ldots, f_0$ such that the diagram

$$\begin{array}{ccccccccc} X_n & \xrightarrow{\partial_n} & X_{n-1} & \xrightarrow{\partial_{n-1}} & X_{n-2} \to \cdots \to & X_0 & \xrightarrow{\varepsilon} & C \\ \vdots f_n & & \downarrow f_{n-1} & & \downarrow f_{n-2} & \downarrow f_0 & & \downarrow \gamma \\ X'_n & \xrightarrow{\partial'_n} & X'_{n-1} & \xrightarrow{\partial'_{n-1}} & X'_{n-2} \to \cdots \to & X'_0 & \xrightarrow{\varepsilon'} & C' \end{array}$$

is commutative. By this commutativity, $\partial'_{n-1} f_{n-1} \partial_n = f_{n-2} \partial\partial = 0$. Hence $\operatorname{Im}(f_{n-1}\partial_n) \subset \operatorname{Ker} \partial'_{n-1}$. By exactness of the bottom row, this kernel is $\partial'_n X'_n$. Since X_n is projective, the map $f_{n-1}\partial_n$ can be lifted to an f_n with $\partial'_n f_n = f_{n-1}\partial_n$, q.e.d.

The construction of the homotopy is similar; it may be obtained directly or by applying the following lemma, noting that the difference of any two chain transformations $f\colon X \to X'$ lifting the same γ is a chain transformation lifting $0\colon C \to C'$.

Lemma 6.2. *Under the hypotheses of Thm. 6.1, let $f\colon X \to X'$ be a chain transformation lifting $\gamma\colon C \to C'$. Suppose that there is a $t\colon C \to X'_0$ such that $\varepsilon' t = \gamma$. Then there exist homomorphisms $s_n\colon X_n \to X'_{n+1}$ for $n = 0, 1, \ldots$ such that, for all n,*

$$\partial' s_0 + t\varepsilon = f_0, \qquad \partial' s_{n+1} + s_n \partial = f_{n+1}.$$

Proof. First, $\varepsilon'(f_0 - t\varepsilon)\colon X_0 \to C'$ is zero. Hence $f_0 - t\varepsilon$ maps the projective module X_0 into $\operatorname{Ker} \varepsilon' = \operatorname{Im}(X'_1 \to X'_0)$; it can therefore be lifted to a map $s_0\colon X_0 \to X'_1$ with $\partial' s_0 = f_0 - t\varepsilon$. Suppose by induction that we have $t = s_{-1}, s_0, \ldots, s_n$, as desired. We wish to find s_{n+1} with $\partial' s_{n+1} = f_{n+1} - s_n \partial$. Now $\partial'(f_{n+1} - s_n \partial) = f_n \partial - (f_n - s_{n-1}\partial)\partial = 0$ by the induction assumption, so $f_{n+1} - s_n\partial$ maps X_{n+1} into $\operatorname{Ker} \partial' = \partial' X'_{n+2}$; therefore it can be lifted to the desired $s_{n+1}\colon X_{n+1} \to X'_{n+2}$.

Let A be a fixed module; apply the functor $\operatorname{Hom}_R(-, A)$ to a resolution (6.1). Since the functor does not preserve exactness, the resulting complex $\operatorname{Hom}_R(X, A)$ may have non-trivial cohomology

$$H^n(X, A) = H^n(\operatorname{Hom}_R(X, A)).$$

Corollary 6.3. *If X and X' are two projective resolutions of C, while A is any module, then $H^n(X, A) \cong H^n(X', A)$ depends only on C and A.*

Proof. By the first part of Thm. 6.1, there are maps $f\colon X \to X'$ and $g\colon X' \to X$ lifting 1_C; by the second half of the theorem, gf is homotopic to $1\colon X \to X$. Hence $f^*\colon H^n(X', A) \to H^n(X, A)$ and g^* have $g^* f^* = 1 = f^* g^*$, so both are isomorphisms, q.e.d.

We now show that this function $H^n(X, A)$ of A and C is exactly $\operatorname{Ext}^n(C, A)$. For $n = 0$, $X_1 \to X_0 \to C \to 0$ is right exact, so

$$0 \to \operatorname{Hom}(C, A) \xrightarrow{\varepsilon^*} \operatorname{Hom}(X_0, A) \to \operatorname{Hom}(X_1, A)$$

is left exact. This states that $\varepsilon^*\colon \operatorname{Hom}(C, A) \cong H^0(X, A)$. For $n > 0$, each n-fold exact sequence S may be regarded as a resolution of C, zero beyond the term A of degree n, as in the diagram

$$\begin{array}{ccccccccccc} & X_{n+1} & \xrightarrow{\partial} & X_n & \to & X_{n-1} & \to \cdots \to & X_0 & \to & C & \\ & & & \downarrow g_n & & \downarrow & & \downarrow & & \| & \\ S\colon & 0 & \to & A & \to & B_{n-1} & \to \cdots \to & B_0 & \to & C & \to 0. \end{array} \tag{6.2}$$

Theorem 6.4. *If C and A are R-modules and $\varepsilon\colon X \to C$ a projective resolution of C, there is an isomorphism*

$$\zeta\colon \operatorname{Ext}^n(C,A) \cong H^n(X,A), \qquad n=0, 1, \ldots, \tag{6.3}$$

defined for $n>0$ as follows. Regard $S \in \in \operatorname{Ext}^n(C,A)$ as a resolution of C, and lift 1_C to $g\colon X \to S$. Then $g_n\colon X_n \to A$ is a cocycle of X. Define

$$\zeta(\operatorname{cls} S) = \operatorname{cls} g_n \in H^n(X,A). \tag{6.4}$$

This isomorphism ζ is natural in A. It is also natural in C in the following sense: If $\gamma\colon C' \to C$, $\varepsilon'\colon X' \to C'$ is a projective resolution of C', and $f\colon X' \to X$ lifts γ, then

$$\zeta' \gamma^* = f^* \zeta\colon \operatorname{Ext}^n(C,A) \to H^n(X',A). \tag{6.5}$$

Proof. First observe that ζ is well defined. Since $g_n \partial = 0$, g_n is an n-cocycle, as stated. Replace g by any other chain transformation g' lifting 1_C, as in (6.2). By Thm. 6.1, there is a chain homotopy s such that $g'_n - g_n = \partial s_n + s_{n-1}\partial$. But $s_n\colon X_n \to 0$, so $s_n = 0$, $g'_n - g_n = s_{n-1}\partial = (-1)^n \delta s_{n-1}$, this by the definition (II.3.1) of the coboundary in $\operatorname{Hom}(X,A)$. This states that the cocycles g'_n and g_n are cohomologous, so $\operatorname{cls} g'_n = \operatorname{cls} g_n$. Next replace S by any congruent exact sequence S'. According to the description of the congruence relation $S \equiv S'$ given in Prop. 5.2, it will suffice to consider the case when there is a morphism $\Gamma\colon S \to S'$ starting and ending with 1. In this case any $g\colon X \to S$ yields $\Gamma g\colon X \to S'$ with the same cocycle $g_n = (\Gamma g)_n$; hence $\operatorname{cls} g_n$ is well defined as a function of $\operatorname{cls} S$. Thus ζ is defined; its naturality properties as asserted follow at once, using suitable compositions of chain transformations.

Rather than proving directly that ζ is an isomorphism we construct its inverse. Given a resolution X, factor $\partial\colon X_n \to X_{n-1}$ as $X_n \xrightarrow{\partial'} \partial X_n \xrightarrow{\varkappa} X_{n-1}$, with $\varkappa$ the injection; this yields an n-fold exact sequence $S_n(C,X)$ as in

$$\begin{array}{lcccccccc} & & X_{n+1} \xrightarrow{\partial} X_n & & & & & & \\ & & \downarrow \partial' \quad \searrow \partial & & & & & & \\ S_n(C,X)\colon & 0 \to & \partial X_n & \xrightarrow{\varkappa} & X_{n-1} & \xrightarrow{\partial} \cdots \to & X_0 & \twoheadrightarrow & C \\ & & \downarrow h & & & & & & \| \\ h S_n\colon & 0 \to & A & & & & & & C. \end{array} \tag{6.6}$$

Any n-cocycle $X_n \to A$ vanishes on $\partial X_{n+1} = \operatorname{Ker} \partial'$, hence may be written uniquely in the form $h\partial'$ for some $h\colon \partial X_n \to A$. Construct the composite hS_n of h with the n-fold exact sequence S_n; this fills in the bottom row of this diagram. Define η: $H^n(X, A) \to \mathrm{Ext}^n(C, A)$ by setting

$$\eta \operatorname{cls}(h\partial') = \operatorname{cls}(hS_n(C, X)), \quad h\colon \partial X_n \to A. \tag{6.7}$$

By the distributive law for the composition in Ext, the right hand side is additive in h. Hence to show η well defined it suffices to show that $\eta(\operatorname{cls} h\partial')$ vanishes when $h\partial'$ is the coboundary of some cochain $k\colon X_{n-1} \to A$. But $h\partial' = \delta k = (-1)^n k\partial = (-1)^n k\varkappa\partial'$ means that $h = \pm k\varkappa$ and hence that $hS_n = \pm k\varkappa S_n$, where $\varkappa S_n \equiv 0$ by Prop. 1.7. Hence η is well defined and is a homomorphism. Comparison of the diagrams (6.2) and (6.6) now shows that $\eta = \zeta^{-1}$.

This theorem states that the groups $\mathrm{Ext}^n(C, A)$ may be computed from any projective resolution $\varepsilon\colon X \to C$; in particular, (6.5) shows how to compute induced homomorphisms $\gamma^*\colon \mathrm{Ext}^n(C, A) \to \mathrm{Ext}^n(C', A)$ from resolutions.

Alternatively, many authors *define* the functor Ext^n without using long exact sequences, setting $\mathrm{Ext}^n(C, A) = H^n(X, A) = H^n(\mathrm{Hom}(X, A))$. This gives a covariant functor of A, while for $\gamma\colon C' \to C$ the induced maps $\gamma^*\colon \mathrm{Ext}^n(C, A) \to \mathrm{Ext}^n(C', A)$ are defined by lifting γ to a comparison $X' \to X$.

Another consequence is a "canonical form" for sequences under congruence:

Corollary 6.5. *If $S \in \varepsilon \in \mathrm{Ext}^n(C, A)$ with $n > 1$, then there is a $T \equiv S$ of the form $T\colon 0 \to A \to B_{n-1} \to B_{n-2} \to \cdots \to B_0 \to C \to 0$ in which the modules $B_{n-2}, \ldots, B_0$ are free.*

Proof. Take $T = hS_n(C, X)$ for a suitable $h\colon \partial X_n \to A$, and X any free resolution of C.

Corollary 6.6. *For abelian groups A and C, $\mathrm{Ext}^n_Z(C, A) = 0$ if $n > 1$.*

Proof. Write $C = F/R$ for F free abelian. Since the subgroup R of the free abelian group F is free, $0 \to R \to F \to C \to 0$ is a free resolution which vanishes (with its cohomology) in dimensions above 1.

Consider now the effect of a ring homomorphism $\varrho\colon R' \to R$ (with $\varrho 1 = 1$). Any left R-module A becomes a left R'-module when the operators are defined by $r'a = (\varrho r')a$; we say that A has been *pulled back* along ϱ to become the R'-module ${}_\varrho A$. Any R-module homomorphism $\alpha\colon C \to A$ is also an R'-module homomorphism ${}_\varrho C \to {}_\varrho A$. By the same token, any long exact sequence S of R-modules pulls back to a long exact sequence ${}_\varrho S$ of R'-modules, and congruent sequences remain congruent. Hence $\varrho^\# \alpha = \alpha$, $\varrho^\#(\operatorname{cls} S) = \operatorname{cls} {}_\varrho S$ define homomorphisms

$$\varrho^\#\colon \mathrm{Ext}^n_R(C, A) \to \mathrm{Ext}^n_{R'}({}_\varrho C, {}_\varrho A), \quad n = 0, 1, \ldots \tag{6.8}$$

called *change of rings*. For ϱ fixed, they are natural in C and A.

These homomorphisms may be calculated from projective resolutions $\varepsilon: X \to C$ and $\varepsilon': X' \to {}_\varrho C$ by R and R'-modules, respectively. To exhibit the ring R, write $H^n(\mathrm{Hom}_R(X,A))$ for $H^n(X,A)$.

Theorem 6.7. *The change of rings $\varrho^\#$, via the isomorphism ζ of Thm. 6.4, is given by the composite map*

$$H^n(\mathrm{Hom}_R(X,A)) \xrightarrow{\varrho^*} H^n(\mathrm{Hom}_{R'}({}_\varrho X, {}_\varrho A)) \xrightarrow{f^*} H^n(\mathrm{Hom}_{R'}(X', {}_\varrho A))$$

where ϱ^* is the cohomology map induced by the chain transformation $\varrho^\#$: $\mathrm{Hom}_R \to \mathrm{Hom}_{R'}$ and $f: X' \to {}_\varrho X$ is a chain transformation lifting the identity of ${}_\varrho C$.

Proof. The case $n=0$ is left to the reader. For $n>0$ take any $S \in \varepsilon \mathrm{Ext}_R^n(C,A)$. As in (6.2), 1_C lifts to $g: X \to S$. Since ${}_\varrho X \to {}_\varrho C$ is a resolution of ${}_\varrho C$, the comparison theorem lifts the identity of ${}_\varrho C$ to a chain transformation $f: X' \to {}_\varrho X$. The diagram is

$$\begin{array}{ccccccccc}
\cdots \to & X'_n & \to & X'_{n-1} & \to \cdots \to & X'_0 & \to & {}_\varrho C \\
& \downarrow{\scriptstyle f_n} & & \downarrow & & \downarrow & & \| \\
\cdots \to & X_n & \to & X_{n-1} & \to \cdots \to & X_0 & \to & C \\
& \downarrow{\scriptstyle g_n} & & \downarrow & & \downarrow & & \| \\
S:\ 0 \to & A & \to & B_{n-1} & \to \cdots \to & B_0 & \to & C\,.
\end{array}$$

Now read off the maps: The isomorphism ζ carries $\mathrm{cls}\, S$ to $\mathrm{cls}\, g_n$, $\varrho^\#$ regards g_n as a R'-module homomorphism, f^* maps $\mathrm{cls}\, g_n$ to $\mathrm{cls}(g_n f_n)$, which is exactly $\zeta(\mathrm{cls}\,{}_\varrho S)$ because gf lifts 1. Hence the result, which will be of use in the treatment of products.

Exercises

1. If $\varepsilon: Y \to C$ is a projective complex over C and $\varepsilon': X \to C$ a resolution of C, construct natural homomorphisms

$$\zeta: \mathrm{Ext}_R^n(C,A) \to H^n(Y,A), \qquad \eta: H^n(X,A) \to \mathrm{Ext}_R^n(C,A).$$

2. (Calculation of Yoneda product by resolutions.) If $X \to C$ and $Y \to A$ are projective resolutions, $g \in \mathrm{Hom}^n(X,A)$ and $h \in \mathrm{Hom}^m(Y,D)$ are cocycles, write g as $g_0 \partial'$ for $g_0: \partial X_n \to A$, lift g_0 to f as in the diagram

$$\begin{array}{ccccccc}
X_{m+n} & \to \cdots \to & X_n & \to & \partial X_n & \to & 0 \\
\downarrow{\scriptstyle f} & & \downarrow & & \downarrow{\scriptstyle g_0} & & \\
Y_m & \to \cdots \to & Y_0 & \to & A & \to & 0\,,
\end{array}$$

show hf an $(m+n)$-cocycle of $\mathrm{Hom}^{m+n}(X,D)$, and prove that the Yoneda product $\eta(\mathrm{cls}\, h) \circ \eta(\mathrm{cls}\, g)$ is $\eta(\mathrm{cls}\, hf)$.

3. Given $E = (\varkappa, \sigma): A \rightarrowtail B \twoheadrightarrow C$ exact and maps $\alpha: A \to A'$, $\xi: B \to A'$, show by a diagram that $(\alpha + \xi\varkappa)E \equiv \alpha E$.

4. If $S = E_n \circ \cdots \circ E_1$, show that any morphism $\Gamma\colon S \to S'$ of n-fold exact sequences starting with a map $\alpha\colon A \to A'$ can be factored as

$$S \to (\alpha E_n) \circ E_{n-1} \circ \cdots \circ E_1 \to S'.$$

5. (Another formulation of the congruence relation on exact sequences.) If $S, S' \in\in \operatorname{Ext}^n(C,A)$, show that $S \equiv S'$ if and only if there is a $T \in\in \operatorname{Ext}^n(C,A)$ with morphisms $\Gamma\colon T \to S$, $\Gamma'\colon T \to S'$ starting and ending with 1's. (Use Exs. 3 and 4 and $T = h\,S_n(C,X)$.)

7. Injective Modules

The description of Ext^n by resolutions reads: Resolve the first argument by projective modules and calculate Ext^n by cohomology: $\operatorname{Ext}^n(C,A) \cong H^n(\operatorname{Hom}(X,A))$. We wish a dual statement, using a suitable resolution of the *second* argument A. For this we need the dual of a projective module; it is called an injective module.

A left R-module J is said to be *injective* if a homomorphism α with range J can always be extended; that is, if for each $\alpha\colon A \to J$ and $A \subset B$ there exists $\beta\colon B \to J$ extending α. Equivalently, J is injective if any diagram of the form

$$\begin{array}{ccccc} 0 & \to & A & \xrightarrow{\varkappa} & B \\ & & {\scriptstyle\alpha}\downarrow & \swarrow{\scriptstyle\beta} & \\ & & J & & \end{array} \qquad (7.1)$$

with horizontal row exact can be filled in (on the dotted arrow) so as to be commutative. The characterization of projective modules in Thm. I.6.3 and Thm. 3.5 dualizes at once to give

Theorem 7.1. *The following properties of a module J are equivalent:*

(i) *J is injective;*

(ii) *For each monomorphism $\varkappa\colon A \to B$, $\varkappa^*\colon \operatorname{Hom}(B,J) \to \operatorname{Hom}(A,J)$ is an epimorphism;*

(iii) *Every short exact sequence $J \rightarrowtail B \twoheadrightarrow C$ splits;*

(iv) *For every module C, $\operatorname{Ext}^1(C,J) = 0$.*

The latter characterization can be further specialized.

Proposition 7.2. *A left R-module J is injective if and only if $\operatorname{Ext}_R(R/L, J) = 0$ for every left ideal L in R.*

Proof. This condition is necessary. Conversely, suppose each $\operatorname{Ext}(R/L, J)$ zero. Given $A \subset B$ and $\alpha\colon A \to J$ we must, as in (7.1), construct an extension $\beta\colon B \to J$ of α. Consider all pairs (S, γ) consisting of a submodule S with $A \subset S \subset B$ and an extension $\gamma\colon S \to J$ of the given $\alpha\colon A \to J$. Partly order these pairs by the rule $(S,\gamma) \leqq (S',\gamma')$ when $S \subset S'$ and γ' is an extension of γ. To any linearly ordered collection

(S_i, γ_i) of these pairs there is an upper bound (T, τ) with $(S_i, \gamma_i) \leqq (T, \tau)$, for take T to be the union of the submodules S_i with τ defined for each t by $\tau t = \gamma_i t$ when $t \in S_i$. Hence, by Zorn's lemma, there is a maximal such pair $(S_\infty, \gamma_\infty)$. We need only prove $S_\infty = B$. If not, there is an element $b \in B$ not in S_∞; take the submodule U of B generated by b and S_∞. Then $r \rightarrow rb + S_\infty$ is an epimorphism $R \twoheadrightarrow U/S_\infty$, the kernel of the epimorphism is a left ideal L in R, and $R/L \cong U/S_\infty$. Since the sequence $S_\infty \rightarrowtail U \twoheadrightarrow U/S_\infty$ is exact, so is the sequence

$$\operatorname{Hom}(U, J) \rightarrow \operatorname{Hom}(S_\infty, J) \rightarrow \operatorname{Ext}(U/S_\infty, J). \tag{7.2}$$

But $\operatorname{Ext}(U/S_\infty, J) \cong \operatorname{Ext}(R/L, J) = 0$ by hypothesis, so $\operatorname{Hom}(U, J) \rightarrow \operatorname{Hom}(S_\infty, J)$ is an epimorphism. In other words, each homomorphism $S_\infty \rightarrow J$ can be extended to a homomorphism $U \rightarrow J$; in particular, $\gamma_\infty : S_\infty \rightarrow J$ can be so extended, a contradiction to the maximality of $(S_\infty, \gamma_\infty)$.

Consider now injective modules over special types of rings R. If R is a field, there are no proper left ideals $L \subset R$, while $\operatorname{Ext}_R(R, -)$ is always zero. Hence every module (= vector space) over a field is injective. Take $R = Z$, the ring of integers. Call a Z-module (= abelian group) D *divisible* if and only if there exists to each integer $m \neq 0$ and each $d \in D$ a solution of the equation $mx = d$.

Corollary 7.3. *An abelian group is injective (as a Z-module) if and only if it is divisible.*

Proof. The only ideals in Z are the principal ideals (m), and $Z/(m)$ is the cyclic group of order m. By Prop. 1.1, $\operatorname{Ext}(Z/(m), A) \cong A/mA$, while $A/mA = 0$ for all $m \neq 0$ precisely when A is divisible.

The construction of projective resolutions rested on the fact that any module is a quotient of a free module, hence certainly a quotient of a projective module. To get injective resolutions we need

Theorem 7.4. *Every R-module is a submodule of an injective R-module.*

Proof. Suppose first that $R = Z$. The additive group Z is embedded in the additive group Q of rational numbers, and Q is divisible. Any free abelian group F is a direct sum of copies of Z; it is embedded in the direct sum of corresponding copies of Q, and this direct sum D is divisible. Now represent the arbitrary abelian group as a quotient $A = F/S$ with F free, and embed F in some divisible group D as above; this embeds $A = F/S$ in D/S. An immediate argument shows that any quotient D/S of a divisible group D is divisible, hence injective. The abelian group A is thus embedded in an injective group D/S.

Return now to the case of an arbitrary ring R. For any abelian group G, the additive group $\operatorname{Hom}_Z(R, G)$ is a left R-module when the product sf, for $s \in R$, $f\colon R \to G$, is defined as the homomorphism $sf\colon R \to G$ with

$$(sf)(r) = f(rs), \quad r \in R. \tag{7.3}$$

If C is any left R-module we can define a homomorphism

$$j\colon C \to \operatorname{Hom}_Z(R, C) \tag{7.4}$$

by letting jc, for $c \in C$, be the homomorphism $jc\colon R \to C$ given as

$$(jc)(r) = rc, \quad r \in R. \tag{7.5}$$

To show this a homomorphism of R-modules, take s, $r \in R$, and compute

$$\begin{aligned}[j(sc)](r) = r(sc) = (rs)\,c = (jc)(rs) \qquad & \text{by (7.5)},\\ = [s(jc)]\,(r), \qquad & \text{by (7.3)}.\end{aligned}$$

This gives $j(sc) = s(jc)$. Since $1c = c$, j is a monomorphism.

Now embed the additive group of C in a divisible group D; this induces a monomorphism of R-modules

$$k\colon \operatorname{Hom}_Z(R, C) \to \operatorname{Hom}_Z(R, D).$$

The composite kj embeds C in $\operatorname{Hom}_Z(R, D)$. If we show that $\operatorname{Hom}_Z(R, D) = J$ is injective, we are done. By Thm. 7.1 (ii), it suffices to show that each monomorphism $\varkappa\colon A \to B$ of R-modules induces an epimorphism $\varkappa^* = \operatorname{Hom}_R(\varkappa, 1_J)$. Here $\varkappa^*$ is the top row of the diagram

$$\begin{array}{llcl}\operatorname{Hom}_R(\varkappa, 1_J)\colon & \operatorname{Hom}_R(B, \operatorname{Hom}_Z(R, D)) & \to & \operatorname{Hom}_R(A, \operatorname{Hom}_Z(R, D))\\ & \quad\downarrow \eta_B & & \quad\downarrow \eta_A\\ \operatorname{Hom}_Z(\varkappa, 1_D)\colon & \operatorname{Hom}_Z(B, D) & \to & \operatorname{Hom}_Z(A, D)\end{array}$$

where the vertical maps are isomorphisms, to be established in a lemma below. These isomorphisms are natural, so the diagram commutes. The bottom row refers not to R, but only to Z; since D is a divisible group, this bottom map $\operatorname{Hom}_Z(\varkappa, 1_D)$ is an epimorphism. Since η_B and η_A are isomorphisms, the top map $\operatorname{Hom}_R(\varkappa, 1_J)$ is also an epimorphism.

Lemma 7.5. *If G is an abelian group and A an R-module, there is a natural isomorphism $\eta_A\colon \operatorname{Hom}_R(A, \operatorname{Hom}_Z(R, G)) \cong \operatorname{Hom}_Z(A, G)$.*

Proof. Take an $f \in \operatorname{Hom}_R(A, \operatorname{Hom}_Z(R, G))$. For $a \in A$, $fa\colon R \to G$; that is, $(fa)(r) \in G$. Now regard f as a function of two variables $f(a, r) \in G$. The fact that fa is a Z-homomorphism means that $f(a, r)$ is additive in the argument r. The fact that $f\colon A \to \operatorname{Hom}_Z(R, G)$ is an R-homomorphism means that $f(a, r)$ is additive in a and that $s(fa) = f(sa)$ for each $s \in R$. By the definition (7.3) of the multiplication by s, this means that always

$$[s(fa)](r) = (fa)(rs) = [f(sa)](r);$$

in other words, that $f(a, rs) = f(sa, r)$ always. In particular, $f(a, s) = f(sa, 1)$ so the function f is determined by $g(a) = f(a, 1)$. Clearly $g: A \to G$. Now η_A and its inverse are defined by

$$(\eta_A f)(a) = f(a, 1); \qquad (\eta_A^{-1} g)(a, r) = g(ra).$$

The maps η_A and η_A^{-1} are clearly homomorphisms and natural (in A and in G).

This idea of regarding a function $f(a, r)$ of two variables as a function of a whose values are functions of r will reappear more formally later (V.3), and this lemma will turn out to be a special case of a more general natural isomorphism, called "adjoint associativity". Injective modules will be studied further in §11.

Exercises

1. If R is an integral domain, show that the field of quotients of R is a divisible R-module. If, in addition, R is a principal ideal domain, show that this field is injective as an R-module.

2. If A is a left R-module and L a left ideal in R, each $a \in A$ defines an R-module homomorphism $f_a: L \to A$ by $f_a(l) = la$. Prove that A is injective if and only if, for all L every $f: L \to A$ is f_a for some a.

3. If K is a complex of R-modules and J an injective R-module, show that α of (4.1) yields an isomorphism

$$H^n(\operatorname{Hom}_R(K, J)) \cong \operatorname{Hom}_R(H_n(K), J).$$

8. Injective Resolutions

A complex $\varepsilon: A \to Y$ *under* the module A is a sequence

$$0 \to A \xrightarrow{\varepsilon} Y^0 \xrightarrow{\delta} Y^1 \xrightarrow{\delta} \cdots \to Y^n \xrightarrow{\delta} Y^{n+1} \to \cdots \tag{8.1}$$

such that the composite of any two successive homomorphisms is zero. In other words, Y is a negative complex, positive in upper indices, and $\varepsilon: A \to Y$ a chain transformation. If this sequence is exact, $\varepsilon: A \to Y$ is called a *coresolution* of A; if each Y_n is injective, $\varepsilon: A \to Y$ is an *injective complex* under A. The results of the previous section show that every module A has an injective (co)resolution — by a customary abuse of language, an "injective resolution".

Theorem 8.1. *(Comparison theorem.) If $\alpha: A \to A'$ is a module homomorphism, $\varepsilon: A \to Y$ a coresolution, and $\varepsilon': A' \to Y'$ an injective complex under A', then there is a chain transformation $f: Y \to Y'$ with $\varepsilon'\alpha = f\varepsilon$ and any two such chain transformations are homotopic.*

The proof is exactly dual to that of Thm. 6.1, which used only the categorical properties of projective modules and exact sequences. Again the map f will be said to *lift* α.

For each module C the negative complex Y determines, as in (4.4), a negative complex

$$\operatorname{Hom}(C,Y):\ \operatorname{Hom}(C,Y^0)\to\operatorname{Hom}(C,Y^1)\to\cdots\to\operatorname{Hom}(C,Y^n)\to\cdots \tag{8.2}$$

Its homology gives Ext, as follows

Theorem 8.2. *For each module C and each injective coresolution $\varepsilon: A\to Y$ there is an isomorphism*

$$\check{\zeta}:\ \operatorname{Ext}^n(C,A)\cong H^n(\operatorname{Hom}(C,Y)),\quad n=0,1,\ldots, \tag{8.3}$$

which is natural in C and natural in A, in the sense that if $\alpha: A\to A'$, $\varepsilon': A'\to Y'$ is an injective coresolution, and $f: Y\to Y'$ is any chain transformation lifting α, then $\check{\zeta}\alpha_*=f_*\check{\zeta}$. Here f_* is the induced homomorphism $f_*: H^n(\operatorname{Hom}(C,Y))\to H^n(\operatorname{Hom}(C,Y'))$.

The homomorphism $\check{\zeta}$ is defined as follows. Regard any $S\in\in\operatorname{Ext}^n(C,A)$ as a coresolution of A, zero beyond the term C of (upper) degree n; by Thm. 8.1 construct a cochain transformation as in

$$\begin{array}{cccccccccccccc}
Y: & 0\to & A & \to & Y^0 & \to & Y^1 & \to\cdots\to & Y^{n-1} & \to & Y^n & \to & Y^{n+1} \\
\uparrow g & & \| & & \uparrow & & \uparrow & & \uparrow & & \uparrow g^n & & \\
S: & 0\to & A & \to & B_{n-1} & \to & B_{n-2} & \to\cdots\to & B_0 & \to & C & \to & 0.
\end{array} \tag{8.4}$$

Then $g^n: C\to Y^n$ is a cycle of $\operatorname{Hom}(C,Y)$. Define

$$\check{\zeta}(\operatorname{cls} S)=(\operatorname{cls} g^n)\in H^n(\operatorname{Hom}(C,Y)). \tag{8.5}$$

The rest of the proof, like the definition, is dual to the proof of Thm. 6.4.

We can summarize the theorems of §6 and §8 in the scheme

$$H^n(\operatorname{Hom}(\operatorname{Res}_P C,A))\cong\operatorname{Ext}^n(C,A)\cong H^n(\operatorname{Hom}(C,\operatorname{Res}_J A)),$$

where $\operatorname{Res}_P C$ denotes an arbitrary projective resolution of C, $\operatorname{Res}_J A$ an arbitrary injective coresolution of A. A symmetric formula $\operatorname{Ext}^n(C,A)\cong H^n(\operatorname{Hom}(\operatorname{Res}_P C,\operatorname{Res}_J A))$ can be established (Ex. V.9.3).

Exercises

1. Carry out the construction of g in (8.4) and of the inverse of ζ.
2. State and prove the dual of Lemma 6.2.
3. For direct sums and products establish the isomorphisms

$$\operatorname{Ext}^n(\textstyle\sum C_t,A)\cong\prod\operatorname{Ext}^n(C_t,A),\qquad \operatorname{Ext}^n(C,\prod A_t)\cong\prod\operatorname{Ext}^n(C,A_t).$$

9. Two Exact Sequences for Ext^n

Composition of long exact sequences with a short exact sequence E from A to C yields *connecting homomorphisms*

$$E^*:\ \operatorname{Ext}^k(A,G)\to\operatorname{Ext}^{k+1}(C,G),\qquad E_*:\ \operatorname{Ext}^k(G,C)\to\operatorname{Ext}^{k+1}(G,A).$$

Since E determines A and C, both $\operatorname{Ext}^k(A,G)$ and $\operatorname{Ext}^{k+1}(C,G)$ may be regarded as contravariant functors of the short exact sequence E. Moreover, each morphism $\Gamma=(\alpha,\beta,\gamma): E\to E'$ of short exact sequences gives $\alpha E\equiv E'\gamma$ and hence

$$E^*\alpha^*=\gamma^*E'^*: \operatorname{Ext}^k(A',G)\to\operatorname{Ext}^{k+1}(C,G).$$

This states that E^* is a natural transformation between functors of E, as is E_*. With these connecting homomorphisms, the exact sequences for Hom and $\operatorname{Ext}=\operatorname{Ext}^1$ already found in (3.1) and (3.2) will now be continued to higher dimensions. Observe similarly that an n-fold exact sequence S starting at A and ending at C is a composite of n short exact sequences; hence composition with S yields *iterated connecting homomorphisms*

$$S^*: \operatorname{Ext}^k(A,G)\to\operatorname{Ext}^{k+n}(C,G),\qquad S_*: \operatorname{Ext}^k(G,C)\to\operatorname{Ext}^{k+n}(G,A),$$

which depend only on the congruence class of S.

Theorem 9.1. *If $E=(\varkappa,\sigma): A\rightarrowtail B\twoheadrightarrow C$ is a short exact sequence of modules and G is another module, then the sequences*

$$\cdots\to\operatorname{Ext}^n(C,G)\xrightarrow{\sigma^*}\operatorname{Ext}^n(B,G)\xrightarrow{\varkappa^*}\operatorname{Ext}^n(A,G)\xrightarrow{E^*}\operatorname{Ext}^{n+1}(C,G)\to\cdots \tag{9.1}$$

$$\cdots\to\operatorname{Ext}^n(G,A)\xrightarrow{\varkappa_*}\operatorname{Ext}^n(G,B)\xrightarrow{\sigma_*}\operatorname{Ext}^n(G,C)\xrightarrow{E_*}\operatorname{Ext}^{n+1}(G,A)\to\cdots \tag{9.2}$$

are exact. These sequences start at the left with $0\to\operatorname{Hom}(C,G)=\operatorname{Ext}^0(C,G)$ and with $0\to\operatorname{Hom}(G,A)$, respectively, and continue to the right for all $n=0,1,2,\ldots$. The maps in these sequences are defined for arguments $\varrho\in\operatorname{Ext}^n(C,G)$, $\omega\in\operatorname{Ext}^n(B,G)$, $\tau\in\operatorname{Ext}^n(A,G)$, ... by composition with $\varkappa$, σ, E as follows:

$$\sigma^*\varrho=\varrho\sigma,\qquad \varkappa^*\omega=\omega\varkappa,\qquad E^*\tau=(-1)^n\tau E, \tag{9.3}$$

$$\varkappa_*\varrho'=\varkappa\varrho',\qquad \sigma_*\omega'=\sigma\omega',\qquad E_*\tau'=E\tau'. \tag{9.4}$$

The sign in the last part of (9.3) occurs because $E^*\tau=\tau E$ involves an interchange of an element E of degree 1 with an element τ of degree n.

Proof. First consider (9.2). Take any free resolution X of G and apply the exact cohomology sequence (Thm. II.4.5) for the sequence E of coefficients. Since the cohomology groups $H^n(X,A)$ are $\operatorname{Ext}^n(G,A)$, and so on, this yields an exact sequence with the same terms as (9.2). To show that the maps in this sequence are obtained by composition, as stated in (9.4), we must prove commutativity in the diagram

$$\begin{array}{ccccccc}
\operatorname{Ext}^n(G,A) & \xrightarrow{\varkappa_*} & \operatorname{Ext}^n(G,B) & \xrightarrow{\sigma_*} & \operatorname{Ext}^n(G,C) & \xrightarrow{E_*} & \operatorname{Ext}^{n+1}(G,A)\\
\downarrow\zeta & & \downarrow\zeta & & \downarrow\zeta & & \downarrow\zeta\\
H^n(X,A) & \xrightarrow{\varkappa_*} & H^n(X,B) & \xrightarrow{\sigma_*} & H^n(X,C) & \xrightarrow{(-1)^{n+1}\delta_E} & H^{n+1}(X,A),
\end{array} \tag{9.5}$$

where each ζ is the isomorphism provided by Thm. 6.4, while δ_E is the connecting homomorphism provided in Thm. II.4.5. Since ζ is natural for coefficient homomorphisms $\varkappa$ and σ, the first two squares are commutative. The commutativity of the right-hand square requires a systematic use of the definitions of various maps involved, as follows. For $n>0$ and $S \in \in \operatorname{Ext}^n(G,C)$, regard $E \circ S$ as a resolution of G and construct the commutative diagram

$$\begin{array}{lcccccccccc}
X\colon & X_{n+1} & \xrightarrow{\partial} & X_n & \to & X_{n-1} & \to \cdots \to & X_0 & \to & G \\
\downarrow f & \downarrow f_{n+1} & & \downarrow f_n & & \downarrow f_{n-1} & & \downarrow & & \| \\
E\circ S\colon & A & \xrightarrow{\varkappa} & B & \to & B_{n-1} & \to \cdots \to & B_0 & \to & G \\
 & \downarrow \sigma & & \downarrow \sigma & & \| & & \| & & \| \\
S\colon & 0 & \to & C & \to & B_{n-1} & \to \cdots \to & B_0 & \to & G
\end{array}$$

where f lifts 1_G. By the definition of ζ,

$$\zeta E_* (\operatorname{cls} S) = (\operatorname{cls} f_{n+1}) \in H^{n+1}(X, A).$$

On the other hand, σf is a chain transformation lifting 1_G, so $\zeta(\operatorname{cls} S) = \operatorname{cls}(\sigma f_n)$. Now δ_E is defined by the switchback $\delta_E = \operatorname{cls} \varkappa^{-1} \delta\, \sigma^{-1} \operatorname{cls}^{-1}$ of (II. 4.12) and $\varkappa^{-1} \delta\, \sigma^{-1}(\sigma f_n) = \varkappa^{-1} \delta f_n = (-1)^{n+1} \varkappa^{-1}(f_n \partial) = (-1)^{n+1} \varkappa^{-1}(\varkappa f_{n+1}) = (-1)^{n+1} f_{n+1}$, so that $\delta_E \operatorname{cls}(\sigma f_n) = (-1)^{n+1} \operatorname{cls} f_{n+1} = (-1)^{n+1} \zeta E_* \operatorname{cls} S$. This shows (9.5) commutative.

For $n=0$ the definition of ζ (and the commutativity proof) is correspondingly simpler.

The exactness of the sequence (9.1) of the theorem is proved similarly, using injective coresolutions. Specifically, let $\varepsilon\colon G \to Y$ be an injective coresolution of G. Then $\operatorname{Hom}(A,Y)$ is, as in §8, a negative complex; furthermore each Y^n is injective, so each sequence $\operatorname{Hom}(C,Y^n \rightarrowtail) \operatorname{Hom}(B,Y^n) \twoheadrightarrow \operatorname{Hom}(A,Y^n)$ is exact. Therefore

$$0 \to \operatorname{Hom}(C,Y) \xrightarrow{\sigma^*} \operatorname{Hom}(B,Y) \xrightarrow{\varkappa^*} \operatorname{Hom}(A,Y) \to 0$$

is an exact sequence of complexes. Hence Thm. II.4.1, in the version with upper indices, states that the first row of the following diagram is exact for each n:

$$\begin{array}{ccccccc}
H^n(\operatorname{Hom}(C,Y)) & \xrightarrow{\sigma^*} & H^n(\operatorname{Hom}(B,Y)) & \xrightarrow{\varkappa^*} & H^n(\operatorname{Hom}(A,Y)) & \xrightarrow{\delta_E} & H^{n+1}(\operatorname{Hom}(C,Y)) \\
\uparrow \tilde{\zeta} & & \uparrow & & \uparrow & & \uparrow \\
\operatorname{Ext}^n(C,G) & \xrightarrow{\sigma^*} & \operatorname{Ext}^n(B,G) & \xrightarrow{\varkappa^*} & \operatorname{Ext}^n(A,G) & \xrightarrow{E^*} & \operatorname{Ext}^{n+1}(C,G).
\end{array}$$

The desired proof that the bottom row is exact requires now only the commutativity of the diagram. Note that the connecting homomorphism δ_E is defined by switchback as $\delta_E = \operatorname{cls} \sigma^{*-1} \delta \varkappa^{*-1} \operatorname{cls}^{-1}$, and no trouble with signs occurs. Given this definition, the proof that commutativity holds is now like that given above for the dual case — though since the proof manipulates not only arrows, but also elements, we cannot say that the proof is exactly dual. Thm. 9.1, though formulated in the language of exact sequences, can also be regarded as a statement about annihilators in the "pseudo-ring" Ext_R of Thm. 5.3. Indeed, if $E=(\varkappa,\sigma)$ is a short exact sequence of R-modules, then

$$\varkappa E=0, \qquad \sigma\varkappa=0, \qquad E\sigma=0$$

and these equations indicate the whole left and right annihilator in Ext_R of each of $\varkappa$, E, and σ, as follows. The right annihilator of $\varkappa$ consists of multiples of E; whenever $\varrho\in\mathrm{Ext}_R$ is such that the composite $\varkappa\varrho$ is defined and is 0, then either $\varrho=0$ or $\varrho=E\tau$ for a suitable $\tau\in\mathrm{Ext}_R$. Similarly, $\varrho\varkappa=0$ implies $\varrho=\tau\sigma$ for some τ, etc. In other words, the left annihilator of $\varkappa$ is the principal left ideal $(\mathrm{Ext}_R)\sigma$.

Exercises

1. Given the usual short exact sequence E of modules and given projective resolutions $\varepsilon'\colon X\to A$ and $\varepsilon''\colon Z\to C$ of the end modules A and C, construct a projective resolution $\varepsilon\colon Y\to B$ of the middle module B and chain transformations $f\colon X\to Y$, $g\colon Y\to Z$ lifting $\varkappa$ and σ, respectively, such that $X\rightarrowtail Y\twoheadrightarrow Z$ is an exact sequence of complexes. (Hint: for each n, take $Y_n=X_n\oplus Z_n$ and define ε and ∂ so that (Y,ε) is a complex.)

2. Use the result of Ex. 1 to give a proof of the exactness of (9.1) by projective resolutions.

3. Deduce Prop. 3.7 from Thm. 9.1 and Cor. 6.6.

4. For A a finite abelian group, Q the additive group of rational numbers, prove $\mathrm{Ext}_Z(A,Z)\cong\mathrm{Hom}_Z(A,Q/Z)$.

10. Axiomatic Description of Ext

The properties already obtained for the sequence of functors Ext^n suffice to determine those functors up to a natural equivalence, in the following sense.

Theorem 10.1. *For each $n=0,1,\ldots$, let there be given a contravariant functor $\mathrm{Ex}^n(A)$ of the module A, taking abelian groups as values, and for each n and each short exact sequence $E\colon A\rightarrowtail B\twoheadrightarrow C$ let there be given a homomorphism $E^n\colon \mathrm{Ex}^n(A)\to\mathrm{Ex}^{n+1}(C)$ which is natural for morphisms $\Gamma\colon E\to E'$ of short exact sequences. Suppose that there is a fixed module G such that*

$$\mathrm{Ex}^0(A)=\mathrm{Hom}(A,G) \qquad \textit{for all } A, \tag{10.1}$$

$$\mathrm{Ex}^n(F)=0 \qquad \textit{for } n>0 \textit{ and all free } F, \tag{10.2}$$

and suppose that for each $E=(\varkappa,\sigma)$ the sequence

$$\cdots\to\mathrm{Ex}^n(C)\xrightarrow{\sigma^*}\mathrm{Ex}^n(B)\xrightarrow{\varkappa^*}\mathrm{Ex}^n(A)\xrightarrow{E^n}\mathrm{Ex}^{n+1}(C)\to\cdots \tag{10.3}$$

is exact. Then there is for each A and n an isomorphism φ_A^n: $\operatorname{Ex}^n(A) \cong \operatorname{Ext}^n(A,G)$, with $\varphi_A^0 = 1$, which is natural in A and such that the diagram

$$\begin{array}{ccc} \operatorname{Ex}^n(A) & \xrightarrow{E^n} & \operatorname{Ex}^{n+1}(C) \\ \downarrow{\scriptstyle\varphi^n} & & \downarrow{\scriptstyle\varphi^{n+1}} \\ \operatorname{Ext}^n(A,G) & \xrightarrow{E^*} & \operatorname{Ext}^{n+1}(C,G) \end{array} \tag{10.4}$$

is commutative for all n and all short exact E: $A \rightarrowtail B \twoheadrightarrow C$.

Property (10.4) reads "φ commutes with the connecting homomorphisms". With the naturality of φ, it states that the φ^n provide a morphism of the long sequence (10.3) into the corresponding sequence (9.1) for Ext.

The same theorems holds with "free" in (10.2) replaced by "projective". Since the functors Ext^n clearly satisfy the analogues of (10.1), (10.2), and (10.3), we may regard these three properties as axioms characterizing the sequence of functors Ext^n "connected" by the homomorphisms E^*.

The proof will construct φ^n by induction on n; the case $n=1$ presents the most interest. Represent each module C as a quotient F/K, with F free. This gives a short exact sequence E_C: $K \rightarrowtail F \twoheadrightarrow C$. By (10.2), $\operatorname{Ex}^1(F)=0$, so the sequence (10.3) becomes

$$\operatorname{Hom}(F,G) \xrightarrow{\varkappa^*} \operatorname{Hom}(K,G) \xrightarrow{E_0^1} \operatorname{Ex}^1(C) \longrightarrow 0.$$

Exactness states that $\operatorname{Ex}^1(C) \cong \operatorname{Hom}(K,G)/\varkappa^* \operatorname{Hom}(F,G)$. The sequence (9.1) for Ext^1 shows $\operatorname{Ext}^1(C,G)$ isomorphic to the same group. Combining these isomorphisms yields an isomorphism φ_C^1: $\operatorname{Ex}^1(C) \cong \operatorname{Ext}^1(C,G)$; by its construction, φ_C^1 is characterized by the equation

$$\varphi_C^1 E_C^1 = E_C^*\colon \operatorname{Hom}(K,G) \to \operatorname{Ext}^1(C,G),$$

which is a special case of (10.4). To show that φ_C^1 is natural for any γ: $C \to C'$, pick an exact $E_{C'}$: $K' \rightarrowtail F' \twoheadrightarrow C'$. The comparison theorem lifts γ to β: $F \to F'$, which induces a morphism $\Gamma = (\alpha, \beta, \gamma)$: $E_C \to E_{C'}$. Since both connecting homomorphisms E^1 and E^* are natural with respect to such morphisms Γ, it follows that $\gamma^* \varphi_{C'}^1 E_{C'}^1 = \gamma^* E_{C'}^* = E_C^* \alpha^* = \varphi_C^1 E_C^1 \alpha^* = \varphi_C^1 \gamma^* E_{C'}^1$. But $E_{C'}^1$ is an epimorphism, so

$$\gamma^* \varphi_{C'}^1 = \varphi_C^1 \gamma^*\colon \operatorname{Ex}^1(C') \to \operatorname{Ext}^1(C,G);$$

φ^1 is indeed natural for maps of C. In particular, if E_C and $E_{C'}$ are two free presentations of the same module C ($\gamma = 1_C$), this identity shows that the homomorphism φ_C^1 is independent of the choice of the particular free module F used in its construction. Finally, if E: $A \rightarrowtail B \twoheadrightarrow C$ is any short exact sequence, the comparison theorem (for F free) again lifts

1 to a morphism $(\alpha, \beta, 1): E_C \to E$ and yields

$$\begin{aligned} 1_C^* E^* &= E_C^* \alpha^* && \text{(because } E^* \text{ is natural)}, \\ &= \varphi_C^1 E_C^1 \alpha^* && \text{(definition of } \varphi), \\ &= \varphi_C^1 E^1 && \text{(because } E^1 \text{ is natural)}. \end{aligned}$$

This is the required property (10.4) for $n=1$.

For $n>1$ we proceed in similar vein, choosing again a short exact sequence E_C with middle term free. Then $\mathrm{Ex}^{n-1}(F)=0=\mathrm{Ex}^n(F)$, so the exact sequence (10.3) becomes

$$0 \to \mathrm{Ex}^{n-1}(K) \xrightarrow{E_0^{n-1}} \mathrm{Ex}^n(C) \to 0$$

and $\mathrm{Ex}^n(C) \cong \mathrm{Ex}^{n-1}(K)$. Using the similar sequence for Ext^n, we define φ^n by

$$\varphi_C^n = E_C^* \varphi_K^{n-1} (E_C^{n-1})^{-1} \colon \mathrm{Ex}^n(C) \cong \mathrm{Ext}^n(C,G)$$

and establish naturality, independence of the choice of F, and the commutativity (10.4) much as in the case $n=1$.

There is a dual characterization for $\mathrm{Ext}(C,A)$ as a functor of A using the second exact sequence (9.2).

Theorem 10.2. *For a fixed module G, the covariant functors $\mathrm{Ext}^n(G,A)$ of A, $n=0, 1, \ldots$ together with the natural homomorphisms $E_*\colon \mathrm{Ext}^n(G,C) \to \mathrm{Ext}^{n+1}(G,A)$ defined for short exact sequences E of modules, are characterized up to a natural isomorphism by these three properties:*

$$\mathrm{Ext}^0(G,A) = \mathrm{Hom}(G,A) \qquad \textit{for all } A, \tag{10.5}$$

$$\mathrm{Ext}^n(G,J) = 0 \qquad \textit{for } n>0 \textit{ and all injective } J, \tag{10.6}$$

The sequence (9.2) *is exact for all E.* (10.7)

Proof. Observe first that Ext^n does have the property (10.6), for an injective module J has the injective coresolution $0 \to J \to J \to 0$, which vanishes in all dimensions above 0. Conversely, the proof that these three properties characterize the $\mathrm{Ext}^n(G,A)$ as functors of A is dual to the proof we have just given.

Exercises

1. (S. Schanuel.) Given two short exact sequences $K \rightarrowtail P \twoheadrightarrow C$ and $K' \rightarrowtail P' \twoheadrightarrow C$ with P and P' projective, $K \subset P$, $K' \subset P'$, and the same end module C: Construct an isomorphism $P \oplus P' \cong P \oplus P'$ which maps $K \oplus P'$ isomorphically on $P \oplus K'$.

2. Call two modules C and C' *projectively equivalent* if there are projective modules Q and Q' and an isomorphism $C \oplus Q' \cong C' \oplus Q$. Let $S\colon K \rightarrowtail P_{n-1} \to \cdots \to P_0 \twoheadrightarrow C$ be an n-fold exact sequence with all P_i projective. Using Ex. 1, show that the projective equivalence class of K depends only on that of C and not on the choice of S.

3. For S as in Ex. 2, show that the iterated connecting homomorphism provides an isomorphism S^*: $\operatorname{Ext}^1(K,G) \cong \operatorname{Ext}^{n+1}(C,G)$ for any G.

11. The Injective Envelope

Every R-module A is a submodule of an injective one (Thm. 7.4). We now show that there is a unique "minimal" such injective module for each A.

An extension $A \subset B$ — or a monomorphism $\varkappa: A' \to B$ with image A — is called *essential* if $S \subset B$ and $S \cap A = 0$ always implies $S = 0$. This amounts to the requirement that to each $b \neq 0$ in B there is an $r \in R$ with $rb \neq 0$ and in A. For example, the additive group Q of rational numbers is an essential extension of the group Z of integers. If $A \subset B$ and $B \subset C$ are essential extensions, so is $A \subset C$.

Lemma 11.1. *If* $\varkappa$: $A' \to B$ *is an essential monomorphism, while* $\lambda: A' \to J$ *is a monomorphism with injective range* J*, there is a monomorphism* μ: $B \to J$ *with* $\mu\varkappa = \lambda$.

In other words, an essential extension of A' *can be embedded in any injective extension of* A'.

Proof. Because J is injective, λ: $A' \to J$ extends to a μ with $\mu\varkappa = \lambda$. Let K be the kernel of μ. Since λ is a monomorphism, $K \cap \varkappa A' = 0$; since $\varkappa$ is essential, $K = 0$. Hence μ is a monomorphism.

Proposition 11.2. *A module* J *is injective if and only if* J *has no proper essential extension.*

Proof. If $J \subset B$ with J injective, then J is a direct summand of B, so the extension $J \subset B$ is inessential unless $J = B$. Conversely, if J has no proper essential extensions, we wish to show that any extension $J \subset B$ splits. Consider the set $\mathcal{S}$ of all submodules $S \subset B$ with $S \cap J = 0$. If a subset $\{S_t\}$ of elements of $\mathcal{S}$ is linearly ordered by inclusion, the union $S = \bigcup S_t$ of the sets S_t is a submodule of B with $S \cap J = 0$, hence also in $\mathcal{S}$. Since any linearly ordered subset of $\mathcal{S}$ has an upper bound in $\mathcal{S}$, Zorn's lemma asserts that $\mathcal{S}$ has an element M maximal in the sense that it is properly contained in no S. Then $J \to B \to B/M$ is an essential monomorphism. But J is assumed to have no proper essential extension, so $J \to B/M$ is an isomorphism, $B = J \cup M$ and $J \cap M = 0$. Thus J is a direct summand of any containing B, so is injective.

This suggests that we might construct a minimal injective extension as a maximal essential extension.

Theorem 11.3. *For every module* A *there is an essential monomorphism* $\varkappa$: $A \to J$ *with* J *injective. If* $\varkappa'$: $A \to J'$ *is another such, there is an isomorphism* θ: $J \to J'$ *with* $\theta\varkappa = \varkappa'$.

Proof. By Thm. 7.4 there is an injective module J_0 with $A \subset J_0$. Let $\mathcal{T}$ be the set of all submodules S of J_0 with $A \subset S$ essential. If $\{S_t\}$ is a subset of $\mathcal{T}$ linearly ordered by inclusion, the union $\bigcup S_t$ is an essential extension of A, hence is in $\mathcal{T}$. By Zorn's lemma again, $\mathcal{T}$ has a maximal element, J, and $A \subset J$ is essential. Any proper essential extension of J could by Lemma 11.1 be embedded in J_0, counter to the maximality of J. Hence J is injective, by Prop. 11.2.

Let $\varkappa: A \to J$ be the injection. If $\varkappa': A \to J'$ is another essential monomorphism to an injective J', Lemma 11.1 gives a monomorphism $\mu: J' \to J$ with $\mu\varkappa' = \varkappa$. Since $\mu J'$ is injective, it is a direct summand of J. Since $A \to J$ is essential, $\mu J'$ must be all of J, so μ is an isomorphism, as asserted.

The essential monomorphism $\varkappa: A \to J$ with J injective, unique up to equivalence, is called the *injective envelope* of A. Its existence was established by BAER [1940]; our proof follows ECKMANN-SCHOPF [1953]. For some of its applications, see MATLIS [1958]. A dual construction — of a "least" projective P with an epimorphism $P \to A$ — is not in general possible (Why?).

Notes. The study of extensions developed first for extensions of multiplicative groups (see Chap. IV), with extensions described by factor systems. The systematic treatment by SCHREIER [1926] was influential, though the idea of a factor system appeared much earlier [HÖLDER 1893]. The same factor systems were important in the representation of central simple algebras as crossed product algebras [BRAUER 1928], [HASSE-BRAUER-NOETHER 1932] and hence in class field theory. An invariant treatment of extensions without factor systems was first broached by BAER [1934, 1935]. That the group of abelian group extensions had topological applications was first realized by EILENBERG-MAC LANE [1942] in their treatment of the universal coefficient problem. There Ext^1 was named. Another proof of the universal coefficient theorem and the homotopy classification theorem of § 4 has been given by MASSEY [1958], using the mapping cone.

Resolutions, perhaps without the name, have long been used, for example in HILBERT [1890]. HOPF in 1944 used them explicitly to describe the homology of a group. CARTAN [1950] used them for the cohomology of groups and gave an axiomatic description as in § 10. Ext^n was defined via resolutions by CARTAN-EILENBERG. The definition by long exact sequences is due to YONEDA [1954], who also has [1960] a more general treatment of composites.

Chapter four

Cohomology of Groups

The cohomology of a group Π provides our first example of the functors $\mathrm{Ext}^n_R(C, A)$ — with R the group ring and $C = Z$. These cohomology groups may be defined directly in terms of a standard "bar resolution". In low dimensions they arise in problems of group extensions by Π; in all dimensions they have a topological interpretation (§ 11).

1. The Group Ring

Let Π be a multiplicative group. The free abelian group $Z(\Pi)$ generated by the elements $x \in \Pi$ consists of the finite sums $\sum m(x)\, x$ with integral coefficients $m(x) \in Z$. The product in Π induces a product

$$\left(\sum_x m(x)\, x\right)\left(\sum_y m'(y)\, y\right) = \sum_{x,y} m(x)\, m'(y)\, x y$$

of two such elements, and makes $Z(\Pi)$ a ring, called the *integral group ring* of Π. Thus an element in $Z(\Pi)$ is a function m on Π to Z, zero except for a finite number of arguments $x \in \Pi$; the sum of two functions is defined by $(m+m')(x) = m(x) + m'(x)$, while the product is $(m m')(x) = \sum m(y)\, m'(z)$, where the latter sum is taken over all y and z in Π with $yz = x$. A ring homomorphism $\varepsilon\colon Z(\Pi) \to Z$, called *augmentation*, is defined by setting

$$\varepsilon\left(\sum_x m(x)\, x\right) = \sum_x m(x). \tag{1.1}$$

Let $\mu_0\colon \Pi \to Z(\Pi)$ be the function which assigns to each $y \in \Pi$ the element labelled $1y$ in $Z(\Pi)$; this means more exactly that $\mu_0 y$ is that function on Π to Z for which $(\mu_0 y)(y) = 1$ and $(\mu_0 y)(x) = 0$ for $x \neq y$. Clearly μ_0 is a multiplicative homomorphism, in the sense that $\mu_0(y y') = (\mu_0 y)(\mu_0 y')$ and $\mu_0(1) = 1$. The group ring $Z(\Pi)$, together with this homomorphism μ_0, can be characterized by the following universal property.

Proposition 1.1. *If Π is a multiplicative group, R a ring with identity, and $\mu\colon \Pi \to R$ a function with $\mu(1) = 1$ and $\mu(xy) = (\mu x)(\mu y)$, then there is a unique ring homomorphism $\varrho\colon Z(\Pi) \to R$ such that $\varrho\, \mu_0 = \mu$.*

Proof. We may define $\varrho\left(\sum m(x)\, x\right) = \sum m(x)\, \mu(x)$; this is a ring homomorphism, and the only such with $\varrho\, \mu_0 = \mu$.

In view of this property it would be more suggestive to call $Z(\Pi)$ not the "group ring of Π", but the *free ring over the multiplicative group Π*.

Modules over $Z(\Pi)$ (Π-modules for short) will appear repeatedly.

Proposition 1.2. *An abelian group A is given a unique structure as a left Π-module by giving either*

(i) *A function on $\Pi \times A$ to A, written xa for $x \in \Pi$, $a \in A$, such that always*

$$x(a_1 + a_2) = x a_1 + x a_2, \qquad (x_1 x_2)\, a = x_1 (x_2 a), \qquad 1a = a; \tag{1.2}$$

(ii) *A group homomorphism*

$$\varphi\colon \Pi \to \operatorname{Aut} A. \tag{1.3}$$

Here $\operatorname{Aut} A$ designates the set of all automorphisms of A; that is, of all isomorphisms $\alpha\colon A \to A$. Under composition, $\operatorname{Aut} A$ is a multiplicative group.

The proof is immediate, for (1.2) gives φ by $\varphi(x)\,a = xa$, while $\operatorname{Aut} A$ is contained in the endomorphism ring $\operatorname{Hom}_Z(A, A)$, and φ extends by Prop. 1.1 to $\psi\colon Z(\Pi) \to \operatorname{Hom}_Z(A, A)$, making A a left module with operators $\psi(u)\,a$ for each $u \in Z(\Pi)$.

In particular, any abelian group A can be regarded as a *trivial* Π-module by taking $\varphi x = 1$; then $xa = a$ for all x.

To each Π-module A we construct an additive, but not necessarily abelian, group $A \times_\varphi \Pi$ called the *semi-direct product* of A and Π with operators φ. Its elements are all pairs (a, x) with the addition

$$(a, x) + (a_1, x_1) = (a + x a_1, x x_1), \qquad x a_1 = \varphi(x)\, a_1. \tag{1.4}$$

One proves that this is a group with the "identity" element $0 = (0, 1)$ and inverse $-(a, x) = (-x^{-1} a, x^{-1})$ and that there is a short exact sequence

$$0 \to A \xrightarrow{\varkappa} A \times_\varphi \Pi \xrightarrow{\sigma} \Pi \to 1, \tag{1.5}$$

where $\varkappa$ is the homomorphism given by $\varkappa a = (a, 1)$, σ is $\sigma(a, x) = x$, and 1 denotes the trivial multiplicative group. Also σ has a right inverse ν defined by $\nu x = (0, x)$ for all x; it is a homomorphism of the multiplicative group Π to the additive group $A \times_\varphi \Pi$.

Exercises

1. A holomorphism h of the multiplicative group G is a 1-1 function on G to G with $h(ab^{-1}c) = (ha)(hb)^{-1}(hc)$ for $a, b, c \in G$. Show that the set of all holomorphisms of G under composition form a group, the *holomorph* $\operatorname{Hol} G$. Construct a short exact sequence $(\lambda, \tau)\colon G \rightarrowtail \operatorname{Hol} G \twoheadrightarrow \operatorname{Aut} G$, where $(\lambda g)(a) = ga$, $(\tau h)\, a = h(1)^{-1} h(a)$, and τ has a right inverse.

2. (R. BAER.) Let A be a Π-module and $\operatorname{Hol} A$ the holomorph of its additive group, as in Ex. 1. Form the direct product $(\operatorname{Hol} A) \times \Pi$ with projections π_1 and π_2 upon its factors, show that $A \times_\varphi \Pi$ is isomorphic to the subgroup of $(\operatorname{Hol} A) \times \Pi$ where

$$\tau \pi_1 = \varphi \pi_2 \colon (\operatorname{Hol} A) \times \Pi \to \operatorname{Aut} A,$$

and compare the sequence (1.5) with that of Ex. 1.

2. Crossed Homomorphisms

If A is a Π-module, a *crossed homomorphism* of Π to A is a function f on Π to A such that

$$f(xy) = x f(y) + f(x), \qquad x, y \in \Pi. \tag{2.1}$$

Then necessarily $f(1) = 0$. For example, if A is a trivial Π-module ($xa = a$ always), a crossed homomorphism is just an ordinary homomorphism of the multiplicative group Π to the additive abelian group A. The sum of two crossed homomorphisms f and g, defined by $(f+g)\,x = f(x) + g(x)$, is a crossed homomorphism. Under this addition the set of all crossed homomorphisms of Π to A is an abelian group which will be denoted by $Z^1_\varphi(\Pi, A)$ — here φ records the Π-module structure $\Pi \to \operatorname{Aut} A$ of A. For each fixed $a \in A$ the function f_a defined by $f_a(x) = xa - a$ is a crossed homomorphism. The functions of this form f_a are *principal crossed homomorphisms*. Since $f_a + f_b = f_{(a+b)}$ and $f_{(-a)} = -f_a$, they constitute a subgroup $B^1_\varphi(\Pi, A)$ of Z^1_φ. The first cohomology group of Π over A is defined to be the quotient group

$$H^1_\varphi(\Pi, A) = Z^1_\varphi(\Pi, A)/B^1_\varphi(\Pi, A). \tag{2.2}$$

If A is the multiplicative group of a field and Π a finite group of automorphisms of A (thus determining the Π-module structure of A), a fundamental theorem of Galois Theory (ARTIN 1944, Thm. 21) asserts that $H^1(\Pi, A) = 0$ — that is, in this case, every crossed homomorphism is principal. Another application of crossed homomorphisms is

Proposition 2.1. *The group of all those automorphisms of the semi-direct product $B = A \times_\varphi \Pi$ which induce the identity both on the subgroup A and the quotient group $B/A \cong \Pi$ is isomorphic to the group $Z^1_\varphi(\Pi, A)$ of crossed homomorphisms. Under this isomorphism the inner automorphisms of B induced by elements of A correspond to the principal crossed homomorphisms.*

Proof. An automorphism ω of the sort described must be given by a formula $\omega(a, x) = (a + f(x), x)$ for some function f on Π to A with $f(1) = 0$. The condition that ω be an automorphism is equivalent to the equation (2.1). Composition of automorphisms then corresponds to the addition of the functions f, and inner automorphisms $(b, x) \to (a, 1) + (b, x) - (a, 1)$ to principal crossed homomorphisms, as asserted.

Crossed homomorphisms may be described in terms of the group ring $Z(\Pi)$ and its augmentation $\varepsilon\colon Z(\Pi) \to Z$, as follows.

Proposition 2.2. *A crossed homomorphism of Π to the $Z(\Pi)$-module A is a homomorphism $g\colon Z(\Pi) \to A$ of abelian groups such that always*

$$g(rs) = rg(s) + g(r)\varepsilon(s), \quad r, s \in Z(\Pi). \tag{2.3}$$

The principal homomorphisms are the homomorphisms g_a defined for a fixed $a \in A$ as $g_a(r) = ra - a\varepsilon(r)$.

Proof. In these formulas $\varepsilon(r)$ and $\varepsilon(s)$ are integers which operate on A on the right as multiples; thus $a\varepsilon(r) = \varepsilon(r)a$. Given any function g, as in (2.3), its restriction $f = g|\Pi$ to the elements $x \in \Pi$ is a crossed homomorphism in the previous sense of (2.1), since $\varepsilon(x) = 1$. Conversely, any crossed homomorphism f in the sense of (2.1) may be extended by linearity to a homomorphism $g\colon Z(\Pi) \to A$ of abelian groups; that is, by $g(\sum m_x x) = \sum m_x f(x)$. Then (2.3) follows from (2.1). We identify f with its extension g, and obtain thus the results stated.

The augmentation $\varepsilon\colon Z(\Pi) \to Z$ is a ring homomorphism, hence its kernel $I(\Pi)$ is a two-sided ideal in $Z(\Pi)$ and therefore also a Π-submodule of $Z(\Pi)$. The injection ι gives an exact sequence

$$0 \to I(\Pi) \xrightarrow{\iota} Z(\Pi) \xrightarrow{\varepsilon} Z \to 0 \tag{2.4}$$

of Π-modules, where Z has the trivial module structure. The map $m \to m1$ of Z to $Z(\Pi)$ is a homomorphism of additive groups (not of Π-modules!) which is a right inverse of ε. Hence the sequence (2.4) splits as a sequence of abelian groups. A left inverse $p\colon Z(\Pi) \to I(\Pi)$ of the injection ι is therefore the map defined for $r \in Z(\Pi)$ as $pr = r - \varepsilon(r)1$. It is a homomorphism of abelian groups and a crossed homomorphism of Π to the module $I(\Pi)$.

Proposition 2.3. *For any Π-module A the operation of restricting to $I(\Pi)$ a crossed homomorphism g of the form (2.3) provides an isomorphism*

$$Z^1_\varphi(\Pi, A) \cong \operatorname{Hom}_{Z(\Pi)}(I(\Pi), A). \tag{2.5}$$

The principal homomorphisms correspond to the module homomorphisms $h_a\colon I(\Pi) \to A$ defined for fixed a by the formula $h_a(u) = ua$, $u \in I(\Pi)$.

Proof. When $\varepsilon(s) = 0$ the identity (2.3) for g becomes $g(rs) = rg(s)$, so g restricted to the kernel of ε is a module homomorphism, as stated. Conversely, any module homomorphism $h\colon I(\Pi) \to A$, when composed with the special crossed homomorphism $pr = r - \varepsilon(r)1$, yields a crossed homomorphism hp on $Z(\Pi)$ whose restriction to $I(\Pi)$ is exactly h. Finally, the principal homomorphisms behave as stated.

For Π fixed, $Z^1_\varphi(\Pi, A)$ and H^1_φ are covariant functors of A; for each module homomorphism $\alpha\colon A \to B$, $(\alpha_* f)(x)$ is defined as $\alpha[f(x)]$. For a fixed abelian group A with the trivial Π-module structure one can make Z^1_φ and H^1_φ contravariant functors of Π; for a group homomorphism $\zeta\colon \Pi \to \Pi'$ and a crossed homomorphism f on Π' define the induced map $\zeta^*\colon Z^1_\varphi(\Pi', A) \to Z^1_\varphi(\Pi, A)$ by $(\zeta^* f)(x) = f(\zeta x)$. This will not do when A is a non-trivial Π- or Π'-module. However, if $\zeta\colon \Pi \to \Pi'$ and A' is a Π'-module via $\varphi'\colon \Pi' \to \operatorname{Aut} A'$, then A' is also a Π-module via $\varphi'\zeta\colon \Pi \to \operatorname{Aut} A'$, and we may define induced homomorphisms

$$\zeta^*\colon Z^1_{\varphi'}(\Pi', A') \to Z^1_{\varphi'\zeta}(\Pi, A'), \quad \zeta^*\colon H^1_{\varphi'}(\Pi', A') \to H^1_{\varphi'\zeta}(\Pi, A')$$

by setting $(\zeta^* f)(x) = f(\zeta x)$ for any crossed homomorphism f on Π'. These induced homomorphisms ζ^* behave functorially; that is, $(\zeta'\zeta)^* = \zeta^* \zeta'^*$ and $1^* = 1$.

More formally, regard the triple (Π, A, φ) as a single object in a category $\mathcal{G}^-$ in which the morphisms $\varrho: (\Pi, A, \varphi) \to (\Pi', A', \varphi')$ are *changes of groups*; that is, pairs $\varrho = (\zeta, \alpha)$ of group homomorphisms with

$$\zeta: \Pi \to \Pi', \quad \alpha: A' \to A, \quad x(\alpha a') = \alpha[(\zeta x) a'] \tag{2.6}$$

for all $x \in \Pi$ and $a' \in A'$. Note that α is backwards (from A' to A), and that the third condition states that α is a homomorphism $\alpha: A' \to A$ of Π-modules. If $\varrho' = (\zeta', \alpha'): (\Pi', A', \varphi') \to (\Pi'', A'', \varphi'')$ is another change of groups, the composite $\varrho' \varrho$ is $(\zeta' \zeta, \alpha \alpha')$. For any crossed homomorphism f' on Π' to A' the definition $(\varrho^* f')(x) = \alpha[f'(\zeta x)]$ gives a map $\varrho^*: Z^1_{\varphi'}(\Pi', A') \to Z^1_{\varphi}(\Pi, A)$ which makes Z^1_{φ} and H^1_{φ} contravariant functors on the change of groups category $\mathcal{G}^-$. This map ϱ^* is the composite

$$H^1_{\varphi'}(\Pi', A') \xrightarrow{\zeta^*} H^1_{\varphi'\zeta}(\Pi, A') \xrightarrow{\alpha_*} H^1_{\varphi}(\Pi, A)$$

of the maps ζ^* and α_* previously defined.

3. Group Extensions

A *group extension* is a short exact sequence

$$E: 0 \to G \xrightarrow{\varkappa} B \xrightarrow{\sigma} \Pi \to 1 \tag{3.1}$$

of not necessarily abelian groups; it is convenient to write the group composition in 0, G, and B as addition; that in Π and 1 as multiplication. As before, the statement that E is exact amounts to the assertion that $\varkappa$ maps G isomorphically onto a normal subgroup of B and that σ induces an isomorphism $B/\varkappa G \cong \Pi$ of the corresponding quotient group. The extension E *splits* if σ has a right inverse ν; that is, if there is a homomorphism $\nu: \Pi \to B$ with $\sigma\nu = 1_\Pi$, the identity automorphism of Π. The semi-direct product extension (1.5) splits.

Let $\operatorname{Aut} G$ denote the group of automorphisms of G, with group multiplication the composition of automorphisms. Conjugation in B yields a homomorphism $\theta: B \to \operatorname{Aut} G$ under which the action of each $\theta(b)$ on any $g \in G$ is given by

$$\varkappa[(\theta b) g] = b + (\varkappa g) - b, \quad b \in B, \ g \in G.$$

Suppose $G = A$ abelian; then $\theta(A) = 1$, so that θ induces a homomorphism $\varphi: \Pi \to \operatorname{Aut} A$ with $\varphi\sigma = \theta$. Thus φ is defined by

$$\varkappa[(\varphi\sigma b) a] = b + (\varkappa a) - b, \quad b \in B, \ a \in A. \tag{3.2}$$

We then say that E is an extension of the abelian group A by the group Π *with the operators* $\varphi: \Pi \to \operatorname{Aut} A$. This map φ records the way in which A appears as a normal subgroup in the extension.

The problem of group extensions is that of constructing all E, given A, Π, and φ. Now φ gives A the structure of a Π-module; hence the group extension problem is that of constructing all E, given Π and a Π-module A. There is at least one such extension, the semi-direct product $A \times_\varphi \Pi$.

If E and E' are any two group extensions, a *morphism* $\Gamma: E \to E'$ is a triple $\Gamma = (\alpha, \beta, \gamma)$ of group homomorphisms such that the diagram

$$\begin{array}{llllllll} E: & 0 \to & A & \to & B & \to & \Pi & \to 1 \\ & & \downarrow \alpha & & \downarrow \beta & & \downarrow \gamma & \\ E': & 0 \to & A' & \to & B' & \to & \Pi' & \to 1 \end{array} \tag{3.3}$$

is commutative. If A and A' are abelian, and φ and $\varphi': \Pi' \to \operatorname{Aut} A'$ the associated operators of E and E', one shows readily that always

$$\alpha[\varphi(x)\,a] = (\varphi'\gamma x)\,\alpha a. \tag{3.4}$$

For example, if $A = A'$, and $\alpha = 1_A$, then $(\varphi x)\,a = (\varphi'\gamma x)\,a$: In other words, the Π-module structure on A is determined by the Π'-module structure. If $\Gamma: E \to E'$ and $\Gamma': E' \to E''$ are morphisms of extensions, so is the composite $\Gamma'\Gamma: E \to E''$.

If E and E' are two group extensions of the same module A by the group Π, a *congruence* $\Gamma: E \to E'$ is a morphism $\Gamma = (\alpha, \beta, \gamma)$ with $\alpha = 1_A$ and $\gamma = 1_\Pi$. For A abelian, it follows from (3.4) that $\varphi = \varphi'$; i.e., congruent extensions have the same operators. The (non-commutative!) short five lemma shows that a congruence $\Gamma = (1_A, \beta, 1_\Pi)$ has β an isomorphism, hence that each congruence has an inverse. We may therefore speak of congruence classes of extensions. Let $\operatorname{Opext}(\Pi, A, \varphi)$ denote the set of all congruence classes of extensions of the abelian group A by Π with operators φ. We wish to describe Opext.

Any extension (3.1) with $G = A$ abelian which splits (under $\nu: \Pi \to B$) is congruent to the semi-direct product $A \times_\varphi \Pi$, under the isomorphism $\beta: B \to A \times_\varphi \Pi$ given by $\beta b = (\varkappa^{-1}[b - \nu\sigma b], \sigma b)$. In detail,

$$\begin{aligned} b + b_1 - \nu\sigma(b + b_1) &= (b - \nu\sigma b) + \nu\sigma b + (b_1 - \nu\sigma b_1) - \nu\sigma b \\ &= (b - \nu\sigma b) + \varkappa[(\theta b)\varkappa^{-1}(b_1 - \nu\sigma b_1)], \end{aligned}$$

exactly as in the addition table (1.4) for the semi-direct product.

If Π is a (non-abelian) free group with generators t_k, then any epimorphism $\sigma: B \to \Pi$ has a right inverse given by setting $\nu t_k = b_k$, where b_k is any element of B with $\sigma b_k = t_k$. Hence any extension by a free group splits, and Opext then consists of a single element.

As a more interesting case, take $\Pi = C_m(t)$ cyclic of finite order m with generator t. In any extension E by C_m identify each $a \in A$ with its image $\varkappa a \in B$, so that $A \subset B$. Choose a representative u for t with $\sigma u = t$; as

$\sigma(mu)=t^m=1$, $mu=a_0\in A$. Each element of B can be written uniquely as $a+iu$ for $a\in A$ and $0\leqq i<m$. By the choice of a_0 and (3.2),

$$mu=a_0, \quad u+a=ta+u. \tag{3.5}$$

With these equations, the sum of any two elements of the form $a+iu$ can be put in the same form. By associativity, $u+mu=(m+1)u=mu+u$, so $u+a_0=a_0+u$. Therefore $a_0=ta_0$, so a_0 is "invariant" under t. This element a_0 is not unique; if $u'=a_1+u$ is a different representative for t in B, then, by (3.5) and induction on m,

$$mu'=m(a_1+u)=a_1+ta_1+\cdots+t^{m-1}a_1+mu=N_t a_1+a_0.$$

Here $N_t a_1=a_1+ta_1+\cdots+t^{m-1}a_1$ is the *norm* with respect to t in the C_m-module A; it is a group homomorphism $N_t\colon A\to A$. Since the coset of a_0 modulo $N_t A$ is uniquely determined by the congruence class of the extension, we have established a correspondence

$$\operatorname{Opext}(C_m(t),A,\varphi)\leftrightarrow[a\mid ta=a]/N_t A. \tag{3.6}$$

This is 1-1; given any invariant a_0, take B to be all symbols $a+iu$ with $0\leqq i<m$ and addition given by (3.5). The invariance of a_0 proves this addition associative, and B is an extension of A by C_m with the given operators. In particular, if A has trivial operators ($ta=a$ always), the expression on the right of (3.6) is the group A/mA — in agreement with the result already found in the case of abelian extensions in Prop. III.1.1. In this case, all extensions of A by C_m are abelian.

Again, let $\Pi=C_\infty\times C_\infty$ be the free multiplicative abelian group on two generators t_1 and t_2. In any extension by Π, take representatives u_i of t_i, $i=1,2$. There is then a constant a_0 in A with $u_2+u_1=a_0+u_1+u_2$, all elements of the extension can be written uniquely as $a+m_1u_1+m_2u_2$ with integral coefficients m_1 and m_2, and the addition in B is determined by the addition in A and the rules

$$u_1+a=t_1a+u_1, \quad u_2+a=t_2a+u_2, \quad u_2+u_1=a_0+u_1+u_2.$$

This addition is always associative and makes the collection of elements $a+m_1u_1+m_2u_2$ a group. If the representatives u_1 and u_2 are replaced by any other $u_1'=a_1+u_1$, $u_2'=a_2+u_2$, for $a_1, a_2\in A$, the constant a_0 is replaced by $a_0+a_2-t_1a_2-a_1+t_2a_1$. Hence, if S is the subgroup of A generated by all sums $a_2-t_1a_2-a_1+t_2a_1$, we have a 1-1 correspondence,

$$\operatorname{Opext}(C_\infty\times C_\infty, A, \varphi)\leftrightarrow A/S. \tag{3.7}$$

Exercises

1. Describe $\operatorname{Opext}(C_\infty\times C_\infty\times C_\infty, A, \varphi)$.
2. Describe $\operatorname{Opext}(C_m\times C_n, A, \varphi)$.
3. Show that Prop. 2.1 holds if $A\times_\varphi \Pi$ is replaced by any extension of (Π, A, φ).

4. Factor Sets

The calculations just made suggest that Opext(Π, A, φ), like Ext, is a group. This group structure can be described by means of certain factor sets.

Let E be an extension (3.1) in Opext(Π, A, φ): For convenience, identify each a with $\varkappa a$. To each x in Π choose a "representative" $u(x)$ in B; that is, an element $u(x)$ with $\sigma u(x) = x$. In particular, choose $u(1) = 0$. Now each coset of A in B contains exactly one $u(x)$, and the elements of B can be represented uniquely as $a + u(x)$ for $a \in A$, $x \in \Pi$. We write the operators as $\varphi(x) a = xa$: Then (3.2) for $b = u(x)$ becomes

$$u(x) + a = xa + u(x). \tag{4.1}$$

On the other hand, the sum $u(x) + u(y)$ must lie in the same coset as $u(xy)$, so there are unique elements $f(x, y) \in A$ such that always

$$u(x) + u(y) = f(x, y) + u(xy). \tag{4.2}$$

Since $u(1) = 0$, we also have

$$f(x, 1) = 0 = f(1, y), \qquad x, y \in \Pi. \tag{4.3}$$

The function f is called a *factor set* of the extension E. With this factor set and the data (Π, A, φ), the addition in B is determined, for the sum of any two elements $a + u(x)$ and $a_1 + u(y)$ of B can be calculated, by (4.1) and (4.2), as

$$[a + u(x)] + [a_1 + u(y)] = (a + xa_1 + f(x, y)) + u(xy). \tag{4.4}$$

By this rule form the triple sums

$$[u(x) + u(y)] + u(z) = f(x, y) + f(xy, z) + u(xyz),$$
$$u(x) + [u(y) + u(z)] = xf(y, z) + f(x, yz) + u(xyz).$$

Their equality (associative law!) gives

$$xf(y, z) + f(x, yz) = f(x, y) + f(xy, z), \qquad x, y, z \in \Pi. \tag{4.5}$$

The factor set f for an extension depends on a choice of representatives; if $u'(x)$ is a second set of representatives with $u'(1) = 0$, then $u'(x)$ and $u(x)$ lie in the same coset, so there is a function g on Π to A with $g(1) = 0$ such that $u'(x) = g(x) + u(x)$. Thus

$$u'(x) + u'(y) = g(x) + xg(y) + u(x) + u(y) = g(x) + xg(y) + f(x, y) + u(xy).$$

The new factor set is $f'(x, y) = \delta g(x, y) + f(x, y)$, where δg is the function

$$(\delta g)(x, y) = xg(y) - g(xy) + g(x), \qquad x, y \in \Pi. \tag{4.6}$$

One verfies that this function δg does satisfy the identity (4.5), with f replaced by δg there.

These observations suggest the following definitions. Let $Z^2_\varphi(\Pi, A)$ denote the set of all functions f on $\Pi \times \Pi$ to A which satisfy the identity (4.5) and the normalization condition (4.3). This set is an abelian group under the termwise addition $(f+f')(x,y)=f(x,y)+f'(x,y)$. Let $B^2_\varphi(\Pi, A)$ denote the subset of Z^2_φ which consists of all functions f of the form $f=\delta g$, where δg is defined as in (4.6) from any function g on Π to A with $g(1)=0$. The factor group

$$H^2_\varphi(\Pi, A)=Z^2_\varphi(\Pi, A)/B^2_\varphi(\Pi, A)$$

is called the 2-dimensional cohomology group of Π over A. Our discussion has suggested

Theorem 4.1. *Given $\varphi\colon \Pi \to \operatorname{Aut} A$, A abelian, the function ω which assigns to each extension of A by Π with operators φ the congruence class of one of its factor sets is a 1-1-correspondence*

$$\omega\colon \operatorname{Opext}(\Pi, A, \varphi) \leftrightarrow H^2_\varphi(\Pi, A) \tag{4.7}$$

between the set Opext *of all congruence classes of such extensions and the 2-dimensional cohomology group. Under this correspondence the semi-direct product corresponds to the zero element of H^2_φ.*

Since H^2_φ is an abelian group, this correspondence ω imposes the desired group structure on Opext. This group structure can also be described conceptually via the Baer product, as set forth in the exercises below.

Proof. Since the factor set of an extension is well defined modulo the subgroup B^2_φ, and since congruent extensions have the same factor sets, we know that the correspondence ω is well defined. The semi-direct product $A \times_\varphi \Pi$ clearly has the trivial function $f(x,y)=0$ as one of its factor sets. If two extensions yield factor sets whose difference is some function $\delta g(x,y)$, then a change of representatives in one extension will make the factor sets equal and the extensions congruent. Therefore (4.7) is a 1-1 correspondence of Opext with part of H^2. Finally, given any f satisfying (4.5) and (4.3), one may define a group B to consist of pairs (a,x) with a sum given as in (4.4) by

$$(a,x)+(a_1,y)=(a+xa_1+f(x,y), xy), \qquad a, b \in A.$$

The module rules and the condition (4.5) show that this composition is associative; it clearly yields an extension with representatives $u(x)=(0,x)$ and factor set f. This completes the proof of the theorem.

If A is abelian, a *central group extension* of A by Π is an extension E as in (3.1) in which $\varkappa A$ is in the center of B. In other words, a central extension is one with operators $\varphi=1$. This theorem thus includes the fact that the set of congruence classes of central extensions of A by Π

is in 1-1 correspondence with the group $H^2(\Pi, A)$, where the abelian group A is taken with trivial operators φ. If Π is abelian, every abelian extension is central, so there is a monomorphism $\operatorname{Ext}^1_Z(\Pi, A) \to H^2(\Pi, A)$.

We can regard the cohomology groups H^2_φ and H^1_φ as the cohomology groups of a suitable complex

$$X_0 \leftarrow X_1 \leftarrow X_2 \leftarrow X_3$$

of free Π-modules. Take X_2 to be the free Π-module generated by all pairs $[x, y]$ of elements $x \neq 1$, $y \neq 1$ of Π. In order to define $[x, y] \in X_2$ for all $x, y \in \Pi$, set also $[1, y] = 0 = [x, 1]$ and $[1, 1] = 0$. A 2-dimensional cochain f of $\operatorname{Hom}_\Pi(X, A)$ is thus a Π-homomorphism $f\colon X_2 \to A$; it is determined by its values $f[x, y]$ on the free generators of X_2; hence is in effect a function on $\Pi \times \Pi$ to A with $f(x, 1) = 0 = f(1, y)$. Next take X_3 to be the free Π-module generated by all triples $[x, y, z]$ of elements not 1 in Π, with $\partial\colon X_3 \to X_2$ given by

$$\partial[x, y, z] = x[y, z] - [xy, z] + [x, yz] - [x, y]; \tag{4.8}$$

the condition that f be a cocycle ($f\partial = 0$) is exactly the identity (4.5). Finally, take X_1 to be the free module generated by all $[x]$ with $x \neq 1$ and set $[1] = 0$. A 1-dimensional cochain is a module homomorphism $X_1 \to A$, and is hence determined by its values on $[x]$, so is, in effect, a function g on Π to A with $g(1) = 0$. If we now define $\partial\colon X_2 \to X_1$ by

$$\partial[x, y] = x[y] - [xy] + [x], \tag{4.9}$$

then $\partial\partial = 0$, and the coboundary of g is the function given by the formula (4.6). Thus $H^2_\varphi(\Pi, A)$ is $H^2(\operatorname{Hom}_{Z(\Pi)}(X, A))$. We get the analogous result for H^1_φ if we take X_0 to be $Z(\Pi)$ and set $\partial[x] = x - 1 \in Z(\Pi)$.

This complex also defines a 0-dimensional cohomology group as $H^0_\varphi(\Pi, A) = H^0(\operatorname{Hom}_{Z(\Pi)}(X, A))$. A 0-dimensional cochain is a module homomorphism $f\colon Z(\Pi) \to A$; it is determined by its value $f(1) = a \in A$. It is a cocycle if $-(\delta f)[x] = f\partial[x] = f(x - 1) = xa - a$ is zero. Hence the 0-cocycles correspond to the elements $a \in A$ invariant under Π ($xa = a$ for all x):

$$H^0_\varphi(\Pi, A) = A^\Pi, \qquad A^\Pi = [a \mid xa = a \text{ for } x \in \Pi]. \tag{4.10}$$

Exercises

The Baer sum, introduced for extensions of modules in Chap. III, can also be applied to group extensions, as indicated in the following sequence of exercises.

1. Prove: If E is an extension of G by Π and $\gamma\colon \Pi' \to \Pi$, there exists an extension E' of G by Π' and a morphism $\Gamma = (1_G, \beta, \gamma)\colon E' \to E$. The pair (Γ, E') is unique up to a congruence of E'. If G is abelian and has operators $\varphi\colon \Pi \to \operatorname{Aut} G$, then E' has operators $\varphi\gamma$. Define $E\gamma = E'$.

2. Under the hypotheses of Ex. 1, prove that each morphism $(\alpha_1, \beta_1, \gamma_1): E_1 \to E$ of extensions with $\gamma_1 = \gamma$ can be factored uniquely through Γ.

3. For $E \in \operatorname{Opext}(\Pi, A, \varphi)$, $\varphi': \Pi \to \operatorname{Aut} A'$, and $\alpha: A \to A'$ a Π-module homomorphism, prove that there exists an extension $E' \in \operatorname{Opext}(\Pi, A', \varphi')$ and a morphism $\Theta = (\alpha, \beta, 1_\Pi): E \to E'$, unique up to a congruence of E'. Define αE to be E'.

4. Under the hypotheses of Ex. 3, prove that for $E_1 \in \operatorname{Opext}(\Pi_1, A', \varphi_1')$ each morphism $(\alpha_1, \beta_1, \gamma_1): E \to E_1$ with $\alpha_1 = \alpha$ and $\varphi_1' \gamma_1 = \varphi'$ can be factored uniquely through Θ.

5. For α, γ, and E as in Exs. 1 and 3, with $G = A$ abelian, prove that $\alpha(E\gamma)$ is congruent to $(\alpha E)\gamma$.

6. Using Exs. 1, 3, and 5, show that Opext is a contravariant functor on the category $\mathscr{G}^-$ of changes of groups.

7. Show that $\operatorname{Opext}(\Pi, A, \varphi)$ is an abelian group under the Baer sum defined by $E_1 + E_2 = \nabla_A (E_1 \times E_2) \Delta_\Pi$, and show that this composition agrees with that given by factor sets.

5. The Bar Resolution

The boundary formulas (4.8) and (4.9) for the complex X of the last section can be generalized to higher dimensions. Specifically, for any group Π we construct a certain chain complex of Π-modules $B_n(Z(\Pi))$. Take B_n to be the free Π-module with generators $[x_1| \dots | x_n]$ all n-tuples of elements $x_1 \neq 1, \dots, x_n \neq 1$ of Π. Operation on a generator with an $x \in \Pi$ yields an element $x[x_1| \dots | x_n]$ in B_n, so B_n may be described as the free abelian group generated by all $x[x_1| \dots | x_n]$. To give a meaning to every symbol $[x_1| \dots | x_n]$, set

$$[x_1| \dots | x_n] = 0 \quad \text{if any one} \quad x_i = 1; \tag{5.1}$$

this is called the *normalization* condition. In particular, B_0 is the free module on one generator, denoted $[\,]$, so is isomorphic to $Z(\Pi)$, while $\varepsilon[\,] = 1$ is a Π-module homomorphism $\varepsilon: B_0 \to Z$, with Z the trivial Π-module.

Homomorphisms $s_{-1}: Z \dashrightarrow B_0$, $s_n: B_n \dashrightarrow B_{n+1}$ of abelian groups are defined by

$$s_{-1} 1 = [\,], \qquad s_n x[x_1| \dots | x_n] = [x | x_1| \dots | x_n]. \tag{5.2}$$

Define Π-module homomorphisms $\partial: B_n \to B_{n-1}$ for $n > 0$ by

$$\left.\begin{aligned} \partial[x_1| \dots | x_n] = x_1[x_2| \dots | x_n] + \sum_{i=1}^{n-1} (-1)^i [x_1| \dots | x_i x_{i+1}| \dots | x_n] + {}\\ + (-1)^n [x_1| \dots | x_{n-1}]; \end{aligned}\right\} \tag{5.3}$$

in particular $\partial[x] = x[\,] - [\,]$, $\partial[x|y] = x[y] - [xy] + [x]$. Note that formula (5.3) holds even when some $x_i = 1$, for then the terms numbered $i-1$ and i on the right cancel, and the remaining terms are zero. All told, we have a diagram

$$Z \underset{s_{-1}}{\overset{\varepsilon}{\leftrightarrows}} B_0 \underset{s_0}{\overset{\partial}{\leftrightarrows}} B_1 \leftrightarrows \cdots \leftrightarrows B_{n-1} \underset{s_{n-1}}{\overset{\partial}{\leftrightarrows}} B_n \leftrightarrows \cdots, \tag{5.4}$$

with solid arrows module homomorphisms and dotted arrows group homomorphisms. Call $B=B(Z(\Pi))$ the *bar resolution*.

Theorem 5.1. *For any group Π the bar resolution $B(Z(\Pi))$ with augmentation ε is a free resolution of the trivial Π-module Z.*

Proof. The B_n are free modules, by construction, so we must show that the sequence of solid arrows in (5.4), with zero adjoined on the left, is exact. We will prove more: That this sequence is a complex of abelian groups with s as a contracting homotopy. The latter statement means that

$$\varepsilon s_{-1}=1, \quad \partial s_0+s_{-1}\varepsilon=1, \quad \partial s_n+s_{n-1}\partial=1, \quad (n>0). \tag{5.5}$$

Each of these equations is immediate from the definition; for example, by (5.3), $\partial s_n(x[x_1|\ldots|x_n])$ starts with $x[x_1|\ldots|x_n]$ while the remaining terms are those of $s_{n-1}\partial x[x_1|\ldots|x_n]$, each with sign changed; this proves the last equation of (5.5). Moreover, these equations determine ε and $\partial_{n+1}\colon B_{n+1}\to B_n$ uniquely by recursion on n, for B_{n+1} is generated as a Π-module by the subgroup $s_n B_n$, and the equations (5.5) give ∂_{n+1} on this subgroup as $\partial_{n+1}s_n=1-s_{n-1}\partial_n$; thus the formula (5.3) for ∂ can be deduced from (5.5) and (5.2) for s. By the same recursion argument it follows that $\varepsilon\partial_1=0$ and $\partial_n\partial_{n+1}=0$, for

$$\partial_n\partial_{n+1}s_n=\partial_n(1-s_{n-1}\partial_n)=\partial_n-(\partial_n s_{n-1})\,\partial_n=\partial_n-\partial_n+s_{n-2}\partial_{n-1}\partial_n$$

gives $\partial^2=0$ by induction. This can also be proved, directly but laboriously, from the formula (5.3) for ∂. Either argument shows $B(Z(\Pi))$ a complex and a resolution of Z, as stated in the theorem.

The same theorem holds for the "non-normalized" bar resolution $\beta(Z(\Pi))$. Here β_n is the free Π-module generated by all the n-tuples $x_1\otimes\cdots\otimes x_n$ of elements of Π (no normalization condition) and ε, ∂, s are given by the same formulas as for B. Thus $B_n\cong\beta_n/D_n$, where D_n is the submodule generated by all $x_1\otimes\cdots\otimes x_n$ with one $x_i=1$. The symbol $\otimes$ is used here because this description makes β_n the $(n+1)$-fold "tensor product" $Z(\Pi)\otimes\cdots\otimes Z(\Pi)$ of the abelian groups $Z(\Pi)$; these tensor products are defined in Chap. V and applied to the bar resolution in Chap. IX.

For any Π-module A we define the *cohomology groups* of Π with coefficients A by the formula

$$H^n(\Pi,A)=H^n(B(Z(\Pi)),A), \tag{5.6}$$

in keeping with the special cases treated in the previous section (where the subscript φ was used to record explicitly the structure of A as a Π-module). The cohomology groups $H^n(\Pi, A)$ are thus those of the cochain complex $B(\Pi, A) = \mathrm{Hom}_\Pi[B(Z(\Pi)), A]$, where Hom_Π is short for $\mathrm{Hom}_{Z(\Pi)}$. Since B_n is a free module with generators $[x_1|\dots|x_n]$ (no $x_i = 1$), an n-cochain $f\colon B_n \to A$ is a Π-module homomorphism which is uniquely determined by its values on these generators. Therefore the group $B^n(\Pi, A)$ of n-cochains may be identified with the set of all those functions f of n arguments x_i in Π, with values in A, which satisfy the "normalization" conditions

$$f(x_1, \dots, x_{i-1}, 1, x_{i+1}, \dots, x_n) = 0, \qquad i = 1, \dots, n. \tag{5.7}$$

The sum of two cochains f_1 and f_2 is given by addition of values as

$$(f_1 + f_2)(x_1, \dots, x_n) = f_1(x_1, \dots, x_n) + f_2(x_1, \dots, x_n).$$

Under this addition the set B^n of all such f is an abelian group. The coboundary homomorphism $\delta\colon B^n \to B^{n-1}$ is defined by

$$\left.\begin{aligned} &\delta f(x_1, \dots, x_{n+1}) = (-1)^{n+1}[x_1 f(x_2, \dots, x_{n+1}) + \\ &\quad + \sum_{i=1}^{n} (-1)^i f(x_1, \dots, x_i x_{i+1}, \dots, x_{n+1}) + (-1)^{n+1} f(x_1, \dots, x_n)].\end{aligned}\right\} \tag{5.8}$$

$H^n(\Pi, A)$ is the n-th cohomology group of this complex $B(\Pi, A)$.

As a functor, $H^n(\Pi, A)$ is contravariant in the objects (Π, A, φ), for if $\varrho = (\zeta, \alpha)$ is a change of groups as in (2.6), the induced map $\varrho^*\colon H^n(\Pi', A') \to H^n(\Pi, A)$ is defined for any $f' \in B'^n$ by

$$(\varrho^* f')(x_1, \dots, x_n) = \alpha[f'(\zeta x_1, \dots, \zeta x_n)], \quad \zeta\colon \Pi \to \Pi', \quad \alpha\colon A' \to A. \tag{5.9}$$

In particular, for Π fixed, $H^n(\Pi, A)$ is a covariant functor of the Π-module A.

Corollary 5.2. *For any Π-module A there is an isomorphism*

$$\theta\colon \mathrm{Ext}^n_{Z(\Pi)}(Z, A) \cong H^n(\Pi, A)$$

which is natural in A.

Since B is a free resolution of the trivial Π-module Z, the result is immediate, by Thm. III.6.4; it shows that the cohomology of a group is a special case of the functor Ext_R, for R the group ring.

For a short exact sequence $E\colon A \rightarrowtail B \twoheadrightarrow C$ of Π-modules, Cor. 5.2 and the usual exact sequence for Ext yield an exact sequence

$$\cdots \to H^n(\Pi, A) \to H^n(\Pi, B) \to H^n(\Pi, C) \xrightarrow{E_*} H^{n+1}(\Pi, A) \to \cdots.$$

The connecting homomorphisms E_* are natural in E. For fixed Π, the cohomology groups $H^n(\Pi, A)$ are covariant functors of A which may be characterized, with these connecting homomorphisms, by three axioms like those for Ext (III.10): The sequence above is exact, $H^0(\Pi, A) \cong A^\Pi$, and $H^n(\Pi, J)=0$ if $n>0$ and J is an injective Π-module.

For Π finite, the coboundary formula gives an amusing result:

Proposition 5.3. *If Π is a finite group of order k, every element of $H^n(\Pi, A)$ for $n>0$ has order dividing k.*

Proof. For each n-cochain f define an $(n-1)$-cochain g by

$$g(x_1, \ldots, x_{n-1}) = \sum_{x \in \Pi} f(x_1, \ldots, x_{n-1}, x).$$

Add the identities (5.8) for all $x = x_{n+1}$ in Π. The last term is independent of x; in the next to the last term, for x_n fixed,

$$\textstyle\sum_x f(\ldots, x_{n-1}, x_n x) = \sum_x f(\ldots, x_{n-1}, x) = g(\ldots, x_{n-1}).$$

Hence the result is

$$\sum_{x \in \Pi} \delta f(x_1, \ldots, x_n, x) = -\delta g(x_1, \ldots, x_n) + k f(x_1, \ldots, x_n).$$

For $\delta f = 0$ this gives $kf = \delta g$ a coboundary, hence the result.

Corollary 5.4. *If Π is finite, while the divisible abelian group D with no elements of finite order is a Π-module in any way, then $H^n(\Pi, D) = 0$ for $n>0$.*

Proof. For g as above, there is an $(n-1)$-cochain h with $g = kh$. Then $kf = \pm k\,\delta h$; since D has no elements of finite order, $f = \pm\delta h$, and the cocycle f is a coboundary.

Corollary 5.5. *If Π is finite, P is the additive group of real numbers, mod 1, and P and Z are trivial Π-modules,* $H^2(\Pi, Z) \cong \operatorname{Hom}(\Pi, P)$.

The (abelian) group $\operatorname{Hom}(\Pi, P)$ of all group homomorphisms $\Pi \to P$ is the character group of Π.

Proof. The additive group R of reals is divisible, with no elements of finite order. The short exact sequence $Z \rightarrowtail R \twoheadrightarrow P$ of trivial Π-modules yields the exact sequence

$$H^1(\Pi, R) \to H^1(\Pi, P) \to H^2(\Pi, Z) \to H^2(\Pi, R).$$

By Cor. 5.4, the two outside groups vanish; since P has trivial module structure, $H^1(\Pi, P) = \operatorname{Hom}(\Pi, P)$. Hence the connecting homomorphism is the desired isomorphism.

To illustrate the use of resolutions, consider the operation of conjugation by a fixed element $t \in \Pi$. Let $\theta_t \colon \Pi \to \Pi$ denote the inner automorphism $\theta_t x = t^{-1} x t$, while, for any Π-module A, $\alpha_t \colon A \to A$ is the automorphism given by $\alpha_t a = ta$. Then $x(\alpha_t a) = xta = t(t^{-1}xta) = \alpha_t[(\theta_t x)a]$, so $(\theta_t, \alpha_t)\colon (\Pi, A, \varphi) \to (\Pi, A, \varphi)$ is a change of groups in the sense of (2.6). The induced map on cohomology is necessarily an isomorphism, but more is true:

Proposition 5.6. *For any Π-module A, conjugation by a fixed $t \in \Pi$ induces the identity isomorphism*

$$(\theta_t, \alpha_t)^*\colon H^n(\Pi, A) = H^n(\Pi, A).$$

Proof. A module homomorphism $g_t\colon B_n(Z(\Pi)) \to B_n(Z(\Pi))$ is given by

$$g_t(x[x_1|\dots|x_n]) = xt[t^{-1}x_1t|\dots|t^{-1}x_it|\dots|t^{-1}x_nt].$$

Observation shows $g_t\partial = \partial g_t$, so g_t is a chain transformation of resolutions which lifts the identity $Z \to Z$. By the comparison theorem for resolutions, g_t is homotopic to the identity, so the induced map on cohomology is the identity. But this induced map carries any n-cochain f into $g_t^* f$ where

$$(g_t^* f)(x_1, \dots, x_n) = f g_t[x_1|\dots|x_n] = tf(t^{-1}x_1t, \dots, t^{-1}x_nt).$$

The cochain on the right is $(\theta_t, \alpha_t)^* f$, as defined by (5.9), hence the conclusion. Note that the comparison theorem has saved us the trouble of constructing an explicit homotopy $g_t \simeq 1$.

This theorem may be read as stating that each n-cocycle f is cohomologous to the cocycle $g_t^* f$ defined above. Like many results in the cohomology of groups, this result was discovered in the case $n=2$ from properties of group extensions (Ex. 3 below).

In the bar resolution, $B_n(Z(\Pi))$ is the free abelian group with free generators all symbols $x[x_1|\dots|x_n]$ with all $x \in \Pi$ and none of $x_1, \dots, x_n$ equal to $1 \in \Pi$. We call these symbols the *nonhomogeneous generators* of B. Now the string of elements $x, x_1, \dots, x_n$ in Π determines and is determined by the string of elements $y_0 = x, y_1 = xx_1, y_2 = xx_1x_2, \dots, y_n = xx_1 \dots x_n$ in Π, and the condition $x_i = 1$ becomes $y_{i-1} = y_i$. Hence the generators of B_n may be labelled by the elements $y_i \in \Pi$, in symbols

$$(y_0, y_1, \dots, y_n) = y_0[y_0^{-1}y_1|y_1^{-1}y_2|\dots|y_{n-1}^{-1}y_n], \tag{5.10}$$

while conversely

$$x[x_1|\dots|x_n] = (x, xx_1, xx_1x_2, \dots, xx_1\dots x_n). \tag{5.11}$$

Translating the boundary formula to this notation proves

Proposition 5.7. *The abelian group $B_n(Z(\Pi))$ contains the elements $(y_0, \dots, y_n)$ of (5.10) for all $y_i \in \Pi$. If $y_{i-1} = y_i$, $(y_0, \dots, y_n) = 0$. The* *remaining such elements are free generators of* B_n. *The* Π*-module structure is given by*

$$y(y_0, y_1, \ldots, y_n) = (y y_0, y y_1, \ldots, y y_n) \tag{5.12}$$

and the boundary $\partial\colon B_n \to B_{n-1}$ *is determined by*

$$\partial(y_0, y_1, \ldots, y_n) = \sum_{i=0}^{n} (-1)^i (y_0, \ldots, \hat{y}_i, \ldots, y_n), \tag{5.13}$$

where the $\wedge$ *over* y_i *indicates that* y_i *is to be omitted.*

Note that this formulation of $B(Z(\Pi))$ uses the multiplication of Π only in the definition (5.12) of the module structure. In view of the form of this definition, the symbols $(y_0, \ldots, y_n)$ are called the *homogeneous* generators of B_n. They have a geometrical flavor. If we regard $(y_0, y_1, \ldots, y_n)$ as an n-simplex σ with the element $y_i \in \Pi$ as a label on the i-th vertex, then $(y_0, \ldots, \hat{y}_i, \ldots, y_n)$ is the $n-1$ simplex which consists of the i-th face of σ with its labels, and the boundary formula (5.13) is the usual formula for the boundary of a simplex as the alternating sum of its $(n-1)$-dimensional faces.

The non-homogeneous generators may be similarly read as a system of edge labels. On the simplex, label the edge from the vertex i to vertex j by $z_{ij} = y_i^{-1} y_j$, so that the simplices σ and $y\sigma$ have the same edge labels, and $z_{ij} z_{jk} = z_{ik}$. Hence the edge labels $x_i = z_{i-1,i}$ determine all the edge labels by composition. The non-homogeneous generator $x[x_1| \ldots |x_n]$ simply records these edge labels x_i and the label $x = y_0$ on the initial vertex, as in the figure

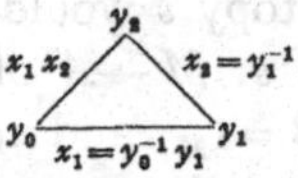

The non-homogeneous boundary formula (5.3) may be read off from these edge labels. This schematic description can be given an exact geometrical meaning when Π is the fundamental group of a space (Eilenberg-MacLane 1945).

Exercises

1. Show that $\beta(Z(\Pi))$ — the non-normalized bar resolution — with a suitable augmentation is a free Π-module resolution of Z.

2. Deduce that $\operatorname{Opext}(\Pi, A, \varphi)$ can be described by factor sets which satisfy (4.5) but not the normalization condition (4.3). Find the identity element in the group extension given by such a non-normalized factor set.

3. For $n=2$ in Prop. 5.6, show explicitly that the cohomologous factor sets f and $g_1^* f$ determine congruent elements of $\operatorname{Opext}(\Pi, A, \varphi)$.

4. Show that $\operatorname{Ext}^n_{Z(\Pi)}(Z, A)$ is a contravariant functor on the category $\mathcal{G}^-$ of changes of groups, and prove the isomorphism θ of Cor. 5.2 natural.

6. The Characteristic Class of a Group Extension

For $n=2$, Cor. 5.2 provides an isomorphism

$$\theta\colon \operatorname{Ext}^2_{Z(\Pi)}(Z,A) \cong H^2(\Pi,A). \tag{6.1}$$

Hence each group extension E of A by Π,

$$E\colon 0 \to A \xrightarrow{\varkappa} B \xrightarrow{\sigma} \Pi \to 1,$$

with given operators φ must determine a two-fold Π-module extension of A by the trivial module Z. It is instructive to construct this module extension

$$\chi(E)\colon 0 \to A \xrightarrow{\alpha} M \xrightarrow{\beta} Z(\Pi) \xrightarrow{\varepsilon} Z \to 0 \tag{6.2}$$

directly. To do so, take $Z(\Pi)$ to be the group ring of Π, with ε its augmentation. Take M to be the quotient module $M=F/L$, where F is the free Π-module on generators $[b]$, one for each $b \neq 0$ in B, with the convention that $[0]=0$ in F, while L is the submodule of F generated by all $[b_1+b_2]-(\sigma b_1)[b_2]-[b_1]$ for $b_1, b_2 \in B$. The module homomorphisms α and β of (6.2) may then be given by $\alpha a = [\varkappa a]+L$, $\beta([b]+L) = \sigma b - 1 \in Z(\Pi)$. Clearly $\beta\alpha=0$ and $\varepsilon\beta=0$, so the sequence $\chi(E)$ of (6.2) may be regarded as a complex of Π-modules. The exactness of this sequence is a consequence of

Lemma 6.1. *As a chain complex of abelian groups,* (6.2) *has a contracting homotopy.*

Proof. A contracting homotopy s would consist of homomorphisms $s\colon Z \to Z(\Pi)$, $s\colon Z(\Pi) \to M$, and $s\colon M \to A$ of abelian groups such that $\varepsilon s = 1_Z$, $\beta s + s\varepsilon = 1_{Z(\Pi)}$, $\alpha s + s\beta = 1_M$, and $s\alpha = 1_A$. The first condition is satisfied by setting $s1=1$, and the second by $sx = [u(x)]+L$, where $u(x) \in B$ is a representative of x in B with $\sigma u(x) = x$ and $u(1)=0$. For all x and b, $u(x)+b-u(x(\sigma b))$ is in the kernel of σ, so there are elements $h(x,b) \in A$ with

$$u(x)+b = \varkappa h(x,b) + u(x(\sigma b)).$$

A homomorphism $s\colon M \to A$ may be defined by $s(x[b]+L) = h(x,b)$. The proof is completed by showing that $\alpha s + s\beta = 1$, $s\alpha = 1$.

The given short exact sequence E of groups thus determines an exact sequence $\chi(E)$ of modules, hence an element of $\operatorname{Ext}^2_{Z(\Pi)}(Z,A)$, called the *characteristic class* of E. That the correspondence

$$\chi\colon \operatorname{Opext}(\Pi,A,\varphi) \to \operatorname{Ext}^2_{Z(\Pi)}(Z,A)$$

is an isomorphism will follow by composing it with the θ of (6.1) and applying

Theorem 6.2. *The composite correspondence*

$$\mathrm{Opext}(\Pi, A, \varphi) \xrightarrow{\chi} \mathrm{Ext}^2_{Z(\Pi)}(Z, A) \xrightarrow{\theta} H^2(\Pi, A) \qquad (6.3)$$

is the isomorphism which assigns to each E the cohomology class of one of its factor sets.

Sketch of proof. To apply the definition of θ we must find a chain transformation of the bar resolution, regarded as a free resolution of the trivial module Z, into the sequence $\chi(E)$, regarded also as a resolution of Z. Such a chain transformation

$$\begin{array}{ccccccccc} \cdots \to B_2(Z(\Pi)) & \to & B_1(Z(\Pi)) & \to & B_0(Z(\Pi)) & \to & Z & \to & 0 \\ \downarrow g_2 & & \downarrow g_1 & & \| & & \| & & \\ 0 \to A & \to & M & \to & Z(\Pi) & \to & Z & \to & 0 \end{array}$$

may be specified in terms of representatives $u(x)$ of x in B, with the usual factor set $f(x,y)$ for the $u(x)$, by the module homomorphisms

$$g_1[x] = [u(x)] + L, \quad g_2[x|y] = f(x,y). \qquad (6.4)$$

The cohomology class belonging to $\chi(E)$ is then the cohomology class of g_2, regarded as a cocycle on $B(Z(\Pi))$; that is, is the cohomology class of a factor set f for the extension E, as asserted in our theorem.

This construction may be reversed. The bar resolution provides a 2-fold exact sequence starting with ∂B_2 and ending in Z. Left multiplication of this sequence by the cocycle f produces the sequence $\chi(E)$.

Exercises

1. Show that α and β as defined for (6.2) are indeed module homomorphisms.
2. Complete the proof of Lemma 6.1, in particular showing that the function h there introduced satisfies $h(x, b_1 + b_2) = h(x(\sigma b_1), b_2) + h(x, b_1)$ and hence that $s: M \to A$ is well defined.
3. Express the function h in terms of the factor set f.
4. Verify that (6.4) gives a chain transformation as claimed.

7. Cohomology of Cyclic and Free Groups

Since $H^2(\Pi, A) \cong \mathrm{Ext}^2_{Z(\Pi)}(Z, A)$, we may calculate the cohomology of a particular group Π by using a Π-module resolution of Z suitably adapted to the structure of the group Π.

Let $\Pi = C_m(t)$ be the multiplicative cyclic group of order m with generator t. The group ring $\Gamma = Z(C_m(t))$ is the ring of all polynomials $u = \sum_{i=0}^{m-1} a_i t^i$ in t with integral coefficients a_i, taken modulo the relation $t^m = 1$. Two particular elements in Γ are

$$N = 1 + t + \cdots + t^{m-1}, \quad D = t - 1. \qquad (7.1)$$

Clearly $ND=0$, while, if $u=\sum a_i t^i$ is any element of Γ,

$$Nu=\sum_{j=0}^{m-1}\left(\sum_{i=0}^{m-1} a_i\right)t^j, \qquad Du=\sum_{j=1}^{m}(a_{j-1}-a_j)\,t^j, \qquad a_m=a_0.$$

If $Du=0$, then $a_0=a_1=\cdots=a_{m-1}$ and $u=Na_0$. If $Nu=0$, then $\sum a_i=0$ and $u=-D[a_0+(a_1+a_0)t+\cdots+(a_{m-1}+\cdots+a_0)t^{m-1}]$. This means that the sequence of Π-modules

$$\Gamma\xleftarrow{D_*}\Gamma\xleftarrow{N_*}\Gamma\xleftarrow{D_*}\Gamma, \qquad D_*u=Du, \qquad N_*u=Nu,$$

is exact. The augmentation $\varepsilon\colon \Gamma\to Z$ is $\varepsilon u=\sum a_i$, hence $\varepsilon u=0$ implies that $u=Dv$ for some v. All told, the long exact sequence

$$0\leftarrow Z\xleftarrow{\varepsilon}\Gamma\xleftarrow{D_*}\Gamma\xleftarrow{N_*}\Gamma\xleftarrow{D_*}\Gamma\leftarrow\cdots \tag{7.2}$$

thus provides a free resolution of Z. This resolution is customarily denoted by W, especially in algebraic topology, where it is of considerable use in calculating cohomology operations (STEENROD [1953]).

For any Π-module A the isomorphism $\operatorname{Hom}_\Pi(\Gamma, A)\cong A$ sends any $f\colon \Gamma\to A$ into $f(1)$. Hence the cochain complex $\operatorname{Hom}_\Pi(W, A)$, with the usual signs $\delta f=(-1)^{n+1}f\partial$ for the coboundary, becomes

$$A\xrightarrow{-D^*}A\xrightarrow{N^*}A\xrightarrow{-D^*}A\longrightarrow\cdots,$$

starting with dimension zero, where $N^*a=Na$, $D^*a=Da=(t-1)a$. The kernel of D^* is the subgroup $[a\,|\,ta=a]$ of all elements of A invariant under the action of $t\in C_m$, while the kernel of N^* is the subgroup of all a in A with $a+ta+\cdots+t^{m-1}a=0$. The cohomology groups of C_m are those of this cochain complex, hence

Theorem 7.1. *For a finite cyclic group C_m of order m and generator t and a C_m-module A, the cohomology groups are*

$$H^0(C_m,A)=[a\,|\,ta=a],$$
$$H^{2n}(C_m,A)=[a\,|\,ta=a]/N^*A, \qquad n>0,$$
$$H^{2n+1}(C_m,A)=[a\,|\,Na=0]/D^*A, \qquad n\geqq 0.$$

Note that these groups for $n>0$ repeat with period two.

Next we consider free groups.

Lemma 7.2. *If F is a free group on free generators e_i, for $i\in J$, then $Z^1_\varphi(F, A)$ is isomorphic to the cartesian product $\prod A_i$ of copies $A_i\cong A$ of A, under the correspondence which sends each crossed homomorphism f to the family $\{fe_i\}$ of its values on the generators.*

Proof. By definition, the *free group* F consists of 1 and the *words* $x=e_{i_1}^{\varepsilon_1}\ldots e_{i_k}^{\varepsilon_k}$ in the generators, with exponents $\varepsilon_j=\pm 1$. If we assume that the word is reduced (i.e., $\varepsilon_j+\varepsilon_{j+1}\neq 0$, when $i_j=i_{j+1}$), then this representation is unique. The product of two words is obtained by juxtaposition and subsequent cancellation. Now a crossed homomorphism f satisfies the equation $f(xy)=xf(y)+f(x)$ and hence also $f(1)=0$ and $f(x^{-1})=-x^{-1}f(x)$. Therefore f is completely determined by its values $f(e_i)=a_i\in A$ on the free generators e_i. Conversely, given constants a_i in A, we may set $f(e_i)=a_i$ and define $f(x)$ by induction on the length of the reduced word x by the formulas

$$f(e_i x)=e_i f(x)+a_i, \qquad f(e_i^{-1}x)=e_i^{-1}f(x)-e_i^{-1}a_i.$$

We verify that these formulas hold even when the word $e_i x$ or $e_i^{-1}x$ is not reduced, and hence that the f so defined is a crossed homomorphism. This completes the proof.

Consider now the exact sequence (2.4) of $Z(F)$-modules

$$0\to I(F)\xrightarrow{j} Z(F)\xrightarrow{\varepsilon} Z\to 0 \tag{7.3}$$

with p the crossed homomorphism from F to $I(F)$ given by $px=x-1$. By Prop. 2.3 the crossed homomorphisms f on F to A correspond one-one to the module homomorphisms $h\colon I(F)\to A$, indeed each h determines an $f=hp$. In particular $fe_i=hpe_i=h(e_i-1)$. Thus the lemma above states that the module homomorphisms h are determined in one-one fashion by their values on $e_i-1\in I(F)$. This means that $I(F)$ is a free F-module on the generators e_i-1. Hence (7.3) is a free resolution of the trivial F-module Z, and may thus be used to calculate the cohomology of F. Since this resolution is zero in dimensions beyond 1, we conclude

Theorem 7.3. *For a free group F, $H^n(F,A)=0$, for $n>1$.*

Exercises

1. Describe $H^1(F,A)$ for F free.
2. Without using crossed homomorphisms, prove $I(F)$ a free module.
3. Find a resolution for Z as a trivial module over the free abelian group Π on two generators, and calculate the cohomology of Π.
4. Determine the Yoneda products for the cohomology groups $H^k(C_m,Z)$, showing that

$$S^{2n}\colon 0\to Z\xrightarrow{N_*}\Gamma\xrightarrow{D_*}\Gamma\to\cdots\to\Gamma\xrightarrow{\varepsilon} Z\to 0$$

is an exact sequence with $2n$ intermediate terms Γ and maps alternately multiplication by N and by D, that, for $n>0$, $H^{2n}(C_m,Z)=\operatorname{Ext}^{2n}(Z,Z)=Z/mZ$ has an additive generator of order m given by the congruence class of the sequence S^{2n}, and that the composite $S^{2n}S^{2k}$ is $S^{2(n+k)}$.

5. If E_0 is the exact sequence $Z\rightarrowtail Z\twoheadrightarrow C_m$, where the map $Z\to Z$ is multiplication by m, show that the characteristic class $\chi(E_0)$ in the sense of §6 is the sequence S^2 of Ex. 4. Deduce that $\operatorname{Opext}(C_m,Z)$ is the cyclic group of order m generated by the extension E_0.

6. Let $\zeta: C_m \to C_k$ be a homomorphism of cyclic groups. For A a trivial Π-module in Thm. 7.1, calculate the induced map ζ^* in cohomology.

8. Obstructions to Extensions

The 3-dimensional cohomology groups appear in the study of extensions of a non-abelian group G. We write the composition in G as addition, even though G is not abelian.

For any element h of G, denote by $\mu(h)$ or μ_h the *inner automorphism* $\mu_h g = h + g - h$ given by conjugation with h. The map $\mu: G \to \operatorname{Aut} G$ is a homomorphism of the additive group G to the multiplicative group $\operatorname{Aut} G$ of all automorphisms of G; its image μG is the group $\operatorname{In} G$ of inner automorphisms of G. This image is a normal subgroup of $\operatorname{Aut} G$, for if $\eta \in \operatorname{Aut} G$, then always

$$\eta(\mu_h g) = \eta(h + g - h) = \eta h + \eta g - \eta h = \mu_{\eta h}(\eta g)$$

and hence

$$\eta \mu_h \eta^{-1} = \mu_{\eta h}, \quad \mu_h = \text{conjugation by } h. \tag{8.1}$$

The factor group $\operatorname{Aut} G/\operatorname{In} G$ is called the group of *automorphism classes* or of *outer automorphisms* of G; it is the cokernel of μ: $G \to \operatorname{Aut} G$. The kernel of μ is the *center* C of G; it consists of all $c \in G$ such that $c + g = g + c$ for all G. The sequence

$$0 \to C \to G \xrightarrow{\mu} \operatorname{Aut} G \to \operatorname{Aut} G/\operatorname{In} G \to 1 \tag{8.2}$$

is therefore exact.

Any group extension

$$E\colon 0 \to G \xrightarrow{\varkappa} B \xrightarrow{\sigma} \Pi \to 1$$

of G by Π determines, via conjugation in the additive group B, a homomorphism θ: $B \to \operatorname{Aut} G$ for which $\theta(\varkappa G) \subset \operatorname{In} G$. It hence determines an induced homomorphism ψ: $\Pi \to \operatorname{Aut} G/\operatorname{In} G$. In other words, for each $b \in B$ the automorphism $g \to b + g - b$ of G is in the automorphism class $\psi(\sigma b)$. We say that the extension E has *conjugation class* ψ: thus ψ records the fashion in which G appears as a normal subgroup of B. Conversely, call a pair of groups Π, G together with a homomorphism ψ: $\Pi \to \operatorname{Aut} G/\operatorname{In} G$ an *abstract kernel*. The general problem of group extensions is that of constructing all extensions E to a given abstract kernel (Π, G, ψ); that is, of constructing all short exact sequences E with given end groups G and Π and given conjugation class ψ. As in § 3, congruent extensions have the same conjugation class.

A given extension E may be described as follows. Identify each $g \in G$ with $\varkappa g \in B$. To each $x \in \Pi$ choose $u(x) \in B$ with $\sigma u(x) = x$, choosing in particular $u(1)=0$. Then conjugation by $u(x)$ yields an automorphism $\varphi(x)\in\psi(x)$ of G with

$$u(x)+g=[\varphi(x)g]+u(x), \qquad x\in\Pi,\ g\in G. \tag{8.3}$$

The sum $u(x)+u(y)$ equals $u(xy)$ up to a summand in G, which we may denote as $f(x,y)\in G$,

$$u(x)+u(y)=f(x,y)+u(xy), \qquad x,y\in\Pi. \tag{8.4}$$

The associativity law for $u(x)+u(y)+u(z)$ implies that

$$[\varphi(x)\,f(y,z)]+f(x,yz)=f(x,y)+f(xy,z). \tag{8.5}$$

If the group G containing the values of f were abelian, this identity would state that $\delta f=0$. Also, conjugation by the left and by the right side of (8.4) must yield the same automorphism of G; hence the identity

$$\varphi(x)\,\varphi(y)=\mu[f(x,y)]\,\varphi(xy), \tag{8.6}$$

which states that μf measures the extent to which φ deviates from a homomorphism $\varphi\colon\Pi\to\operatorname{Aut}G$.

Conversely, these conditions may be used to construct an extension as follows:

Lemma 8.1. *Given Π, G, and functions φ on Π to* Aut G, *f on $\Pi\times\Pi$ to G which satisfy the identities* (8.5) *and* (8.6) *and the (normalization) conditions $\varphi(1)=1$, $f(x,1)=0=f(1,y)$, the set $B_0[G,\varphi,f,\Pi]$ of all pairs (g,x) under the sum defined by*

$$(g,x)+(g_1,y)=(g+\varphi(x)\,g_1+f(x,y),xy) \tag{8.7}$$

is a group. With the homomorphisms $g\to(g,1)$ and $(g,x)\to x$, $G\rightarrowtail B_0\twoheadrightarrow\Pi$ is an extension of G by Π with conjugation class given by the automorphism class of φ.

Proof. A routine calculation shows that (8.5) and (8.6) yield the associative law. Because of the normalization condition, $(0,1)$ is the zero, while $(-f(x^{-1},x)-\varphi(x^{-1})\,g,\ x^{-1})$ is the negative of the element (g,x).

We call the group $B_0=[G,\varphi,f,\Pi]$ so constructed a *crossed product* group and the resulting extension a *crossed product extension*. Our analysis just before the previous lemma showed that any extension was isomorphic to such a crossed product, in the following explicit sense.

Lemma 8.2. *If $\varphi(x)\in\psi(x)$ has $\varphi(1)=1$, then any extension E of the abstract kernel (Π,G,ψ) is congruent to a crossed product extension $[G,\varphi,f,\Pi]$ with the given function φ.*

Proof. In the given extension E the representatives $u(x)$ can be chosen so that $g \to u(x)+g-u(x)$ is any automorphism in the automorphism class $\psi(x)$. Make this choice so that the automorphism is $\varphi(x)$. Each element of B then has a unique representation as $g+u(x)$, and the addition rules (8.3) and (8.4) yield a sum which corresponds under $g+u(x) \to (g, x)$ to that in the crossed product (8.7). This correspondence is a congruence. This proves the lemma.

Suppose now that just the abstract kernel (Π, G, ψ) is given. In each automorphism class $\psi(x)$ choose an automorphism $\varphi(x)$, taking care to pick $\varphi(1)=1$. Since ψ is a homomorphism into $\operatorname{Aut} G/\operatorname{In} G$, $\varphi(x)\varphi(y)\varphi(xy)^{-1}$ is an inner automorphism. To each $x, y \in \Pi$ choose an element $f(x,y)$ in G yielding this inner automorphism, in particular picking $f(x,1)=0=f(1,y)$; then $\varphi(x)\,\varphi(y)=\mu[f(x,y)]\,\varphi(xy)$. This is (8.6); we would like (8.5) to hold, but this need not be so. The associative law for $\varphi(x)\,\varphi(y)\,\varphi(z)$ shows only that (8.5) holds after μ is applied to both sides. The kernel of μ is the center C of G; hence there is for all x, y, z an element $k(x,y,z) \in C$ such that

$$[\varphi(x)\, f(y,z)]+f(x,yz)=k(x,y,z)+f(x,y)+f(xy,z)\,. \tag{8.5'}$$

Clearly $k(1,y,z)=k(x,1,z)=k(x,y,1)=0$, so that this function k may be regarded as a normalized 3-cochain of Π with coefficients in C.

The abelian group $C=\text{center}\,(G)$ may be regarded as a Π-module, for each automorphism $\varphi(x)$ of G carries C into C and yields for $c \in C$ an automorphism $c \to \varphi(x)\,c$ independent of the choice of $\varphi(x)$ in its class $\psi(x)$. We may thus write xc for $\varphi(x)\,c$.

We call the cochain k of (8.5′) an *obstruction* of the abstract kernel (Π, G, ψ). There are various obstructions to a given kernel, depending on the choice of $\varphi(x) \in \psi(x)$ and of f satisfying (8.6), but when there is an extension E we have shown in (8.5) that there is an obstruction $k=0$; hence

Lemma 8.3. *An abstract kernel (Π, G, ψ) has an extension if and only if one of its obstructions is the cochain identically* 0.

Next we prove

Lemma 8.4. *Any obstruction k of a kernel (Π, G, ψ) is a non-homogeneous 3-dimensional cocycle of $B(Z(\Pi))$.*

We must prove $\delta k=0$. This is plausible, for if only G were abelian and φ a homomorphism, the definition (8.5′) of k would read $k=\delta f$, hence would give $\delta k=\delta\delta f=0$. The proof consists in showing that $\delta\delta$ is still 0 in the non-abelian case. In detail, for x, y, z, t in Π we calculate the expression

$$L=\varphi(x)\,[\varphi(y)\, f(z,t)+f(y,zt)]+f(x,yzt)$$

in two ways. In the first way, apply (8.5′) to the inside terms beginning $\varphi(y)f(z,t)$; upon application of the homomorphism $\varphi(x)$ to the result there are terms $\varphi(x)f(y,z)$ and $\varphi(x)f(yz,t)$ to each of which (8.5′) may again be applied. When the terms k in the center are put in front, the result reads

$$L=[xk(y,z,t)]+k(x,yz,t)+k(x,y,z)+U, \tag{8.8}$$

where U is an abbreviation for the expression

$$U=f(x,y)+f(xy,z)+f(xyz,t).$$

In the second way of calculation, the automorphisms $\varphi(x)\,\varphi(y)$ on the first terms in the brackets in L may be rewritten by (8.6) to give

$$L=f(x,y)+\varphi(xy)\,f(z,t)-f(x,y)+\varphi(x)\,f(y,zt)+f(x,yzt).$$

Using (8.5′) on each term involving φ, the fact that all values of k lie in the center gives

$$L=k(xy,z,t)+k(x,y,zt)+U \tag{8.9}$$

with U as before. But the terms added to U in (8.8) and (8.9) are respectively the positive and the negative terms in $\delta k(x,y,z,t)$; hence comparison of (8.8) and (8.9) gives $\delta k=0$, q.e.d.

We now investigate the effect of different choices of φ and f in the construction of an obstruction to a given kernel.

Lemma 8.5. *For given $\varphi(x)\in\psi(x)$, a change in the choice of f in* (8.6) *replaces k by a cohomologous cocycle. By suitably changing the choice of f, k may be replaced by any cohomologous cocycle.*

Proof. Since the kernel of μ is the center C of G, any other choice of the function f in (8.6) must have the form

$$f'(x,y)=h(x,y)+f(x,y), \qquad h(x,1)=0=h(1,y) \tag{8.10}$$

where the function h has values in C, hence may be viewed as a 2-dimensional normalized cocycle of Π with values in C. Now the definition (8.5′) states essentially that the obstruction k is the coboundary $k=\delta f$. The obstruction k' of f' is thus $k'=\delta(h+f)$. The values of h lie in the center, so we may write $\delta(h+f)=(\delta h)+(\delta f)$; the new obstruction thus has the asserted form; since in (8.10) h may be chosen arbitrarily in C, we can indeed replace the obstruction k by any cohomologous cocycle.

Lemma 8.6. *A change in the choice of the automorphisms $\varphi(x)$ may be followed by a suitable new selection of f such as to leave the obstruction cocycle k unchanged.*

Proof. Let $\varphi(x)\in\psi(x)$ be replaced by automorphisms $\varphi'(x)\in\psi(x)$ with $\varphi'(1)=1$. Since $\varphi(x)$ and $\varphi'(x)$ lie in the same automorphism class there must be elements $g(x) \in G$ with $g(1)=0$ such that $\varphi'(x) = [\mu g(x)]\varphi(x)$. Using (8.1) and (8.6), calculate that

$$\varphi'(x)\,\varphi'(y) = \mu[g(x)+\varphi(x)\,g(y)+f(x,y)-g(xy)]\,\varphi'(xy).$$

As the new function $f'(x,y)$ we may then select the expression in brackets. We write this definition as

$$f'(x,y)+g(xy) = g(x)+\varphi(x)\,g(y)+f(x,y). \tag{8.11}$$

This definition has the form $f'=(\delta g)+f$, so that we should have $\delta f' = (\delta\delta g)+\delta f = \delta f$, modulo troubles with commutativity. If one, in fact, successively transforms the expression $\varphi'(x)\,f'(y,z)+f'(x,yz)+g(xyz)$ by (8.11) and (8.6) one obtains $k(x,y,z)+f'(x,y)+f'(xy,z)+g(xyz)$, which shows that the obstruction k is the same as before.

These results may be summarized as follows.

Theorem 8.7. *In any abstract kernel (Π, G, ψ), interpret the center C of G as a Π-module with operators $xc=\varphi(x)\,c$ for any choice of automorphisms $\varphi(x) \in \psi(x)$. The assignment to this kernel of the cohomology class of any one of its obstructions yields a well defined element* $\operatorname{Obs}(\Pi, G, \psi) \in H^3(\Pi, C)$. *The kernel (Π, G, ψ) has an extension if and only if* $\operatorname{Obs}(\Pi, G, \psi) = 0$.

Indeed, when the cohomology class of k is zero, any obstruction k has the form $k=\delta h$. By Lemma 8.5, there is a new choice f' for f which makes the obstruction identically zero; with this factor set f' the extension may be constructed as the crossed product $[G, \varphi, f', \Pi]$.

To complete the study of the extension problem we have the following result on the manifold of extensions.

Theorem 8.8. *If the abstract kernel (Π, G, ψ) has an extension, then the set of congruence classes of extensions is in 1-1 correspondence with the set $H^2(\Pi, C)$, where C is the center of G with module structure as in Thm. 8.7.*

We shall actually show more: That the group $H^2(\Pi, C)$ operates as a group of transformations on the set $\operatorname{Opext}(\Pi, G, \psi)$ and that this operation is simply transitive, in that from any one extension E_0 we obtain all congruence classes of extensions, each once, by operation with the elements of $H^2(\Pi, C)$.

Proof. Write any extension $E \in \operatorname{Opext}(\Pi, G, \psi)$ as a crossed product $[G, \varphi, f, \Pi]$. Hold φ fixed. Represent each element of $H^2(\Pi, C)$ by a factor set (2-cocycle) h. The required operation is $[G, \varphi, f, \Pi] \to [G, \varphi, h+f, \Pi]$. The stated properties of this operation follow. In particular, to show that any extension E' is so obtained from E, write E', as in Lemma 8.2, in the form of a crossed product $[G, \varphi, f', \Pi]$ with the same function φ.

Two applications of (8.6) give

$$\mu[f(x,y)]=\varphi(x)\,\varphi(y)\,\varphi(xy)^{-1}=\mu[f'(x,y)].$$

This states that $f(x,y)-f'(x,y)$ lies in the kernel of μ, that is, in the center of G. If h is defined as $h(x,y)=-f(x,y)+f'(x,y)$, then (8.5) for f and f' shows $\delta h=0$, hence h a cocycle with $f'=h+f$, as desired.

The operations of H^2 on Opext may also be defined in invariant terms, without using factor sets. Represent an element of $H^2(\Pi,C)$, according to Thm. 4.1, as an extension D of C by Π with the indicated operators. Let $C\times G$ be the cartesian product of the groups C and G. Define a "codiagonal" map ∇: $C\times G\to G$ by setting $\nabla(c,g)=c+g$; since C is the center of G, this is a homomorphism. The result of operating with D on an extension E in $\operatorname{Opext}(\Pi,G,\psi)$ may then be written as $\nabla(D\times E)\,\Delta_\Pi$. Exactly as in the case of the Baer sum (Ex. 4.7) this does yield an extension of G by Π with the operators ψ; if we calculate the factor set for this extension we find that it is given, just as above, by a map $f\to h+f$.

9. Realization of Obstructions

We have proved that the obstruction to an extension problem is an element of $H^3(\Pi,C)$. If $C=0$, the obstruction vanishes, hence the extension problem has a solution. The result is

Theorem 9.1. *If the (additive, non-abelian) group G has center* 0, *then any abstract kernel (Π,G,ψ) has an extension.*

This simple result is worth a direct proof. Since G is centerless, $G\rightarrowtail \operatorname{Aut}G\twoheadrightarrow \operatorname{Aut}G/\operatorname{In}G$ is an extension E_0; the induced extension $E_0\psi$ of Ex. 4.1 is the desired extension of G by Π with operators ψ.

In other cases the extension problem may not have a solution. By § 7 there are cases (e.g., with Π finite cyclic) where $H^3(\Pi,C)\neq 0$; the obstruction theory above then produces abstract kernels with no extension, provided that we know that every 3-cocycle can be realized as an obstruction. This fact, which is also of interest in showing that the cohomology of groups "fits" the extension problem, may be stated as follows.

Theorem 9.2. *Given Π not cyclic of order* 2, *a Π-module C, and any cohomology class $\bar{k}$ of $H^3(\Pi,C)$ there exists a group G with center C and a homomorphism ψ: $\Pi\to\operatorname{Aut}G/\operatorname{In}G$ inducing the given Π-module structure on C and such that* $\operatorname{Obs}(\Pi,G,\psi)=\bar{k}$.

The theorem is true for all Π (cf. Eilenberg-Mac Lane [1947]); a special proof is required when Π is cyclic of order 2.

The proof is obtained by reversing the considerations leading to the definition of the obstruction in such a way as to construct a "free" kernel with a given 3-cocycle k as obstruction.

Take $G=C\times F$, where C is the given Π-module and F is the free (non-abelian) group with generators all symbols $[x,y]$ for $x\neq 1$ and $y\neq 1$ in Π. Write the composition in F and G as addition. Define a function f on $\Pi\times\Pi$ to $G\supset F$ by $f(x,1)=f(1,y)=0$, and $f(x,y)=[x,y]$ for $x\neq 1\neq y$. For each $x\in\Pi$ define an endomorphism $\beta(x)\colon G\to G$ by setting $\beta(x)\,c=xc$ (under the module structure of C) and

$$\beta(x)\,[y,z]=k(x,y,z)+f(x,y)+f(xy,z)-f(x,yz) \tag{9.1}$$

for each generator $[y,z]$ of F. Since k is normalized (i.e., $k(x,y,1)=k(x,1,z)=k(1,y,z)=0$ always), this equation also holds with $[y,z]$ replaced by $f(y,z)$; that is, with y or $z=1$. The equation thus asserts that $k=\delta f$, in the same "non-abelian" sense as in the definition (8.5') of the obstruction.

By this definition, $\beta(1)$ is the identity automorphism. We now assert that always

$$\beta(x)\,\beta(y)=\mu[f(x,y)]\,\beta(xy)\colon\ G\to G. \tag{9.2}$$

Both sides have the same effect on an element c of the Π-module C; hence it suffices to prove that the endomorphism on each side of (9.2) has the same effect on any one of the generators $[z,t]$ of F. First calculate $\beta(x)\,\beta(y)\,[z,t]$ by repeated applications of the definition (9.1), once for $\beta(y)$ and three times for $\beta(x)$. The terms in k all lie in C, which is surely contained in the center of G, so can be collected. These terms in k include all the terms of $\delta k(x,y,z,t)$ except for the term $-k(xy,z,t)$. Since $\delta k=0$, we can replace the terms in k by the one term $k(xy,z,t)$. The result is

$$\begin{aligned}\beta(x)\beta(y)\,[z,t]&=f(x,y)+k(xy,z,t)+f(xy,z)+f(xyz,t)-f(xy,zt)-f(x,y)\\&=f(x,y)+\beta(xy)\,[z,t]-f(x,y)\\&=\mu[f(x,y)]\,\beta(xy)\,[z,t].\end{aligned}$$

This proves (9.2).

We claim that each $\beta(x)$ is an automorphism of G. Indeed, (9.2) proves that $\beta(x)\,\beta(x^{-1})=\mu[f(x,x^{-1})]\,\beta(1)=\mu[f(x,x^{-1})]$ is an inner automorphism. Hence $\beta(x^{-1})$ has kernel 0 and $\beta(x)$ has image G. Since x is arbitrary, this gives the result.

Denote by $\psi(x)$ the automorphism class containing $\beta(x)$. By (9.2), ψ is a homomorphism $\psi\colon \Pi\to\operatorname{Aut}G/\operatorname{In}G$, hence (Π,G,ψ) is an abstract kernel. Since Π is not cyclic of order 2, we can assume that Π contains more than two elements. The free group F then has more than one generator, hence is centerless, so that C is exactly the center of $G = C \rtimes F$. Our construction has been designed precisely to yield the given cocycle k as the obstruction of this kernel, hence the theorem.

A many-one correspondence of abstract kernels with center C to the group $H^3(\Pi, C)$ has now been established. This correspondence can be so decorated as to become a group isomorphism; one first defines a relation of similarity between abstract kernels such that two kernels are similar if and only if they have the same obstruction; with a suitable product of kernels the group of similarity classes of kernels (Π, G, ψ) with fixed Π and fixed Π-module C as center is then isomorphic to $H^3(\Pi, C)$. The details are given in EILENBERG-MAC LANE [1947].

No reasonable analogous interpretation of $H^4(\Pi, C)$ or of higher dimensional cohomology groups is known.

10. Schur's Theorem

We now apply factor sets to a problem in group theory.

For any set S the collection $\mathrm{Aut}\, S$ of all 1-1 mappings of S onto itself is a group under composition. A (multiplicative) group G is said to *act* on the set S if a homomorphism $\mu: G \to \mathrm{Aut}\, S$ is given. Equivalently, to each $g \in G$ and to each "point" $s \in S$ there is given a unique point $gs = \mu(g)s \in S$ so that always $(g_1 g_2)s = g_1(g_2 s)$ and $1s = s$. The *orbit* of a point $s_0 \in S$ under the action of G is the set of all gs_0 for $g \in G$; any other point in this subset has the same orbit. The whole set S is the union of disjoint orbits. The set of all $h \in G$ such that $hs_0 = s_0$ is a subgroup H of G called the group *fixing* s_0. The correspondence $gH \to gs_0$ is a 1-1 mapping of the left cosets of H in G onto the orbit of s_0. By definition, the number of such cosets is the index $[G:H]$; if the index is finite, it is therefore the number of points in the orbit. Thus when a finite group G acts on a set S, the number of points in each orbit is a divisor of the order of G.

Take S to be the set of all subgroups U of a given group G. The correspondence $U \to gUg^{-1}$ defines an action of G on S; one says that G *acts* on S by *conjugation*. Similarly G (or any subgroup of G) acts by conjugation on the set of elements of G.

Theorem 10.1. *(Cauchy's Theorem.) If the order n of a finite group G is divisible by a prime number p, then G contains an element of order p.*

The proof is by induction on n. Let G act on itself by conjugation. The orbit of an element c consists of c alone when always $gcg^{-1} = c$; that is, precisely when c is in the center C of G. Let m denote the order of C and $k_i > 1$, the number of points in the i-th orbit outside C, $i = 1, \ldots, t$. Since G is the union of disjoint orbits, $n = m + k_1 + \cdots + k_t$.

If m is divisible by p, write the abelian group C as a direct sum of cyclic groups; one of these summands then has order divisible by p, hence contains an element of order p. Otherwise m is prime to p, so also at least one of the integers k_i. But k_i is the number of points in some orbit, hence equals the index $[G:H]$ of some subgroup. Since p does not divide k_i it must divide the order of the subgroup H. By the induction assumption, H contains an element of order p.

A p-group is a group in which every element has order some power of the prime p. By Cauchy's Theorem, a finite p-group may also be described as a group of order some power of p.

Theorem 10.2. *Any finite p-group $\neq 1$ has a center $C \neq 1$.*

Proof. Let the p-group act on itself by conjugation. Each orbit consists of p^{m_i} points for some exponent $m_i \geqq 0$; together the orbits exhaust the p^n elements of the group. Since the orbit of 1 consists of itself only, $p^n = 1 + \sum p^{m_i}$. Therefore at least $p-1$ other orbits consist of one element c only. These elements lie in the center C, so $C \neq 1$.

A *maximal p-subgroup* of G is a p-group $P \subset G$ which is contained in no larger p-subgroup of G. By Cauchy's Theorem, a finite group of order n has at least one maximal p-subgroup $\neq 1$ for each prime p which divides n.

A subgroup U of G is said to *normalize* a subgroup V if $uVu^{-1} = V$ for all $u \in U$; that is, if V is a one-point orbit under the action of U on subgroups of G.

Lemma 10.3. *If P and Q are maximal p-subgroups of G such that P normalizes Q, then $P = Q$.*

Proof. Let PQ denote the subgroup of G generated by P and Q. Since P normalizes Q, Q is a normal subgroup of PQ. Since P is a p-group, so is its quotient $P/P \cap Q \cong PQ/Q$. Thus PQ is an extension of the p-group Q by the p-group $P/P \cap Q$, hence is itself a p-group. Since P is contained in no larger p-subgroup, $P = PQ$, so $P \supset Q$. Since Q is contained in no larger p-subgroup, $P = Q$.

Any conjugate of a maximal p-subgroup is itself a maximal p-subgroup. Moreover

Theorem 10.4. *Any two maximal p-subgroups of a finite group are conjugate.*

Proof. Let S be the set of all conjugates in G of some maximal p-subgroup P, and let P act on S by conjugation. By the lemma, a point $P' \in S$ is a one-point orbit exactly when $P' = P$. The number of points in any other orbit is the index of a subgroup of P, hence is divisible by p. Therefore the number of points in S is congruent to 1, modulo p.

Any maximal p-subgroup Q of G acts by conjugation on S; under this action each orbit again has either one point or a number of points divisible by p. The congruence above thus shows that there is a one-point orbit P'. In other words, Q normalizes some conjugate P' of P, so, by the lemma, $Q=P'$ and is itself a conjugate of P.

Theorem 10.5. (SCHUR-ZASSENHAUS) *If the integers m and n are relatively prime, any extension of a group of order m by one of order n splits.*

Proof. Let $G \rightarrowtail B \overset{\sigma}{\twoheadrightarrow} \Pi$ be such an extension, with G of order m and Π of order n. This extension *splits* if σ has a right inverse; that is, if B contains a subgroup (also of order n) mapped by σ isomorphically on Π.

Suppose first that G is abelian. The given extension is then an element $e \in H^2(\Pi, G)$. By Prop. 5.3, $ne=0$; trivially, $me=0$. Since m and n are relatively prime, $e=0$; the extension splits.

For G not abelian the proof will be by induction on the order m of G. It suffices to prove that the extension B contains a subgroup of order n, for such a subgroup is mapped by $B \to \Pi$ isomorphically upon Π.

Take a prime p dividing m and a maximal p-subgroup P of B. The normalizer N of P in B is defined to be the set of all b with $bPb^{-1}=P$. The index $[B:N]$ is then the number of conjugates of P in B. All these conjugates must lie in G and are maximal p-subgroups there. By Thm. 10.4 they are all conjugate in G. Now $G \cap N$ is the normalizer of P in G, so the index $[G: G\cap N]$ is the number of these conjugates and is therefore equal to $[B:N]$. This index equality (see the diagram) proves also that $n=[B:G]=[N: G\cap N]$. Now P and $G\cap N$ are normal subgroups of N, and N/P is an extension of the group $(G\cap N)/P$, of order some proper divisor of m, by the group $N/G\cap N$ of order n. By the induction assumption, N/P thus contains a subgroup of order n, which may be written in the form H/P for some H with $P \subset H \subset N$ and $[H:P]=n$.

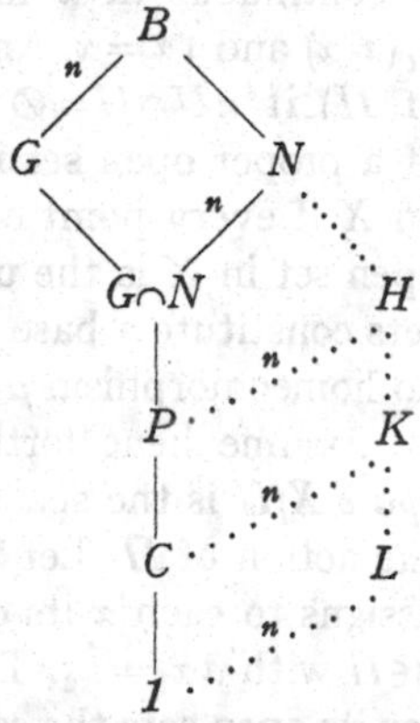

The center C of P is, by Thm. 10.2, not 1. Conjugation by elements of $H \subset N$ maps P onto itself and hence C onto itself, so that C and P are normal in H. Thus H/C is an extension of the p-group P/C by the group H/P of order n prime to p. Since $C \neq 1$, the order of P/C is less than m, so the induction assumption provides a subgroup $K/C \subset H/C$ of order n. This group K is an extension of the abelian p-subgroup C

by a group K/C of order n, hence splits by the abelian case already reated. This splitting provides a subgroup $L \subset K$ of order n, and the subgroup L splits the original extension B.

Exercises

1. (The first Sylow Theorem.) If the order of a finite group G is divisible by p^k, p a prime, then G contains a subgroup of order p^k.

2. If p^n is the highest power of p dividing the order of G, every maximal p-subgroup of G has order p^n.

3. If the order of the finite group Π is prime to the order of the finite abelian group A, prove that $H^n(\Pi, A) = 0$ for $n > 0$ and any Π-module structure on A.

4. Let $\sigma: B \to \Pi$ be an extension of an abelian group G of order m by Π of order n, with $(n, m) = 1$, as in the Schur-Zassenhaus Theorem. If S and T are two subgroups of B isomorphic to Π under σ, show S and T conjugate under conjugation by an element of G (use $H^1(\Pi, G) = 0$).

11. Spaces with Operators

The geometrical meaning of the cohomology groups of a group will now be illustrated by an examination of spaces with operators.

For any topological space X, let $\mathrm{Aut}(X)$ denote the group of all homeomorphisms of X with itself. A group Π *operates* on the space X if a homomorphism $\mu: \Pi \to \mathrm{Aut}(X)$ is given. Equivalently, to each $a \in \Pi$ and each $x \in X$ a unique point $ax = \mu(a)x \in X$ is given such that ax is continuous in x for each fixed a and such that always $(a_1 a_2)x = a_1(a_2 x)$ and $1x = x$. An open set U in X is called *proper* (under the action of Π) if $aU \cap U = \emptyset$ (the null set) whenever $a \neq 1$. Any open subset of a proper open set is proper. The group Π is said to *operate properly* on X if every point of X is contained in a proper open set; then every open set in X is the union of proper open sets, so that the proper open sets constitute a base for the topology of X. When Π operates properly, no homeomorphism $\mu(a)$ with $a \neq 1$ can leave a point x fixed.

Assume henceforth that Π operates properly on X. The *quotient space* X/Π is the space whose points are the orbits of points of X under the action of Π. Let the projection $p: X \to X/\Pi$ be the function which assigns to each x its orbit px. Thus $px_1 = px_2$ if and only if there is an $a \in \Pi$ with $ax_1 = x_2$. The topology of X/Π is defined by taking as a base for its open sets the sets pU, where U is a proper open set of X under Π; these sets $V = pU$ are called *proper* in X/Π.

Proposition 11.1. *The map $p: X \to X/\Pi$ is continuous. The space X/Π is covered by proper open sets V; each $p^{-1}V$ is the union of disjoint open sets U_α such that each restriction $p|U_\alpha$ is a homeomorphism $U_\alpha \cong V$.*

This proposition asserts that X is a "covering space" for X/Π under the map p. The U_α are the *sheets* of X over V.

Proof. If U is proper and $V=pU$, then $p^{-1}V$ is the union of the sets aU for $a\in\Pi$. These sets are disjoint by the assumption that U is proper. Each aU is mapped by p onto V, and the open sets aU are all proper, and map onto proper open sets in X/Π, so that $p|aU$ is indeed a homeomorphism.

For example, let X be the real line E^1 and Π the infinite (multiplicative) cyclic group with generator c, acting on E^1 by the rule $c^k x = x+k$ for any integer k. Then open intervals of length less than 1 on the line are proper, so Π acts properly. The quotient E^1/Π is homeomorphic to the unit circle S^1. If we identify E^1/Π with S^1, $p: E^1\to S^1$ becomes the map $px=e^{2\pi i x}$ which wraps the line E^1 around the circle S^1. Similarly, the free abelian group on two generators b and c acts properly on the Euclidean plane E^2 by $b^k c^l(x,y)=(x+k, y+l)$; here b is horizontal translation and c vertical translation, each by one unit. The quotient space E^2/Π is the 2-dimensional torus $S^1\times S^1$. Again, the cyclic group of order 2 operates properly on the 2-sphere S^2 by mapping each point into its diametrically opposite point, and S^2/Π is the real projective plane. In each of these cases X is the "universal covering space" of X/Π, and Π is the "fundamental group" of X/Π [Hu 1959].

Now consider the singular homology of X, as defined in Chap. II.

Lemma 11.2. *If the group Π operates properly on X, then the singular complex $S(X)$ is a complex of free Π-modules.*

Proof. The group $S_n(X)$ of n-chains is the free abelian group generated by the singular n-simplices $T: \Delta^n\to X$. For each $a\in\Pi$ the composite aT is also a singular n-simplex; the operators $T\to aT$ make $S_n(X)$ a Π-module. If d_iT denotes the i-th face of T, then $a(d_iT)=d_i(aT)$, hence $\partial=\sum(-1)^i d_i: S_n\to S_{n-1}$ is a Π-module homomorphism. Thus $S(X)$ is a complex of Π-modules. To show $S_n(X)$ free, pick any subset $X_0\subset X$ (a "fundamental domain") containing exactly one point from each orbit of X under Π. Then those singular n-simplices T with initial vertex in X_0 constitute a set of free generators for $S_n(X)$ as a module.

Lemma 11.3. *If the group Π operates properly on the space X, any $T: \Delta^n\to X/\Pi$ can be written as $T=pT'$ for some $T': \Delta^n\to X$. With suitable choice of one T' for each T, these T' are free generators of $S_n(X)$ as a Π-module.*

We say that T can be lifted to T'; the possibility of such a lifting is actually a consequence of a more general fact on the lifting of maps in a covering space.

Proof. If T is "small" in the sense that $T(\Delta^n)$ is contained in a proper open subset V of X/Π, and if U is any sheet over V then T can be lifted to $T'=(p|U)^{-1}T$ in U. The general case can then be handled by subdividing Δ^n into small pieces, lifting T in succession on these pieces. It is technically easier to do this by replacing Δ^n by the n-cube $I^n = I \times \cdots \times I$ (n factors), where I is the unit interval. Since Δ^n is homeomorphic to I^n, it will suffice to lift any $T: I^n \to X/\Pi$. The cube I^n is covered by the inverse images $T^{-1}(V)$ of proper open sets of X/Π. Since I^n is a compact metric space, the Lebesgue Lemma provides a real $\varepsilon > 0$ such that any subset of diameter less than ε lies in one of the $T^{-1}(V)$. Now subdivide I^n into congruent n-cubes, with sides parallel to the axes, each of diameter less than ε. Then T can be lifted in succession on the cubes of this subdivision, beginning with the cubes on the bottom layer. When we come to lift T on any one cube, the continuous lifting T' will already be defined on a certain connected set of faces of this cube and will lie in one sheet U over some proper V; the rest of the cube is then lifted by $(p|U)^{-1}$. This completes the proof.

Proposition 11.4. *If Π operates properly on X, while the abelian group A has the trivial Π-module structure, then $p: X \to X/\Pi$ induces an isomorphism $p^*: \operatorname{Hom}_Z(S(X/\Pi), A) \cong \operatorname{Hom}_\Pi(S(X), A)$ of chain complexes and hence an isomorphism*

$$p^*: H^n(X/\Pi, A) \cong H^n(\operatorname{Hom}_\Pi(S(X), A)). \tag{11.1}$$

Proof. A cochain $f: S_n(X/\Pi) \to A$ is uniquely determined by its values on the n-simplices T of X/Π, while a cochain f' of $S(X)$, as a module homomorphism $f': S_n(X) \to A$, is uniquely determined by its values on the free module generators T' of $S_n(X)$. Since these generators are in 1-1 correspondence $T' \to pT'$ by Lemma 11.3 and since $(p^* f)\, T' = f(pT')$, the result follows.

More generally, when A is any Π-module, the cohomology of $\operatorname{Hom}_\Pi(S(X), A)$ is known as the *equivariant cohomology* of X with coefficients A; in this general circumstance the theorem would still hold if $H^n(X/\Pi, A)$ were interpreted as the cohomology of X/Π with "local coefficients" A, defined as in Eilenberg [1947] and Eilenberg-Mac Lane [1949]. The main result now is

Theorem 11.5. *If a group Π operates properly on an acyclic space X and if A is an abelian group with trivial Π-module structure, there is an isomorphism*

$$H^n(X/\Pi, A) \cong H^n(\Pi, A), \quad n = 0, 1, \ldots, \tag{11.2}$$

natural in A, between the cohomology groups of the quotient space X/Π and those of the group Π.

Proof. The hypothesis that X is acyclic means that $H_n(S(X)) = 0$ for $n > 0$ and $H_0(S(X)) \cong Z$. This latter isomorphism yields an epimorphism $S_0(X) \to Z$ with kernel $\partial S_1(X)$. Thus the exact sequence of Π-modules

$$\cdots \to S_1(X) \to S_0(X) \to Z \to 0$$

is a free resolution of the trivial module Z. Hence the equivariant cohomology of $S(X)$ is $\mathrm{Ext}^n_{Z(\Pi)}(Z, A)$, which is $H^n(\Pi, A)$ by Cor. 5.2.

The system (X, Π) consisting of a topological space with a proper group of operators Π may be regarded as an object in a category where the morphisms ϱ: $(X, \Pi) \to (X', \Pi')$ are pairs $\varrho = (\xi, \gamma)$, with $\xi: X \to X'$ a continuous map and $\gamma: \Pi \to \Pi'$ a group homomorphism such that always $\xi(a x) = (\gamma a)\, \xi x$ for $a \in \Pi$. The isomorphism (11.2) is natural for these maps.

This theorem provides a geometric interpretation of all the cohomology groups of a group Π. Assuming some concepts from homotopy theory, let Y be a pathwise connected topological space with fundamental group $\Pi = \pi_1(Y)$. Then, if Y has suitable local connectivity, one can construct its universal covering space; this is a space upon which Π properly operates in such wise that Y is homeomorphic to X/Π. Suppose, in particular, that Y is aspherical (all higher homotopy groups vanish). One can then prove that the universal covering space X is acyclic. Thm. 11.5 thus applies to show that the cohomology of the aspherical space Y is, in fact, isomorphic to the cohomology of the fundamental group of Y.

Notes. The fact that the cohomology of an aspherical space Y depends only on the fundamental group was proved by Hurewicz [1935], while the expression of this dependence via the cohomology of groups was discovered by Eilenberg-Mac Lane [1943, 1945b], and later but independently by Eckmann [1945–1946]. There is a corresponding result expressing the homology of Y by the homology of Π, found by Hopf [1945] and independently by Freudenthal [1946]. All these investigations were stimulated by the prior study of Hopf [1942] on the influence of the fundamental group on the second homology group of any space. This line of investigation provided the justification for the study of cohomology of groups in all dimensions and was the starting point of homological algebra. The 1-dimensional cohomology groups (crossed homomorphisms) had been long known; the 2-dimensional cohomology groups, in the guise of factor sets, had appeared long since in the study of group extensions by Schreier [1926], Baer [1934, 1935], Hall [1938], and Fitting [1938]. Earlier, Schur had considered projective representations ϱ of a group Π. Each ϱ is a homomorphism of Π to the group of projective collineations of complex projective n-space, hence may be represented by a set of $(n+1) \times (n+1)$ non-singular complex matrices A_x for $x \in \Pi$ with $A_x A_y = f(x,y)\, A_{xy}$, where $f(x,y)$ is a non-zero complex number. This f is a factor set for Π in the multiplicative group C^* of non-zero complex numbers. Hence Schur's "multiplicator", which is the cohomology group $H^2(\Pi, C^*)$, with trivial Π-module structure for C^*. (For recent literature Asano-Shoda [1935], Frucht [1955], Kochendörffer [1956].) Projective representations of infinite groups have been studied by Mackey [1958]. The 3-dimensional cohomology groups of a group were first considered by Teichmüller [1940] in a study of simple algebras over a number field. The cohomology of groups has been applied extensively in class-field theory: HOCHSCHILD [1950], TATE [1952], ARTIN-TATE [1960].

Exercises

1. Show that a set V is open in X/Π if and only if $p^{-1}V$ is open in X. (This asserts that X/Π has the standard "quotient space" topology.)

2. Construct an explicit homeomorphism of Δ^n with I^n.

Chapter five

Tensor and Torsion Products

1. Tensor Products

Let G be a right R-module and A a left R-module — a situation we may indicate as G_R, $_RA$. Their *tensor product* $G \otimes_R A$ is the abelian group generated by the symbols $g \otimes a$ for $g \in G$ and $a \in A$ subject to the relations

$$(g+g') \otimes a = g \otimes a + g' \otimes a, \quad g \otimes (a+a') = g \otimes a + g \otimes a', \quad (1.1)$$

$$gr \otimes a = g \otimes ra, \quad a \in A, \ r \in R, \ g \in G. \quad (1.2)$$

More formally, this statement describes $G \otimes_R A$ as a factor group $(G \bigcirc A)/S$, where $G \bigcirc A$ is the free abelian group with generators all symbols $g \bigcirc a$, while S is the subgroup of $G \bigcirc A$ generated by all elements $(g+g') \bigcirc a - g \bigcirc a - g' \bigcirc a$, $g \bigcirc (a+a') - g \bigcirc a - g \bigcirc a'$, and $gr \bigcirc a - g \bigcirc ra$. Then $g \otimes a$ denotes the coset $(g \bigcirc a) + S$ in $(G \bigcirc A)/S$.

The intention is that $G \otimes_R A$ be a group in which an element of G can be "multiplied" by an element of A to give a "product" $g \otimes a$; one wishes the product to be distributive, as assured by (1.1), and associative, as in (1.2). More formally, let $G \times A$ be the cartesian product of the sets G and A, while M is any abelian group. Call a function f on $G \times A$ to M *biadditive* if always

$$f(g+g', a) = f(g, a) + f(g', a), \quad f(g, a+a') = f(g, a) + f(g, a') \quad (1.3)$$

and *middle associative* if always

$$f(gr, a) = f(g, ra). \quad (1.4)$$

If f satisfies both conditions, call f *middle linear*. Now $g \otimes a$ is middle linear by definition, and $G \otimes_R A$ is the universal range for any middle linear f, in the following sense.

Theorem 1.1. *Given modules G_R and $_RA$ and a middle linear function f on $G \times A$ to an abelian group M, there is a unique homomorphism $\omega: G \otimes_R A \to M$ of abelian groups with $\omega(g \otimes a) = f(g, a)$.*

Proof. The formula $\omega(g\otimes a)=f(g,a)$ defines ω on the generators of $G\otimes_R A$; the assumption that f is middle linear implies that ω "preserves" the relations (1.1) and (1.2) defining $G\otimes_R A$; hence ω is a homomorphism; it is manifestly the only such. This proof is a shorter statement of the following argument: Since $G\bigcirc A$ is free abelian on the generators $g\bigcirc a$, there is a unique homomorphism ω': $G\bigcirc A\to M$ with $\omega'(g\bigcirc a)=f(g,a)$. The assumptions on f show that ω' maps the subgroup S above into zero; hence ω' factors as $G\bigcirc A\to(G\bigcirc A)/S\to M$; the second factor is the desired ω.

This theorem has a variety of uses. First, it gives a universal property of $M_0=G\otimes_R A$ which characterizes this group and the middle linear function $\otimes$: $G\times A\to M_0$ uniquely (up to an isomorphism of M_0). Hence the theorem may be taken as a conceptual definition of the tensor product. Next, the theorem states that every middle linear f can be obtained from one such function $\otimes$ followed by a group homomorphism ω; in this sense, the theorem reduces middle linear functions to homomorphisms. Finally, the theorem states that a homomorphism ω with domain the tensor product $G\otimes_R A$ is uniquely defined by giving the images of the symbols $g\otimes a$ under ω, provided only that these images are additive in g and a and middle associative in R. This last version we shall use repeatedly to construct maps ω.

For example, if γ: $G_R\to G'_R$ and α: ${}_RA\to{}_RA'$ are R-module homomorphisms, then in $G'\otimes_R A'$ we can form the expression $\gamma g\otimes\alpha a$, which is middle associative and additive in $g\in G$ and $a\in A$. Hence there is a homomorphism $\gamma\otimes\alpha$: $G\otimes_R A\to G'\otimes_R A'$ with $(\gamma\otimes\alpha)(g\otimes a)=\gamma g\otimes\alpha a$. Clearly $1_G\otimes 1_A=1$, and, for matching maps, $\gamma\gamma'\otimes\alpha\alpha'=(\gamma\otimes\alpha)(\gamma'\otimes\alpha')$; hence $G\otimes_R A$ is a covariant bifunctor of A and G. Moreover

$$\gamma\otimes(\alpha+\beta)=\gamma\otimes\alpha+\gamma\otimes\beta,\qquad (\gamma_1+\gamma_2)\otimes\alpha=\gamma_1\otimes\alpha+\gamma_2\otimes\alpha. \tag{1.5}$$

These identities can be applied to a direct sum diagram to give an isomorphism

$$\zeta\colon G\otimes_R(A\oplus B)\cong(G\otimes_R A)\oplus(G\otimes_R B). \tag{1.6}$$

Alternatively, since $(g\otimes a,\ g\otimes b)$ is middle linear as a function of g and (a,b), we can construct ζ directly by Thm. 1.1 as that homomorphism ζ: $G\otimes_R(A\oplus B)\to(G\otimes_R A)\oplus(G\otimes_R B)$ with $\zeta[g\otimes(a,b)]=(g\otimes a,\ g\otimes b)$; ζ^{-1} may also be constructed from $g\otimes a\to g\otimes(a,0)$ and $g\otimes b\to g\otimes(0,b)$.

The ring R may be considered as either a left or a right module over itself. For modules G_R and ${}_RA$ one has isomorphisms (of abelian groups)

$$G\otimes_R R\cong G,\qquad R\otimes_R A\cong A, \tag{1.7}$$

given by $g\otimes r\to gr$, $r\otimes a\to ra$.

If $\varrho: S \to R$ is a homomorphism of rings, each right R-module G becomes a right S-module G_ϱ when the operators are defined by $gs = g(\varrho s)$. Similarly, each left R-module A becomes a left S-module ${}_\varrho A$; this is "pull-back along ϱ", as in III.6. If $\varrho': T \to S$ is a second ring homomorphism, $G_{(\varrho\varrho')} = (G_\varrho)_{\varrho'}$, while ${}_{(\varrho\varrho')}A = {}_{\varrho'}({}_\varrho A)$, in opposite order.

Lemma 1.2. *(The Pull-back Lemma.) For a ring homomorphism* $\varrho: S \to R$ *and modules* G_R, ${}_R A$, ${}_R C$ *there are natural homomorphisms*

$$\varrho_\#: (G_\varrho) \otimes_S ({}_\varrho A) \to G \otimes_R A, \qquad \varrho^\#: \operatorname{Hom}_R(C, A) \to \operatorname{Hom}_S({}_\varrho C, {}_\varrho A). \qquad (1.8)$$

If ϱ *is an epimorphism, both* $\varrho_\#$ *and* $\varrho^\#$ *are isomorphisms.*

Proof. For $g \in G$, $a \in A$, and $s \in S$,

$$gs \otimes_R a = g\varrho(s) \otimes_R a = g \otimes_R \varrho(s)\, a = g \otimes_R sa,$$

so $g \otimes_R a$ is S-middle associative. Thus $\varrho_\#(g \otimes_S a) = g \otimes_R a$ determines a homomorphism, by Thm. 1.1. If $\varrho(R) = S$, $\varrho_\#$ has an inverse $g \otimes_R a \to g \otimes_S a$. Similarly each R-module homomorphism $f: C \to A$ is an S-module homomorphism, and conversely if $\varrho(S) = R$.

We normally write the modules G_ϱ, ${}_\varrho A$ without the subscript ϱ when this is indicated by the context $G \otimes_S A$.

An abelian group A is a module over the ring Z of integers, so our definition of tensor product includes that of the tensor product $G \otimes A$ of two abelian groups (here $\otimes$ is short for $\otimes_Z$). In this case, any bi-additive function $f(g, a)$ is automatically middle associative, for, with m any positive integer,

$$\begin{aligned} f(mg, a) &= f(g + \cdots + g, a) = f(g, a) + \cdots + f(g, a) \\ &= f(g, a + \cdots + a) = f(g, ma). \end{aligned}$$

This holds also for negative m, since $f(-g, a) = -f(g, a) = f(g, -a)$. Hence the middle associativity condition (1.2) may be omitted in defining $\otimes_Z$.

Tensor products of finite abelian groups can be explicitly computed. For each positive integer m, let $Z_m(g_0)$ be the cyclic group with generator g_0 of order m, while mA denotes that subgroup of A which consists of all multiples ma, $a \in A$. We claim that an isomorphism

$$\eta: A/mA \cong Z_m(g_0) \otimes A \qquad (1.9)$$

is given by setting $\eta(a + mA) = g_0 \otimes a$. Indeed, since $g_0 \otimes ma = mg_0 \otimes a = 0$, the product $g_0 \otimes a$ depends only on the coset of a, modulo mA, hence η is a homomorphism $A/mA \to Z_m \otimes A$. To construct an inverse for η, note that any generator of the tensor product has the form $kg_0 \otimes a$, for some $k \in Z$; since the product ka is distributive in both factors, the formula $\psi(kg_0 \otimes a) = ka + mA$ provides a homomorphism from right to left in (1.9). Clearly $\psi\eta = 1$, while $\eta\psi(kg_0 \otimes a) = g_0 \otimes ka = kg_0 \otimes a$, so also $\eta\psi = 1$. Therefore η and ψ are mutually reciprocal isomorphisms, proving (1.9).

We also have $Z \otimes A \cong A$ by (1.7). Since any finitely generated abelian group is a direct sum of cyclic groups Z and Z_m, these formulas, with (1.6), provide a calculation of $G \otimes A$ for G finitely generated. Note also that $G \otimes A \cong A \otimes G$.

Exercises

1. Prove $Z_m \otimes Z_n \cong Z_{(m,n)}$, where (m,n) is the g.c.d. of m and n.
2. Show that $G \otimes_R \sum A_t \cong \sum G \otimes_R A_t$.
3. Determine the tensor product of two free modules.
4. If Q is the additive group of rational numbers, $Q \otimes Q \cong Q$.

2. Modules over Commutative Rings

The meaning of tensor products may be illustrated by examining other special cases. If K is a commutative ring (as usual, with an identity), then any left K-module A can be regarded as a *right* K-module, simply by defining the multiple ak, with $k \in \mathrm{K}$ on the right, as ka. The rule $a(kk') = (ak)k'$ then follows, because K is commutative, by the calculation $a(kk') = (k'k)a = k'(ka) = (ak)k'$; the other axioms for a right module follow even more directly. With this observation, it is fruitless to distinguish between left and right modules over K; instead we speak simply of *modules* and write scalar multiples on either side, as may be convenient.

For modules A and B over a commutative K, the tensor product $A \otimes_{\mathrm{K}} B$ is not just an abelian group, but is also a K-module, with multiples defined (on the generators) as

$$k(a \otimes b) = (ka) \otimes b, \qquad (\text{or } = a \otimes kb). \tag{2.1}$$

This definition leads to a variant of Thm. 1.1. Let A, B, and M be K-modules. Call a function f on $A \times B$ to M *K-bilinear* if $f(a,b)$ is K-linear in each argument when the other is fixed (e.g., $f(k_1a_1 + k_2a_2, b) = k_1 f(a_1, b) + k_2 f(a_2, b)$). Thus $a \otimes b$ is a K-bilinear function f on $A \times B$ to $A \otimes_{\mathrm{K}} B$, and Thm. 1.1 implies that any K-bilinear function f on $A \times B$ to M can be written as $f(a,b) = \omega(a \otimes b)$, for a unique homomorphism $\omega\colon A \otimes_{\mathrm{K}} B \to M$ of K-modules.

Since $A \otimes_{\mathrm{K}} B$ is still a K-module, one may form iterated tensor products such as $(A \otimes_{\mathrm{K}} B) \otimes_{\mathrm{K}} C$; this iterated product is associative and commutative, in the sense that $\xi[(a \otimes b) \otimes c] = a \otimes (b \otimes c)$ and $\tau(a \otimes b) =$

$b\otimes a$ define natural isomorphisms

$$\xi\colon (A\otimes B)\otimes C\cong A\otimes(B\otimes C),\qquad \tau\colon A\otimes B\cong B\otimes A \tag{2.2}$$

of K-modules, with $\otimes$ short for $\otimes_{\mathsf K}$. The function $(a\otimes b)\otimes c$ is K-*trilinear* (i.e., K-linear in each argument separately) and is universal for K-trilinear functions on $A\times B\times C$ to a K-module. The same holds for K-multilinear functions in any number of arguments.

Similarly, (cf. I.6) $\mathrm{Hom}_{\mathsf K}(A,B)$ becomes a K-module if to each $f\colon A\to B$ the multiple $kf\colon A\to B$ is defined as $(kf)(a)=k(fa)$.

A module over a field F is simply a vector space V, and $\mathrm{Hom}_F(V,W)$ is the vector space of all linear transformations $f\colon V\to W$. Suppose that V and W have finite bases $\{e_1,\ldots,e_m\}$ and $\{h_1,\ldots,h_n\}$, respectively. This means that V is a direct sum $\sum Fe_i$ of copies Fe_i of the field F. Since "Hom" carries finite direct sums to direct sums, $\mathrm{Hom}_F(V,W)$ is a vector space of dimension mn, as in the usual representation of linear transformations $f\colon V\to W$ by $m\times n$ matrices. Since the tensor product is additive, $V\otimes_F W$ has a basis of mn vectors $e_i\otimes h_j$, hence has dimension mn. In particular, any vector u of $V\otimes_F V$ has a unique expression as $u=\sum x^{ij}(e_i\otimes e_j)$; the m^2 constants $x^{ij}\in F$ are known as the "components" of the "tensor" u relative to the basis $\{e_i\}$. From a change of bases one calculates the corresponding change in these components x^{ij}. Classical tensor analysis, lacking a proper conceptual definition of the tensor product, described twice covariant tensors (elements u of $V\otimes_F V$) strictly in terms of such components and their transformations under change of basis. A tensor with one covariant and one contravariant index is, by definition, an element of $V\otimes_F V^*$, where $V^*=\mathrm{Hom}_F(V,F)$ is the dual space. Now the given basis $\{e_i\}$ for V determines a dual basis $\{e^i\}$ for V^*. Any tensor in $V\otimes_F V^*$ has a unique representation as a sum $\sum x^i{}_j(e_i\otimes e^j)$, so is determined by components $x^i{}_j$, for $i,j=1,\ldots,n$.

Exercises

1. If a new basis $\{e_i'\}$ in the finite dimensional space V is given by the formulas $e_i'=\sum_j t_i{}^j e_j$, calculate the resulting transformation in the components of
   a) a twice covariant tensor in $V\otimes_F V$;
   b) a tensor in $V\otimes_F V^*$.
2. Describe the transformation of components for tensors covariant in r indices and contravariant in s indices.

3. Bimodules

If R and S are two rings, an R-S-bimodule A — in symbols ${}_RA_S$ — is an abelian group which is both a left R-module and a right S-module, with always $(ra)s=r(as)$. For example, any ring R is an R-R-bimodule;

any left R-module can be regarded as an R-Z-bimodule; any K-module, with K commutative, is a K-K-bimodule; etc. If A and B are R-S-bimodules, we denote by $\mathrm{Hom}_{R\text{-}S}(A,B)$ the abelian group of all bimodule homomorphisms $f\colon A\to B$; that is, of all those group homomorphisms f with $r(fa)s=f(ras)$ always. A bimodule ${}_RA_S$ can be *pulled back* by ring homomorphisms $\varrho\colon R'\to R$ and $\sigma\colon S'\to S$ to give an R'-S'-bimodule ${}_\varrho A_\sigma$.

The functors Hom and $\otimes$ carry suitable bimodules into bimodules. To show how this takes place, let T, R, and S be any three rings. Then we have the implication

$$ {}_TG_R \;\&\; {}_RA_S \;\Rightarrow\; {}_T[G\otimes_R A]_S, \tag{3.1}$$

where the bimodule structure indicated on the right is defined on generators, according to Thm. 1.1, by $t(g\otimes a)s=tg\otimes as$. Note that the formula $t(g\otimes a)=tg\otimes a$ which makes $G\otimes A$ a left module over T is essentially the same as the formula $\gamma(g\otimes a)=(\gamma g)\otimes a$ which makes $G\otimes_R A$ a covariant functor of G. Similarly there is an implication

$$ {}_SC_R \;\&\; {}_TA_R \;\Rightarrow\; {}_T[\mathrm{Hom}_R(C,A)]_S, \tag{3.2}$$

where the bimodule structure on the right is defined for each $f\colon C\to A$ by $(tfs)(c)=t[f(sc)]$. The reader should show that this does produce a T-S-bimodule, noting that the given bimodule associativities $s(cr)=(sc)r$ and $t(ar)=(ta)r$ are used to insure that tfs is indeed a homomorphism of right R-modules when f is one. Observe also how the contravariance of Hom_R in C changes left operators of S on C into right operators of S on $\mathrm{Hom}_R(C,A)$. In case $S=T$, the group $\mathrm{Hom}_{S\text{-}R}(C,A)$ of bimodule homomorphisms can be described as the set of all those f in the S-S-bimodule $\mathrm{Hom}_R(C,A)$ with $sf=fs$. For left-module homomorphisms, the analogue of (3.2) is the implication

$$ {}_RC_S \;\&\; {}_RA_T \;\Rightarrow\; {}_S[\mathrm{Hom}_R(C,A)]_T. \tag{3.3}$$

An *endomorphism* of the right R-module A is by definition an R-module homomorphism $f\colon A\to A$. Under addition and composition the set of all R-endomorphisms of A form a ring $\mathrm{End}_R(A)=\mathrm{Hom}_R(A,A)$ with identity element 1_A. The equation $(fa)r=f(ar)$ which states that f is a homomorphism of right R-modules also states that A is an $\mathrm{End}_R(A)$-R-bimodule. If ${}_SA_R$ is a bimodule, the left multiplication l_s by $s\in S$, defined by $l_sa=sa$, is an R-endomorphism of A, and the correspondence $s\to l_s$ is a ring homomorphism $S\to\mathrm{End}_RA$. Conversely, given A_R and a ring homomorphism $S\to\mathrm{End}_RA$, pull-back along this homomorphism yields a bimodule ${}_SA_R$. In our treatment of $\mathrm{Ext}_R(C,A)$ (Chap. III), we showed how to multiply an element $S_0\in\mathrm{Ext}^n_R(C,A)$ on the left by a homomorphism $\alpha\colon A\to A'$ and on the right by a $\gamma\colon C'\to C$, and we proved (Lemma III.1.6) the congruence $(\alpha S_0)\gamma \equiv \alpha(S_0\gamma)$. For endomorphisms α and γ, this means that $\mathrm{Ext}^n_R(C,A)$ is an $\mathrm{End}_R(A)$-$\mathrm{End}_R(C)$-bimodule. If we pull back this bimodule structure along $T \to \mathrm{End}_R C$ and $S \to \mathrm{End}_R A$, we get the implication

$$_TC_R \;\&\; _SA_R \;\Rightarrow\; _S[\mathrm{Ext}^n_R(C,A)]_T, \qquad (3.4)$$

exactly as in (3.2) for $n=0$.

A function of two variables $f(a,b)$ can be turned into a function ηf of the first variable a whose values are functions of the second variable, according to the formula $[(\eta f)\,a]\,b = f(a,b)$. This change of simultaneous arguments to successive arguments appears in many connections; for example, in the treatment of the topology of function spaces. In the present context it takes the following form, which we call the *adjoint associativity* of Hom and $\otimes$:

Theorem 3.1. *If R and S are rings with A, B, and C modules in the situation A_R, $_RB_S$, C_S, there is a natural isomorphism*

$$\eta\colon \mathrm{Hom}_S(A\otimes_R B, C) \cong \mathrm{Hom}_R(A, \mathrm{Hom}_S(B,C)) \qquad (3.5)$$

of abelian groups defined for each $f\colon A\otimes_R B \to C$ by

$$[(\eta f)\,a]\,(b) = f(a\otimes b), \quad a\in A,\ b\in B. \qquad (3.6)$$

The proof is mechanical. First check that (3.6) assigns to each $a\in A$ and each S-module homomorphism $f\colon A\otimes_R B\to C$ a function $F = [(\eta f)\,a]$ which, as a function of b, is an S-homomorphism $[(\eta f)\,a]\colon B\to C$. Next check that ηf, as a function of a, is an R-module homomorphism of A into $\mathrm{Hom}_S(B,C)$. Finally check that $\eta(f_1+f_2) = \eta f_1 + \eta f_2$, so that η is a group homomorphism, as asserted.

To show that η is an isomorphism, construct an inverse map ζ. To this end, take any right R-module homomorphism $g\colon A\to \mathrm{Hom}_S(B,C)$, and consider the function $(ga)\,b$ of $a\in A$, $b\in B$. Any r in R operates on a on the right and on b on the left, and

$$[g(ar)]\,(b) = [(ga)\,r]\,b = (ga)\,(rb),$$

this because g is an R-module homomorphism and because of the way an operation of r on a homomorphism $ga\colon B\to C$ was defined. This equation is the "middle associative" property for the function $(ga)\,b$ of the elements a and b. Hence, by Thm. 1.1, a homomorphism $\zeta g\colon A\otimes_R B\to C$ is defined by setting

$$(\zeta g)\,(a\otimes b) = (ga)\,b.$$

One checks that $\zeta\colon \mathrm{Hom}_R(A, \mathrm{Hom}_S(B,C)) \to \mathrm{Hom}_S(A\otimes_R B, C)$, and that both composites $\zeta\eta$ and $\eta\zeta$ are the identity. Both domain and range of ζ are functors of A, B, and C, covariant in C but contravariant in A and B. Moreover ζ and η are natural homomorphisms between these functors.

Corollary 3.2. *If U, R, S, and T are rings and ${}_UA_R$, ${}_RB_S$, ${}_TC_S$ are bimodules, then the map η of (3.5) is a isomorphism of T-U-bimodules. If $U=T$, it induces a natural isomorphism*

$$\eta': \operatorname{Hom}_{T\text{-}S}(A\otimes_R B, C)\cong\operatorname{Hom}_{T\text{-}R}(A, \operatorname{Hom}_S(B,C)). \tag{3.7}$$

Proof. The right U-module structure on the terms of (3.5) and the description of these terms as functors of A are given by identical formulas. Hence the fact that η is natural (in A and C) implies that η is a module homomorphism in U and T. In case $U=T$, this yields (3.7).

As another application we prove

Corollary 3.3. *If P_R is projective as an R-module, while the bimodule ${}_RP'_S$ is projective as an S-module, then $P\otimes_R P'$ is a projective S-module.*

Proof. To say that P' is S-projective means that to each epimorphism $B\twoheadrightarrow C$ of S-modules the induced map $\operatorname{Hom}_S(P', B)\to\operatorname{Hom}_S(P', C)$ is an epimorphism (of R-modules). Since P is projective as an R-module,

$$\operatorname{Hom}_R(P, \operatorname{Hom}_S(P', B))\to\operatorname{Hom}_R(P, \operatorname{Hom}_S(P', C))$$

is an epimorphism. Application of adjoint associativity to each side gives the statement that $P\otimes_R P'$ is S-projective.

A simpler analogue of adjoint associativity is the associativity of the tensor product. In the situation A_R, ${}_RB_S$, ${}_SC$, the correspondence $(a\otimes b)\otimes c\to a\otimes(b\otimes c)$ yields the natural isomorphism

$$(A\otimes_R B)\otimes_S C\cong A\otimes_R(B\otimes_S C). \tag{3.8}$$

If in addition ${}_UA_R$ and ${}_SC_T$, this is an isomorphism of U-T-bimodules. We normally identify the two sides of (3.8) by this isomorphism.

For modules A_R, ${}_RB$ we also make the identifications

$$A\otimes_R R=A, \quad R\otimes_R B=B \tag{3.9}$$

by the natural isomorphisms $a\otimes r\to ar$, $r\otimes b\to rb$.

Exercises

1. If A and B are left R-modules, show that $\operatorname{Hom}_Z(A,B)$ is an R-R-bimodule, and that the subgroup $\operatorname{Hom}_R(A,B)$ consists of those group homomorphisms $f: A\to B$ with $rf=fr$.

2. For G_R, ${}_RA$ show that $G\otimes_R A$ is an $\operatorname{End}_R(G)$-$\operatorname{End}_R(A)$-bimodule.

3. For R-S-bimodules C and A define the group $\operatorname{Ext}_{R\text{-}S}(C,A)$ of bimodule extensions of A by C.

4. Establish a "permuted" adjoint associativity

$$\mathrm{Hom}(A\otimes B, C)\cong \mathrm{Hom}(B, \mathrm{Hom}(A,C)).$$

Deduce that if ${}_U P_R$ is a projective U-module and ${}_R P'$ a projective R-module, then $P\otimes_R P'$ is a projective U-module.

5. In the situation A_K, B_K, C_K, with K commutative, establish the natural isomorphism $\mathrm{Hom}_K(A, \mathrm{Hom}_K(B,C))\cong \mathrm{Hom}_K(B, \mathrm{Hom}_K(A,C))$ of K-modules.

4. Dual Modules

The *dual* or *conjugate* of a left R-module A is the right R-module $A^* = \mathrm{Hom}_R(A,R)$. Thus an element of A^* is an R-module map $f\colon A\to R$, while $fr\colon A\to R$ is the R-module map defined for each $a\in A$ by $(fr)a = (fa)r$. The dual of an R-module homomorphism $\alpha\colon A\to A'$ is $\alpha^* = \mathrm{Hom}(\alpha, 1)\colon A'^*\to A^*$, so the dual is a contravariant functor on left modules to right modules. Similarly, the dual of a right R-module G is a left R-module G^*.

For left modules A and B there is a natural isomorphism

$$(A\oplus B)^*\cong A^*\oplus B^*. \tag{4.1}$$

Indeed, in the direct sum diagram $A\rightleftarrows A\oplus B\leftrightarrows B$ take the dual of each object and each map; the result is still a direct sum diagram, with the injections $\iota_A\colon A\to A\oplus B$ and ι_B converted into projections $\iota_A^*\colon (A\oplus B)^*\to A^*$ and ι_B^*.

By the properties of Hom, $A\rightarrowtail B\twoheadrightarrow C$ short exact gives $C^*\rightarrowtail B^*\to A^*$ (left) exact. In other words, if $A\subset B$, then $(B/A)^*\cong C^*$ is isomorphic to that submodule of B^* which consists of all those $f\colon B\to R$ which vanish on A. Call this submodule the *annihilator* of A, in symbols $\mathrm{Annih}\,A$; thus

$$(B/A)^*\cong \mathrm{Annih}\,A\subset B^*, \qquad B^*/\mathrm{Annih}\,A\rightarrowtail A^*. \tag{4.2}$$

For each left R-module A, there is a natural homomorphism

$$\varphi_A\colon A\to A^{**} = \mathrm{Hom}_R(\mathrm{Hom}_R(A,R),R), \tag{4.3}$$

which assigns to each $a\in A$ the map $\varphi a\colon A^*\to R$ with $(\varphi a)f = f(a)$. In other words, for fixed a, regard the expression $f(a)$ as a linear function of the element $f\in A^*$.

Theorem 4.1. *If L is a finitely generated and projective left R-module, then L^* is a finitely generated projective right R-module. For such L, $\varphi\colon L\to L^{**}$ is a natural isomorphism.*

Proof. If F is free on the generators $e_1, \ldots, e_n$, we may define elements e^j in F^* by

$$e^j(e_i) = 1, \quad \text{if } i=j,$$
$$= 0, \quad \text{if } i\neq j.$$

Any $f\colon F\to R$ is uniquely determined by the elements $fe_i=r_i\in R$, hence $f=\sum e^j r_j$, and F^* is free on the generators $e^1,\ldots,e^n$. They are said to be the *dual basis* to $e_1,\ldots,e_n$. Now φ_F maps the e_i to the basis elements dual to the e^j, so $\varphi_F\colon F\to F^{**}$ is an isomorphism.

If L is finitely generated and projective, there is such an F with $F\twoheadrightarrow L$ and $F\cong L\oplus L'$; L' is also finitely generated and projective. Hence $F^*\cong L^*\oplus L'^*$, and $L\oplus L'\cong F\cong F^{**}\cong L^{**}\oplus L'^{**}$; this isomorphism carries L onto L^{**} by φ_L, whence the conclusion.

For example, if R is a field, any finitely generated module V (i.e., any finite dimensional vector space) is free. For such spaces $V^{**}\cong V$, and, for $V\supset W$,

$$(V/W)^*\cong \operatorname{Annih} W;\qquad V^*/\operatorname{Annih} W\cong W^*.$$

For left modules A and C there is a natural homomorphism

$$\zeta\colon A^*\otimes_R C\to \operatorname{Hom}_R(A,C) \tag{4.4}$$

defined for each $f\colon A\to R$ and each $c\in C$ by $[\zeta(f\otimes c)]a=f(a)c$ for all a. One checks that $\zeta(f\otimes c)$ is a module homomorphism $A\to C$, and that this homomorphism is biadditive and middle associative in f and c.

Proposition 4.2. *If L is a finitely generated projective left R-module, then ζ is a natural isomorphism $\zeta=\zeta_L\colon L^*\otimes_R C\cong \operatorname{Hom}_R(L,C)$.*

For example, if V and W are finite dimensional vector spaces, take $L=V^*$ and $C=W$. Then $L^*\cong V$, so ζ gives $V\otimes W\cong \operatorname{Hom}(V^*,W)$. Thus tensor products of such vector spaces may be defined via Hom and the dual. Alternatively, $V\otimes W$ is the dual of the space of bilinear maps of $V\times W$ to the base field.

Proof. First suppose that $L=F$ is free on the generators $e_1,\ldots,e_n$. With the dual basis $e^1,\ldots,e^n$, each element of $F^*\otimes C$ has a unique representation as $\sum e^i\otimes c_i$ for constants $c_i\in C$. But $\zeta(\sum e^i\otimes c_i)=f$ is that homomorphism $f\colon F\to C$ with $f(e_j)=c_j$, $j=1,\ldots,n$. Since F is free, any $f\colon F\to C$ is uniquely determined by its values $f(e_j)$ for all j. Hence ζ_F is an isomorphism. The case when L is a finitely generated projective is now treated as in the proof of Thm. 4.1.

Proposition 4.3. *For modules L and B over a commutative ring* K, *with L finitely generated and projective, there is a natural isomorphism $\psi\colon L^*\otimes B^*\cong (L\otimes B)^*$.*

Proof. For any two K-modules A and B a natural homomorphism $\psi\colon A^*\otimes B^*\to (A\otimes B)^*$ is defined for $f\in A^*$, $g\in B^*$ by setting

$$[\psi(f\otimes g)](a\otimes b)=f(a)\,g(b)\in \mathrm{K}.$$

This map ψ is the composite

$$A^*\otimes B^*\xrightarrow{\zeta}\operatorname{Hom}(A,B^*)=\operatorname{Hom}(A,\operatorname{Hom}(B,\mathrm{K}))\cong \operatorname{Hom}(A\otimes B,\mathrm{K})$$

of ζ of (4.4) with adjoint associativity. The latter is always an isomorphism, and so is ζ when $A=L$ is finitely generated and projective.

Note. For further discussions of duality see DIEUDONNÉ [1958], MORITA [1958], BASS [1960], or JANS [1961].

Exercises

1. For each ${}_RA$, show that $\theta(a\otimes f)=fa$ is a bimodule homomorphism $\theta: A\otimes_Z A^* \to R$.

2. For modules ${}_RA, G_R$, a bimodule homomorphism $\psi: A\otimes_Z G\to R$ is called a *pairing*. Show that ψ determines $\psi_G: G\to A^*$ such that $\psi=\theta(1\otimes\psi_G)$, for θ as in Ex 1.

3. For each R-module A, prove that the composite

$$A^* \xrightarrow{\varphi(A^*)} A^{***} \xrightarrow{(\varphi_A)^*} A^*$$

is the identity.

5. Right Exactness of Tensor Products

The tensor product preserves short right exact sequences:

Theorem 5.1. *If G is a right R-module, while $D\xrightarrow{\beta} B\xrightarrow{\sigma}\!\!\!\to C$ is an exact sequence of left R-modules, then*

$$G\otimes_R D\xrightarrow{1\otimes\beta} G\otimes_R B\xrightarrow{1\otimes\sigma} G\otimes_R C\to 0 \tag{5.1}$$

is an exact sequence (of abelian groups).

Proof. With the cokernel L of $1\otimes\beta$ manufacture the exact sequence

$$G\otimes_R D\xrightarrow{1\otimes\beta} G\otimes_R B\xrightarrow{\eta} L\to 0.$$

Compare this with (5.1). The composite $(1\otimes\sigma)(1\otimes\beta)=1\otimes\sigma\beta$ is zero, so $1\otimes\sigma$ factors as $\sigma'\eta$ for some $\sigma': L\to G\otimes_R C$. Since $\sigma(B)=C$, there is to each c in C a b with $\sigma b=c$. By exactness at B, each $\eta(g\otimes b)$ depends only on $g\in G$ and $c\in C$, but not on the choice of b. Moreover, $\eta(g\otimes b)$ is biadditive and middle associative. Hence Thm. 1.1 gives $\omega: G\otimes_R C\to L$ with $\omega(g\otimes c)=\eta(g\otimes b)$, and $\sigma'\omega=1$, $\omega\sigma'=1$. Thus $\omega: G\otimes_R C\cong L$ makes (5.1) isomorphic to the manufactured sequence and hence exact.

Corollary 5.2. *The tensor product of two epimorphisms is an epimorphism.*

Proof. By the theorem, if τ and σ are epimorphisms, so are $\tau\otimes 1$ and $1\otimes\sigma$, hence also their composite $(\tau\otimes 1)(1\otimes\sigma)=\tau\otimes\sigma$. For the kernel of $\tau\otimes\sigma$, see Lemma VIII.3.2 or Ex. 3 below.

In Thm. 5.1 it would not be true to state that a short exact sequence $(\varkappa,\sigma): A\rightarrowtail B\twoheadrightarrow C$ yields a short exact sequence like (5.1) because when $\varkappa: A\to B$ is a monomorphism $1\otimes\varkappa: G\otimes_R A\to G\otimes_R B$ need not be a monomorphism. To illustrate this, take $R=Z$, $A=2Z$ (the group of even integers), $B=Z$, $\varkappa$ the injection, and $G=Z_2(g)$ a cyclic group of order 2 with generator g. Then, as calculated in (1.9), $Z_2(g)\otimes(2Z)$ is the cyclic group of order 2 with generator $g\otimes 2$, while

$$(1\otimes\varkappa)(g\otimes 2)=g\otimes 2=2g\otimes 1=0\otimes 1=0,$$

so $g\otimes 2$ is in the kernel of $1\otimes\varkappa$.

This example can be reformulated thus. For a submodule $A\subset B$ one cannot assume that $G\otimes A\subset G\otimes B$, because an element $g\otimes a$ of $G\otimes A$ may be non-zero while the "same" element $g\otimes a$ becomes zero in $G\otimes B$. For these and related reasons we have insisted from the start that the inclusion $A\subset B$ be represented by a map $\varkappa: A\to B$.

In this example, the integer 2 can be replaced by any integer m. Thus we can describe certain elements in $\operatorname{Ker}(1\otimes\varkappa)$ for $R=Z$ and $(\varkappa,\sigma): A\rightarrowtail B\twoheadrightarrow C$ any short exact sequence of abelian groups. These elements $g\otimes a$ arise whenever there is an element b with both $\varkappa a=mb$ and $mg=0$ for the same integer m, for then

$$(1\otimes\varkappa)(g\otimes a)=g\otimes\varkappa a=g\otimes mb=mg\otimes b=0\otimes b=0.$$

Now $\varkappa a$, and hence a, is determined by b, while $g\otimes a$ depends only on $\sigma b\in C$. Indeed, $\sigma b=\sigma b'$ by exactness implies $b'=b+\varkappa a_0$ for some a_0, whence $\varkappa(a+ma_0)=mb'$ and $g\otimes(a+ma_0)=g\otimes a+g\otimes ma_0=g\otimes a$. The kernel element $g\otimes a$ depends on g, $m\in Z$, and $\sigma b=c$; furthermore $mc=m(\sigma b)=\sigma(mb)=\sigma\varkappa a=0$, by exactness. By way of notation, set

$$k(g,m,c)=g\otimes a\in\operatorname{Ker}(1\otimes\varkappa),\quad mg=0=mc; \tag{5.2}$$

here a is any element of A such that $\varkappa a=mb$, $\sigma b=c$ for some b; that is, a is obtained by "switchback" as $a=\varkappa^{-1}m\,\sigma^{-1}c$. In the next section, we shall show that the elements $k(g,m,c)$ of (5.2) generate $\operatorname{Ker}(1\otimes\varkappa)$.

These elements $k(g,m,c)$ satisfy certain identities. They are additive in g and in c; for example, additivity in c means that

$$k(g,m,c_1+c_2)=k(g,m,c_1)+k(g,m,c_2) \tag{5.3}$$

whenever $mc_1=0=mc_2$. For any two integers m and n, one calculates that

$$k(g,mn,c)=k(g,m,nc) \tag{5.4}$$

whenever $mg=0$, $mnc=0$, and that

$$k(g,mn,c)=k(gm,n,c) \tag{5.5}$$

whenever $gmn=0$, $nc=0$. Here we have written gm for mg because we can consider the abelian group G as a *right* module over Z. These relations will now be used to define a new group.

Exercises

1. If $A \rightarrowtail B$, show that each element in $\operatorname{Ker}(A\otimes Z_m \to B\otimes Z_m)$ has the form $k(c, m, 1)$ for 1 the generator of Z_m.

2. If J is a two-sided ideal in the ring R show that the map $a\otimes(r+J)\to ar$ for $a\in J$ yields an epimorphism $J\otimes_R(R/J)\twoheadrightarrow J/J^2$ of R-modules. If $J^2 \neq J$ prove that the injection $J\to R$ induces a map $J\otimes_R(R/J)\to R\otimes_R(R/J)$ which is not a monomorphism.

3. If $(\gamma,\tau)\colon G\rightarrowtail H\twoheadrightarrow K$ and $(\beta,\sigma)\colon D\rightarrowtail B\twoheadrightarrow C$ are exact, then $\tau\otimes\sigma\colon H\otimes_R B\twoheadrightarrow K\otimes_R C$ has kernel $\gamma_*(G\otimes B)\cup\beta_*(H\otimes D)$.

6. Torsion Products of Groups

For abelian groups A and G we define the torsion product $\operatorname{Tor}(G,A)$ as that abelian group which has generators all symbols $\langle g, m, a\rangle$, with $m\in Z$, $gm=0$ in G, and $ma=0$ in A, subject to the relations ("additivity" and "slide" rules for factors m, n)

$$\langle g_1+g_2, m, a\rangle=\langle g_1, m, a\rangle+\langle g_2, m, a\rangle, \qquad g_i m=0=ma, \tag{6.1}$$

$$\langle g, m, a_1+a_2\rangle=\langle g, m, a_1\rangle+\langle g, m, a_2\rangle, \qquad gm=0=ma_i, \tag{6.2}$$

$$\langle g, mn, a\rangle=\langle gm, n, a\rangle, \qquad gmn=0=na, \tag{6.3}$$

$$\langle g, mn, a\rangle=\langle g, m, na\rangle, \qquad gm=0=mna. \tag{6.4}$$

Each relation is imposed whenever both sides are defined; in each case this amounts to the requirement that the symbols on the right hand side be defined. The additivity relations (6.1) and (6.2) imply that $\langle 0, m, a\rangle=0=\langle g, m, 0\rangle$. Hence $\operatorname{Tor}(G,A)=0$ when A has no elements (except 0) of finite order. Also $\operatorname{Tor}(A,G)\cong\operatorname{Tor}(G,A)$.

If $\alpha\colon A\to A'$, the definition $\alpha_*\langle g, m, a\rangle=\langle g, m, \alpha a\rangle$ makes $\operatorname{Tor}(G,A)$ a covariant functor of A. It is likewise covariant in G. From (6.2) one deduces $(\alpha+\beta)_*=\alpha_*+\beta_*$ and hence the isomorphism $\operatorname{Tor}(G, A_1\oplus A_2)\cong\operatorname{Tor}(G,A_1)\oplus\operatorname{Tor}(G,A_2)$. Thus to calculate $\operatorname{Tor}(G,A)$ for finitely generated groups it will suffice to make a calculation for G finite cyclic.

For $G=Z_q(g_0)$ a cyclic group of order q and generator g_0, there is an isomorphism

$$\zeta\colon {}_qA\cong\operatorname{Tor}(Z_q(g_0), A), \tag{6.5}$$

where ${}_qA$ denotes the subgroup of those elements $a\in A$ for which $qa=0$. Indeed, each $a\in{}_qA$ yields an element $\zeta a=\langle g_0, q, a\rangle$ in $\operatorname{Tor}(Z_q, A)$; by (6.2), ζ is a homomorphism. To find a homomorphism η in the reverse direction, write each element of Z_q as $g_0 k$ for some $k\in Z$; each generator of the torsion product then has the form $\langle g_0 k, m, a\rangle$ where $ma=0$ and $mk\equiv 0 \pmod q$. With $n=mk/q$, (6.3) and (6.4) give

$$\langle g_0 k, m, a\rangle=\langle g_0, km, a\rangle=\langle g_0, q, na\rangle.$$

This suggests that η be defined by $\eta\langle g_0 k, m, a\rangle = (mk/q)a$. The reader should verify that this definition respects the defining relations (6.1) to (6.4), in the sense that η carries elements defined to be equal in Tor into equal elements of ${}_qA$. This shows that η yields a homomorphism $\eta\colon \operatorname{Tor}(Z_q, A) \to {}_qA$. Furthermore $\eta\zeta a = a$, while the calculation displayed just above shows that $\zeta\eta = 1$. Therefore η and ζ are reciprocal isomorphisms, as asserted.

For a fixed cyclic group, the isomorphism (6.5) is natural in A, but depends on the choice of the generator for the cyclic group Z_q.

The torsion product, born of the inexactitude of $\otimes$, does measure that inexactitude as follows.

Theorem 6.1. *If $E = (\varkappa, \sigma)\colon A \rightarrowtail B \twoheadrightarrow C$ is an exact sequence of abelian groups, then each abelian group G gives an exact sequence*

$$\left.\begin{array}{l} 0 \to \operatorname{Tor}(G,A) \to \operatorname{Tor}(G,B) \xrightarrow{\sigma_*} \operatorname{Tor}(G,C) \\ \quad \xrightarrow{E_*} G\otimes A \xrightarrow{1\otimes\varkappa} G\otimes B \to G\otimes C \to 0. \end{array}\right\} \tag{6.6}$$

The maps are those induced by $\varkappa$ and σ except for E_*, which is defined on each generator of Tor (G, C) *by the formula*

$$E_*\langle g, m, c\rangle = k(g, m, c) \tag{6.7}$$

for k as in (5.2). This map E_* is natural when its arguments are considered as bifunctors of E and G.

Proof. E_* is a homomorphism because the identities already noted in (5.3), (5.4), and (5.5) for k match exactly the defining identities for Tor. Naturality is readily proved. Since each $k(g,m,c)$ lies in $\operatorname{Ker}(1\otimes\varkappa)$, one has $(1\otimes\varkappa)E_* = 0$, and one also verifies that $E_*\sigma_* = 0$. As usual, the crux of the exactness proof lies in the demonstration that each kernel is contained in the corresponding image.

It suffices to prove this, we claim, in the case when G has a finite number of generators. As a sample consider exactness at $G\otimes A$. An element $u = \sum g_i \otimes a_i$ of $G\otimes A$ involves only a finite number of elements of G. If its image $(1\otimes\varkappa)u = \sum g_i \otimes \varkappa a_i$ is zero in $G\otimes B$, it is zero because of a finite number of defining relations for $G\otimes B$; these relations again involve but a finite list $h_1, \ldots, h_m$ of elements of G. Now take G_0 to be the subgroup of G generated by all the elements $g_1, \ldots, g_n, h_1, \ldots, h_m$ which have occurred, and let $\iota\colon G_0 \to G$ be the injection. Then $u_0 = \sum g_i \otimes a_i$ is an element of $G_0 \otimes A$ with $(\iota\otimes 1)u_0 = u$. By naturality the diagram

$$\begin{array}{ccccc} \operatorname{Tor}(G_0, C) & \xrightarrow{E_*} & G_0\otimes A & \xrightarrow{\varkappa_* = 1\otimes\varkappa} & G_0\otimes B \\ \downarrow{\iota_*} & & \downarrow{\iota\otimes 1} & & \downarrow{\iota\otimes 1} \\ \operatorname{Tor}(G, C) & \xrightarrow{E_*} & G\otimes A & \xrightarrow{\varkappa_* = 1\otimes\varkappa} & G\otimes B \end{array}$$

commutes, and we are allowed to assume the top row exact. Since G_0 contains all the elements $h_j \in G$ used to show $\varkappa_* u = 0$, these same elements will show $\varkappa_* u_0 = 0$ in $G_0 \otimes B$. By exactness of the top row, there is a $t_0 \in \operatorname{Tor}(G_0, C)$ with $E_* t_0 = u_0$. But $E_* \iota_* t_0 = (\iota \otimes 1) E_* t_0 = (\iota \otimes 1) u_0 = u$; this proves the bottom row exact at $G \otimes A$.

This argument depends not on the particular form of the definitions of Tor and $\otimes$, but only on the fact that these groups were described by generators and relations.

Return to the proof of exactness. Now G is finitely generated, hence representable as a direct sum of cyclic groups. Since both Tor and $\otimes$ carry direct sums into direct sums, the sequence (6.6) is the direct sum of the corresponding sequences for cyclic groups G. If $G = Z$ is cyclic infinite, the torsion products are all zero and the sequence is isomorphic to the given sequence E. If $G = Z_q$ is finite cyclic, the various terms have been calculated in (1.9) and (6.5); the calculations amount to a diagram in which the central portion is

$$\begin{array}{ccccccc} \cdots \to & \operatorname{Tor}(Z_q, B) & \to & \operatorname{Tor}(Z_q, C) & \xrightarrow{E_*} & Z_q \otimes A & \to \cdots \\ & \uparrow \zeta & & \uparrow \zeta & & \uparrow \eta & \\ \cdots \to & {}_qB & \to & {}_qC & \xrightarrow{E_\#} & A/qA & \to \cdots . \end{array}$$

In the second row, define $E_\#$ by the switchback rule $E_\# c = \varkappa^{-1} q \sigma^{-1} c + qA$; with this definition this diagram is readily seen to be commutative. Since η of (1.9) and ζ of (6.5) are isomorphisms, the exactness of the top row is now reduced to the exactness of the bottom row, which reads in full

$$0 \to {}_qA \to {}_qB \to {}_qC \xrightarrow{E_\#} A/qA \to B/qB \to C/qC \to 0.$$

Exactness here may be verified from the definitions of the terms and the exactness of E. For example, if $\varkappa(a + qA) = 0$ in B/qB, then $\varkappa a = qb$ for some $b \in B$. Thus $\sigma(qb) = 0$, hence $\sigma b = c \in {}_qC$; the very definition of the switchback yields $E_\# c = a + qA$.

We leave the reader to prove

Theorem 6.2. *The following conditions on an abelian group G are equivalent:*

(i) *G has no elements of finite order, except* 0;

(ii) $\operatorname{Tor}(G, A) = 0$ *for every abelian group A*;

(iii) *If $\varkappa\colon A \to B$ is a monomorphism, so is $1 \otimes \varkappa\colon G \otimes A \to G \otimes B$*;

(iv) *Any short exact sequence remains exact upon tensor multiplication by G*;

(v) *Any exact sequence remains exact when tensored with G.*

Such a group G is said to be *torsion-free* (condition (i)).

A different description of the generators of $\operatorname{Tor}(G,A)$ is useful for generalization. The triple $\langle g, m, a\rangle$ determines three homomorphisms

$$G \xleftarrow{\mu} Z \xleftarrow{\partial} Z, \qquad \nu\colon Z = Z^* \to A$$

by $\mu 1 = g$, $\partial 1 = m$, $\nu 1 = a$. Regard $L\colon Z \leftarrow Z$ as a chain complex, zero except in dimensions $L_0 = Z = L_1$; since $\mu\partial = 0$, $\mu\colon L \to G$ is a chain complex over G. Regard the dual L^* as a chain complex $\partial^*\colon L_0^* \to L_1^*$ over A via $\nu\colon L_1^* \to A$. The triple $\langle g, m, a\rangle$ has become a triple (μ, L, ν), where

L and its dual L^* are chain complexes (of "length" 1),

$\mu\colon L \to G$ and $\nu\colon L^* \to A$ are chain transformations.

The slide rules (6.3) and (6.4) can be written as one rule

$$\langle g' n_0, m, a\rangle = \langle g', m', n_1 a\rangle; \quad n_0 m = m' n_1, \quad g' m' = 0 = m a. \tag{6.8}$$

If m and m' determine chain complexes L and L', $n_0 m = m' n_1$ makes

$$\begin{array}{cccc} L\colon & Z & \xleftarrow{m} & Z \\ \varrho\downarrow & \downarrow n_0 & & \downarrow n_1 \\ L'\colon & Z & \xleftarrow{m'} & Z \end{array}$$

commutative, hence $\varrho\colon L \to L'$ a chain transformation. Now g' and a determine $\mu'\colon L_0' = Z \to G$ and $\nu\colon L_1^* \to A$ by $\mu' 1 = g'$, $\nu 1 = a$ and $\mu'\varrho 1 = g' n_0$, $\nu\varrho^* 1 = n_1 a$. In this notation, the slide rule (6.8) becomes

$$(\mu'\varrho, L, \nu) = (\mu', L', \nu\varrho^*), \qquad \varrho\colon L \to L'.$$

Exercises

1. For both sides defined, prove $\langle g, m_1 + m_2, a\rangle = \langle g, m_1, a\rangle + \langle g, m_2, a\rangle$.

2. Let $Q > Z$ be the additive group of rational numbers, and let $T(A)$ (the "torsion subgroup") be the subgroup of A consisting of all elements of finite order in A. Establish a natural isomorphism $\operatorname{Tor}(Q/Z, A) \cong T(A)$.

3. If Q_p is that subgroup of Q consisting of all rational numbers with denominator some power of p, describe $\operatorname{Tor}(Q_p/Z, A)$.

4. Investigate $\operatorname{Tor}(G, A)$ when its arguments are infinite direct sums.

5. If A and B are finite abelian groups prove that $A \otimes B \cong \operatorname{Tor}(A, B)$ (The isomorphism is not natural).

6. Show that $\operatorname{Tor}(G, A) = 0$ if for each element a of finite order k in A and each g of finite order l in G one always has k and l relatively prime.

7. For modules G_R, ${}_R A$, let $T(G, A)$ be defined by generators $\langle g, r, a\rangle$ for $gr = 0 = ra$ and relations (6.1) through (6.4). Show that the sequence (6.6) with Tor replaced by T need not be exact at $G \otimes_R A$.

7. Torsion Products of Modules

For fixed $n \geqq 0$ we consider chain complexes L of length n,

$$L\colon L_0 \xleftarrow{\partial} L_1 \xleftarrow{\partial} \cdots \xleftarrow{\partial} L_{n-1} \xleftarrow{\partial} L_n,$$

with each L_k a finitely generated projective right R-module. The dual $L^* = \operatorname{Hom}_R(L, R)$ can also be regarded as a chain complex L^*, with L_k^* as the chains of dimension $n-k$,

$$L^*\colon L_n^* \xleftarrow{\delta} L_{n-1}^* \xleftarrow{\delta} \cdots \xleftarrow{\delta} L_1^* \xleftarrow{\delta} L_0^*.$$

Each L_k^* is also a finitely generated projective left module, and $\delta_k\colon L_k^* \to L_{k+1}^*$ is defined from $\partial_{k+1}\colon L_{k+1} \to L_k$ as $\delta_k = (-1)^{k+1} \partial_{k+1}^*$. Here and below, we could equally well require the L_k to be finitely generated and free; the same then holds for the L_k^*.

If G is a right R-module, regarded as a trivial chain complex, a chain transformation $\mu\colon L \to G$ is a module homomorphism $\mu_0\colon L_0 \to G$ with $\mu_0 \partial = 0\colon L_1 \to G$, while a chain transformation $\nu\colon L^* \to {}_R C$ is a module homomorphism $\nu\colon L_n^* \to C$ with $\nu\delta = 0$. For given modules G_R and ${}_R C$ we take as the elements of $\operatorname{Tor}_n^R(G, C)$ all the triples

$$t = (\mu, L, \nu), \quad \mu\colon L \to G, \quad \nu\colon L^* \to C,$$

where L has length n and μ, ν are chain transformations, as above. If L' is a second such complex and $\varrho\colon L \to L'$ a chain transformation, then so is the dual $\varrho^*\colon L'^* \to L^*$. Given $\mu'\colon L' \to G$ and $\nu\colon L^* \to C$, we propose that

$$(\mu' \varrho, L, \nu) = (\mu', L', \nu \varrho^*). \tag{7.1}$$

These maps may be exhibited by a pair of commutative diagrams

$$\begin{array}{ccccccc} G \leftarrow & L_0 & \leftarrow \cdots \leftarrow & L_n & & L_0^* & \to \cdots \to & L_n^* & \to C \\ \| & \downarrow \varrho_0 & & \downarrow \varrho_n & & \varrho_0^* \uparrow & & \varrho_n^* \uparrow & \| \\ G \leftarrow & L_0' & \leftarrow \cdots \leftarrow & L_n', & & L_0'^* & \to \cdots \to & L_n'^* & \to C, \end{array}$$

resembling the definition of the congruence relation on Ext^n by morphisms of long exact sequences. Formally, the equality relation on Tor_n is to be the weakest equivalence relation in which (7.1) holds; this means that two triples in Tor_n are equal if the second is obtained from the first by a finite succession of applications of the rule (7.1). This describes Tor_n as a set.

This set is a functor. Indeed, for maps $\eta\colon G \to G'$, $\gamma\colon C \to C'$ the rules

$$\eta_*(\mu, L, \nu) = (\eta\mu, L, \nu), \quad \gamma_*(\mu, L, \nu) = (\mu, L, \gamma\nu) \tag{7.2}$$

preserve the equality (7.1) and make Tor_n a covariant bifunctor.

Two triples t_1 and t_2 in $\mathrm{Tor}_n(G,C)$ have as direct sum the triple

$$(\mu_1, L^1, \nu_1) \oplus (\mu_2, L^2, \nu_2) = (\mu_1 \oplus \mu_2, L^1 \oplus L^2, \nu_1 \oplus \nu_2)$$

in $\mathrm{Tor}_n(G\oplus G, C\oplus C)$. If $t_1 = t_1'$ and $t_2 = t_2'$ according to (7.1), then $t_1 \oplus t_2 = t_1' \oplus t_2'$. If ω_G is the automorphism of $G\oplus G$ given by $\omega(g_1, g_2) = (g_2, g_1)$, then $(\omega_G)_*(t_1\oplus t_2) - (\omega_C)_*(t_2 \oplus t_1)$, as one sees by applying (7.1) with ϱ: $L^1\oplus L^2 \to L^2 \oplus L^1$ the map interchanging the summands.

Now $\mathrm{Tor}_n(G,C)$ is an abelian group when the addition is defined by

$$t_1 + t_2 = (\nabla_G)_* (\nabla_C)_* (t_1 \oplus t_2) \in \mathrm{Tor}_n(G,C), \tag{7.3}$$

with ∇_G: $G\oplus G \to G$ and ∇_C the codiagonal maps (III.2.1'). The proof of the group axioms is direct. The associative law follows from the associativity of the codiagonal maps. The commutative law follows from $(\omega_G)_*(t_1\oplus t_2) = (\omega_C)_*(t_2 \oplus t_1)$ and $\nabla_G \omega_G = \nabla_G$. As a zero for the addition we may take $(0,0,0)$, where the middle zero designates the zero complex of length n, while the inverse $-(\mu, L, \nu)$ is $(-\mu, L, \nu)$. The maps η_* and γ_*, defined as in (7.2), respect this addition, so Tor_n is a bifunctor to the category of abelian groups. The same formulas (7.2) show that if the modules G and C are bimodules ${}_T G_R$, ${}_R C_S$ for other rings T and S, then Tor_n is a bimodule ${}_T(\mathrm{Tor}_n)_S$, much as in (3.1).

Proposition 7.1. *The symbols* (μ, L, ν) *in* Tor_n *are additive in* μ *and* ν; *e.g.*,

$$(\mu_1 + \mu_2, L, \nu) = (\mu_1, L, \nu) + (\mu_2, L, \nu). \tag{7.4}$$

Proof. Recall (III.2.2) that $\mu_1 + \mu_2 = \nabla_G(\mu_1 \oplus \mu_2)\Delta_L$. The dual of the diagonal map Δ_L: $L \to L\oplus L$ is the codiagonal ∇_{L^*}: $L^* \oplus L^* \to L^*$. Hence the equality rule (7.1) and the definition (7.3) yield (7.4) as

$$\begin{aligned}(\mu_1 + \mu_2, L, \nu) &= (\nabla_G(\mu_1\oplus\mu_2)\Delta_L, L, \nu) = (\nabla_G(\mu_1 \oplus \mu_2), L\oplus L, \nu \nabla_{L^*})\\ &= (\nabla_G(\mu_1\oplus\mu_2), L\oplus L, \nabla_C(\nu\oplus\nu)) = (\mu_1, L, \nu) + (\mu_2, L, \nu).\end{aligned}$$

Proposition 7.2. *Every element of* $\mathrm{Tor}_n(G,C)$ *has the form* (μ, F, ν) *where* μ: $F\to G$, ν: $F^* \to C$, *and* F *is a chain complex of length* n *of finitely generated free right modules. Hence the functor* Tor_n *defined using complexes of finitely generated free modules* F_i *is naturally isomorphic to the functor* Tor_n *defined using complexes of finitely generated projective modules* L_i.

Proof. The construction above, using only free modules instead of projectives, yields a functor $\mathrm{Torf}_n(G,C)$. Since each free complex F of length n is also projective, each element (μ, F, ν) of Torf_n is also an element of Tor_n. This map Torf $\to$ Tor has a two-sided inverse. For take any $(\mu, L, \nu) \in \mathrm{Tor}_n$. Each L_k can be written as a direct summand of some finitely generated free module $F_k = L_k \oplus M_k$. Make F a complex with boundary $\partial \oplus 0$: $L_k \oplus M_k \to L_{k-1} \oplus M_{k-1}$. The injection ι: $L \to F$

and the projection $\pi\colon F\to L$ are chain transformations with $\pi\iota=1$, $\iota^*\pi^*=1$. By our equality rule,

$$(\mu, L, \nu)=(\mu, L, \nu\iota^*\pi^*)=(\mu\pi, F, \nu\iota^*);$$

this is an element of Torf, for F is a free complex of length n. By this process, triples equal in the sense (7.1) are turned into equal triples of Torf; hence the natural isomorphism $\mathrm{Tor}_n\cong\mathrm{Torf}_n$.

For $n=0$, Tor_0 may be identified with $\otimes$:

Theorem 7.3. *There is a natural isomorphism* $G\otimes_R C\cong\mathrm{Tor}_0^R(G,C)$.

Proof. Each $g\in G$ determines a map $\mu_g\colon R\to G$ of right R-modules by $\mu_g(r)=gr$; similarly each $c\in C$ determines a map $\nu_c\colon R=R^*\to C$ of left R-modules by $\nu_c(1)=c$. The triple $(\mu_g, R, \nu_c)\in\mathrm{Tor}_0(G,C)$ is additive in g and c and middle linear, so $g\otimes c\to(\mu_g, R, \nu_c)$ is a homomorphism $G\otimes C\to\mathrm{Tor}_0(G,C)$ of abelian groups. It is natural. This homomorphism takes each element $\sum g_i\otimes c_i$ of $G\otimes C$ into the triple (μ, F, ν), where F is free on generators e_i, $\mu e_i=g_i$, and $\nu e^i=c_i$.

To construct an inverse map Θ, use Prop. 7.2 to write each element of $\mathrm{Tor}_0(G,C)$ as (μ, F, ν) where $\mu\colon F\to G$, $\nu\colon F^*\to C$, and F is a finitely generated free module. Choose any free generators $e_1, \ldots, e_m$ for F, use the dual basis $e^1, \ldots, e^m$ of F^*, and set

$$\Theta(\mu, F, \nu)=\sum_{i=1}^{m}\mu(e_i)\otimes\nu(e^i)\in G\otimes_R C.$$

To express the equality in Tor_0, write $\varrho\colon F\to F'$, in terms of bases e_i and e'_j, as $\varrho e_i=\sum_j e'_j r_{ji}$ with a matrix $\{r_{ji}\}$ of elements from R. Then $\varrho^* e'^j=\sum_i r_{ji}e^i$ and

$$\begin{aligned}\Theta(\mu'\varrho, F, \nu)&=\sum_i\Big(\sum_j\mu'(e'_j r_{ji})\otimes\nu e^i\Big)\\ &=\sum_j\Big(\mu' e'_j\otimes\sum_i\nu(r_{ji}e^i)\Big)=\Theta(\mu', F', \nu\varrho^*).\end{aligned}$$

This shows Θ well defined for the equality in Tor; also, if $F=F'$ and e'_j is a different basis in F, it shows the definition of Θ independent of the choice of the basis in F. Since Θ is a two-sided inverse of the previous map, the proof is complete.

Corollary 7.4. *For L a finitely generated and projective right R-module a natural isomorphism* $\xi\colon\mathrm{Hom}_R(L^*, C)\cong L\otimes_R C$ *is defined by* $\xi(\nu)=(1_L, L, \nu)$. *Hence each element t of* $\mathrm{Tor}_0(L,C)$ *has a unique representation as* $t=(1_L, L, \nu)$ *for some* $\nu\colon L^*\to C$.

Proof. By additivity (Prop. 7.1), ξ is a natural homomorphism. To show it an isomorphism, it suffices to prove the composite

$$L\otimes_R C\cong L^{**}\otimes_R C\xrightarrow{\zeta}\mathrm{Hom}_R(L^*, C)\xrightarrow{\xi}\mathrm{Tor}_0(L,C)$$

the identity, where ζ is the isomorphism defined in Prop. 4.2. If $L=R$, the definitions show that $\xi\zeta$ is the identity; since all functors are additive, this makes $\xi\zeta$ the identity for L finitely generated and free. Any L is a direct summand of a finitely generated free module F. Since ξ and ζ are natural, their composite maps direct summands to direct summands, hence is the identity for any L.

The torsion products are symmetric in G and A. To show this, construct from the ring R its *opposite ring* R^{op}: The additive group of R^{op} is an isomorphic copy of that of R, under an isomorphism $r \to r^{\mathrm{op}}$; the product in R^{op} is defined by $r^{\mathrm{op}} s^{\mathrm{op}} = (sr)^{\mathrm{op}}$. Each right R-module G becomes a left R^{op}-module via the definition $r^{\mathrm{op}} g = gr$; symmetrically, each left R-module A is a right R^{op}-module.

Proposition 7.5. *The correspondence* $(\mu, L, \nu) \to (\nu, L^*, \mu)$ *is an isomorphism*

$$\mathrm{Tor}_n^R(G,A) \cong \mathrm{Tor}_n^{R^{\mathrm{op}}}(A,G), \quad n=0,1,\ldots.$$

Proof. The complex L^* consists of finitely generated projective R^{op}-modules. Hence the correspondence is well defined; it is clearly an isomorphism.

For a short exact sequence $E=(\varkappa,\sigma)\colon A \rightarrowtail B \twoheadrightarrow C$ and an element $t=(\mu,L,\nu) \in \mathrm{Tor}_n(G,C)$ with $n>0$ a product $Et \in \mathrm{Tor}_{n-1}(G,A)$ may be defined. Regard $\nu\colon L^* \to C$ and E as complexes over C, the first projective and the second exact. By the comparison theorem, there is a chain map φ:

$$\begin{array}{ccccccc} \cdots \to & L^*_{n-1} & \xrightarrow{\partial} & L^*_n & \xrightarrow{\nu} & C & \\ & \downarrow{\scriptstyle \varphi_{n-1}} & & \downarrow{\scriptstyle \varphi_n} & & \| & \\ E\colon 0 \to & A & \to & B & \to & C & \to 0. \end{array} \tag{7.5}$$

Let ${}^{n-1}_{\;0}L$ designate the chain complex of length $n-1$ formed by removing the last module L_n from L, and set

$$E(\mu,L,\nu) = (\mu, {}^{n-1}_{\;0}L, \varphi_{n-1}). \tag{7.6}$$

Theorem 7.6. *For* $E \in \mathrm{Ext}^1(C,A)$ *and* $t \in \mathrm{Tor}_n(G,C)$ *the product* Et *is a well defined element of* $\mathrm{Tor}_{n-1}(G,A)$ *which satisfies the associative laws*

$$\alpha(Et) = (\alpha E)\, t, \qquad (E\gamma)\, t' = E(\gamma_* t'), \qquad E(\eta_* t) = \eta_*(Et), \tag{7.7}$$

for $\alpha\colon A \to A'$, $\gamma\colon C' \to C$, $\eta\colon G \to G'$, *and* $t' \in \mathrm{Tor}_n(G,C')$. *It provides a homomorphism*

$$\mathrm{Ext}^1(C,A) \otimes_Z \mathrm{Tor}_n(G,C) \to \mathrm{Tor}_{n-1}(G,A), \qquad n=1,2,\ldots. \tag{7.8}$$

Proof. Any different choice φ' for the chain transformation φ of (7.5) is homotopic to φ, so there is an $s\colon L^*_n \to A$ with $\varphi'_{n-1} = \varphi_{n-1} + s\partial$.

The product Et defined via φ' is

$$(\mu, {}^{n-1}_{0}L, \varphi'_{n-1}) = (\mu, {}^{n-1}_{0}L, \varphi_{n-1}) + (\mu, {}^{n-1}_{0}L, s\delta).$$

But let ${}^{n}_{1}L$ be L with the first module L_0 removed; then $\partial: {}^{n}_{1}L \to {}^{n-1}_{0}L$ is a chain transformation, and the second term above is $(\pm\mu\partial, {}^{n}_{1}L, s) = (0, {}^{n}_{1}L, s) = 0$. The product Et is thus independent of the choice of φ. If $(\mu'\varrho, L, \nu) = (\mu', L', \nu\varrho^*)$ is an equality in Tor_n for some $\varrho: L \to L'$, then $\varphi\varrho^*$ is a chain transformation, and the products

$$E(\mu'\varrho, L, \nu) = (\mu'\varrho, {}^{n-1}_{0}L, \varphi_{n-1}) = (\mu', {}^{n-1}_{0}L', \varphi_{n-1}\varrho^*_{n-1}) = E(\mu', L', \nu\varrho^*)$$

are equal. Hence Et is well defined.

Consider the associative laws (7.7). If $\alpha: A \to A'$, attach the morphism $E \to \alpha E$ to the bottom of the diagram (7.5). It gives $(\alpha E)(\mu, L, \nu) = (\mu, {}^{n-1}_{0}L, \alpha\varphi_{n-1}) = \alpha(\mu, {}^{n-1}_{0}L, \varphi_{n-1}) = \alpha[E(\mu, L, \nu)]$, proving the first rule of (7.7). A similar large diagram for $\gamma: C' \to C$, $E\gamma \in \operatorname{Ext}^1(C', A)$, and $t' \in \operatorname{Tor}_n(G, C')$ proves the second of (7.7); the third rule is immediate.

To say that (7.8) is a homomorphism is to say that the product Et is additive in each factor E and t separately. But $E(t+t') = Et + Et'$ follows at once by the definition (7.3) of the addition in Tor_n. The other rule $(E_1 + E_2)t = E_1 t + E_2 t$ derives from the definition $E_1 + E_2 = \nabla_A (E_1 \oplus E_2) \Delta_C$ of the addition of extensions. The proof is complete.

Given E and G, a map $E_*: \operatorname{Tor}_n(G, C) \to \operatorname{Tor}_{n-1}(G, A)$ is defined as $E_* t = Et$; hence the long sequence

$$\left.\begin{array}{l} \cdots \to \operatorname{Tor}_n(G,A) \to \operatorname{Tor}_n(G,B) \to \operatorname{Tor}_n(G,C) \xrightarrow{E_*} \\ \qquad \operatorname{Tor}_{n-1}(G,A) \to \operatorname{Tor}_{n-1}(G,B) \to \cdots. \end{array}\right\} \tag{7.9}$$

Its exactness will be proved in the next section by homological means.

An element $S \in \sigma \in \operatorname{Ext}^m(C, A)$ is a long exact sequence which may be written as a composite $S = E_1 E_2 \ldots E_m$ of short exact sequences. Define the product σt to be $E_1(E_2 \ldots (E_m t))$. By (7.7), the result is unchanged by a congruence $(E''\gamma) \circ E' \equiv E'' \circ (\gamma E')$ of long exact sequences, hence gives a well-defined "composite" connecting homomorphism

$$\operatorname{Ext}^m(C, A) \otimes \operatorname{Tor}_n(G, C) \to \operatorname{Tor}_{n-m}(G, A), \qquad n \geqq m. \tag{7.10}$$

Exact sequences in the first argument of Tor_n yield symmetric results. For $E' \in \operatorname{Ext}^1(K, G)$ and $t \in \operatorname{Tor}_n(K, A)$ a product $E't \in \operatorname{Tor}_{n-1}(G, A)$ is defined, with properties as in Thm. 7.6, and yielding composite connecting homomorphisms

$$\operatorname{Ext}^m(K, G) \otimes \operatorname{Tor}_n(K, A) \to \operatorname{Tor}_{n-m}(G, A), \qquad n \geqq m. \tag{7.11}$$

Multiplications by E and E' commute in the following sense:

Theorem 7.7. *Let* $E=(\varkappa,\sigma): A \rightarrowtail B \twoheadrightarrow C$, *and* $E'=(\lambda,\tau): G \rightarrowtail H \twoheadrightarrow K$ *be short exact sequences of left and right modules, respectively, while* $t \in \operatorname{Tor}_n(K,C)$. *If* $n \geqq 2$, $EE't = -E'Et \in \operatorname{Tor}_{n-2}(G,A)$.

Proof. Take $t=(\mu, L, \nu)$. The products Et and $E't$ are calculated from the diagrams

$$\begin{array}{ccccc} L^*_{n-1} & \xrightarrow{\partial} & L^*_n & \xrightarrow{\nu} & C \\ \downarrow{\scriptstyle \varphi_{n-1}} & & \downarrow{\scriptstyle \varphi_n} & & \| \\ A & \xrightarrow{\varkappa} & B & \xrightarrow{\sigma} & C, \end{array} \qquad \begin{array}{ccccc} L_1 & \xrightarrow{\partial} & L_0 & \xrightarrow{\mu} & K \\ \downarrow{\scriptstyle \psi_1} & & \downarrow{\scriptstyle \psi_0} & & \| \\ G & \xrightarrow{\lambda} & H & \xrightarrow{\tau} & K \end{array}$$

as $Et=(\mu, {}^{n-1}_{\;0}L, \varphi_{n-1})$ and $E't=(\psi_1, {}^{n}_{1}L, \nu)$. If $n \geqq 2$, the diagrams do not overlap, so we may calculate $EE't$ from the first diagram if we note that ∂ for L^* and ∂ for $({}^{n}_{1}L)^*$ have opposite signs. Changing the sign of φ in the diagram gives $EE't=(\psi_1, {}^{n-1}_{\;1}L, -\varphi_{n-1})$. Similarly, but without sign trouble, $E'Et=(\psi_1, {}^{n-1}_{\;1}L, \varphi_{n-1})$. Hence $E'E=-EE'$, as asserted.

Exercises

1. By taking L free with a given basis, show that the elements of $\operatorname{Tor}_1(G,C)$ can be taken to be symbols $((g_1, \ldots, g_m), \varkappa, (c_1, \ldots, c_n))$ with $g_i \in G$, $c_j \in C$ and $\varkappa$ an $m \times n$ matrix of entries from R such that $(g_1, \ldots, g_m)\,\varkappa = 0 = \varkappa\,(c_1, \ldots, c_n)'$; here the prime denotes the transpose. Describe the addition of such symbols and show that the equality of such is given by sliding matrix factors of $\varkappa$ right and left.

2. Obtain a similar definition of $\operatorname{Tor}_n(G,C)$.

3. Prove that $\operatorname{Tor}_n(P,C)=0$ for $n>0$ and P projective. (Hint: show first that it suffices to prove this when P is finitely generated.)

The exactness of (7.9) can be proved directly (i.e., without homology) as in the following sequence of exercises.

4. Show that the composite of two successive maps in (7.9) is zero and that the exactness of (7.9) for G finitely generated implies that for all G.

5. For $E'=(\lambda,\tau): G \rightarrowtail H \twoheadrightarrow K$ exact with H free show that $E'_*: \operatorname{Tor}_n(K,C) \to \operatorname{Tor}_{n-1}(G,C)$ is an isomorphism for $n>1$ and a monomorphism with image $\operatorname{Ker}(\lambda \otimes 1_C)$ for $n=1$. (Hint: construct an inverse map.) Show that E'_* maps the displayed portion of (7.9) for $n=1$ isomorphically onto the Ker-Coker sequence of the 2×3 diagram with rows $G \otimes E$ and $H \otimes E$.

6. Prove by induction on n that the displayed portion of (7.9) is exact.

8. Torsion Products by Resolutions

The functor $\operatorname{Ext}^n(C,A)$ can be calculated (Thm. III.6.4) from a projective resolution X of C as $H^n(\operatorname{Hom}_R(X,A))$. There is an analogous calculation for $\operatorname{Tor}_n(G,A)$. If $\varepsilon: X \to G$ is a projective resolution by right R-modules, $X \otimes_R A$ is a complex of abelian groups, with boundary $\partial \otimes 1_A: X_n \otimes A \to X_{n-1} \otimes A$. The comparison theorem for resolutions shows the homology $H_n(X \otimes_R A)$ independent of the choice of X, hence a function of G and A which we call, for the moment,

$$\operatorname{Top}_n^R(G,A)=H_n(X\otimes_R A).$$

It is clearly a functor of A, and also a functor of G. For, given $\eta\colon G\to G'$, choose a projective resolution $\varepsilon'\colon X'\to G'$, lift η to a chain transformation $f\colon X\to X'$, and construct the induced map $f_*\colon H_n(X\otimes A)\to H_n(X'\otimes A)$. By the comparison theorem, any two such f's are homotopic, so f_* depends only on η and gives $\eta_*\colon \operatorname{Top}_n(G,A)\to\operatorname{Top}_n(G',A)$. Thus Top_n is a covariant bifunctor, which we now identify with Tor_n. (Often Tor_n is defined to be what we have called Top_n.)

Theorem 8.1. *For a resolution $\varepsilon\colon X\to G$ of the module G_R and a module ${}_RA$ there is a homomorphism*

$$\omega\colon \operatorname{Tor}_n^R(G,A)\to H_n(X\otimes_R A),\qquad n=0,1,\ldots,\tag{8.1}$$

natural in A. If X is a projective resolution, ω is an isomorphism, natural in G and A.

Sketch. Each (μ,L,ν) of Tor_n consists of a projective complex $\mu\colon L\to G$ of length n over G and an n-cycle $(1,L_n,\nu)\in\operatorname{Tor}_0(L_n,A)$ of the complex $L\otimes A$, hence determines a homology class in $H_n(L\otimes A)$. A comparison $L\to X$ gives a homology class in $H_n(X\otimes A)$, thus an element of Top_n.

Proof. Take $t=(\mu,L,\nu)$ in $\operatorname{Tor}_n(G,A)$. The comparison theorem yields a chain transformation $h\colon L\to X$ of the projective complex L over G (via μ) to the exact complex X over G (via ε). Set

$$\omega(\mu,L,\nu)=\operatorname{cls}(h_n,L_n,\nu)\in H_n(X\otimes A).$$

This makes sense, for $h_n\colon L_n\to X_n$, $\nu\colon L_n^*\to A$, so $(h_n,L_n,\nu)\in\operatorname{Tor}_0(X_n,A)$ $=X_n\otimes A$. It is a cycle there, as

$$\begin{aligned}\partial(h_n,L_n,\nu)&=(\partial h_n,L_n,\nu)=(h_{n-1}\partial,L_n,\nu)\\&=(h_{n-1},L_{n-1},\nu\,\partial^*)=(h_{n-1},L_{n-1},0)=0.\end{aligned}$$

The homology class of this cycle is unique, for if $h'\colon L\to X$ is another chain transformation lifting 1_G, there is a homotopy s with $h'_n=h_n+\partial s_n+s_{n-1}\partial$. Then

$$\begin{aligned}(h'_n,L_n,\nu)&=(h_n,L_n,\nu)+(\partial s_n,L_n,\nu)+(s_{n-1}\partial,L_n,\nu)\\&=(h_n,L_n,\nu)+\partial(s_n,L_n,\nu)+(s_{n-1},L_{n-1},0),\end{aligned}$$

which is the original cycle (h_n,L_n,ν) plus a boundary. Furthermore, if $t=(\mu'\varrho,L,\nu)$ and $t'=(\mu',L',\nu\,\varrho^*)$ for some $\varrho\colon L\to L'$ are equal elements according to the definition of Tor_n, while $h'\colon L'\to X$, then $h'\varrho\colon L\to X$ and $\omega t=\omega t'$ in Tor_0.

To show ω a homomorphism, note that two chain transformations $h^i: L^i \to X$ yield $\nabla(h^1 \oplus h^2): L^1 \oplus L^2 \to X$, and hence that

$$\begin{aligned}\omega[(\mu_1, L^1, \nu_1) + (\mu_2, L^2, \nu_2)] &= \omega[\nabla(\mu_1 \oplus \mu_2), L^1 \oplus L^2, \nabla(\nu_1 \oplus \nu_2)] \\ &= \operatorname{cls}[\nabla(h_n^1 \oplus h_n^2), L_n^1 \oplus L_n^2, \nabla(\nu_1 \oplus \nu_2)] = \omega(\mu_1, L^1, \nu_1) + \omega(\mu_2, L^2, \nu_2).\end{aligned}$$

That ω is natural in A is immediate, while the asserted naturality in G follows by observing that a chain transformation $f: X \to X'$ lifting $\eta: G \to G'$ composes with an $h: L \to X$ to give $fh: L \to X'$.

It suffices to show ω an isomorphism when the resolution X is free. Any homology class in $X \otimes A$ is the class of a cycle in some $X' \otimes A$, where X' is a suitable finitely generated subcomplex of X. By Cor. 7.4, this cycle can be written as $(1, X_n', \nu)$ for some $\nu: X_n'^* \to A$. If the complex X', with $\varepsilon': X' \to G$, is cut off beyond dimension n, it is one of the complexes L used in the definition of Tor_n, so $t = (\varepsilon', X', \nu)$ is an element of $\operatorname{Tor}_n(G, A)$. The injection $\iota: X' \to X$ shows $\omega t = \operatorname{cls}(\iota, X_n', \nu)$. Hence ω is an epimorphism.

It remains to prove ω a monomorphism. Suppose $\omega t = 0$ for some t. This means that the cycle (h_n, L_n, ν) is a boundary in $X \otimes A$, hence also in some $X' \otimes A$ for $X' \subset X$ a finitely generated free subcomplex of X. Choose X' to contain $h(L)$. Then $h: L \to X$ cut down yields $h': L \to X'$ with $(h_n', L_n, \nu) = (1, X_n', \nu h_n'^*)$ the boundary of some $(n+1)$-chain of $X' \otimes A$. By Cor. 7.4, write this chain as $(1, X_{n+1}', \zeta)$ for some $\zeta: X_{n+1}'^* \to A$. Now

$$(1, X_n', \nu h_n'^*) = \partial(1, X_{n+1}', \zeta) = (1, X_n', \zeta \partial^*),$$

so the uniqueness assertion of Cor. 7.4 yields $\nu h_n'^* = \zeta \partial^*$. Let ${}_0^n X'$ be the part of X' from X_0' to X_n' inclusive and ${}^{n+1}_{1}X'$ the part from X_1' to X_{n+1}', so that $h': L \to {}_0^n X'$ and $\partial: {}^{n+1}_{1}X' \to {}_0^n X'$ are chain transformations. The original element t of $\operatorname{Tor}_n(G, A)$ becomes

$$\begin{aligned}(\mu, L, \nu) &= (\varepsilon' h', L, \nu) = (\varepsilon', {}_0^n X', \nu h'^*) \\ &= (\varepsilon', {}_0^n X', \zeta \partial^*) = (\varepsilon' \partial, {}^{n+1}_{1}X', \zeta) = (0, -, -) = 0,\end{aligned}$$

and $t = 0$, as desired. The proof is complete.

It is convenient to have a homomorphism "converse" to ω.

Corollary 8.2. *If $\eta: Y \to G$ is a projective complex over G, there is a homomorphism*

$$\tau: H_n(Y \otimes_R A) \to \operatorname{Tor}_n^R(G, A) \qquad (8.2)$$

natural in A. If Y is a resolution, $\tau = \omega^{-1}$.

Proof. Let X be a projective resolution of G. The comparison theorem lifts 1_G to a chain transformation $f: Y \to X$ such that $f_*: H_n(Y \otimes_R A) \to H_n(X \otimes_R A)$ is independent of the choice of f. Set $\tau = \omega^{-1} f_*$.

The connecting maps $t \to Et$ may also be calculated from resolutions:

Proposition 8.3. *Let* $\varepsilon: X \to G$ *be a projective resolution. Each short exact sequence* $E: A \rightarrowtail B \twoheadrightarrow C$ *of left R-modules yields an exact sequence* $X \otimes E: X \otimes A \rightarrowtail X \otimes B \twoheadrightarrow X \otimes C$ *of complexes with a connecting homomorphism* $\partial_{X \otimes E}: H_n(X \otimes C) \to H_{n-1}(X \otimes A)$. *For each* $t \in \operatorname{Tor}_n(G, C)$,

$$\omega(Et) = (-1)^n \partial_{X \otimes E} \omega t;$$

that is, the isomorphism ω *of Thm. 8.1 commutes with connecting homomorphisms.*

The proof applies the relevant definitions directly and is left to the reader. The exactness of the homology sequence for the sequence $X \otimes E$ of complexes now implies

Theorem 8.4. *A short exact sequence* $E: A \rightarrowtail B \twoheadrightarrow C$ *of left R-modules and a right R-module G yield a long exact sequence*

$$\cdots \to \operatorname{Tor}_n(G, A) \to \operatorname{Tor}_n(G, B) \to \operatorname{Tor}_n(G, C) \xrightarrow{E_*} \operatorname{Tor}_{n-1}(G, A) \to \cdots, \qquad (8.3)$$

ending with $\operatorname{Tor}_0(G, C) = G \otimes C \to 0$. *The map* E_* *is left multiplication by E.*

For a projective module $A = P$, the exactness of a resolution X makes $H_n(X \otimes P)$ and hence $\operatorname{Tor}_n(G, P)$ zero for $n > 0$. Much as for Ext (III.10) we can now characterize Tor by axioms, as follows.

Theorem 8.5. *For a fixed right R-module G the covariant functors* $\operatorname{Tor}_n(G, A)$ *of* A, $n = 0, 1, \ldots$, *taken together with the homomorphisms* $E_*: \operatorname{Tor}_n(G, C) \to \operatorname{Tor}_{n-1}(G, A)$, *natural for short exact sequences E of modules, are characterized up to natural isomorphisms by the properties*

(i) $\operatorname{Tor}_0(G, A) = G \otimes_R A$ *for all A,*

(ii) $\operatorname{Tor}_n(G, F) = 0$ *for* $n > 0$, *and all free F,*

(iii) *The sequence* (8.3) *is exact for all E.*

By symmetry (Prop. 7.5), $\operatorname{Tor}_n(G, A)$ will also yield a long exact sequence when the first argument G is replaced by a short exact sequence; this gives a corresponding characterization of the Tor_n as functors of G for fixed A. For $R = Z$, it follows that Tor_1 for abelian groups agrees with the functor Tor defined by generators and relations in § 6.

Theorem 8.6. *The following properties of a right R-module G are equivalent:*

(i) *For every left R-module C,* $\operatorname{Tor}_1(G, C) = 0$;

(ii) *Whenever* $\varkappa: A \to B$ *is a monomorphism, so is* $1 \otimes \varkappa: G \otimes A \to G \otimes B$;

(iii) *Every exact sequence of left R-modules remains exact upon tensor multiplication by G;*

(iv) *If A is a left module and $G' \rightarrowtail G'' \twoheadrightarrow G$ is exact so is the sequence* $G' \otimes A \rightarrowtail G'' \otimes A \twoheadrightarrow G \otimes A$;

(v) *For every* $_RC$ *and every* $n > 0$, $\operatorname{Tor}_n(G, C) = 0$.

Proof. Clearly (iii) $\Rightarrow$ (ii). Conversely, given (ii), Thm. 5.1 implies that any short exact sequence remains exact upon tensor multiplication by G; since a long exact sequence is a composite of short ones, this gives (iii).

Given (i), the long exact sequences for Tor_n yield (ii) and (iv). Conversely, given (ii), represent C as a quotient $C \cong P/A$ of a projective P, so that

$$0 = \operatorname{Tor}_1(G, P) \to \operatorname{Tor}_1(G, C) \to G \otimes A \to G \otimes P$$

is exact with $G \otimes A \to G \otimes P$ a monomorphism by (ii), and therefore $\operatorname{Tor}_1(G, C) = 0$. The proof that (iv) $\Rightarrow$ (i) is analogous.

Finally, (v) $\Rightarrow$ (i); conversely $C \cong P/A$ and exactness of

$$0 = \operatorname{Tor}_n(G, P) \to \operatorname{Tor}_n(G, C) \to \operatorname{Tor}_{n-1}(G, A)$$

show by induction on n that (i) $\Rightarrow$ (v).

A module G with the equivalent properties listed in Thm. 8.6 is said to be *flat*. Note the analogy: P projective means that the functor $\operatorname{Hom}(P, -)$ preserves exact sequences; G flat means that the functor $G \otimes -$ preserves exact sequences. Every projective module is clearly flat. When $R = Z$, Thm. 6.2 shows that a flat Z-module is just a torsion-free abelian group. Hence a flat module need not be projective.

Exercises

1. If $\eta\colon Y \to A$ is a projective resolution, establish an isomorphism $\omega'\colon \operatorname{Tor}_n^R(G, A) \cong H_n(G \otimes_R Y)$. For $E'\colon G \rightarrowtail H \twoheadrightarrow K$ exact prove that $\omega' E' = \partial_{E \otimes Y} \omega'$.

2. For a projective resolution X of G let $S_n(G, X)$ be the n-fold exact sequence $0 \to \partial X_n \to X_{n-1} \to \cdots \to G \to 0$. Show that the isomorphism ω of Thm. 8.1 is $\omega t = \operatorname{cls} \partial^{-1}[S_n(G, X) t]$.

9. The Tensor Product of Complexes

If K_R and $_RL$ are chain complexes of right and left R-modules, respectively, their *tensor product* $K \otimes_R L$ is the chain complex of abelian groups with

$$(K \otimes_R L)_n = \sum_{p+q=n} K_p \otimes_R L_q, \tag{9.1}$$

with boundary homomorphisms defined on the generators $k \otimes l$ by

$$\partial(k \otimes l) = \partial k \otimes l + (-1)^{\deg k} k \otimes \partial l. \tag{9.2}$$

If K and L are positive complexes, so is $K \otimes L$, and the direct sum in (9.1) is finite, with p running from 0 to n. The boundary formula (9.2) resembles that for the derivative of a product of two functions; the sign $(-1)^{\deg k}$ appears in accord with the *standard sign commutation rule*: Whenever two symbols u and v with degrees are interchanged, affix the sign $(-1)^e$ with $e=(\text{degree } u)(\text{degree } v)$. In the second term of (9.2), ∂ of degree -1 has been moved past k, hence the sign. With this sign, one checks that $\partial\partial=0$.

If $f\colon K \to K'$ and $g\colon L \to L'$ are chain transformations, the definition $(f \otimes g)(k \otimes l) = fk \otimes gl$ gives a chain transformation $f \otimes g\colon K \otimes L \to K' \otimes L'$; in this way the tensor product is a covariant bifunctor of complexes. For chain homotopies one has

Proposition 9.1. *If $f_1 \simeq f_2\colon K \to K'$ and $g_1 \simeq g_2\colon L \to L'$, then $f_1 \otimes g_1 \simeq f_2 \otimes g_2$. In detail, chain homotopies $s\colon f_1 \simeq f_2$ and $t\colon g_1 \simeq g_2$ yield a homotopy*

$$u\colon f_1 \otimes g_1 \simeq f_2 \otimes g_2\colon K \otimes L \to K' \otimes L' \tag{9.3}$$

given as $u = s \otimes g_1 + f_2 \otimes t$; that is, by

$$u(k \otimes l) = sk \otimes g_1 l + (-1)^{\deg k} f_2 k \otimes tl.$$

This is in accord with the sign convention, since t of degree 1 has been commuted past k.

Proof. First, s and t give homotopies $s \otimes 1\colon f_1 \otimes 1 \simeq f_2 \otimes 1\colon K \otimes L \to K' \otimes L$ and $1 \otimes t\colon 1 \otimes g_1 \simeq 1 \otimes g_2$. Composing these two homotopies (by Prop. II.2.3) gives the result.

Corollary 9.2. *If $f\colon K \to K'$ and $g\colon L \to L'$ are chain equivalences, so is $f \otimes g\colon K \otimes L \to K' \otimes L'$.*

As a first application of the tensor product of complexes, we show that the torsion products can be computed from resolutions of both arguments, as follows.

Theorem 9.3. *If $\varepsilon\colon X \to G$ and $\eta\colon Y \to A$ are projective resolutions of the modules G_R and ${}_RA$, respectively, then $\varepsilon \otimes 1\colon X \otimes Y \to G \otimes Y$ induces an isomorphism $H_n(X \otimes_R Y) \cong H_n(G \otimes_R Y)$, and hence an isomorphism*

$$H_n(X \otimes_R Y) \cong \operatorname{Tor}_n^R(G, A), \qquad n = 0, 1, \ldots. \tag{9.4}$$

Proof. Let F^k, for $k=0, 1, \ldots$, be the subcomplex of $X \otimes Y$ spanned by all $X_i \otimes Y_j$ with $j \leqq k$, while M^k is the subcomplex of $G \otimes Y$ consisting of all $G \otimes Y_j$ with $j \leqq k$. Then

$$\left.\begin{array}{l} 0 = F^{-1} \subset F^0 \subset F^1 \subset \cdots \subset X \otimes Y, \\ 0 = M^{-1} \subset M^0 \subset M^1 \subset \cdots \subset G \otimes Y, \end{array}\right\} \tag{9.5}$$

and $\varepsilon \otimes 1$ maps F^k into M^k. Since $\partial(x\otimes y) = \partial x \otimes y \pm x \otimes \partial y$ in $X \otimes Y$, the quotient complex F^k/F^{k-1} is isomorphic to

$$X_0 \otimes Y_k \xleftarrow{\partial\otimes 1} X_1 \otimes Y_k \leftarrow X_2 \otimes Y_k \leftarrow \cdots.$$

Similarly M^k/M^{k-1} consists only of the chain group $G \otimes Y_k$ in dimension k. Because each Y_k is projective and $0 \leftarrow G \leftarrow X_0 \leftarrow X_1 \leftarrow \cdots$ is exact, the sequence

$$0 \leftarrow G \otimes Y_k \leftarrow X_0 \otimes Y_k \leftarrow X_1 \otimes Y_k \leftarrow \cdots$$

is exact. This amounts to the statement that $\varepsilon \otimes 1 : F^k/F^{k-1} \to M^k/M^{k-1}$ induces an isomorphism in homology for all k. On the other hand, $\varepsilon \otimes 1$ maps the exact sequence $F^{k-1} \rightarrowtail F^k \twoheadrightarrow F^k/F^{k-1}$ into the corresponding sequence for the M's, as in the commutative diagram

$$\begin{array}{ccccccccc} H_{n+1}(F^k/F^{k-1}) & \to & H_n(F^{k-1}) & \to & H_n(F^k) & \to & H_n(F^k/F^{k-1}) & \to & H_{n-1}(F^{k-1}) \\ \downarrow & & \downarrow & & \downarrow & & \downarrow & & \downarrow \\ H_{n+1}(M^k/M^{k-1}) & \to & H_n(M^{k-1}) & \to & H_n(M^k) & \to & H_n(M^k/M^{k-1}) & \to & H_{n-1}(M^{k-1}). \end{array}$$

We claim that $H_n(F^k) \to H_n(M^k)$ is an isomorphism for all n and k. This is true for k negative and all n. Now suppose by induction that this is so for smaller k and all n. Thus the four outside vertical maps in the diagram are isomorphisms, so the Five Lemma makes the middle vertical map an isomorphism. This completes the induction.

In dimension n every cycle or boundary of $X \otimes Y$ will appear within F^{n+1}. Hence the isomorphism $H_n(F^k) \cong H_n(M^k)$ for large k (specifically, for $k \geqq n+1$) implies the desired isomorphism $H_n(X \otimes Y) \cong H_n(G \otimes Y)$. Now $H_n(G \otimes Y) \cong \operatorname{Tor}_n(G, A)$ by the symmetric case of Thm. 8.1; hence the result.

A sequence of subcomplexes F^k of $X \otimes Y$ arranged as in (9.5) is called a *filtration* of $X \otimes Y$. The method here used of comparing two complexes via filtrations of each will be formulated in general terms in Chap. XI.

Exercises

1. For complexes K, L, M over a commutative ring, establish the adjoint associativity $\operatorname{Hom}(K \otimes L, M) \cong \operatorname{Hom}(K, \operatorname{Hom}(L, M))$.

2. Let $f: K \to L$ be a chain transformation, F^k a filtration of K and M^k one of L with $f(F^k) \subset M^k$. If $f_*: H_n(F^k/F^{k-1}) \to H_n(M^k/M^{k-1})$ is an isomorphism for all n and k, while for each n there is a k such that the injections induce isomorphisms $H_n(F^k) \cong H_n(K)$, $H_n(M^k) \cong H_n(L)$, show $f_*: H_n(K) \to H_n(L)$ an isomorphism for all n.

3. If $\varepsilon: X \to C$ is a projective resolution and $\eta: A \to Y$ an injective coresolution prove that $\operatorname{Ext}^n(C, A) \cong H^n(\operatorname{Hom}(X, Y))$.

10. The KÜNNETH Formula

The tensor product of complexes corresponds to the cartesian product of spaces X and Y, in the sense that the singular complex $S(X \times Y)$ can be proved (VIII.8) chain equivalent to $S(X) \otimes S(Y)$. This suggests the problem of the present section: To determine the homology of $K \otimes L$ in terms of the homologies of K and L.

The boundary formula (9.2) shows that the tensor product $u \otimes v$ of cycles is a cycle in $K \otimes L$ and that the tensor product of a cycle by a boundary is a boundary. Hence for cycles u and v in K and L, respectively,

$$p(\operatorname{cls} u \otimes \operatorname{cls} v) = \operatorname{cls}(u \otimes v) \tag{10.1}$$

is a well defined homology class in $K \otimes L$, so yields a homomorphism

$$p\colon H_m(K) \otimes_R H_q(L) \to H_{m+q}(K \otimes_R L)$$

of abelian groups, called the (external) *homology product*. The direct sum $\sum H_m \otimes H_q$ for $m+q=n$ is thereby mapped into $H_n(K \otimes_R L)$, and the image gives all of $H_n(K \otimes L)$ under stringent conditions on the modules $B_m(K)$, $C_m(K)$, and $H_m(K)$ of boundaries, cycles, and homology classes of K, respectively:

Theorem 10.1 *(The* KÜNNETH *Tensor Formula.) If L is a complex of left R-modules while K is a complex of right R-modules satisfying*

(i) *$C_n(K)$ and $H_n(K)$ are projective modules, for all n,*
then, for each n, the homology product is an isomorphism

$$p\colon \sum_{m+q=n} H_m(K) \otimes_R H_q(L) \cong H_n(K \otimes_R L). \tag{10.2}$$

This is a consequence of a more general theorem, which among other things shows that the image of p does not usually exhaust $H(K \otimes_R L)$.

Theorem 10.2. *(The* KÜNNETH *Formula.) If L is a complex of left R-modules and K a complex of right R-modules satisfying*

(ii) *$C_n(K)$ and $B_n(K)$ are flat modules, for all n,*
then there is for each dimension n a short exact sequence

$$0 \to \sum_{m+q=n} H_m(K) \otimes_R H_q(L) \xrightarrow{p} H_n(K \otimes_R L) \xrightarrow{\beta} \sum_{m+q=n-1} \operatorname{Tor}_1^R(H_m(K), H_q(L)) \to 0 \tag{10.3}$$

where p is the homology product and β a natural homomorphism.

Neither complex K, L need be positive.

This implies the previous theorem. Indeed, since $H_n(K) \cong C_n(K)/B_n(K)$, the hypothesis (i) that $H_n(K)$ is projective implies that $C_n \twoheadrightarrow H_n$ splits, hence that $B_n(K)$ is a direct summand of the projective module C_n, so is itself projective. Now every projective module is flat (Thm. 8.6), so C_n and B_n are flat, as required for (ii). Moreover, H_m flat makes $\operatorname{Tor}_1(H_m, H_q)=0$, so (10.3) reduces to (10.2).

Before proving Thm. 10.2 we treat the special case when the boundary in K is zero. It suffices to set $K=G$.

Lemma 10.3. *If G is a flat right R-module, $p: G\otimes H_n(L)\cong H_n(G\otimes L)$.*

Proof. Set $H_n=H_n(L)$, $C_n=C_n(L)$, $B_n=B_n(L)$. To say that H_n is the n-th homology group of L is to say that the commutative diagram

$$\begin{array}{ccccccccc} & & 0 & & 0 & & & & \\ & & \uparrow & & \downarrow & & & & \\ 0 & \to & B_n & \to & C_n & \to & H_n & \to & 0 \\ & & \uparrow & & \downarrow & & & & \\ & & L_{n+1} & \xrightarrow{\partial} & L_n & & & & \\ & & & & \downarrow{\scriptstyle\partial} & & & & \\ & & & & L_{n-1} & & & & \end{array}$$

has exact rows and columns. Indeed, the exactness of the long column states that C_n is the kernel of $\partial: L_n\to L_{n-1}$, while the exactness of the short column gives B_n as ∂L_{n+1}, and the exact row defines H_n as C_n/B_n. Now take the tensor product of this diagram with G. Since G is flat, the new diagram is exact, and states that $G\otimes H_n$ is isomorphic, under p, to the homology group $H_n(G\otimes L)$, thus proving the lemma.

To prove Thm. 10.2, we regard the families $C_n=C_n(K)$ and $D_n=K_n/C_n\cong B_{n-1}(K)$ as complexes of flat modules with zero boundary, so that $C\rightarrowtail K\twoheadrightarrow D$ is an exact sequence of complexes. As $D_n\cong B_{n-1}(K)$ is flat by hypothesis, $\operatorname{Tor}_1(D_n, L_q)=0$, so the sequence $E: C\otimes L\rightarrowtail K\otimes L\twoheadrightarrow D\otimes L$ is also an exact sequence of complexes. The usual exact homology sequence for E reads

$$H_{n+1}(D\otimes L)\xrightarrow{E_{n+1}} H_n(C\otimes L)\to H_n(K\otimes L)\to H_n(D\otimes L)\xrightarrow{E_n} H_{n-1}(C\otimes L),$$

with connecting homomorphisms E_n. Equivalently, the sequence

$$0\to \operatorname{Coker} E_{n+1}\to H_n(K\otimes L)\to \operatorname{Ker} E_n\to 0 \qquad (10.4)$$

is exact. We wish to compare this with the sequence (10.3) of the theorem, which also has $H_n(K\otimes L)$ as middle term. Let ∂' denote the map $D_{m+1}\to C_m$ induced by ∂.

The homology module $H_m(K)$ can be described by a short exact sequence S: $D_{m+1}\rightarrowtail C_m\twoheadrightarrow H_m(K)$. Take the tensor product of this sequence with $H_q(L)$. Since C_m is flat, $\operatorname{Tor}_1(C_m, H_q(L))=0$, so the long exact sequence for the torsion product, summed over $m+q=n$, becomes

$$\begin{array}{ccccccc} \sum \operatorname{Tor}_1(H_m(K), H_q(L)) & \xrightarrow{S_*} & \sum D_{m+1}\otimes H_q(L) & \xrightarrow{\partial'\otimes 1} & \sum C_m\otimes H_q(L) & \twoheadrightarrow & \sum H_m(K)\otimes H_q(L) \\ & & \downarrow{\scriptstyle p} & & \downarrow{\scriptstyle p} & & \\ & & H_{n+1}(D\otimes L) & \xrightarrow{E_{n+1}} & H_n(C\otimes L). & & \end{array} \qquad (10.5)$$

Since D_{m+1} and C_m are flat, the vertical maps are isomorphisms by the lemma. If we know that the square part of the diagram is commutative, we get $\mathrm{Ker}(E_{n+1})$ from $\mathrm{Ker}(\partial'\otimes 1)$; explicitly, $\mathrm{Ker}(E_{n+1})\cong \mathrm{Ker}(\partial'\otimes 1)\cong \sum \mathrm{Tor}_1(H(K), H(L))$, and $\mathrm{Coker}\, E_{n+1}\cong \mathrm{Coker}(\partial'\otimes 1)\cong \sum H(K)\otimes H(L)$. Thereby (10.4) becomes the desired sequence (10.3). One checks also that the first map of (10.3) is indeed the homology product, while the second map β is described by the commutativity of the diagram

$$\begin{array}{ccc} H_n(K\otimes L) & \xrightarrow{\beta} & \sum_{m+q=n} \mathrm{Tor}_1(H_{m-1}(K), H_q(L)) \\ \downarrow & & \downarrow S_* \\ H_n((K/C)\otimes L) & \xleftarrow{p} & \sum_{m+q=n} (K/C)_m \otimes H_q(L) \end{array} \tag{10.6}$$

with $K\to K/C$ the canonical projection, p an isomorphism, S_m the short exact sequence S_m: $K_m/C_m \rightarrowtail C_{m-1} \twoheadrightarrow H_{m-1}(K)$, and S_* the sum of the corresponding connecting homomorphisms on Tor_1. This shows β natural, but note that its definition is *not* symmetric in K and L; if $C_n(L)$ and $B_n(L)$ are also flat, symmetric arguments on L will produce a possibly different map β'. We show below (Prop. 10.6) that $\beta=\beta'$ for complexes of abelian groups; we conjecture that this should hold in general.

It remains to show the square in (10.5) commutative. An element $\sum d_i\otimes \mathrm{cls}\, v_i$ in $D_{m+1}\otimes H_q(L)$ is mapped by p to $\mathrm{cls}(\sum d_i\otimes v_i)$. The definition of the connecting homomorphism E_{n+1} reads: Pull the cycle $\sum d_i\otimes v_i$ of $D\otimes L$ back to a chain $\sum k_i\otimes v_i$ in $K\otimes L$, take its boundary $\sum \partial' d_i\otimes v_i$ pulled back to $C\otimes L$ and the homology class of the result. This gives $\mathrm{cls}(\sum \partial' d_i\otimes v_i)=p(\partial'\otimes 1)(\sum d_i\otimes \mathrm{cls}\, v_i)$, hence the commutativity.

In the case of complexes of abelian groups we can say more.

Theorem 10.4. *(The* KÜNNETH *Formula for Abelian Groups.) For (not necessarily positive) chain complexes K and L of abelian groups where no K_n has elements of finite order except* 0, *the sequence* (10.3) *is exact and splits by a homomorphism which is not natural.*

Proof. Since K_n torsion-free implies that its subgroups C_n and B_n are also torsion-free and hence flat (as Z-modules), the previous theorem gives the exact sequence (10.3). It remains to show that it splits. First suppose that both K and L are complexes of free abelian groups K_m and L_q. Then $D_m\cong \partial K_m \subset K_{m-1}$ is a subgroup of a free abelian group, hence is free, so that K_m splits as an extension of C_m by D_m, with $K_m\cong C_m\oplus D_m$. The homomorphism cls: $C_m\to H_m(K)$ can thus be extended to a map φ_m: $K_m\to H_m(K)$ with $\varphi_m c=\mathrm{cls}\, c$ for each cycle c. There is a similar ψ_q: $L_q\to H_q(L)$ for the free complex L. The tensor product of these group homomorphisms yields a map $\varphi\otimes\psi\colon (K\otimes L)_n\to\sum H_m(K)\otimes H_q(L)$; since φ and ψ vanish on boundaries, so does $\varphi\otimes\psi$. There is thus an induced map $(\varphi\otimes\psi)_*\colon H_n(K\otimes L)\to\sum H_m(K)\otimes H_q(L)$. For cycles u and v, $(\varphi\otimes\psi)_*\,p(\operatorname{cls} u\otimes\operatorname{cls} v)=(\varphi\otimes\psi)_*\operatorname{cls}(u\otimes v)=\varphi u\otimes\psi v=\operatorname{cls} u\otimes \operatorname{cls} v$ so $(\varphi\otimes\psi)_*\,p=1$; $(\varphi\otimes\psi)_*$ is a left inverse of p, splitting our sequence.

Now consider any complexes K and L (with K torsion-free). Just below we will show that one can choose a free complex K' and a chain transformation $f\colon K'\to K$ such that each $f_*\colon H_p(K')\to H_p(K)$ is an isomorphism. With a similar choice $g\colon L'\to L$, the naturality of p and β means that the diagram

$$\begin{array}{ccccccccc}
0\to & \sum H(K')\otimes H(L') & \xrightarrow{p} & H(K'\otimes L') & \xrightarrow{\beta} & \sum \operatorname{Tor}(H(K'),H(L')) & \to 0\\
& \downarrow{\scriptstyle f_*\otimes g_*} & & \downarrow{\scriptstyle (f\otimes g)_*} & & \downarrow{\scriptstyle \operatorname{Tor}(f_*,g_*)} & \\
0\to & \sum H(K)\otimes H(L) & \xrightarrow{p} & H(K\otimes L) & \xrightarrow{\beta} & \sum \operatorname{Tor}(H(K),H(L)) & \to 0
\end{array}$$

commutes. By the choice of f and g, the outside vertical maps are isomorphisms. Hence by the Short Five Lemma the middle vertical map is also an isomorphism. The bottom exact sequence is thereby isomorphic to the top exact sequence, which has just been shown to split. Therefore the bottom sequence splits.

This proof, due to A. DOLD, depends on the following useful lemma.

Lemma 10.5. *If K is a complex of abelian groups there exists a complex X of free abelian groups and a chain transformation $f\colon X\to K$ such that $f_*\colon H_n(X)\to H_n(K)$ is an isomorphism for each dimension n.*

Proof. It suffices to take X the direct sum of complexes $X^{(n)}$ with chain transformations $f^{(n)}\colon X^{(n)}\to K$ such that $(f^{(n)})_*\colon H_n(X^{(n)})\cong H_n(K)$ and $H_q(X^{(n)})=0$ for $q\neq n$. For fixed n, construct a diagram

$$\begin{array}{ccccc}
0\to & R_{n+1} & \xrightarrow{j} & F_n & \to 0\\
& \vdots\downarrow{\scriptstyle\eta} & & \downarrow{\scriptstyle\xi} & \\
& K_{n+1} & \to & K_n. &
\end{array}$$

First write the group C_n of n-cycles of K as a quotient of a free group F_n; this gives $\xi\colon F_n\to C_n\subset K_n$. Next take $R_{n+1}=\xi^{-1}B_n$ and $j\colon R_{n+1}\to F_n$ the injection. Since R_{n+1} is free and $\xi j R_{n+1}=\partial K_{n+1}$, ξj lifts to a map η which makes the diagram commute. The top row is now a complex $X^{(n)}$ with homology $F_n/R_{n+1}\cong C_n/B_n=H_n(K)$ in dimension n and all other homology groups zero. The vertical maps constitute a chain transformation which is a homology isomorphism in dimension n, as required.

Thm. 10.4 shows that the homology of $K\otimes L$ is spanned by two types of cycles. Type I is a cycle $u\otimes v$ built from cycles $u\in K$, $v\in L$; in the theorem, $\operatorname{Im} p$ is spanned by the classes of Type I cycles. Secondly, consider a triple $\langle \operatorname{cls} u, m, \operatorname{cls} v\rangle$ in $\operatorname{Tor}_1(H(K), H(L))$; there are then chains k and l with $\partial k = mu$, $\partial l = mv$ for the same integer m; thus

$$(1/m)\partial(k \otimes l) = u \otimes l + (-1)^{n+1} k \otimes v, \qquad \dim u = n$$

is a (Type II) cycle. One may verify that its homology class is determined, modulo $\operatorname{Im} p$, by $\operatorname{cls} u$ and $\operatorname{cls} v$. This yields an expression for β in the KÜNNETH formula, as follows:

Proposition 10.6. *For* $t = \langle \operatorname{cls} u, m, \operatorname{cls} v\rangle \in \operatorname{Tor}_1(H_n(K), H(L))$ *with* $\partial k = mu$, $\partial l = mv$ *the formula* $\gamma t = (-1)^{n+1} \operatorname{cls}(1/m)\,\partial(k \otimes l)$ *defines a homomorphism*

$$\gamma\colon \operatorname{Tor}_1(H(K), H(L)) \to H(K \otimes L)/p[H(K) \otimes H(L)].$$

Under the hypotheses of Thm. 10.4, γ *is an isomorphism and its inverse induces* β.

Proof. Since $D = K/C$, the map $H(K \otimes L) \to H(D \otimes L)$ carries $\gamma t = \operatorname{cls}[(-1)^{n+1} u \otimes l + k \otimes v]$ into $\operatorname{cls}[(k + C) \otimes v]$. The maps $p S_*$ of (10.6) carry t into $(k + C) \otimes \operatorname{cls} v$ and thence into $\operatorname{cls}[(k + C) \otimes v]$. The identity of these two results proves that γ induces β, as stated.

Exercises

1. Show that Thm. 10.1 holds with (i) replaced by either (iii) $C_n(K)$, $B_n(K)$, and $H_n(K)$ are flat modules, for all n, or (iv) $C_n(K)$, $B_n(K)$, and $H_n(L)$ are flat modules, for all n.

2. For K and L finitely generated complexes of free abelian groups, calculate the Betti numbers and the torsion coefficients of $K \otimes L$ from those of K and L. (Cf. II.6; this version gives the original theorem of KÜNNETH [1923, 1924].)

3. Prove Thm 10.4 as follows. It suffices to take K finitely generated, hence to take K elementary (Ex. II.2.2). In this case every cycle of $K \otimes L$ can be written as a sum of cycles of types I and II; deduce that p is a monomorphism and γ of Prop. 10.2 an isomorphism (EILENBERG-MAC LANE [1954, § 12]).

4. State a KÜNNETH formula for $K \otimes L \otimes M$.

5. Using this, establish for abelian groups the isomorphisms

$$\operatorname{Tor}(A, \operatorname{Tor}(B, C)) \cong \operatorname{Tor}(\operatorname{Tor}(A, B), C),$$
$$\operatorname{Ext}(A, \operatorname{Ext}(B, C)) \cong \operatorname{Ext}(\operatorname{Tor}(A, B), C).$$

11. Universal Coefficient Theorems

The various homologies of a complex may now be listed. If K is a complex of right R-modules, while ${}_RA$ and G_R are modules, regard A as a complex (with trivial grading $A = A_0$ and boundary $\partial = 0$), so that

$$K, \quad K \otimes_R A, \quad \operatorname{Hom}_R(K, R), \quad \operatorname{Hom}_R(K, G)$$

are complexes derived from K. The homology groups

$$H_n(K \otimes_R A), \qquad H^n(K, G) = H^n(\mathrm{Hom}_R(K, G))$$

are known, respectively, as the n-dimensional homology of K with coefficients A and the n-dimensional cohomology of K with coefficients G. According to our rules for shifting indices up or down, $H^n(\mathrm{Hom}_R(K, G))$ is $H_{-n}(\mathrm{Hom}_R(K, G))$. When K is a positive complex, $H_n(K \otimes_R A) = 0$ for $n < 0$, while $H^n(K, G) = 0$ for $n < 0$; hence the custom of writing the homology index down, the cohomology index up. For K positive, $H_n(K \otimes_R A)$ is sometimes written as $H_n(K, A)$. Warning: Do not shift *this* index up, where it would have a different meaning $H^{-n}(K, A) = H_n(\mathrm{Hom}(K, A))$.

Consider complexes of abelian groups ($R = Z$). If each K_n is free, the universal coefficient theorem (Thm. III.4.1) is an exact sequence

$$0 \to \mathrm{Ext}(H_{n-1}(K), G) \to H^n(K, G) \to \mathrm{Hom}(H_n(K), G) \to 0.$$

We now have a corresponding homology theorem:

Theorem 11.1. *If K is a (not necessarily positive) complex of abelian groups with no elements of finite order and A is an abelian group, there is for each dimension n a split exact sequence of groups*

$$0 \to H_n(K) \otimes A \xrightarrow{p} H_n(K \otimes A) \to \mathrm{Tor}(H_{n-1}(K), A) \to 0 \qquad (11.1)$$

with both homomorphisms natural and p defined for a cycle u of K by $p(\mathrm{cls}\, u \otimes a) = \mathrm{cls}(u \otimes a)$. If K is a complex of vector spaces over some field and V a vector space over the same field, then p: $H_n(K) \otimes V \cong H_n(K \otimes V)$.

This is a corollary to the previous Thm. 10.4. A direct proof is easy when K is free. Write ∂_n for $\partial \otimes 1 \colon K_n \otimes A \to K_{n-1} \otimes A$. The exact sequence

$$0 \to C_n \to K_n \to C_{n-1} \to H_{n-1} \to 0$$

is a free resolution of H_{n-1}; its tensor product with A then has homology 0 in dimension 2, $\mathrm{Tor}(H_{n-1}, A)$ in dimension 1, and $H_{n-1} \otimes A$ in dimension zero. The first states that $C_n \otimes A$ can be regarded as a subgroup of $K_n \otimes A$; indeed

$$\mathrm{Im}\, \partial_{n+1} \subset C_n \otimes A \subset \mathrm{Ker}\, \partial_n \subset K_n \otimes A.$$

The second states that $\mathrm{Ker}\, \partial_n / C_n \otimes A \cong \mathrm{Tor}(H_{n-1}, A)$; the third (with n replaced by $n+1$) that $C_n \otimes A / \mathrm{Im}\, \partial_{n+1} \cong H_n \otimes A$. Therefore $H_n(K \otimes A) = \mathrm{Ker}\, \partial_n / \mathrm{Im}\, \partial_{n+1}$ is an extension of $H_n \otimes A$ by $\mathrm{Tor}(H_{n-1}, A)$, as asserted by the exact sequence (11.1).

Corollary 11.2. *If K and K' are complexes of abelian groups, each with no elements of finite order, while $f\colon K \to K'$ is a chain transformation with $f_*\colon H_n(K) \cong H_n(K')$ an isomorphism for each n, then $f_*\colon H_n(K \otimes A) \to H_n(K' \otimes A)$ is an isomorphism for every abelian group A and every n.*

Proof. Write the sequences (11.1) for K and K' and apply the Five Lemma, as in the proof of Cor. III.4.6.

These universal coefficient theorems express the homology and cohomology of K with any coefficients in terms of the so-called "integral" homology $H_n(K)$, at least when the K_n are free. If the K_n are free and finitely generated abelian groups, there are corresponding expressions in terms of the "integral" cohomology $H^n(K, Z)$, as in Ex. 2 below.

Exercises

1. For abelian groups K and A construct natural homomorphisms $\mathrm{Hom}(K,Z) \otimes A \to \mathrm{Hom}(K,A)$ and $K \otimes A \to \mathrm{Hom}(\mathrm{Hom}(K,Z),A)$. Show them isomorphisms when K is a finitely generated free group, and chain transformations when K is a complex.

2. Let K be a complex of abelian groups, with each K_n a finitely generated free abelian group. Write $H^n(K)$ for $H^n(\mathrm{Hom}(K,Z))$. Using Ex. 1 and the universal coefficient theorems, establish natural exact sequences

$$0 \to H^n(K) \otimes G \to H^n(K,G) \to \mathrm{Tor}(H^{n+1}(K),G) \to 0,$$

$$0 \to \mathrm{Ext}(H^{n+1}(K),A) \to H_n(K \otimes A) \to \mathrm{Hom}(H^n(K),A) \to 0.$$

3. If K is a complex of finitely generated free abelian groups, show that the n-th Betti number b_n of K (II.2) is the dimension of the vector space $H_n(K \otimes Q)$, where Q is the field of rational numbers.

4. For K as in Ex. 3, and Z_p the field of integers modulo p, calculate the dimension of the vector space $H_n(K \otimes Z_p)$ from the Betti numbers and torsion coefficients of K.

5. If K is a complex of vector spaces over a field F, write K^* for its dual $\mathrm{Hom}(K, F)$. If each K_n is finite dimensional, establish the natural isomorphisms $H^n(K^*) \cong [H_n(K)]^*$.

Notes: Tensor products were long used implicitly; for example, via $G \otimes_R \sum Re_i \cong \sum Ge_i$ or $V \otimes W \cong \mathrm{Hom}(V^*, W)$. Their central role in multilinear algebra was highlighted by Bourbaki's [1948] treatise on this subject. The tensor product for abelian groups was first explicitly defined by Whitney [1938]. The universal coefficient theorem 11.1 was first proved by Čech [1935] who thereby first introduced (but did not name) the torsion product Tor_1. Cartan-Eilenberg used resolutions to define the higher torsion products. The description (§ 6) of Tor_1 for abelian groups by generators and relations (Eilenberg-MacLane [1954, § 12]) is useful in treating the Bockstein spectrum of a complex K of abelian groups (the various $H_n(K, Z_m)$ and their interrelations — Bockstein [1958]; Palermo [1957]). A similar description (Ex. 7.1, 7.2) of Tor_n by generators and relations (Mac Lane [1955]) involves some rather mysterious new functors, the "slide products" (e.g., T in Ex. 6.7) and leads to the conceptual characterization (§ 7) of the elements of Tor_n as triples $\langle \mu, L, \nu \rangle$.

Chapter six

Types of Algebras

1. Algebras by Diagrams

This chapter studies the formal properties of various types of algebras over a fixed commutative ring K, with $\otimes$ short for $\otimes_{\mathsf{K}}$, Hom for $\mathrm{Hom}_{\mathsf{K}}$.

A K-algebra Λ is a ring which is also a K-module such that always

$$k(\lambda_1\lambda_2)=(k\lambda_1)\lambda_2=\lambda_1(k\lambda_2), \quad k\in\mathsf{K}, \quad \lambda_1, \lambda_2\in\Lambda.$$

If 1_Λ is the identity element of Λ, then $I(k)=k1_\Lambda$ defines a ring homomorphism I: $\mathsf{K}\to\Lambda$. Indeed, a K-algebra may be described as a ring Λ together with a ring homomorphism I: $\mathsf{K}\to\Lambda$ such that always $(Ik)\lambda=\lambda(Ik)$; that is, with $I\mathsf{K}$ in the center of Λ.

The product $\lambda_1\lambda_2$ is left and right distributive, so is a K-bilinear function. Hence $\pi(\lambda_1\otimes\lambda_2)=\lambda_1\lambda_2$ determines a K-module homomorphism π: $\Lambda\otimes\Lambda\to\Lambda$. In these terms a K-algebra may be described as a K-module Λ equipped with two homomorphisms

$$\pi=\pi_\Lambda\colon \Lambda\otimes\Lambda\to\Lambda, \quad I=I_\Lambda\colon \mathsf{K}\to\Lambda \tag{1.1}$$

of K-modules such that the diagrams

$$\begin{array}{ccc} \Lambda\otimes\Lambda\otimes\Lambda & \xrightarrow{\pi\otimes 1} & \Lambda\otimes\Lambda \\ \downarrow{\scriptstyle 1\otimes\pi} & & \downarrow{\scriptstyle \pi} \\ \Lambda\otimes\Lambda & \xrightarrow{\pi} & \Lambda \end{array}, \qquad \begin{array}{ccccc} \mathsf{K}\otimes\Lambda & \cong & \Lambda & \cong & \Lambda\otimes\mathsf{K} \\ \downarrow{\scriptstyle I\otimes 1} & & \| & & \downarrow{\scriptstyle 1\otimes I} \\ \Lambda\otimes\Lambda & \xrightarrow{\pi} & \Lambda & \xleftarrow{\pi} & \Lambda\otimes\Lambda \end{array} \tag{1.2}$$

are commutative. Indeed, the first diagram asserts that the product is associative, while the left and right halves of the second diagram state that $I(1_{\mathsf{K}})$ is a left and right identity element for the product in Λ and that $\pi(Ik\otimes\lambda)=k\lambda=\pi(\lambda\otimes Ik)$.

In case K is the ring Z of integers, a Z-algebra is simply a ring, so this gives a diagrammatic definition of a ring, via tensor products of abelian groups. The dual diagrams define a "coring" or a "coalgebra". Algebras may be graded by degrees such that $\deg(\lambda_1\lambda_2)=\deg\lambda_1+\deg\lambda_2$, or may have a differential ∂, with $\partial(\lambda_1\lambda_2)=(\partial\lambda_1)\lambda_2+\lambda_1(\partial\lambda_2)$. This chapter will give a uniform treatment of these various types of algebras and the modules over them. As an illustration of algebras with a differential, we first consider certain resolutions over a polynomial ring.

Let $P=F[x]$ be the usual ring of polynomials in an indeterminate x with coefficients in a field F; actually, P can be regarded as an F-algebra, but for the moment we consider it just as a commutative ring. Since $F=F[x]/(x)$ is the quotient of P by the principal ideal (x) of all multiples of x, we can regard F as a P-module so that $\varepsilon(x)=0$ defines a P-module homomorphism ε: $P\to F$. Form the sequence

$$0\leftarrow F\xleftarrow{\varepsilon}P\xleftarrow{\partial}Pu\leftarrow 0 \tag{1.3}$$

of P-modules, where Pu is the free P-module on one generator u and ∂ the P-module homomorphism with $\partial u = x$. The sequence is exact, so is a free resolution of F. For any P-module A, the group $\operatorname{Ext}^1_P(F,A)$ may be calculated from this resolution as the first cohomology group of the complex

$$\operatorname{Hom}_P(P,A) \to \operatorname{Hom}_P(Pu,A) \to 0.$$

Under the isomorphism $\operatorname{Hom}_P(P,A) \cong A$, this is the complex $\delta\colon A \to A$ with $\delta a = -xa$, so $\operatorname{Ext}^1_P(F,A) \cong A/(x)A$. By taking the tensor product of the resolution (1.3) with a module B, we find $\operatorname{Tor}^P_1(F, B)$ to be the submodule of B consisting of all $b \in B$ with $xb = 0$. For example, $\operatorname{Ext}^1_P(F,F) \cong F$, and $\operatorname{Tor}^P_1(F,F) \cong F$.

Similarly, let $P = F[x, y]$ be the ring of polynomials in two indeterminates x and y over F. If (x, y) denotes the ideal generated by x and y, then $F = P/(x, y)$ is again a P-module and $\varepsilon\colon P \to F$ a P-module homomorphism with $\varepsilon(x) = 0 = \varepsilon(y)$. The kernel of ε can be written as the image of the free P-module on two generators u and v under the module homomorphism $\partial_1\colon Pu \oplus Pv \to P$ with $\partial_1 u = x$, $\partial_1 v = y$. The kernel of this map ∂_1 consists of all $fu + gv$ for polynomials $f, g \in P$ such that $fx + gy = 0$; by the unique factorization of polynomials we must then have $f = -hy$ and $g = hx$ for some polynomial h. This kernel is therefore the image of the free module $P(uv)$ on one generator uv under the homomorphism ∂_2 with $\partial_2(huv) = (hx)v - (hy)u = fu + gv$. Since P has no divisors of zero, ∂_2 is a monomorphism. We have thereby shown that the sequence

$$0 \leftarrow F \overset{\varepsilon}{\leftarrow} P \overset{\partial_1}{\leftarrow} Pu \oplus Pv \overset{\partial_2}{\leftarrow} P(uv) \leftarrow 0 \qquad (1.4)$$

is exact. From this resolution one calculates that $\operatorname{Ext}^1_P(F,F) \cong F \oplus F \cong \operatorname{Tor}^P_1(F,F)$, and $\operatorname{Ext}^2_P(F,F) \cong F \cong \operatorname{Tor}^P_2(F,F)$.

In the resolution (1.4) omit F and write $E = P \oplus Pu \oplus Pv \oplus P(uv)$. Now set $vu = -uv$, $u^2 = 0$, $v^2 = 0$; this makes E a ring, with 1_P acting as the identity and products given, for example, by $(fu)(gv) = (fg)(uv) = -(gv)(fu)$. It is called the "exterior" ring over P in two generators u and v. Its elements may be "graded" by assigning dimensions as $\dim 1_P = 0$, $\dim u = 1 = \dim v$, and $\dim uv = \dim u + \dim v = 2$, in accordance with the usual dimensions for the resolution (1.4). The dimension of a product is then the sum of the dimensions of its factors. Furthermore, the boundary homomorphism in the resolution is now a module homomorphism $\partial\colon E \to E$ of degree -1 with $\partial u = x$, $\partial v = y$, and $\partial(uv) = (\partial u)v - u(\partial v)$. This implies a formula for the differential of a product of two elements e_1, e_2 of E as

$$\partial(e_1 e_2) = (\partial e_1)e_2 + (-1)^{\dim e_1} e_1(\partial e_2). \qquad (1.5)$$

This "Leibniz formula" is typical for a ring which is also a complex. Other examples are found in the next chapter, which may be read in parallel with this one.

Exercises

1. Prove that the three definitions given for a K-algebra are equivalent.
2. If J is an ideal in K, show that K/J is a K-algebra.
3. For $P=F[x,y]$ and A any P-module, show that $\mathrm{Ext}_P^2(F,A)$ is the quotient $A/(xA\cup yA)$, while

$$\mathrm{Ext}_P^1(F,A)\cong[(a_1,a_2)\mid a_1,a_2\in A,\ xa_2=ya_1]/[(xa,\ ya)\mid a\in A].$$

4. Obtain a similar formula for $\mathrm{Tor}_1^P(F,F)$ when $P=F[x,y]$.
5. Obtain a free resolution for F as a module over the polynomial ring $F[x,y,z]$ in three indeterminates.

2. Graded Modules

An (externally) *Z-graded* K-*module* is a family $M=\{M_n, n=0, \pm 1, \pm 2, \ldots\}$ of K-modules M_n; an element m of M_n is also said to be an element of *degree* n in M (briefly, $\deg m=n$). A *graded submodule* $S\subset M$ is a family of submodules $S_n\subset M_n$, one for each n. For two Z-graded modules L and M a *homomorphism* $f\colon L\to M$ of degree d is a family $f=\{f_n\colon L_n\to M_{n+d};\ n\in Z\}$ of K-module homomorphisms f_n. The set of all $f\colon L\to M$ of fixed degree d is a K-module $\mathrm{Hom}_d(L,M)$. The composite of homomorphisms of degrees d and d' has degree $d+d'$. A Z-graded module M may also be written with upper indices as $M^n=M_{-n}$; in particular, $\mathrm{Hom}^d(L,M)=\mathrm{Hom}_{-d}(L,M)$.

A *graded* K-module M is a Z-graded module with $M_n=0$ for $n<0$. These graded modules are of most frequent occurrence, and will be studied below, leaving the reader to formulate the corresponding facts for Z-graded modules. Warning: Many authors use "graded" for our Z-graded and "positively graded" for our graded modules.

A *trivially graded* module M has $M_n=0$ for $n\neq 0$.

The graded K-modules M, with morphisms $\hom(L,M)=\mathrm{Hom}_0(L,M)$ the homomorphisms of degree 0, form a category. Each $f\colon L\to M$ of degree 0 has kernel, image, cokernel, and coimage defined as expected (i.e., termwise for each n); they are graded modules with the usual properties. For fixed degree d, $\mathrm{Hom}_d(L,M)$ is a bifunctor on this category, contravariant in L and covariant in M. Alternatively, the family $\mathrm{Hom}(L,M)=\{\mathrm{Hom}_d(L,M)\}$ is a bifunctor on this category to the category of Z-graded K-modules. Both bifunctors are left exact, in the sense of Thm. I.6.1 and Thm. I.6.2.

The tensor product of two graded modules L and M is the graded module given by

$$(L\otimes M)_n=\sum_{p+q=n} L_p\otimes M_q; \tag{2.1}$$

in brief, the grading in the tensor product is defined by $\deg(l \otimes m) = \deg l + \deg m$. If $f: L \to L'$ and $g: M \to M'$ are homomorphisms of degrees d and e, respectively, then $f \otimes g: L \otimes M \to L' \otimes M'$ is the homomorphism of degree $d+e$ defined by

$$(f \otimes g)(l \otimes m) = (-1)^{(\deg l)(\deg g)} (fl \otimes gm) \tag{2.2}$$

in accord with the sign convention (interchange g and l). For $\deg f = \deg g = 0$, this makes $L \otimes M$ a covariant bifunctor on graded modules to graded modules. It is right exact, as in Thm. V.5.1.

The tensor product for graded modules satisfies the same formal identities as in the ungraded (= trivially graded) case; that is, there are natural isomorphisms of degree 0

$$\alpha: L \otimes (M \otimes N) \cong (L \otimes M) \otimes N, \qquad \alpha[l \otimes (m \otimes n)] = (l \otimes m) \otimes n, \tag{2.3}$$

$$\tau: L \otimes M \cong M \otimes L, \qquad \tau[l \otimes m] = (-1)^{(\deg l)(\deg m)} m \otimes l, \tag{2.4}$$

$$\mathrm{K} \otimes M \cong M \cong M \otimes \mathrm{K}, \qquad k \otimes m \to km \leftarrow m \otimes k. \tag{2.5}$$

Here $\otimes = \otimes_{\mathrm{K}}$, and the ground ring K is regarded as the trivially graded module K with $\mathrm{K}_0 = \mathrm{K}$, $\mathrm{K}_n = 0$ for $n \neq 0$. We regard these isomorphisms as identities. This we can do because they are manifestly consistent with each other: Given any two iterated tensor products of the same modules $M_1, \ldots, M_s$, a suitable combination of these isomorphisms provides a canonical map of the first tensor product into the second—deleting or adding factors K at will, and with sign according to the sign conventions, as in (2.4).

The same properties of $\mathrm{Hom}_{\mathrm{K}}$ and $\otimes_{\mathrm{K}}$ hold for Z-graded modules and in a variety of other cases, as follows.

A *bigraded* K-module B is a family $B = \{B_{p,q} \mid p, q \in Z\}$ of K-modules with $B_{p,q} = 0$ when $p < 0$ or $q < 0$; a homomorphism $f: B \to B'$ of *bidegree* (d, e) is a family $\{f_{p,q}: B_{p,q} \to B'_{p+d,q+e}\}$ of K-module homomorphisms. For example, the tensor product of two graded modules L and M is initially a bigraded module $\{L_p \otimes M_q\}$, which the summation (2.1) has turned into a singly graded module. Similarly, the tensor product of two bigraded modules B and C is a 4-graded module which yields a bigraded module by

$$(B \otimes C)_{m,n} = \sum_{p+q=m} \sum_{r+s=n} B_{p,r} \otimes C_{q,s}. \tag{2.6}$$

An element of $B_{p,q}$ is said to have *total* degree $p+q$. The natural isomorphisms (2.3), (2.4), and (2.5) hold for bigraded modules when the total degrees are used in the sign of the transposition τ.

Trigraded modules, Z-bigraded modules, and the like are defined similarly.

An *internally graded* K-module A is a K-module with a given direct sum decomposition $A=\sum A_n$; in other words, A and its submodules A_n, $n=0, 1, \ldots$, are given so that each element $a\neq 0$ in A has a unique representation as a finite sum of non-zero elements from different submodules A_n. The elements of A_n are said to be *homogeneous* elements of A, of degree n. Each internally graded module A determines an externally graded module $\{A_n\}$. Conversely, each externally graded module $M=\{M_n\}$ determines an associated internally graded module $M_*=\sum M_n$. Moreover, $(L\otimes M)_*=L_*\otimes M_*$, but $\operatorname{Hom}(L_*, M_*)$ is larger than $[\operatorname{Hom}(L, M)]_*$, because a K-module homomorphism $f: L_*\to M_*$ need not be a sum of a finite number of homogeneous homomorphisms.

In much of the literature, "graded module" means internally graded module. Following a suggestion of JOHN MOORE, we have chosen to work with external gradings. This choice has the advantage that in either event we always operate with the homogeneous elements and not with the sums $m_0+\cdots+m_n$ of elements of different degrees. Similarly, one needs only the homogeneous homomorphisms $L\to M$, not the arbitrary homomorphisms $L_*\to M_*$. Moreover, our choice dispenses with the use of infinite direct sums, so that we can define a graded object M over any category $\mathcal{M}$ to be a family $\{M_n\}$ of objects in $\mathcal{M}$, with morphisms of various degrees, just as for modules. For example, a *graded set* S is a family of sets $\{S_n, n=0, 1, 2, \ldots\}$.

3. Graded Algebras

A *graded* K-*algebra* Λ is a graded K-module equipped with two K-module homomorphisms $\pi=\pi_\Lambda: \Lambda\otimes\Lambda\to\Lambda$ and $I=I_\Lambda: \mathrm{K}\to\Lambda$, each of degree 0, which render commutative the diagrams

$$\begin{array}{ccc} \Lambda\otimes\Lambda\otimes\Lambda & \xrightarrow{\pi\otimes 1} & \Lambda\otimes\Lambda \\ \downarrow{\scriptstyle 1\otimes\pi} & & \downarrow{\scriptstyle\pi} \\ \Lambda\otimes\Lambda & \xrightarrow{\pi} & \Lambda \end{array}, \qquad \begin{array}{ccccc} \mathrm{K}\otimes\Lambda & = & \Lambda & = & \Lambda\otimes\mathrm{K} \\ \downarrow{\scriptstyle I\otimes 1} & & \| & & \downarrow{\scriptstyle 1\otimes I} \\ \Lambda\otimes\Lambda & \xrightarrow{\pi} & \Lambda & \xleftarrow{\pi} & \Lambda\otimes\Lambda. \end{array} \tag{3.1}$$

The first asserts that the "product" $\lambda\mu=\pi(\lambda\otimes\mu)$ is associative, and the second that $I_\Lambda(1_{\mathrm{K}})=1_\Lambda$ is a two-sided identity for this product. A *homomorphism* $f: \Lambda\to\Lambda'$ between two graded algebras over the same K is a homomorphism of degree 0 of graded K-modules such that the diagrams

$$\begin{array}{ccc} \Lambda\otimes\Lambda & \xrightarrow{\pi_\Lambda} & \Lambda \\ \downarrow{\scriptstyle f\otimes f} & & \downarrow{\scriptstyle f} \\ \Lambda'\otimes\Lambda' & \xrightarrow{\pi_{\Lambda'}} & \Lambda', \end{array} \qquad \begin{array}{ccc} \mathrm{K} & \xrightarrow{I_\Lambda} & \Lambda \\ \| & & \downarrow{\scriptstyle f} \\ \mathrm{K} & \xrightarrow{I_{\Lambda'}} & \Lambda' \end{array} \tag{3.2}$$

are commutative.

These definitions may be restated in terms of elements. A graded algebra Λ is a family of K-modules $\{\Lambda_n, n=0, 1, \ldots\}$ with a distinguished element $1 \in \Lambda_0$ and a function which assigns to each pair of elements λ, μ a product $\lambda\mu$ which is K-bilinear and such that always

$$\deg(\lambda\mu) = \deg\lambda + \deg\mu,$$

$$\lambda(\mu\nu) = (\lambda\mu)\nu, \quad 1\lambda = \lambda = \lambda 1.$$

Similarly, an algebra homomorphism $f\colon \Lambda \to \Lambda'$ is a function carrying elements of Λ to those of Λ' so as to preserve all the structures involved:

$$\left.\begin{array}{lll} f(\lambda+\mu) = f\lambda + f\mu, \quad f(k\lambda) = k(f\lambda), & & \text{(module structure)},\\ \deg(f\lambda) = \deg\lambda, & & \text{(grading)},\\ f(\lambda\mu) = (f\lambda)(f\mu), \quad f(1_\Lambda) = 1_{\Lambda'} & & \text{(product)}. \end{array}\right\} \quad (3.3)$$

We emphasize that each homomorphism f takes the identity to the identity.

As for rings, we also assume for algebras that $1 \neq 0$.

A *graded subalgebra* $\Sigma \subset \Lambda$ is a graded submodule of Λ such that $1_\Lambda \in \Sigma$ and $\sigma, \sigma' \in \Sigma$ imply $\sigma\sigma' \in \Sigma$. Thus Σ is itself a graded algebra, with the same identity as Λ, and the injection $i\colon \Sigma \to \Lambda$ is a monomorphism of graded K-algebras. If $f\colon \Lambda \to \Lambda'$ is an algebra homomorphism, the image $f(\Sigma)$ is a graded subalgebra of Λ'.

A graded *left ideal* $L \subset \Lambda$ is a graded submodule of Λ such that $\Lambda L \subset L$ (i.e., $\lambda \in \Lambda$ and $l \in L$ imply $\lambda l \in L$). Thus L is closed under products, but need not be a subalgebra since it may not contain the identity 1_Λ. If $a_1, \ldots, a_s$ are elements of Λ, the smallest graded left ideal containing all a_i is often denoted by $\Lambda(a_1, \ldots, a_s)$ or simply by $(a_1, \ldots, a_s)$, with Λ understood. In degree n, it consists of all sums $\sum \lambda_i a_i$ with $\lambda_i \in \Lambda$ of degree $n - \deg a_i$. A graded right ideal $R \subset \Lambda$ is similarly defined by the condition that $R\Lambda \subset R$.

A *graded* (two-sided) *ideal* J of Λ is a graded submodule which is both a left and a right graded ideal of Λ. The quotient module Λ/J is a graded algebra with a product determined by the condition that the projection $\eta\colon \Lambda \to \Lambda/J$ is a homomorphism of graded algebras. This quotient algebra, with the map η, is characterized up to isomorphism by the fact that any homomorphism $f\colon \Lambda \to \Lambda'$ of graded algebras with $f(J) = 0$ has a unique factorization as $f = g\eta$ for some algebra homomorphism $g\colon \Lambda/J \to \Lambda'$. Moreover, the kernel of any homomorphism $f\colon \Lambda \to \Lambda'$ of graded algebras is an ideal of Λ. (Note: In case $J = \Lambda$, the quotient "ring" $\Lambda/J = 0$ has $1 = 0$, counter to our convention $1 \neq 0$.)

A graded algebra Λ is commutative (some authors say *skew-commutative* or *anti-commutative*) if always

$$\lambda\mu = (-1)^{\deg\lambda \deg\mu} \mu\lambda; \qquad (3.4)$$

that is, if $\pi_\Lambda = \pi_\Lambda \tau: \Lambda \otimes \Lambda \to \Lambda$, with τ the transposition (2.4). In consequence, the elements of even degree commute in the ordinary sense. If, in addition, $\lambda^2 = 0$ for every element λ of odd degree, Λ is called *strictly commutative*. If the ground ring K is a field of characteristic not 2, then any commutative graded K-algebra is strictly commutative, for (3.4) with deg λ odd gives $\lambda\lambda = -\lambda\lambda$, $2\lambda^2 = 0$, so 2^{-1} in K implies $\lambda^2 = 0$.

For example, a *graded polynomial algebra* $P = P_{\mathsf{K}}[x]$ may be defined for an "indeterminate" x of any degree $d \geqq 0$. If $d = 0$, P is the ordinary ring of polynomials in x with coefficients in K. For $d > 0$, P is the graded module with $P_n = 0$ for $n \not\equiv 0 \pmod{d}$, while P_{qd} is the free K-module on one generator x^q for each $q \geqq 0$; the product is defined by $x^p x^q = x^{p+q}$. If d is even, this polynomial algebra is commutative. P is characterized up to isomorphism by the fact that it is free in x: For any graded K-algebra Λ with a selected element λ_d of degree d there is a unique homomorphism $f: P \to \Lambda$ of graded algebras with $fx = \lambda_d$.

The *exterior algebra* $E = E_{\mathsf{K}}[u]$ on one symbol u of odd degree d is constructed from the free K-module Ku with one generator u as the graded algebra E with $E_0 = \mathsf{K}$, $E_d = \mathsf{K}u$, $E_n = 0$ for $0 \neq n \neq d$, and with product determined by $1u = u = u1$, $u^2 = 0$. It is strictly commutative. We may also define E as the quotient algebra $P_{\mathsf{K}}[x]/(x^2)$, where x is an indeterminate of degree d and (x^2) denotes the (two-sided) ideal in P generated by x^2. The algebra E may be characterized as the strictly commutative algebra free on u: Given any strictly commutative Λ with a selected element $\lambda_d \in \Lambda_d$, there is a unique homomorphism $f: E_{\mathsf{K}}[u] \to \Lambda$ of graded algebras with $f(u) = \lambda_d$.

The *tensor algebra* $T(M)$ of a K-module M is the graded K-module

$$T_0(M) = \mathsf{K}, \quad T_n(M) = M^n = M \otimes \cdots \otimes M \qquad (n \text{ factors}),$$

with product given by the identification map $\pi: M^p \otimes M^q \cong M^{p+q}$. In other words, the product is formed by juxtaposition, as in

$$(m_1 \otimes \cdots \otimes m_p)(m_1' \otimes \cdots \otimes m_q') = m_1 \otimes \cdots \otimes m_p \otimes m_1' \otimes \cdots \otimes m_q'.$$

Clearly T is a covariant functor on K-modules to graded K-algebras. More generally, if M is a graded K-module, a tensor algebra $T(M)$ is defined similarly, with

$$T_0(M) = \mathsf{K} \oplus \sum_{p=1}^{\infty} (M_0)^p, \qquad T_n(M) = \sum M_{d_1} \otimes \cdots \otimes M_{d_t},$$

where the second sum is taken over all d_i with $d_1 + \cdots + d_t = n$. For $M = M_1$ this includes the previous case. The graded algebra $T(M)$ with the obvious K-module injection $M \to T(M)$ (of degree 0) is characterized up to isomorphism by the following "universal" property:

Proposition 3.1. *If M is a graded modue land Λ a graded algebra over* K, *each homomorphism* $g: M \to \Lambda$ *of graded modules, of degree zero, extends to a unique homomorphism* $f: T(M) \to \Lambda$ *of graded algebras.*

Proof. Set $f(m_1 \otimes \cdots \otimes m_p) = (g m_1) \ldots (g m_p)$.

In particular, if M is the free graded K-module F on n free generators $x_1, \ldots, x_n$, each of a given degree, $T(F)$ is the *free graded algebra* on these generators, in the sense that any set map $\xi: \{x_1, \ldots, x_n\} \to \Lambda$ of degree zero extends to a unique homomorphism $T(F) \to \Lambda$ of graded algebras. When F has just one generator x, $T(F)$ is the polynomial algebra on the indeterminate x; when V is a vector space over a field K, $T(V)$ is the tensor algebra of V over K, consisting of all covariant tensors in any number of indices (cf. V.2).

The ground ring K itself is a graded K-algebra, with trivial grading. An *augmented* graded algebra is a graded algebra Λ together with a homomorphism $\varepsilon: \Lambda \to$ K of graded algebras. The polynomial, exterior, and tensor algebras each have an evident such augmentation. An augmented algebra has been called a "supplemented" algebra (CARTAN-EILENBERG). In the present book, an "augmentation" of an object C in a category $\mathscr{C}$ will always mean a morphism $\varepsilon: C \to B$ into some fixed "base" object B of $\mathscr{C}$. In the category of K-algebras, the base object is the algebra K; in the category of chain complexes of abelian groups, it is the trivial complex Z, and so on.

Starting with graded K-modules, we have defined graded K-algebras by the product and identity element morphisms π and I which make the diagrams (3.1) commutative. By starting with other types of modules, we get the corresponding types of algebras. Thus, the diagrams (1.2) for (ungraded) K-modules define K-algebras; call them *ungraded* K-algebras when a distinction is necessary. Similarly Z-graded modules yield *Z-graded algebras*, bigraded modules, *bigraded algebras*, and internally graded modules yield *internally graded algebras*. As before, internally and externally graded algebras are equivalent: Each graded algebra Λ determines an internally graded algebra $\Lambda_* = \sum \Lambda_n$, with product given by bilinearity as in

$$(\lambda_0 + \cdots + \lambda_p)(\mu_0 + \cdots + \mu_q) = \sum \lambda_i \mu_j, \quad \lambda_i \in \Lambda_i, \ \mu_j \in \Lambda_j.$$

Note that a graded algebra isn't an algebra, but that an internally graded algebra may be regarded simply as an algebra (ignore the grading). The internally graded ideals, defined as above, are usually called *homogeneous ideals*; they are among the ideals of the associated ungraded algebra.

Exercises

1. Describe the free graded K-module on any graded set of generators.

2. Describe the bigraded tensor algebra of a bigraded module, and prove the analogue of Prop. 3.1.

3. Let S be a set of elements in a graded algebra Λ. Show that the set of all homogeneous sums of products $\lambda s \lambda'$, for $s \in S$, is a graded ideal in Λ and is the smallest ideal containing S. It is called the *ideal generated* by S (or, *spanned* by S).

4. Show that a graded K-algebra may be described as a graded ring R equipped with a homomorphism I: $\mathrm{K} \to R$ of graded rings such that always $(I k) r = r (I k)$.

4. Tensor Products of Algebras

The *tensor product* of two graded K-algebras Λ and Σ is their tensor product $\Lambda \otimes \Sigma$, as graded modules, with product map defined as the composite

$$(\Lambda \otimes \Sigma) \otimes (\Lambda \otimes \Sigma) \xrightarrow{1 \otimes \tau \otimes 1} \Lambda \otimes \Lambda \otimes \Sigma \otimes \Sigma \xrightarrow{\pi \otimes \pi} \Lambda \otimes \Sigma, \tag{4.1}$$

where τ is the (signed) transposition (2.4) of Σ and Λ, and with identity-element map given by $I \otimes I$: $\mathrm{K} = \mathrm{K} \otimes \mathrm{K} \to \Lambda \otimes \Sigma$. In terms of elements, the product is given by

$$(\lambda \otimes \sigma)(\lambda' \otimes \sigma') = (-1)^{\deg \sigma \deg \lambda'} \lambda \lambda' \otimes \sigma \sigma'$$

and the identity of $\Lambda \otimes \Sigma$ is $1_\Lambda \otimes 1_\Sigma$. The axioms for a graded algebra all hold. If f: $\Lambda \to \Lambda'$ and g: $\Sigma \to \Sigma'$ are homomorphisms of graded algebras, so is $f \otimes g$: $\Lambda \otimes \Sigma \to \Lambda' \otimes \Sigma'$. Also, $\lambda \to \lambda \otimes 1_\Sigma$, $\sigma \to 1_\Lambda \otimes \sigma$ define homomorphisms

$$\Lambda \to \Lambda \otimes \Sigma \leftarrow \Sigma$$

of graded algebras. With these mappings, the tensor product $\Lambda \otimes \Sigma$ is characterized up to isomorphism by the following property:

Proposition 4.1. *If f: $\Lambda \to \Omega$ and g: $\Sigma \to \Omega$ are homomorphisms of graded K-algebras such that always*

$$(f \lambda)(g \sigma) = (-1)^{\deg \lambda \deg \sigma} (g \sigma)(f \lambda), \tag{4.2}$$

there is a unique homomorphism h: $\Lambda \otimes \Sigma \to \Omega$ of graded algebras with $h(\lambda \otimes 1) = f(\lambda)$, $h(1 \otimes \sigma) = g(\sigma)$.

The proof is left to the reader (set $h(\lambda \otimes \sigma) = f(\lambda)\, g(\sigma)$).

If Ω is commutative, condition (4.2) holds automatically. Thus in the category of commutative graded algebras, $\Lambda \to \Lambda \otimes \Sigma \leftarrow \Sigma$ is a universal diagram with ends Λ and Σ. In the category of all (not necessarily commutative) algebras, the universal diagram requires a free product [Cohn 1959], the couniversal diagram a direct product $\Lambda \times \Sigma$ as defined below in (VII.5.1).

The tensor product of algebras, with this characterization, applies in all the other relevant cases: The tensor product of K-algebras (trivial grading); of rings ($\mathrm{K} = Z$); of bigraded algebras. In each case the tensor product of algebras is commutative (τ: $\Lambda \otimes \Sigma \cong \Sigma \otimes \Lambda$) and associative, and satisfies $K \otimes \Lambda \cong \Lambda$; in other words, the natural isomorphisms (2.3) — (2.5) hold for algebras. The tensor product of algebras is also called their KRONECKER *product* or, in older literature, their "direct" product.

We also assign to each graded algebra Λ a graded *opposite algebra* Λ^{op}. This is defined to be the graded K-module Λ with the same identity element and the new product $\pi_\Lambda \tau \colon \Lambda \otimes \Lambda \to \Lambda$ (transpose the order of the factors, with the appropriate sign, then multiply). To avoid the inconvenience of writing two different products for the same pair of elements, we also say that the underlying graded module of Λ^{op} is an isomorphic copy of that of Λ, under an isomorphism $\lambda \to \lambda^{\mathrm{op}}$, and with product defined by $\lambda^{\mathrm{op}} \mu^{\mathrm{op}} = (-1)^{\deg\lambda \deg\mu} (\mu\lambda)^{\mathrm{op}}$. This product is clearly associative. For example, if Λ is the (trivially graded) algebra of $n \times n$ matrices with entries in K, with the usual "row-by-column" matrix multiplication, then Λ^{op} is the ring of $n \times n$ matrices with "column-by-row" multiplication. The same construction of an opposite applies to rings (as already noted in V.7), to bigraded algebras, etc., and in each case there are natural isomorphisms

$$(\Lambda^{\mathrm{op}})^{\mathrm{op}} \cong \Lambda, \qquad (\Lambda \otimes \Sigma)^{\mathrm{op}} \cong \Lambda^{\mathrm{op}} \otimes \Sigma^{\mathrm{op}}. \tag{4.3}$$

The tensor product may be used to construct various examples of algebras, as follows.

Let $P_K[x_i]$, $i=1, \ldots, n$, be the graded polynomial algebra (§ 3) on the indeterminate x_i of even degree $d_i \geqq 0$. The commutative graded algebra

$$P_K[x_1, \ldots, x_n] = P_K[x_1] \otimes \cdots \otimes P_K[x_n] \tag{4.4}$$

is called the *graded* polynomial algebra on the given x_i. In each dimension m, $P_K[x_1, \ldots, x_n]$ is the free K-module on all

$$x_1^{e_1} \otimes \cdots \otimes x_n^{e_n} \quad \text{with} \quad e_1 d_1 + \cdots + e_n d_n = m \quad (\text{if } e_i = 0, \text{ read } x_i^0 \text{ as } 1_K);$$

two such generators are multiplied by adding the corresponding exponents. This polynomial algebra is the *free* commutative algebra on the generators x_i of even degree, in the sense of the following characterization.

Proposition 4.2. *If Λ is a commutative graded algebra, any set map $\xi\colon \{x_1, \ldots, x_n\} \to \Lambda$ with $\deg(\xi x_i) = \deg x_i$ for all i extends to a unique homomorphism $f\colon P_K[x_1, \ldots, x_n] \to \Lambda$ of graded algebras.*

Proof. Since $P_K[x_i]$ is free on x_i, the correspondence $x_i \to \xi x_i$ extends to an algebra homomorphism $f_i\colon P_K[x_i] \to \Lambda$. Since Λ is commutative, these f_i combine by Prop. 4.1 to give a unique $f\colon P_K \to \Lambda$.

If all x_i have the same degree, this property shows that a change in the order of the indeterminates simply replaces the polynomial algebra by an isomorphic algebra; hence the order of the x_i is irrelevant. If all the x_i have degree zero, $P_{\mathsf{K}}[x_1, \ldots, x_n]$ is trivially graded. We may regard it as an ungraded algebra, and denote it as $\mathsf{K}[x_1, \ldots, x_n]$; it is the ordinary polynomial algebra in n indeterminates over K. For n given constants $k_i \in \mathsf{K}$, Prop. 4.2 yields a unique homomorphism $f\colon P_{\mathsf{K}} \to \mathsf{K}$ with $f x_i = k_i$, $i = 1, \ldots, n$. This is the homomorphism obtained by the familiar process of "substituting k_i for x_i, $i = 1, \ldots, n$".

We next construct a similar free strictly commutative algebra with generators u_i of odd degree (degree 1 will suffice). For n letters $u_1, \ldots, u_n$, each of degree 1, the tensor product (over K)

$$E_{\mathsf{K}}[u_1, \ldots, u_n] = E_{\mathsf{K}}[u_1] \otimes \cdots \otimes E_{\mathsf{K}}[u_n]$$

is a strictly commutative graded algebra, called the *exterior algebra* over K with generators $u_1, \ldots, u_n$. As before,

Proposition 4.3. *The exterior algebra $E = E_{\mathsf{K}}[u_1, \ldots, u_n]$ is in degree 1 the free K-module E_1 on the generators $u_1, \ldots, u_n$. If Λ is any strictly commutative graded K-algebra, each module homomorphism $\beta\colon E_1 \to \Lambda_1$ extends to a unique homomorphism $f\colon E \to \Lambda$ of graded algebras.*

The product of two elements e and e' in the exterior algebra is often written as $e \wedge e'$. Clearly E is the free module on all products (in order) of generators u_i; the products of degree $p > 0$ are

$$u_{i_1} u_{i_2} \ldots u_{i_p} = u_{i_1} \wedge u_{i_2} \wedge \ldots \wedge u_{i_p},$$

with $1 \leqq i_1 < i_2 < \cdots < i_p \leqq n$. The number of such products is $(p, n-p)$, where

$$(p, q) = (p+q)!/(p!\,q!) = \binom{p+q}{p} = \binom{p+q}{q} \tag{4.5}$$

is our notation for the binomial coefficient. Any permutation σ of the marks $1, \ldots, p$ can be written as the composite of $\operatorname{sgn}\sigma$ transpositions of adjacent marks, where $\operatorname{sgn}\sigma \equiv 1$ or $0 \pmod 2$ according as the permutation σ is odd or even, so the commutation rule yields

$$u_{i\sigma 1} u_{i\sigma 2} \ldots u_{i\sigma p} = (-1)^{\operatorname{sgn}\sigma} u_{i_1} u_{i_2} \ldots u_{i_p}.$$

The tensor product $\mathsf{K} \otimes_Z \mathsf{K}'$ of two commutative rings is a commutative ring, and the definition of E shows that

$$E_{\mathsf{K}}[u] \otimes_Z E_{\mathsf{K}'}[u'] = E_{\mathsf{K} \otimes \mathsf{K}'}[u, u']. \tag{4.6}$$

There are similar isomorphisms for more u's, more factors, or for E replaced by P. The polynomials on n commuting indeterminates with coefficients in a not necessarily commutative (ungraded) K-algebra Λ may be defined as

$$P_\Lambda[x_1, \ldots, x_n] = \Lambda \otimes_{\mathsf{K}} P_{\mathsf{K}}[x_1, \ldots, x_n]. \tag{4.7}$$

Exercises

1. In any graded algebra Λ let $C = C(\Lambda)$ be the ideal spanned (cf. Ex. 3.3) by all differences $\lambda\mu - (-1)^{mn}\mu\lambda$ for $m = \deg\lambda$, $n = \deg\mu$. Show that Λ/C is commutative, and that any homomorphism of Λ into a commutative algebra factors uniquely through the projection $\Lambda \to \Lambda/C$.

2. The *symmetric algebra* $S(M)$ is defined from the tensor algebra as $S(M) = T(M)/C(T(M))$, for C as in Ex. 1. Show that Prop. 3.1 holds when Λ is commutative and $T(M)$ is replaced by $S(M)$, and that, for M free on a finite set of generators of even degrees, $S(M)$ is the polynomial algebra.

3. Make a similar construction of the exterior algebra on any graded K-module M consisting of elements all of degree 1.

4. In Ex. 2, show $S(M \oplus N) \cong S(M) \otimes S(N)$.

5. Show that a free strictly commutative graded algebra on any finite graded set of generators may be constructed as a tensor product of polynomial and exterior algebras.

6. If $P = \mathrm{K}[x]$ and $Q = P[y]$, show that Q, as an (ungraded) K-algebra, is isomorphic to $\mathrm{K}[x, y]$. Extend this result to the graded case with more indeterminates.

5. Modules over Algebras

Let Λ be a graded K-algebra. A *left Λ-module A* is a graded K-module together with a homomorphism $\pi_A : \Lambda \otimes A \to A$ of graded K-modules, of degree zero, such that the diagrams

$$\begin{array}{ccc} \Lambda\otimes\Lambda\otimes A & \xrightarrow{\pi_\Lambda\otimes 1} & \Lambda\otimes A \\ \downarrow{\scriptstyle 1\otimes\pi_A} & & \downarrow{\scriptstyle \pi_A} \\ \Lambda\otimes A & \xrightarrow{\pi_A} & A \end{array}, \qquad \begin{array}{ccc} \mathrm{K}\otimes A & = & A \\ \downarrow{\scriptstyle I_\Lambda\otimes 1} & & \| \\ \Lambda\otimes A & \xrightarrow{\pi_A} & A \end{array} \tag{5.1}$$

commute. Alternatively, a left Λ-module is a graded abelian group A together with a function assigning to each $\lambda \in \Lambda$ and $a \in A$ an element $\lambda a \in A$ with $\deg(\lambda a) = \deg\lambda + \deg a$ such that always (for $\deg\lambda_1 = \deg\lambda_2$, $\deg a_1 = \deg a_2$)

$$(\lambda_1 + \lambda_2)a = \lambda_1 a + \lambda_2 a, \qquad \lambda(a_1 + a_2) = \lambda a_1 + \lambda a_2, \tag{5.2}$$

$$(\lambda\mu)a = \lambda(\mu a), \qquad 1_\Lambda a = a. \tag{5.3}$$

Indeed, given these conditions, the definition $ka = (k 1_\Lambda)a$ makes A a graded K-module. By (5.3), $(k\lambda)a = k(\lambda a) = \lambda(ka)$ holds. With (5.2) this makes the function λa K-bilinear, so defines π_A as $\pi_A(\lambda \otimes a) = \lambda a$. Finally, (5.3) is a restatement of the commutativity (5.1).

If C and A are left Λ-modules, a *Λ-module homomorphism* $f\colon C \to A$ of degree d is a homomorphism of graded K-modules, of degree d, such that

$$f\pi_C = \pi_A(1 \otimes f)\colon \Lambda \otimes C \to A\,; \tag{5.4}$$

in other words, such that always

$$f(\lambda c) = (-1)^{(\deg f)(\deg\lambda)}\lambda(fc)\,; \tag{5.4'}$$

the usual sign arises from the definition (2.2) of $1 \otimes f$. The set of all such f of degree d is a K-module which we denote as $\operatorname{Hom}^{-d}(C, A)$.

The class ${}_{\Lambda}\mathcal{M}$ of all left Λ-modules is a category with morphisms $\hom_{\Lambda}(C, A) = \operatorname{Hom}^0_{\Lambda}(C, A)$ those of degree 0. In ${}_{\Lambda}\mathcal{M}$, direct sums, sub- and quotient-modules, kernel, image, coimage, and cokernel are defined as expected, with the usual properties. For each n, $\operatorname{Hom}^n(C, A)$ is an additive bifunctor on ${}_{\Lambda}\mathcal{M}$ to K-modules, contravariant in C and covariant in A. The family $\operatorname{Hom}_{\Lambda}(C, A) = \{\operatorname{Hom}^n_{\Lambda}(C, A), n = 0, \pm 1, \pm 2, \ldots\}$ is a similar bifunctor on ${}_{\Lambda}\mathcal{M}$ to Z-graded K-modules. According to the definition (5.4) of a Λ-module homomorphism, we can also describe $\operatorname{Hom}_{\Lambda}(C, A)$ as that Z-graded K-module which is the kernel of the natural homomorphism

$$\psi\colon \operatorname{Hom}(C, A) \to \operatorname{Hom}(\Lambda \otimes C, A), \qquad \operatorname{Hom} = \operatorname{Hom}_{K}, \tag{5.5}$$

of Z-graded K-modules defined by

$$\psi f = \pi_A (1 \otimes f) - f \pi_C\colon \Lambda \otimes C \to A.$$

Proposition 5.1. *The functor* $\operatorname{Hom}_{\Lambda}$ *is left exact; that is, if* $D \to B \to C \to 0$ *is a short right exact sequence in* ${}_{\Lambda}\mathcal{M}$, *then the induced sequence*

$$0 \to \operatorname{Hom}_{\Lambda}(C, A) \to \operatorname{Hom}_{\Lambda}(B, A) \to \operatorname{Hom}_{\Lambda}(D, A) \tag{5.6}$$

is exact, with the corresponding result when A is replaced by a short left exact sequence.

Proof. Construct the commutative 3×3 diagram

$$\begin{array}{ccccccc}
0 \to & \operatorname{Hom}_{\Lambda}(C,A) & \to & \operatorname{Hom}_{\Lambda}(B,A) & \to & \operatorname{Hom}_{\Lambda}(D,A) \\
 & \downarrow & & \downarrow & & \downarrow \\
0 \to & \operatorname{Hom}(C,A) & \to & \operatorname{Hom}(B,A) & \to & \operatorname{Hom}(D,A) \\
 & \downarrow{\scriptstyle\psi} & & \downarrow{\scriptstyle\psi} & & \downarrow{\scriptstyle\psi} \\
0 \to & \operatorname{Hom}(\Lambda \otimes C,A) & \to & \operatorname{Hom}(\Lambda \otimes B,A) & \to & \operatorname{Hom}(\Lambda \otimes D,A).
\end{array}$$

By right exactness of the tensor product (Thm. V.5.1), $\Lambda \otimes D \to \Lambda \otimes B \to \Lambda \otimes C \to 0$ is right exact. The left exactness of Hom_K makes the last two rows left exact; by the definition (5.5), all three columns are left exact (when starting with $0 \to \cdots$). The 3×3 lemma (in the strong form of Ex. II.5.4) now shows the first row left exact, q.e.d.

Right Λ-modules G are treated similarly. A homomorphism $\gamma\colon G \to G'$ of right Λ-modules must satisfy $\gamma(g\lambda) = (\gamma g)\lambda$; no sign is needed (as in (5.4')), because the homomorphism and the module operations act on opposite sides of $g \in G$. A right Λ-module G may also be described as a left Λ^{op}-module, with operators switched by $\lambda^{\mathrm{op}} g = (-1)^{(\deg \lambda)(\deg g)} g \lambda$; this definition insures that $\lambda^{\mathrm{op}}(\mu^{\mathrm{op}} g) = (\lambda^{\mathrm{op}} \mu^{\mathrm{op}}) g$.

Given modules G_Λ and ${}_\Lambda A$, their *tensor product* over Λ is a graded K-module. It is defined to be the cokernel of the map φ of graded K-modules

$$G\otimes\Lambda\otimes A\xrightarrow{\varphi} G\otimes A\to G\otimes_\Lambda A\to 0 \qquad (5.7)$$

given as $\varphi(g\otimes\lambda\otimes a)=g\lambda\otimes a-g\otimes\lambda a$. This amounts to stating that each $(G\otimes_\Lambda A)_n$ is the K-module quotient of $(G\otimes A)_n$ by the submodule generated by all differences $g\lambda\otimes a-g\otimes\lambda a$ in $(G\otimes A)_n$. This tensor product is characterized via the middle linear functions to a graded K-module M, just as in Thm. V.1.1:

Theorem 5.2. *If f is a family of* K*-bilinear functions $f_{p,q}$ on $G_p\times A_q$ to M_{p+q} which is Λ-middle associative in the sense that always $f(g\lambda,a)=f(g,\lambda a)$, there is a unique homomorphism ω: $G\otimes_\Lambda A\to M$ of graded* K*-modules, with* $\deg\omega=0$, *such that always $\omega(g\otimes a)=f(g,a)$.*

Proof. Each $f_{p,q}$ is bilinear, hence determines $\omega'_{p,q}$: $G_p\otimes A_q\to M_{p+q}$ with $\omega'(g\otimes a)=f_{p,q}(g,a)$; the middle associativity insures that ω' vanishes on the image of the map φ of (5.7), and hence that ω' induces a map ω on the cokernel $G\otimes_\Lambda A$ of φ, as desired.

This result implies that Λ-module homomorphisms γ: $G\to G'$ and α: $A\to A'$ of degrees d and e, respectively, determine a homomorphism $\gamma\otimes\alpha$: $G\otimes_\Lambda A\to G'\otimes_\Lambda A'$ of degree $d+e$ of graded K-modules by the formula

$$(\gamma\otimes\alpha)(g\otimes a)=(-1)^{(\deg\alpha)(\deg g)}\gamma g\otimes\alpha a, \qquad (5.8)$$

with the expected rules for composing $(\gamma\otimes\alpha)$ with $(\gamma'\otimes\alpha')$ — with a sign $(-1)^{(\deg\alpha)(\deg\gamma')}$. In particular, $G\otimes_\Lambda A$ is a covariant and biadditive bifunctor on the categories $\mathcal{M}_\Lambda$ and ${}_\Lambda\mathcal{M}$ of right and left Λ-modules to graded K-modules. From the definition (5.7) it follows as in Prop. 5.1 that this functor carries right exact sequences (in G or A) into right exact sequences.

Modules over other types of algebras (Z-graded, bigraded, etc.) are correspondingly defined. Note that each Λ-module A automatically carries the same type of structure as Λ (e.g., graded when Λ is graded, bigraded when Λ is bigraded). We may introduce modules with added structure; thus a graded module over an ungraded algebra Λ means a module over Λ, regarded as a trivially graded algebra — exactly as for graded modules over the commutative ring K.

If Λ and Σ are two graded K-algebras, a Λ-Σ-bimodule A — in symbols ${}_\Lambda A_\Sigma$ — is a graded K-module which is both a left Λ-module and a right Σ-module such that always $(\lambda a)\sigma=\lambda(a\sigma)$. This condition amounts to the commutativity of an appropriate diagram. Note that $ka=(k1_\Lambda)a=a(k1_\Sigma)$, so that the same given K-module structure on A comes from the left Λ-module structure by pull-back along I: $\mathrm{K}\to\Lambda$, or from the right Σ-module structure by pull-back along $I: \mathsf{K} \to \Sigma$. For example, any graded algebra Λ is a Λ-Λ-bimodule. Since Λ is a left Λ-module and Σ a right Σ-module, the tensor product $\Lambda \otimes_{\mathsf{K}} \Sigma$ is a Λ-Σ-bimodule; in fact, the free bimodule on one generator $1 \otimes 1$. Similarly, for modules ${}_{\Lambda}A$ and B_{Σ} the tensor product $A \otimes_{\mathsf{K}} B$ is canonically a Λ-Σ-bimodule via $\lambda(a \otimes b)\sigma = \lambda a \otimes b\sigma$.

The typographical accident that a letter has two sides could hardly mean that modules are restricted to one-sided modules and bimodules. Indeed, we naturally reach trimodules; for example, modules ${}_{\Lambda}A$ and ${}_{\Sigma}B_{\Omega}$ will have a tensor product $A \otimes_{\mathsf{K}} B$ which is canonically a right Ω- and left Λ-Σ-module. Here we have called C a left Λ-Σ-module if it is both a left Λ- and a left Σ-module such that always

$$\lambda(\sigma c) = (-1)^{(\deg \lambda)(\deg \sigma)} \sigma(\lambda c).$$

Fortunately we can reduce trimodules to bimodules or even to left modules over a single algebra. By setting $(\lambda \otimes \sigma)c = \lambda(\sigma c)$, each left Λ-Σ-module may be regarded as a left $(\Lambda \otimes \Sigma)$-module, or conversely. Similarly we have the logical equivalences

$$B_{\Sigma} \Leftrightarrow {}_{\Sigma^{\mathrm{op}}}B, \qquad {}_{\Lambda}A_{\Sigma} \Leftrightarrow {}_{(\Lambda \otimes \Sigma^{\mathrm{op}})}A \tag{5.9}$$

via $\sigma^{\mathrm{op}} b = (-1)^{(\deg \sigma)(\deg b)} b\sigma$, $(\lambda \otimes \sigma^{\mathrm{op}})a = (-1)^{(\deg \sigma)(\deg a)} \lambda a \sigma$. This reduction carries with it the definitions of Hom and $\otimes$ for bimodules. Thus for bimodules ${}_{\Sigma}G_{\Lambda}$ and ${}_{\Lambda}A_{\Sigma}$ the *bimodule tensor product*

$$G \otimes_{\Lambda-\Sigma} A = G \otimes_{(\Lambda \otimes \Sigma^{\mathrm{op}})} A \tag{5.10}$$

is by (5.7) the quotient of $G \otimes_{\mathsf{K}} A$ by the graded K-submodule spanned by all

$$g\lambda \otimes a - g \otimes \lambda a, \quad \sigma g \otimes a - (-1)^{(\deg \sigma)(\deg g + \deg a)} g \otimes a\sigma.$$

The vanishing of the first expression is Λ-middle associativity; that of the second is *Σ-outside associativity*. Similarly the graded K-module of bimodule homomorphisms of ${}_{\Lambda}C_{\Sigma}$ into ${}_{\Lambda}A_{\Sigma}$ is written $\mathrm{Hom}_{\Lambda-\Sigma}(C, A) = \mathrm{Hom}_{(\Lambda \otimes \Sigma^{\mathrm{op}})}(C, A)$.

Exercises

1. An (ungraded) K-algebra Λ is a ring R equipped with a ring homomorphism $I: \mathsf{K} \to R$ with $I(\mathsf{K})$ in the center of R. Show that a left Λ-module A is just a left R-module, with the K-module structure of A given by pull-back along I. Show also that $\mathrm{Hom}_{\Lambda}(C, A) = \mathrm{Hom}_{R}(C, A)$ and $G \otimes_{\Lambda} A = G \otimes_{R} A$.

2. As in Ex. 1, reduce modules over the graded algebra Λ to modules over Λ, regarded as a graded ring (cf. Ex. 3.4).

6. Cohomology of free Abelian Groups

As an illustration of tensor products of algebras we calculate the cohomology of a free abelian group.

For the group ring $Z(\Pi_1 \times \Pi_2)$ of the cartesian product of two multiplicative groups Π_1 and Π_2 there is a natural isomorphism

$$Z(\Pi_1 \times \Pi_2) \cong Z(\Pi_1) \otimes Z(\Pi_2). \tag{6.1}$$

For $Z(\Pi_i)$ is characterized (Prop. IV.1.1) by the fact that any multiplicative map μ_i of Π_i into a ring S, with $\mu_i(1) = 1_S$, extends to a ring homomorphism $Z(\Pi_i) \to S$. By Prop. 4.1, a multiplicative map $\mu: \Pi_1 \times \Pi_2 \to S$ with $\mu(1) = 1$ then extends to a unique ring homomorphism $Z(\Pi_1) \otimes Z(\Pi_2) \to S$, so that $Z(\Pi_1) \otimes Z(\Pi_2)$ satisfies this characterization of the group ring $Z(\Pi_1 \times \Pi_2)$.

Let C_∞ be the infinite (multiplicative) cyclic group with generator t, and $R = Z(C_\infty)$ its group ring. Any element of R is a polynomial in positive, negative, and zero powers of t, hence may be written as $t^m p(t)$ where p is an (ordinary) polynomial in positive powers of t with integral coefficients. The kernel of the augmentation $\varepsilon: R \to Z$ is the set of all multiples of $t-1$, hence the exact sequence

$$0 \leftarrow Z \overset{\varepsilon}{\leftarrow} R \overset{\partial}{\leftarrow} Ru \leftarrow 0 \tag{6.2}$$

with Ru the free R-module with one generator u and $\partial u = t-1$. Thus $\partial: R \leftarrow Ru$ is a free R-module resolution of ${}_\varepsilon Z$; it is a special case of the resolution found in (IV.7.3) for any free group, and is analogous to the resolution (1.3) for a polynomial ring. For any R-module A, $H^1(C_\infty, A)$ may be calculated from this resolution to be the factor group $A/[ta - a \mid a \in A]$, while $H^n(C_\infty, A) = 0$ if $n > 1$.

The free abelian group Π on n generators $t_1, \ldots, t_n$ is the cartesian product of n infinite cyclic groups. By (6.1) the group ring $Z(\Pi)$ is $R^1 \otimes \cdots \otimes R^n$, where each R^i is the group ring $Z(C_\infty(t_i))$, while the augmentation $\varepsilon: Z(\Pi) \to Z$ is the tensor product $\varepsilon^1 \otimes \cdots \otimes \varepsilon^n$ of the augmentations $\varepsilon^i: R^i \to Z$. For each index i form the R^i-projective resolution $X^i: R^i \leftarrow R^i u_i$ as in (6.2). Form the tensor product complex

$$X = X^1 \otimes_Z X^2 \otimes_Z \cdots \otimes_Z X^n;$$

it is a chain complex of free $R^1 \otimes \cdots \otimes R^n = Z(\Pi)$-modules.

On the one hand, each X^i is a complex of free abelian groups. The iterated Künneth tensor formula (Thm. V.10.1) shows that the homology product

$$\sum H_{m_1}(X^1) \otimes \cdots \otimes H_{m_n}(X^n) \to H_m(X^1 \otimes \cdots \otimes X^n)$$

is an isomorphism in each dimension $m = m_1 + \cdots + m_n$. But $H_{m_i}(X^i) = 0$ unless $m_i = 0$, while $\varepsilon_i: H_0(X^i) \cong Z$, so $H_m(X) = 0$ for m positive, while $\varepsilon: H_0(X) \cong Z$. This proves that X is a free resolution of Z as a Π-module.

On the other hand, each X^i is the exterior algebra $E_{R^i}[u^i]$; as in (4.6), X is the exterior algebra $E_{Z(\Pi)}[u_1, \ldots, u_n]$, so has the form of an exact sequence

$$0 \leftarrow Z \leftarrow X_0 \leftarrow X_1 \leftarrow \cdots \leftarrow X_n \leftarrow 0$$

of Π-modules, with each X_p free on the generators $u_{i_1} \otimes \cdots \otimes u_{i_p}$ with $1 \leqq i_1 < \cdots < i_p \leqq n$. Since $\partial u_i = t_i - 1$, the boundary formula (V.9.2) for the tensor product gives

$$\partial (u_{i_1} \otimes \cdots \otimes u_{i_p}) = \sum_{k=1}^{p} (-1)^{k-1} (t_{i_k} - 1) u_{i_1} \otimes \cdots \otimes \hat{u}_{i_k} \otimes \cdots \otimes u_{i_p}, \quad (6.3)$$

where the $\hat{}$ indicates omission. The cohomology of Π may be computed from this resolution. For any Π-module A,

$$\operatorname{Ext}^p_\Pi(Z, A) \cong H^p(\Pi, A) = 0, \qquad p > n. \quad (6.4)$$

For $p \leqq n$, a p-cochain $f\colon X_p \to A$, as a module homomorphism, is determined by $(p, n-p)$ arbitrary elements $f(u_{i_1} \otimes \cdots \otimes u_{i_p}) \in A$, and

$$\delta f (u_{i_1} \otimes \cdots \otimes u_{i_{p+1}}) = \sum_{k=1}^{p+1} (-1)^{k-1} (t_{i_k} - 1) f (u_{i_1} \otimes \cdots \otimes \hat{u}_{i_k} \otimes \cdots \otimes u_{i_{p+1}}).$$

In particular, if A is an abelian group regarded as a trivial Π-module ($t_i a = a$ for all i), then δf is always zero, so $H^p(\Pi, A)$ is simply the direct sum of $(p, n-p)$ copies of A.

Exercises

1. For Π free abelian as above, show that $H^1(\Pi, A)$ is the quotient L/M, where L is the subgroup of $A \oplus \ldots \oplus A$ (n summands) consisting of all $(a_1, \ldots, a_n)$ with $t_i a_j - t_j a_i = a_j - a_i$ always, while M is all $(t_1 a - a, \ldots, t_n a - a)$ for $a \in A$. Interpret this result in terms of classes of crossed homomorphisms.

2. Obtain a similar formula for $H^2(\Pi, A)$ and compare this with the result found for two generators in IV.3.7.

3. Determine $H^n(\Pi, A)$ for Π free abelian on n generators.

7. Differential Graded Algebras

The resolution X of the last section is both a complex and an algebra, with the boundary of a product given by the Leibniz formula (1.5). Such we call a *DG*-algebra. Further examples of *DG*-algebras will appear in the next chapter; they will be used extensively in Chap. X, to which the following systematic development will be relevant.

A positive complex $X = (X, \partial)$ of K-modules is a graded K-module $X = \{X_n\}$ equipped with a K-module homomorphism $\partial = \partial_X\colon X \to X$ of degree -1 such that $\partial^2 = 0$. A positive complex will thus also be called a *differential graded module* (*DG*-module for short); the homology of

X is the graded K-module $H(X)=\{H_n(X)\}$. A chain transformation ($=DG$-module homomorphism) $f\colon X\to X'$ is just a homomorphism of graded modules, of degree 0, with $\partial_{X'}f=f\partial_X$. The set of all such f is an abelian group $\hom(X, X')$; with these morphisms the DG-modules form a category. Similarly, a not necessarily positive complex of K-modules is a differential Z-graded module (DG_Z-module).

The tensor product $X\otimes Y$ of two DG-modules is the tensor product over K of the graded modules X and Y equipped with the differential $\partial=\partial_X\otimes 1+1\otimes\partial_Y$. According to the definition (2.2) of $1\otimes\partial_Y$ this gives

$$\partial(x\otimes y)=\partial x\otimes y+(-1)^{\deg x}x\otimes\partial y, \tag{7.1}$$

in agreement with the previous definition (V.9.2) of the tensor product of chain complexes. This tensor product of DG-modules satisfies the standard natural isomorphisms (2.3), (2.4), and (2.5); in the latter, the ground ring K is regarded as a DG-module with trivial grading and differential $\partial=0$.

For DG-modules X and Y the Z-graded module $\operatorname{Hom}(X,Y)=\{\operatorname{Hom}^n(X,Y)\}$ has a differential defined for each $f\in\operatorname{Hom}^n$ as $\partial_H f=\partial_Y f+(-1)^{n+1}f\partial_X$, as in (III.4.4). Thus $\operatorname{Hom}(X,Y)$ is a DG_Z-module. Note especially that $\operatorname{Hom}(X,Y)$ with capital H in "Hom" stands for homomorphisms of graded modules of all degrees, while $\hom(X,Y)$, with lower case h, includes only the homomorphisms of DG-modules, of degree 0.

A *DG-algebra* $U=(U,\partial)$ over K is a graded algebra U equipped with a graded K-module homomorphism $\partial\colon U\to U$ of degree -1 with $\partial^2=0$, such that the Leibniz formula

$$\partial(u_1u_2)=(\partial u_1)u_2+(-1)^{\deg u_1}u_1(\partial u_2) \tag{7.2}$$

always holds. Similarly, a homomorphism $f\colon U\to U'$ of DG-algebras is a homomorphism of graded algebras (conditions (3.3)) with $\partial f=f\partial$. With these morphisms, the DG-algebras form a category.

By the Leibniz formula the product of two cycles is a cycle, and the product of a boundary ∂u_1 by a cycle u_2 is a boundary $\partial(u_1u_2)$. Hence a product of homology classes in $H(U)$ may be defined by $(\operatorname{cls} u_1)(\operatorname{cls} u_2)=\operatorname{cls}(u_1u_2)$; this makes $H(U)$ a graded algebra. Any homomorphism $f\colon U\to U'$ of DG-algebras induces a homomorphism $f_*\colon H(U)\to H(U')$ of graded algebras.

The *tensor product* $U\otimes U'$ of two DG-algebras is their tensor product as graded algebras, with the differential given by (7.1). The analogue of Prop. 4.1 holds. The *opposite* U^{op} of a DG-algebra is the opposite of U, as a graded algebra, with the same differential.

A left U-module $X=(X,\partial)$ is a left module over the graded algebra U equipped with a graded K-module homomorphism $\partial\colon X\to X$ of degree

-1 with $\partial^2=0$, such that the formula

$$\partial(ux)=(\partial u)x+(-1)^{\deg u}u(\partial x) \tag{7.3}$$

always holds. Equivalently, a left U-module X is a DG-module over K equipped with a homomorphism $U\otimes_{\mathsf{K}}X\to X$ of DG-modules, of degree 0, written $u\otimes x\to ux$, such that the standard conditions

$$(u_1+u_2)x=u_1x+u_2x,\quad u(x_1+x_2)=ux_1+ux_2,$$
$$(u_1u_2)x=u_1(u_2x),\qquad 1x=x,$$

always hold, as in the diagrams (5.1). If X and Y are U-modules, a morphism $\xi\colon X\to Y$ is a homomorphism of the whole structure: A homomorphism of DG-modules, of degree 0, which is also a homomorphism of modules over the graded algebra U; in other words, ξ is additive and

$$\xi(kx)=k(\xi x),\quad \xi(\partial x)=\partial(\xi x),\quad \xi(ux)=u(\xi x),\quad \deg(\xi x)=\deg x. \tag{7.4}$$

The K-module of all such morphisms ξ is written $\hom_U(X,Y)$. With these morphisms, the left U-modules form a category in which sub- and quotient modules, kernels, images, coimages, and cokernels are defined as usual. Right U-modules are treated similarly.

On this category we define bifunctors Hom_U and $\otimes_U$. For U-modules X and Y, a *graded* U-module homomorphism $f\colon X\to Y$ of degree $-n$ is a homomorphism of X to Y, regarded just as modules over the graded algebra U; in other words, f is additive and

$$f(kx)=k(fx),\quad f(ux)=u(fx),\quad \deg(fx)=\deg x-n, \tag{7.5}$$

but f need not commute with ∂. The set of all such f is a K-module $\mathrm{Hom}_U^n(X,Y)$. The family $\mathrm{Hom}_U(X,Y)=\{\mathrm{Hom}_U^n(X,Y)\}$ becomes a DG_Z-module over K when the differential $\partial_H\colon \mathrm{Hom}^n\to\mathrm{Hom}^{n+1}$ is defined by the usual formula

$$\partial_H f=\partial_Y f+(-1)^{n+1}f\,\partial_X. \tag{7.6}$$

Thus Hom_U with capital "H" differs from $\hom_U$, with lower case "h":

$\mathrm{Hom}_U(X,Y)$ is a DG_Z-module over K; elements all $f\colon X\to Y$;

$\hom_U(X,Y)$ is an (ungraded) K-module; elements all $\xi\colon X\to Y$.

Moreover, $\hom_U$ is the K-module of cycles of degree 0 in the complex Hom_U.

Let X be a right U-module, Y a left U-module. Considered just as modules over the graded algebra U, they define a graded K-module $X\otimes_U Y$ which becomes a DG-module over K when the differential is defined by (7.1); for, by that formula, $\partial(xu\otimes y)=\partial(x\otimes uy)$ (U-middle associativity). Thus the *elements* of Hom_U and $\otimes_U$ are defined from the grading and module structure of X and Y; the *differentials* on Hom_U and $\otimes_U$ come from the differentials on X and Y.

For two *DG*-algebras U and U' a U-U'-bimodule (X, ∂) has *one* differential ∂ which satisfies (7.3) for $\partial(ux)$ and the corresponding rule for $\partial(xu')$ — just as a bimodule has just one K-module structure induced from U or from U'.

The augmented case is relevant. A *differential graded augmented algebra* U (*DGA*-algebra, for short) is a *DG*-algebra together with an augmentation $\varepsilon: U \to \mathsf{K}$ which is a homomorphism of *DG*-algebras. Here the ground ring K is regarded as a *DG*-algebra with trivial grading ($\mathsf{K}_0 = \mathsf{K}$) and differential ($\partial = 0$). Such an augmentation is entirely determined by its component of degree 0, which is a homomorphism $\varepsilon_0: U_0 \to \mathsf{K}$ of (ungraded) K-modules with

$$\varepsilon_0 1 = 1, \quad \varepsilon_0(u_0 u_0') = (\varepsilon_0 u_0)(\varepsilon_0 u_0'), \quad \varepsilon_0 \partial = 0: U_1 \to \mathsf{K}.$$

A *DG*-algebra U is *connected* if $U_0 = \mathsf{K}$ and $\partial: U_1 \to U_0$ is zero; this implies $H_0(U) \cong \mathsf{K}$ (hence the choice of the term "connected": A topological space X is path-connected precisely when $H_0(X) \cong Z$). A connected *DG*-algebra has a canonical augmentation $\varepsilon_0 = 1: U_0 \to \mathsf{K}$.

Next some examples of *DG*-algebras. Take the polynomial algebra $P_{\mathsf{K}}[x]$ in an indeterminate x of degree 1, select some $k_0 \in \mathsf{K}$, and set $\partial x = k_0$; with this, $\partial x^{2m} = 0$, $\partial x^{2m+1} = k_0 x^{2m}$, and P is a *DG*-algebra. Similarly, the exterior algebra $E_{\mathsf{K}}[u]$, with u of degree 1, has a unique differential with $\partial u = k_0$, and is a *DG*-algebra.

If X is a *DG*-module over K, the tensor algebra $T(X)$ has a unique *DG*-algebra structure such that the injection $X \to T(X)$ is a chain transformation; the requisite differential in $T(X)$ is given by

$$\partial(x_1 \otimes \cdots \otimes x_p) = \sum_{i=1}^{p} (-1)^{\eta_i} x_1 \otimes \cdots \otimes \partial x_i \otimes \cdots \otimes x_p$$

with $\eta_i = \deg x_1 + \cdots + \deg x_{i-1}$, in accord with the sign convention. The analogue of Prop. 3.1 holds for this $T(X)$.

One may construct universal *DG*-algebras on given generators. Thus if x has degree 2 and u degree 1, there is exactly one *DG*-algebra structure on $V = P[x] \otimes E[u]$ for which $\partial x = u$, for by the Leibniz rule (7.2) the differential is given on the free K-module generators of the algebra V as

$$\partial(x^m \otimes 1) = m x^{m-1} \otimes u, \quad \partial(x^m \otimes u) = 0. \tag{7.7}$$

If u_2 is a selected element of degree 2 in any strictly commutative *DG*-algebra U, there is a unique homomorphism $f: V \to U$ of *DG*-algebras with $fx = u_2$ (and hence with $fu = \partial u_2$).

Similar considerations will define differential internally graded and differential Z-graded algebras.

Exercises

1. For DG-modules over K, show that the exact homology sequence (Thm. II.4.1) for a short exact sequence $E: W \rightarrowtail X \twoheadrightarrow Y$ of DG-homomorphisms $\varkappa$ and σ takes the form of an *exact triangle*

$$\begin{array}{ccc} H(W) & \xrightarrow{\varkappa_*} & H(X) \\ & {}_{\partial_E}\nwarrow \quad \swarrow_{\sigma_*} & \\ & H(Y) & \end{array}$$

(kernel = image at each vertex), with $\varkappa_*$ and σ_* homomorphisms of graded modules of degree 0, while the connecting homomorphism ∂_E has degree -1. [The usual long exact sequence spirals around this triangle, dropping one level with each ∂_E.]

2. Prove that a DG-algebra U is a DG-module over K with homomorphisms $\pi: U \otimes U \to U$ and $I: K \to U$ of DG-modules, of degree 0, satisfying (3.1). Give a similar definition of U-modules by (5.1), and show that Hom_U and $\otimes_U$ may be obtained from Hom_K and $\otimes_K$, for DG-modules, by the analogues of (5.5) and (5.7).

3. For V as in (7.7) determine the graded homology algebra $H(V)$ when $K = Z$ and when $K = Z_p$ (the field of integers modulo p).

4. Construct a universal strictly commutative DG-algebra on a given finite set of generators (of odd and even degrees).

5. For $\deg x_i = 2$, $\deg u_i = 1$, the graded algebra $P[x_1, \ldots, x_n] \otimes E[u_1, \ldots, u_n]$ is isomorphic to the tensor product of n algebras $V_i = P[x_i] \otimes E[u_i]$ like that treated in the text, hence has a unique differential with $\partial x_i = u_i$, $i = 1, \ldots, n$. For any polynominal p in the x_i, show that $\partial p = \sum \frac{\partial p}{\partial x_i} \otimes u_i$, where $\frac{\partial p}{\partial x_i}$ denotes the usual partial derivative. Hence show directly that $\partial^2 p = 0$. Note that ∂p is the usual differential of the function p of n variables if we replace u_i by a symbol $d x_i$.

8. Identities on Hom and ⊗

Consider modules and bimodules over various graded K-algebras Λ, Σ, and Ω (which may equally well be DG-algebras). The functors Hom_Λ and $\otimes_\Lambda$ have inherited module structure as follows

$${}_\Sigma C_\Lambda \,\&\, {}_\Omega A_\Lambda \Rightarrow {}_\Omega[\mathrm{Hom}_\Lambda(C, A)]_\Sigma,$$

$${}_\Sigma G_\Lambda \,\&\, {}_\Lambda A_\Omega \Rightarrow {}_\Sigma(G \otimes_\Lambda A)_\Omega,$$

defined for $f: C \to A$ by $(\omega f \sigma)(c) = \omega[f(\sigma c)]$ and for $g \otimes a$ by $\sigma(g \otimes a)\omega = \sigma g \otimes a \omega$, just as in (V.3.2) and (V.3.1).

There are several natural isomorphisms for iterated tensor products. Thus

$$\Lambda \otimes_\Lambda A \cong A, \qquad ({}_\Lambda A) \tag{8.1}$$

is given by $\lambda \otimes a \to \lambda a$. The commutative law

$$G \otimes_\Lambda A \cong A \otimes_{\Lambda^{\mathrm{op}}} G, \qquad (G_\Lambda, {}_\Lambda A) \tag{8.2}$$

is given by $g\otimes a\to(-1)^{(\deg g)(\deg a)}a\otimes g$. The associative law

$$\alpha\colon A\otimes_{\Lambda\otimes\Omega}(B\otimes_\Sigma C)\cong(A\otimes_\Lambda B)\otimes_{\Sigma\otimes\Omega}C,\quad (A_{\Lambda-\Omega},\ {}_\Lambda B_\Sigma,\ {}_{\Sigma-\Omega}C)\tag{8.3}$$

is given by $\alpha[a\otimes(b\otimes c)]=(a\otimes b)\otimes c$; here $B\otimes_\Sigma C$ is regarded as a left Ω-module with operators $\omega(b\otimes c)=(-1)^{(\deg b)(\deg\omega)}b\otimes\omega c$. To show this map α well-defined, observe first that for fixed a the function $(a\otimes b)\otimes c$ is bilinear and Σ-middle associative in b and c. By Thm. 5.2 there is for each a a unique homomorphism $F(a)\colon B\otimes_\Sigma C\to(A\otimes_\Lambda B)\otimes_{\Sigma\otimes\Omega}C$ which satisfies $F(a)(b\otimes c)=(a\otimes b)\otimes c$. The function $F(a)(b\otimes c)$ is again bilinear and $(\Lambda\otimes\Omega)$-middle associative in its arguments in A and $B\otimes_\Sigma C$. By Thm. 5.2 again there is a unique homomorphism α with $\alpha[a\otimes(b\otimes c)]=(a\otimes b)\otimes c$. The inverse of α is constructed similarly. The associativity law also holds in simpler cases; e.g., with Ω omitted (set $\Omega=\mathsf{K}$ in (8.3)). A general version of the commutative law is the *middle four interchange*

$$\tau\colon(A\otimes_\Lambda B)\otimes_{\Lambda'\otimes\Sigma'}(C\otimes_\Sigma D)\cong(A\otimes_{\Lambda'}C)\otimes_{\Lambda\otimes\Sigma}(B\otimes_{\Sigma'}D)\tag{8.4}$$

defined for modules $A_{\Lambda-\Lambda'}$, ${}_\Lambda B_{\Sigma'}$, ${}_{\Lambda'}C_\Sigma$, ${}_{\Sigma-\Sigma'}D$ by setting

$$\tau[(a\otimes b)\otimes(c\otimes d)]=(-1)^{(\deg b)(\deg c)}(a\otimes c)\otimes(b\otimes d).$$

In the DG-case, all of these natural isomorphisms are isomorphisms of DG-modules over K, as one verifies by showing that each of the given isomorphisms commutes with the differential which we have defined on $\otimes_U$.

For the functor Hom_Λ alone, we have the natural isomorphism

$$\mathrm{Hom}_\Lambda(\Lambda,A)\cong A\qquad({}_\Lambda A)\tag{8.5}$$

given by $f\to f(1)$ and the natural homomorphism

$$\mathsf{K}\to\mathrm{Hom}_\Lambda(A,A),\qquad({}_\Lambda A)\tag{8.6}$$

given by mapping 1_K into the identity homomorphism $1_A\colon A\to A$.

Adjoint associativity is the natural isomorphism

$$\eta\colon\mathrm{Hom}_{\Omega-\Sigma}(A\otimes_\Lambda B,C)\cong\mathrm{Hom}_{\Omega-\Lambda}(A,\mathrm{Hom}_\Sigma(B,C))\tag{8.7}$$

for modules ${}_\Omega A_\Lambda$, ${}_\Lambda B_\Sigma$, and ${}_\Omega C_\Sigma$, defined for $f\colon A\otimes_\Lambda B\to C$ by $[(\eta f)a]b=f(a\otimes b)$, just as in the case of rings (V.3.5). In the DG-case one checks that η commutes with the differentials defined on both sides, hence is an isomorphism of DG_Z-modules over K. In particular, taking the cycles of degree zero on each side of (8.7) gives the natural isomorphism

$$\mathrm{hom}_{\Omega-\Sigma}(A\otimes_\Lambda B,C)\cong\mathrm{hom}_{\Omega-\Lambda}(A,\mathrm{Hom}_\Sigma(B,C))\tag{8.8}$$

— where, as above, hom with lower case h denotes homomorphisms of degree zero of the full DG_Z-structure. In this case, since A has no elements of negative degree, the Z-graded $\mathrm{Hom}(B, C)$ on the right may be replaced by the graded module $\{\mathrm{Hom}^{-n}(B, C), n=0, 1, \ldots\}$.

Composition of homomorphisms yields a map

$$\mathrm{Hom}_\Lambda(B, C) \otimes_\Omega \mathrm{Hom}_\Lambda(A, B) \to \mathrm{Hom}_\Lambda(A, C) \qquad (A_\Lambda, {}_\Omega B_\Lambda, C_\Lambda) \quad (8.9)$$

which is natural in A and C. Another useful natural homomorphism is the Hom-⊗ *interchange* for modules ${}_\Lambda B$, ${}_\Lambda A$, ${}_{\Lambda'}B'$, ${}_{\Lambda'}A'$,

$$\zeta\colon \mathrm{Hom}_\Lambda(B, A) \otimes \mathrm{Hom}_{\Lambda'}(B', A') \to \mathrm{Hom}_{\Lambda\otimes\Lambda'}(B \otimes B', A \otimes A'), \quad (8.10)$$

defined for $f\colon B \to A$ and $f'\colon B' \to A'$ by

$$[\zeta(f \otimes f')](b \otimes b') = (-1)^{(\deg f')(\deg b)} fb \otimes f'b'.$$

In the DG-case, this and composition are homomorphisms of DG-modules.

A notational curiosity emerges here. In this definition, $f \otimes f'$ denotes a typical element in the tensor product shown on the left of (8.10). Previously, in (2.2), we used $f \otimes f'$ to denote the homomorphism $B \otimes B' \to A \otimes A'$ here written as $\zeta(f \otimes f')$. The two symbols $f \otimes f'$ need not agree, because ζ may well have a kernel not zero. This ambiguity is not serious; long ago we observed that the tensor product $a \otimes b$ of two elements has meaning only when the modules in which these elements lie are specified, and may become zero when one or the other module is enlarged.

Various other natural homomorphisms may be defined by composition of these. For example, the evaluation homomorphism

$$e\colon \mathrm{Hom}_\Lambda(A, B) \otimes A \to B \qquad ({}_\Lambda A, {}_\Lambda B) \qquad (8.11)$$

is given for $f\colon A \to B$ by $e(f \otimes a) = f(a)$; i.e., by taking the value of the function f at a. It may be written as the composition

$$\mathrm{Hom}_\Lambda(A, B) \otimes A \to \mathrm{Hom}_\Lambda(A, B) \otimes \mathrm{Hom}_\Lambda(\Lambda, A) \to \mathrm{Hom}_\Lambda(\Lambda, B) \to B$$

of the maps (8.5), (8.9), and (8.5). It would be instructive to know the various identities holding between composites of the assorted natural homomorphisms (8.1)—(8.10) described above.

As an application, consider free and projective Λ-modules over a graded algebra Λ. A left Λ-module P is projective, as usual, if each epimorphism $\sigma\colon B \twoheadrightarrow C$ of left Λ-modules, of degree 0, induces an epimorphism $\mathrm{hom}_\Lambda(P, B) \twoheadrightarrow \mathrm{hom}_\Lambda(P, C)$. The *free* Λ-module on the graded set S of generators is the Λ-module C containing S and characterized up to isomorphism by the usual property (Prop. I.5.1) that each set map

$S \to_\Lambda A$ of degree zero extends to a unique Λ-module homomorphism $C \to A$; as usual, a free module is projective. The algebra Λ itself is the free Λ-module on one generator 1 of degree 0; the free Λ-module on any S may be constructed as the direct sum $\Lambda S = \sum \Lambda s$ for $s \in S$. Here Λs denotes the left Λ-module with elements λs of degrees $\deg(\lambda s) = \deg \lambda + \deg s$. Note that $\Lambda S = \Lambda \otimes \mathsf{K} S$, where $\mathsf{K}S = \sum \mathsf{K}s$ is the free graded K-module on the generators S. In other words, each free graded K-module F yields a free Λ-module $\Lambda \otimes F$. Similarly,

Proposition 8.1. *If M is a projective graded K-module and Λ a graded K-algebra, then $\Lambda \otimes M$ is a projective Λ-module.*

The proof, as in Cor. V.3.3, follows from the adjoint associativity

$$\hom_\Lambda(\Lambda \otimes M, B) \cong \hom(M, \operatorname{Hom}_\Lambda(\Lambda, B)) = \hom(M, B).$$

The same associativity proves more generally

Proposition 8.2. *For each graded K-module M define a homomorphism $e: M \to \Lambda \otimes M$ of graded K-modules by $e(m) = 1 \otimes m$. This e is universal: For every left Λ-module A each homomorphism $g: M \to A$ of graded K-modules of degree 0 can be factored uniquely through e as $g = \gamma e$, with $\gamma: \Lambda \otimes M \to A$ a Λ-module homomorphism of degree 0.*

Proof. Observe that γ must have $\gamma(\lambda \otimes m) = \lambda g(m) \in A$; the right side of this formula is K-bilinear in λ and m, hence defines γ uniquely.

For "e is universal" in the sense of this proposition we also say that $\Lambda \otimes M$ is the *relatively free* Λ-module generated by the graded K-module M, or that $\Lambda \otimes M$ is (Λ, K)-free. Similarly, for two graded algebras Λ and Σ, each $\Lambda \otimes M \otimes \Sigma$ is a $(\Lambda\text{-}\Sigma, \mathsf{K})$ relatively free bimodule; if M is K-projective, it is Λ-Σ projective; if M is K-free, it is Λ-Σ-free.

For Thm. X.7.4 we shall need

Proposition 8.3. *If B and B' are free left Λ- and Λ'-modules of finite type, the* Hom-$\otimes$ *interchange is a natural isomorphism*

$$\zeta: \operatorname{Hom}_\Lambda(B, A) \otimes \operatorname{Hom}_{\Lambda'}(B', A') \cong \operatorname{Hom}_{\Lambda \otimes \Lambda'}(B \otimes B', A \otimes A'), \ ({}_\Lambda A, {}_{\Lambda'} A').$$

Proof. By direct sums, this reduces to the case $B = \Lambda$, $B' = \Lambda'$; in this case ζ is the identity $A \otimes A' \cong A \otimes A'$.

Exercises

1. Give a direct proof of the middle four interchange; that is, show that τ as specified is well defined and has an inverse.

2. Deduce the middle four interchange by repeated applications of the associativity (8.3) and $A \otimes_{\mathsf{K}} B \cong B \otimes_{\mathsf{K}} A$.

3. For modules ${}_\Lambda C_\Sigma$ and ${}_\Lambda A_\Omega$ describe the bimodule structure on $\operatorname{Hom}_\Lambda(C, A)$ (Attention to signs!).

4. Describe the behavior of composition (8.9) for a map $B \to B'$.

5. Show that ζ of (8.10) may have non-zero kernel (Hint: use finite cyclic groups).

6. Construct a natural homomorphism

$$A \otimes_\Omega \mathrm{Hom}_\Lambda(B,C) \to \mathrm{Hom}_\Lambda(\mathrm{Hom}_\Omega(A,B),C).$$

9. Coalgebras and Hopf Algebras

A formal dualization of the notion of an algebra yields that of a coalgebra. These coalgebras have recently gained importance from a variety of topological applications; for instance, the singular complex of a topological space turns out to be a coalgebra.

A *graded coalgebra* W over the commutative ground ring K is a graded K-module W with two homomorphisms $\psi: W \to W \otimes W$ and $\varepsilon: W \to \mathrm{K}$ of graded K-modules, each of degree 0, such that the diagrams

$$\begin{array}{ccc} W & \xrightarrow{\psi} & W\otimes W \\ \downarrow{\scriptstyle\psi} & & \downarrow{\scriptstyle 1\otimes\psi} \\ W\otimes W & \xrightarrow{\psi\otimes 1} & W\otimes W\otimes W, \end{array} \qquad \begin{array}{ccccc} W\otimes W & \xleftarrow{\psi} & W & \xrightarrow{\psi} & W\otimes W \\ \downarrow{\scriptstyle\varepsilon\otimes 1} & & \| & & \downarrow{\scriptstyle 1\otimes\varepsilon} \\ \mathrm{K}\otimes W & = & W & = & W\otimes\mathrm{K} \end{array} \tag{9.1}$$

are commutative. The first diagram gives the *associative law* for the *diagonal map* (or coproduct) ψ; the second diagram states that ε is a *counit*. Coalgebras which are not associative or which have no counit are sometimes useful, but will not occur in this book. A *homomorphism* $\mu: W \to W'$ of coalgebras is a K-module homomorphism of degree 0 such that the diagrams

$$\begin{array}{ccc} W & \xrightarrow{\psi} & W\otimes W \\ \downarrow{\scriptstyle\mu} & & \downarrow{\scriptstyle\mu\otimes\mu} \\ W' & \xrightarrow{\psi'} & W'\otimes W', \end{array} \qquad \begin{array}{ccc} W & \xrightarrow{\varepsilon} & \mathrm{K} \\ \downarrow{\scriptstyle\mu} & & \| \\ W' & \xrightarrow{\varepsilon'} & \mathrm{K} \end{array} \tag{9.2}$$

are commutative. If the following diagram is commutative

$$\begin{array}{ccc} W & \xrightarrow{\psi} & W\otimes W \\ \| & & \downarrow{\scriptstyle\tau} \\ W & \xrightarrow{\psi} & W\otimes W, \end{array} \qquad \tau(w_1\otimes w_2) = (-1)^{\deg w_1 \deg w_2} w_2 \otimes w_1, \tag{9.3}$$

we call the graded coalgebra W *commutative*. As usual, our definition includes the special cases of coalgebras (W trivially graded) and graded corings ($\mathrm{K}=Z$). DG-coalgebras may also be defined by the diagram (9.1), for W a DG-K-module. In particular, the ground ring K itself is a (trivially graded) K-coalgebra with diagonal map $\mathrm{K} \to \mathrm{K} \otimes \mathrm{K}$ the canonical isomorphism and counit the identity $\mathrm{K} \to \mathrm{K}$.

If W and W' are graded coalgebras, their tensor product $W \otimes W'$ (as graded modules) is a graded coalgebra with diagonal map the composite

$$W\otimes W' \xrightarrow{\psi\otimes\psi'} W\otimes W\otimes W'\otimes W' \xrightarrow{1\otimes\tau\otimes 1} (W\otimes W')\otimes(W\otimes W'), \tag{9.4}$$

for τ as in (9.3), and with counit $\varepsilon \otimes \varepsilon': W \otimes W' \to \mathrm{K} \otimes \mathrm{K} = \mathrm{K}$.

For completeness, let us also define comodules by dualizing the diagrammatic definition (5.1) of a module over an algebra. A graded left *W-comodule* over the graded coalgebra W is a graded K-module C equipped with a homomorphism $\varphi: C \to W \otimes C$ of degree zero such that the diagrams

$$\begin{array}{ccc} C & \xrightarrow{\varphi} & W\otimes C \\ \downarrow{\scriptstyle\varphi} & & \downarrow{\scriptstyle 1\otimes\varphi} \\ W\otimes C & \xrightarrow{\psi\otimes 1} & W\otimes W\otimes C, \end{array} \qquad \begin{array}{ccc} C & = & \mathsf{K}\otimes C \\ {\scriptstyle\varphi}\downarrow & & \| \\ W\otimes C & \xrightarrow{\varepsilon\otimes 1} & \mathsf{K}\otimes C \end{array} \tag{9.5}$$

both commute.

A *graded Hopf algebra* V is a graded K-module $V = \{V_n\}$ which, with this grading, is *both* a graded algebra for a product map $\pi: V \otimes V \to V$ and unit $I: \mathsf{K} \to V$ and a graded coalgebra for a diagonal ψ and a counit ε, and such that

(i) $I: \mathsf{K} \to V$ is a homomorphism of graded coalgebras;

(ii) $\varepsilon: V \to \mathsf{K}$ is a homomorphism of graded algebras;

(iii) $\pi: V \otimes V \to V$ is a homomorphism of graded coalgebras.

Condition (i) states that $\psi(1) = 1 \otimes 1$ and that $\varepsilon I: \mathsf{K} \to \mathsf{K}$ is the identity. Condition (ii) states that V is an augmented algebra, with augmentation the counit. In view of the definition (9.4), condition (iii) states that the following diagram commutes

$$\begin{array}{ccccc} V\otimes V & \xrightarrow{\psi\otimes\psi} & V\otimes V\otimes V\otimes V & \xrightarrow{1\otimes\tau\otimes 1} & V\otimes V\otimes V\otimes V \\ \downarrow{\scriptstyle\pi} & & & & \downarrow{\scriptstyle\pi\otimes\pi} \\ V & & \xrightarrow{\qquad\psi\qquad} & & V\otimes V\ , \end{array} \tag{9.6}$$

for τ as in (9.3). But $(\pi \otimes \pi)(1 \otimes \tau \otimes 1)$ is the product map in the tensor product algebra $V \otimes V$, so this diagram may equally well be read as

(iii′) $\psi: V \to V \otimes V$ is a homomorphism of graded algebras.

Thus (iii) is equivalent to (iii′).

A homomorphism $\nu: V \to V'$ of Hopf algebras is a K-module homomorphism which is both an algebra and a coalgebra homomorphism.

Let V and V' be graded Hopf algebras over K. A formal argument from the definitions shows that $V \otimes V'$ is a graded Hopf algebra over K, with grading that of the tensor product of graded modules, product and unit that of the tensor product of algebras, coproduct and counit that of the tensor product (9.4) of coalgebras.

Now for some examples of Hopf algebras.

The ground ring K is itself (trivially) a graded Hopf algebra.

Let $E = E_{\mathsf{K}}[u]$ be the exterior algebra on one symbol u of degree 1. Since E is the free strictly commutative algebra on one generator u, there are unique algebra homomorphisms $\varepsilon: E \to \mathsf{K}$, $\psi: E \to E \otimes E$ with

$$\varepsilon(u)=0, \qquad \psi(u)=u\otimes 1+1\otimes u. \tag{9.7}$$

With this structure, we claim that E is a Hopf algebra. To prove it a coalgebra, note that $(\psi\otimes 1)\psi$ and $(1\otimes\psi)\psi$ may both be characterized as the unique homomorphism $\eta: E\to E\otimes E\otimes E$ of algebras with $\eta(u)=u\otimes 1\otimes 1+1\otimes u\otimes 1+1\otimes 1\otimes u$; by a similar argument $(\varepsilon\otimes 1)\psi=1 =(1\otimes\varepsilon)\psi$. Condition (i) for a Hopf algebra is trivial, while conditions (ii) and (iii') follow by the definitions of ε and ψ.

Let $P=P_{\mathsf{K}}[x]$ be the polynomial algebra in one symbol x of even degree. By a similar argument, it is a Hopf algebra with

$$\varepsilon(x)=0, \qquad \psi(x)=x\otimes 1+1\otimes x. \tag{9.8}$$

Since ψ is an algebra homomorphism, $\psi(x^n)=(\psi x)^n$, so

$$\psi(x^n)=\sum_{p+q=n}(p,q)\,x^p\otimes x^q, \qquad (p,q)=(p+q)!/(p!\,q!). \tag{9.9}$$

By taking tensor products of Hopf algebras, it follows that the exterior algebra $E_{\mathsf{K}}[u_1,\ldots,u_n]$ on generators u_i of degree 1 or the polynomial algebra $P_{\mathsf{K}}[x_1,\ldots,x_n]$ on generators x_i of even degrees is a Hopf algebra.

The group ring $Z(\Pi)$ of any multiplicative group is a (trivially graded) Hopf algebra over Z, for if $x\in\Pi$, the function $\psi(x)=x\otimes x$ on Π to $Z(\Pi)\otimes Z(\Pi)$ carries 1 to 1 and products to products, hence (Prop. IV.1.1) extends to a ring homomorphism $\psi: Z(\Pi)\to Z(\Pi)\otimes Z(\Pi)$. With the usual augmentation $\varepsilon: Z(\Pi)\to Z$ this makes $Z(\Pi)$ a coalgebra (condition (9.1)) and a Hopf algebra, with unit $I: Z\to Z(\Pi)$ the injection. Any homomorphism $\zeta: \Pi\to\Pi'$ of groups induces a homomorphism $\mathrm{Z}(\zeta): Z(\Pi)\to Z(\Pi')$ of Hopf algebras.

For any commutative K, the *group algebra* $\mathsf{K}(\Pi)$ is defined as the K-algebra $\mathsf{K}\otimes_Z Z(\Pi)$; equivalently, it is the free K-module with free generators the elements $x\in\Pi$ and product determined by the product in Π. It is a Hopf algebra, with coproduct $\psi(x)=x\otimes x$.

Now consider left modules A, B, C over a graded Hopf algebra V: that is, modules over the graded algebra V. The tensor product $A\otimes_{\mathsf{K}} B$ is a left $(V\otimes V)$-module, but becomes a left V-module by pull-back along the diagonal $\psi: V\to V\otimes V$: We write this module as $A\otimes B= {}_{\psi}(A\otimes_{\mathsf{K}}B)$. The associative law (9.1) for ψ proves the usual associativity law $A\otimes(B\otimes C)=(A\otimes B)\otimes C$ for this tensor product. Moreover, the ground ring K is a left V-module ${}_{\varepsilon}\mathsf{K}$ by pull-back along $\varepsilon: V\to\mathsf{K}$, and the rule $(\varepsilon\otimes 1)\psi=1=(1\otimes\varepsilon)\psi$ gives the isomorphism $\mathsf{K}\otimes A\cong A\cong A\otimes\mathsf{K}$. Using these two isomorphisms, parallel to (2.3) and (2.5), one can define an *algebra over a graded Hopf algebra* V — by exactly the mechanism used to define algebras over K itself. If the coproduct ψ is commutative (9.3), one can obtain the isomorphism $\tau: A \otimes B \cong B \otimes A$ for V-modules, and in this case one can define the tensor product of algebras over V.

Exercises

1. If M is a K-module show that the tensor algebra $T(M)$ has a unique Hopf algebra structure with $\psi(m) = m \otimes 1 + 1 \otimes m$.

2. If Λ is a graded K-algebra with each Λ_q finitely generated and projective as a K-module, show that the dual Λ^* is a coalgebra with diagonal map induced by π^* (use Prop. V.4.3). Under similar hypotheses, show that the dual of a Hopf algebra is a Hopf algebra.

3. Characterize the group algebra $K(\Pi)$ by the analogue of Prop. IV.1.1.

Notes. Originally, a linear associative algebra meant an algebra over a field K which was of finite dimensions as a vector space over that field, and the classical theory dealt with the structure of such algebras (e.g. the Wedderburn Principal Theorem X.3.2). In analysis, algebras of continuous functions were vector spaces of infinite dimension. In topology, the cup product in cohomology (Chap. VIII) introduces graded algebras over a commutative ring not a field. Bourbaki and Chevalley [1956] codified the present general concept of a graded algebra, and emphasized a principle due to E. H. Moore: State theorems in the maximum useful generality; e.g., for graded algebras, not just for rings. Hopf algebras first occurred in H. Hopf's study of the cohomology of a Lie group. Their algebraic structure has been examined by various authors (e.g., Borel [1953], Halpern [1958]); for a systematic treatment see Milnor-Moore [196 ?]. Algebras over Hopf algebras were recently considered by Steenrod [1962].

Chapter seven

Dimension

This chapter is a brief introduction to the extensive applications of homological algebra to ring theory and algebraic geometry. We define various dimensions, use them in polynomial rings and separable algebras, and in the Hilbert theorem on syzygies. Subsequent chapters are independent of this material, except for the description (§ 3) of Ext and Tor for algebras and the direct product and ground ring extensions for algebras.

1. Homological Dimension

For abelian groups C and A, $\mathrm{Ext}_Z^2(C, A)$ is always zero; we say that C, regarded as a module over the ring Z of integers, has homological dimension at most 1. Over any ring R, a projective module P is characterized by the fact that all $\mathrm{Ext}_R^1(P, G)$ vanish; we say that P has homological dimension 0. The general phenomenon may be described as follows.

Theorem 1.1. *For each integer n, the following conditions on a left R-module C are equivalent:*

(i) *For all left R-modules B,* $\operatorname{Ext}^{n+1}(C, B)=0$;

(ii) *Any exact sequence of modules*

$$S\colon 0\to C_n\to X_{n-1}\to\cdots\to X_0\to C\to 0,$$

with the X_i *all projective has the first term* C_n *projective;*

(iii) *C has a projective resolution of length n:*

$$0\to X_n\to X_{n-1}\to\cdots\to X_0\to C\to 0.$$

Here and below we write Ext *for* Ext_R.

Proof. Factor the sequence S of (ii) into short exact sequences $E_i\colon C_{i+1}\rightarrowtail X_i\twoheadrightarrow C_i$. Each gives the standard long exact sequence

$$\operatorname{Ext}^k(X_i, B)\to\operatorname{Ext}^k(C_{i+1}, B)\xrightarrow{E_i^*}\operatorname{Ext}^{k+1}(C_i, B)\to\operatorname{Ext}^{k+1}(X_i, B)$$

of (III.9.1). Since X_i is projective, the outside terms $\operatorname{Ext}^k(X_i, B)$ are zero if $k>0$, so the connecting homomorphism E_i^* is an isomorphism. The iterated connecting homomorphism S^* is the composite $E_0^*\ldots E_{n-1}^*$, hence an isomorphism

$$S^*\colon \operatorname{Ext}^1(C_n, B)\cong\operatorname{Ext}^{n+1}(C, B).$$

Now given $\operatorname{Ext}^{n+1}(C, B)=0$ by (i), this isomorphism makes $\operatorname{Ext}^1(C_n, B)=0$ for all B, hence C_n projective as in (ii). Since C has at least one projective resolution, (ii) implies (iii). Given a resolution of the form (iii), $\operatorname{Ext}^{n+1}(C, B)$ computed thereby is 0, whence (i).

The *homological dimension* of an R-module C is defined by the statement that $\operatorname{h.dim}_R C\leqq n$ when any one of the equivalent conditions of Thm. 1.1 hold. In other words, $\operatorname{h.dim}_R C=n$ means that all $\operatorname{Ext}^{n+1}(C, B)=0$, but that $\operatorname{Ext}^n(C, B)\neq 0$ for at least one module B.

Corollary 1.2. *If* $\operatorname{h.dim}_R C=n$, *then for all modules* ${}_RB$ *and* G_R,

$$\operatorname{Ext}^{n+k}(C, B)=0,\qquad \operatorname{Tor}_{n+k}(G, C)=0,\qquad k>0,$$

while for each $m\leqq n$ *there is a left module* B_m *with* $\operatorname{Ext}^m(C, B_m)\neq 0$.

Proof. The first result follows by (iii). If $\operatorname{Ext}^n(C, B)\neq 0$ for $n>0$, imbed B in an injective module J to get a short exact sequence $B\rightarrowtail J\twoheadrightarrow B'$. The corresponding exact sequence

$$\operatorname{Ext}^{n-1}(C, B')\to\operatorname{Ext}^n(C, B)\to\operatorname{Ext}^n(C, J)=0$$

shows that $\operatorname{Ext}^{n-1}(C, B')\neq 0$.

Similarly, $\operatorname{h.dim} C=\infty$ implies that for each positive integer n there is a module B_n with $\operatorname{Ext}^n(C, B_n)\neq 0$. The homological dimension of a module C can be calculated from any projective resolution

$0 \leftarrow C \leftarrow X_0 \leftarrow X_1 \leftarrow \cdots$ as the first n with $\mathrm{Im}(X_n \to X_{n-1})$ projective (for $n=0$, read X_{-1} as C), or as ∞ when none of these images is projective.

For example, the calculations of VI.6 show that the trivial module Z over the group ring $Z(\Pi)$ of a free abelian group Π on n generators has homological dimension n.

The *left global dimension* of a ring R is defined as

$$\text{l.gl.dim.}\, R = \sup\,(\text{h.dim}\, C),$$

where the supremum is taken over all left R-modules C. For example, $\text{l.gl.dim}\, Z = 1$. For a field F, every module is a vector space, hence free, so $\text{l.gl.dim}\, F = 0$. More generally,

Proposition 1.3. *Each of the following conditions is equivalent to* $\text{l.gl.dim}\, R = 0$:

(i) *Every left R-module is projective;*

(ii) *Every short exact sequence* $A \rightarrowtail B \twoheadrightarrow C$ *of left R-modules splits;*

(iii) *Every left R-module is injective;*

(iv) *Every left ideal of R is injective, as a left R-module;*

(v) *Every left ideal of R is a direct summand of R, as a left R-module.*

Proof. Condition (i) is the definition of $\text{l.gl.dim}\, R = O$. Given (i), each short exact sequence (ii) has C projective, hence splits. Since every such sequence beginning with A splits, each A is injective by Prop. III.7.1. Hence (i) $\Rightarrow$ (ii) $\Rightarrow$ (iii), and the reverse argument shows (iii) $\Rightarrow$ (ii) $\Rightarrow$ (i). Clearly (iii) $\Rightarrow$ (iv) $\Rightarrow$ (v). Given (v) and a left ideal L, the short exact sequence $L \rightarrowtail R \twoheadrightarrow R/L$ splits, so that $\mathrm{Hom}(R,A) \to \mathrm{Hom}(L,A)$ is an epimorphism for each module A. By Prop. III.7.2, A is injective Hence (v) $\Rightarrow$ (iii); the proof is complete.

Theorem 1.4. *For each ring R and each* $n \geqq 0$ *the following conditions are equivalent:*

(i) $\text{l.gl.dim}\, R \leqq n$;

(ii) *Each left R-module has homological dimension* $\leqq n$;

(iii) $\mathrm{Ext}^{n+1} = 0$, *as a functor of left R-modules;*

(iv) $\mathrm{Ext}^k = 0$ *for all* $k > n$;

(v) *Any exact sequence*

$$S\colon 0 \to A \to Y_0 \to \cdots \to Y_{n-1} \to A_n \to 0,$$

with n intermediate modules Y_k *all injective, has* A_n *injective.*

Proof. The first four conditions are equivalent by Thm. 1.1. The sequence S in (v) gives a connecting homomorphism which is an isomorphism $S_*\colon \mathrm{Ext}^1(C, A_n) \cong \mathrm{Ext}^{n+1}(C, A)$ for each C. But $\mathrm{Ext}^1(C, A_n) = 0$

for all C states exactly that A_n is injective; hence the equivalence of (iii) and (v).

Corollary 1.5. (AUSLANDER [1955].) *For any ring* R

$$\text{l.gl.dim}\, R = \sup\{\text{h.dim}\, R/L \mid L \textit{ a left ideal in } R\}.$$

Proof. (MATLIS [1959].) If the supremum is infinite, $\text{l.gl.dim}\, R = \infty$. Hence assume that the supremum is $n < \infty$, so that $\text{Ext}^{n+1}(R/L, A) = 0$ for all left ideals L and all R-modules A. For each S as in (v) above, S_*: $\text{Ext}^1(R/L, A_n) \cong \text{Ext}^{n+1}(R/L, A) = 0$. By Prop. III.7.2, A_n is injective; by the theorem, $\text{l.gl.dim}\, R \leqq n$.

The condition $\text{l.gl.dim}\, R = 0$ is equivalent to the requirement that R be semi-simple, and so is connected with classical representation theory. Indeed, a left R-module A may be regarded as an abelian group A together with the ring homomorphism φ: $R \to \text{End}_Z A$ which gives the left operators of R on A. This φ is a *representation* of R and A is the corresponding *representation module.* The module A is called *simple* (and the corresponding representation *irreducible*) if $A \neq 0$ and A has no submodules except 0 and A. A module A is *semi-simple* if it is a direct sum of simple modules; a ring $R \neq 0$ is *semi-simple* if it is a semi-simple left R-module. Using Zorn's lemma, one can prove (see e.g. CARTAN-EILENBERG, Prop. I.4.1) that a module is semi-simple if and only if every submodule of A is a direct summand of A. By condition (v) of Prop. 1.3 it then follows that R is semi-simple if and only if $\text{l.gl.dim}\, R = 0$, and by (ii) that every left module over a semi-simple ring R is itself semi-simple. It can also be proved that left semi-simplicity of R (as here defined) is equivalent to right semi-simplicity.

Various other dimensions can be introduced. For example, the *left injective dimension* of a module is defined by the analogue of Thm. 1.1 using injective resolutions, so that the equivalence (v)$\Leftrightarrow$(iii) in the theorem above states that the left global dimension of R agrees with its left global injective dimension. *Right dimensions* are defined using right R-modules; KAPLANSKY [1958] has constructed an example of a ring for which the left and right global dimensions differ by 1. AUSLANDER has proved that if R satisfies the ascending chain condition for left ideals and for right ideals, its left and right global dimensions agree (for proof see NORTHCOTT [1960], Thm. 7.20). The *finitistic left global dimension* of R is the supremum of the homological dimensions of all left R-modules C with $\text{h.dim}\, C < \infty$. The *weak dimension* of a module C is defined by replacing the condition that $\text{Ext}^{n+1}(C, A) = 0$ for all A by the weaker condition that $\text{Tor}_{n+1}(G, C) = 0$ for all G_R. For example, C is flat if and only if its weak homological dimension is 0. For the development of these ideas, see BASS [1960].

Exercises

1. State and prove the analogue of Thm. 1.1 for left injective dimensions.

2. If $\text{l.gl.dim}\, R \geqq 1$, then

$$\text{l.gl.dim}\, R = 1 + \sup\{\text{h.dim}\, L \mid L \text{ a left ideal in } R\}.$$

3. If $A \rightarrowtail B \twoheadrightarrow C$ is a short exact sequence of R-modules, then if any two have finite homological dimension, so does the third.

4. In Ex. 3 above, show that $\text{h.dim}\, A < \text{h.dim}\, B$ implies $\text{h.dim}\, C = \text{h.dim}\, B$, $\text{h.dim}\, A = \text{h.dim}\, B$ implies $\text{h.dim}\, C \leqq 1 + \text{h.dim}\, B$, and $\text{h.dim}\, A > \text{h.dim}\, B$ implies $\text{h.dim}\, C = 1 + \text{h.dim}\, A$.

2. Dimensions in Polynomial Rings

In (VI.1.4) the exterior ring provided an explicit resolution for a field F regarded as a module over the polynomial ring $F[x, y]$ in two indeterminates. The same device works if F is replaced by a commutative ring K or the two indeterminates are replaced by n such.

In detail, let $P = \mathsf{K}[x_1, \ldots, x_n]$ be the polynomial ring in n indeterminates x_i, each of degree 0. Then $\varepsilon(x_i) = 0$ defines an augmentation $\varepsilon = \varepsilon_P \colon P \to \mathsf{K}$, while pull-back along ε makes K a P-module ${}_\varepsilon\mathsf{K}$. This amounts to regarding K as the quotient module $P/(x_1, \ldots, x_n)$, where $(x_1, \ldots, x_n)$ denotes the ideal in P generated by all the x_i.

Let $E = E_P[u_1, \ldots, u_n]$ be the exterior algebra over P on n generators u_i, each of degree 1. Thus E_m in each degree m is the free P-module with generators all exterior products of m of the u_i in order. The differential with $\partial u_i = x_i$ makes E a DG-algebra over P with $\partial E_{m+1} \subset E_m$, while ε_P gives an augmentation $E_0 \to \mathsf{K}$. Together these provide a sequence

$$0 \leftarrow {}_\varepsilon\mathsf{K} \overset{\varepsilon}{\leftarrow} P = E_0 \overset{\partial}{\leftarrow} E_1 \leftarrow \cdots \leftarrow E_n \leftarrow 0 \tag{2.1}$$

of P-modules and P-module homomorphisms.

Proposition 2.1. *For P the polynomial ring in n indeterminates over K, the exterior algebra E in n generators over P provides, as in (2.1), a free P-module resolution of ${}_\varepsilon\mathsf{K}$.*

The proof will construct K-module homomorphisms $\eta \colon \mathsf{K} \to E$ and $s \colon E \to E$ of respective degrees 0 and 1 such that η is a chain transformation with $\varepsilon\eta = 1$ while s is a chain homotopy $s \colon 1 \simeq \eta\varepsilon \colon E \to E$. This contracting homotopy will show (2.1) exact as a sequence of K-modules, hence exact as a sequence of P-modules, hence a resolution.

The chain transformation η is defined by $\eta k = k1$; clearly $\varepsilon\eta = 1$. The homotopy s is constructed by induction on n. Set

$$P'' = \mathsf{K}[x_1, \ldots, x_{n-1}], \qquad P' = \mathsf{K}[x_n],$$
$$E'' = E_{P''}[u_1, \ldots, u_{n-1}], \quad E' = E_{P'}[u_n].$$

Thus $P=P''\otimes P'$, E'' and E' are DG-algebras over P'' and P' respectively, their tensor product $E''\otimes E'$ is a $P''\otimes P'$-algebra, and (VI.4.6) gives an isomorphism $E\cong E''\otimes E'$ of DG-algebras over P. Moreover, $\varepsilon=\varepsilon''\otimes\varepsilon'$ and $\eta=\eta''\otimes\eta'$. For $n=1$, the contracting homotopy $s'\colon E'_0\to E'_1$ may be defined on a polynomial $f=\sum a_i x^i$ of degree k in $x=x_n$ with coefficients $a_i\in \mathsf{K}$ by setting

$$s'(a_0+a_1x+\cdots+a_k x^k)=(a_1+a_2x+\cdots+a_k x^{k-1})\,u;$$

then $\partial s'f=f-a_0=f-\eta'\varepsilon' f$ and $s'\partial fu=fu$, so $s'\colon 1\simeq\eta'\varepsilon'$. Note in particular that s', though a homomorphism of K-modules, is not a homomorphism of P-modules.

Now assume by induction that there is a K-chain homotopy $s''\colon 1\simeq\eta''\varepsilon''\colon E''\to E''$. Since we already have s', Prop. V.9.1 gives a K-chain homotopy s on $E=E''\otimes E'$, and so completes the induction.

The resolution (2.1) is known as the *Koszul resolution*; it first occurs explicitly in a study of Lie algebras by KOSZUL [1950].

Theorem 2.2. *If $P=\mathsf{K}[x_1,\ldots,x_n]$ is the ungraded polynomial algebra over a commutative ring* K *in n indeterminates x_i, while $I\colon \mathsf{K}\to P$ is the injection and* K *is a P-module in any way such that ${}_I\mathsf{K}=\mathsf{K}$, then*

$$\text{h.dim}_P\,\mathsf{K}=n,\qquad \text{h.dim}_P(x_1,\ldots,x_n)=n-1,\tag{2.2}$$

$\operatorname{Ext}^m_P(\mathsf{K},\mathsf{K})$ *is the direct sum of $(m, n-m)$ copies of* K, *and* $\operatorname{Tor}^P(\mathsf{K},\mathsf{K})=\{\operatorname{Tor}^P_m(\mathsf{K},\mathsf{K})\}$ *is an exterior algebra over* K *on n generators in* $\operatorname{Tor}_1(\mathsf{K},\mathsf{K})$.

Proof. Suppose first that K is the P-module ${}_\varepsilon\mathsf{K}$. The Koszul resolution (2.1) stops with degree n. Hence the homological dimension of K is at most n.

We may calculate $\operatorname{Tor}^P(\mathsf{K},\mathsf{K})$ from the resolution (2.1) as the homology of the complex

$$\begin{aligned}\mathsf{K}\otimes_P E_P&=\mathsf{K}\otimes_P(P\otimes E_{\mathsf{K}}[u_1,\ldots,u_n])\\&=(\mathsf{K}\otimes_P P)\otimes E_{\mathsf{K}}\cong \mathsf{K}\otimes E_{\mathsf{K}}=E_{\mathsf{K}}[u_1,\ldots,u_n]\end{aligned}$$

with boundary $\partial(k\otimes u_i)=k\otimes x_i$. But under the isomorphisms above, $k\otimes x_i\to(k\otimes_P x_i)\otimes 1\to k\,\varepsilon(x_i)\otimes 1=0$, since by definition $\varepsilon(x_i)=0$. Thus the differential on the complex is zero, so $\operatorname{Tor}^P(\mathsf{K},\mathsf{K})$ is the exterior algebra over K in n generators. In particular, $\operatorname{Tor}^P_n(\mathsf{K},\mathsf{K})\cong\mathsf{K}\neq 0$, so $\text{h.dim}_P\,\mathsf{K}$ is exactly n. Similarly, $\operatorname{Ext}_P(\mathsf{K},\mathsf{K})$ is calculated from the resolution as the cohomology of the complex

$$\begin{aligned}\operatorname{Hom}_P(E_P,\mathsf{K})&\cong\operatorname{Hom}_P(P\otimes E_{\mathsf{K}},\mathsf{K})\\&\cong\operatorname{Hom}_{\mathsf{K}}(E_{\mathsf{K}},\operatorname{Hom}_P(P,\mathsf{K})\cong\operatorname{Hom}_{\mathsf{K}}(E_{\mathsf{K}},\mathsf{K})).\end{aligned}$$

The coboundary in this complex is again zero, so $\operatorname{Ext}_P^m(\mathsf{K}, \mathsf{K}) \cong \operatorname{Hom}_{\mathsf{K}}(E_m, \mathsf{K})$ is the direct sum of $(m, n-m)$ copies of K, as asserted in the Theorem.

Now consider the ideal $J = (x_1, \ldots, x_n) = \operatorname{Ker} \varepsilon$. Since $\partial\colon E_1 \to E_0 = P$ has image exactly J, the Koszul resolution (2.1) yields a resolution

$$0 \leftarrow J \leftarrow E_1 \leftarrow \cdots \leftarrow E_n \leftarrow 0$$

of J, with E_{m+1} in "dimension" m. Hence $\operatorname{Ext}_P^m(J, \mathsf{K}) \cong \operatorname{Hom}_{\mathsf{K}}(E_{m+1}, \mathsf{K})$ for $m > 0$, so $\operatorname{Ext}_P^{n-1}(J, \mathsf{K}) = \mathsf{K} \neq 0$, and J has exact dimension $n-1$, as asserted.

Now let K have some other P-module structure, say by operators $p \circ k$ for $p \in P$. The condition ${}_I\mathsf{K} = \mathsf{K}$ states that this P-module structure, pulled back along the injection $I\colon \mathsf{K} \to P$ with $I(k) = k 1_P$, is the original K-module structure of K; in other words, that $k' \circ k = k'k$. Now $\varrho(p) = p \circ 1_{\mathsf{K}}$ defines an algebra homomorphism $\varrho\colon P \to \mathsf{K}$, because $p \circ k = p \circ (1_{\mathsf{K}} k) = \varrho(p) \circ k = \varrho(p)\, k$. In other words, K is the P-module ${}_\varrho\mathsf{K}$ obtained by pull-back along ϱ. But set $a_i = \varrho\, x_i \in \mathsf{K}$ and $x_i' = x_i - a_i$. Then P can be viewed as the polynomial algebra $\mathsf{K}[x_1', \ldots, x_n']$ and $\varrho\, x_i' = 0$, so ϱ is the corresponding augmentation, and the previous calculations apply.

In conclusion, note that $\operatorname{Tor}^P(\mathsf{K}, \mathsf{K}) = E_{\mathsf{K}}[u_1, \ldots, u_n]$ turns out to be not just a graded P-module — as it should be, on general principles — but actually a graded algebra; to wit, the exterior algebra on the n cycles (homology classes) u_i in Tor_1. This algebra structure on $\operatorname{Tor}^P(\mathsf{K}, \mathsf{K})$ hides a mystery. By our general results we may (and did) compute $\operatorname{Tor}_m^P(\mathsf{K}, \mathsf{K})$ from any convenient resolution. By "accident" the DG-module E which we used as a resolution was in fact a DG-algebra, so $\operatorname{Tor}^P(\mathsf{K}, \mathsf{K})$ inherited "by accident" an algebra structure. We shall show in Chap. VIII that this structure arises intrinsically from the fact that K (as a P-module) is a P-algebra; indeed, the torsion product of two algebras is an algebra.

Exercises

1. Calculate $\operatorname{Tor}^P(J, \mathsf{K})$ and $\operatorname{Ext}_P(J, \mathsf{K})$ for $J = (x_1, \ldots, x_n)$.
2. Show that $\operatorname{h.dim}_P(x_1, \ldots, x_k) = k - 1$.
3. Examine Thm. 2.2 when K is a skew field.

3. Ext and Tor for Algebras

If Λ is an (ungraded!) K-algebra, the usual functors $\operatorname{Ext}_\Lambda$ and $\operatorname{Tor}^\Lambda$ may be regarded as functors with values which are K-modules. For this purpose, as in VI.1, we regard the K-algebra Λ as the composite object $\Lambda = (R, I)$ consisting of a ring R and a ring homomorphism $I\colon \mathsf{K} \to R$

with $(Ik)\,r=r\,(Ik)$; that is, with $I(\mathsf{K})$ in the center of R. A left Λ-module (say as defined by (VI.5.2) and (VI.5.3)) is just a left R-module A; by pull-back along $I\colon \mathsf{K}\to R$ it is a K-module, hence also an R-K-bimodule ${}_R A_{\mathsf{K}}$. A homomorphism $\alpha\colon A\to A'$ of left Λ-modules is defined to be a homomorphism of left R-modules, and then is automatically a homomorphism of R-K-bimodules.

Proposition 3.1. *For $\Lambda=(R,I)$ a K-algebra and C, A left Λ-modules, the abelian group $\operatorname{Ext}_R^n(C,A)$ has two K-module structures induced by the K-module structures of C and A, respectively. These two K-module structures agree; if we write $\operatorname{Ext}_\Lambda^n(C,A)$ for the resulting K-module, then $\operatorname{Ext}_\Lambda^n$ is a bifunctor from Λ-modules to K-modules which satisfies the axioms formulated in Thm.* III.10.1. *For a third Λ-module D, composition is a homomorphism of K-modules*

$$\operatorname{Ext}_\Lambda^k(A,D)\otimes_{\mathsf{K}}\operatorname{Ext}_\Lambda^m(C,A)\to\operatorname{Ext}_\Lambda^{k+m}(C,D). \tag{3.1}$$

Proof. The first K-module structure is that induced on $\operatorname{Ext}_R^n(C,A)$ as a functor of C by the R-module endomorphisms $p_k\colon C\to C$ defined for each $k\in\mathsf{K}$ by $p_k(c)=kc$; the second arises similarly from A. Equivalently, regard C and A as R-K-bimodules; then $\operatorname{Ext}_R^n(C,A)$ is a K-K-bimodule as in V.3.4. The crux of the proof is the demonstration that these two K-module structures agree.

For $n=0$ and $f\in\operatorname{Hom}_R(C,A)$, the first K-module structure defines kf by $(kf)c=f(kc)$, the second by $(fk)c=k(fc)$. Since f is a K-module homomorphism, they agree.

For $n>0$, take a long exact sequence $S\in\in\operatorname{Ext}_R^n(C,A)$. Multiplication by k is a morphism $p_k\colon S\to S$ of sequences of R-modules which agrees on the left end with multiplication in A by k and on the right end with multiplication in C by k. By Prop. III.5.1, $p_k S\equiv S p_k$, and the structures agree. Alternatively, if X is a projective resolution of C and $\operatorname{Ext}_R^n(C,A)$ is calculated as $H^n(\operatorname{Hom}_R(X,A))$, the K-module structure, like the functorial structure, is computed from that of X or of A, which are known to agree in $\operatorname{Hom}_R(X,A)$.

Any R-module homomorphism $\alpha\colon A\to A'$ commutes with the endomorphism p_k, so the induced map $\alpha_*\colon\operatorname{Ext}_R^n(C,A)\to\operatorname{Ext}_R^n(C,A')$ is a K-module homomorphism, and $\operatorname{Ext}^n(C,A)$ is a bifunctor of K-modules. The connecting homomorphisms are also K-module homomorphisms, and the Yoneda composite is K-bilinear; hence (3.1).

The treatment of torsion products is similar.

Proposition 3.2. *If $\Lambda=(R,I)$ and G_Λ, ${}_\Lambda C$ are Λ-modules, then for each $n\geqq 0$ the abelian group $\operatorname{Tor}_n^R(G,C)$ has two K-module structures induced by the K-module structures of G and C, respectively. These two K-module structures agree; if we write $\operatorname{Tor}_n^\Lambda(G,C)$ for the resulting K-module,* *then* Tor_n^Λ *is a covariant bifunctor from* Λ*-modules to* K*-modules which satisfies the axioms formulated in Thm.* V.8.5.

Proof. For $n=0$, $kg\otimes c=g\otimes kc$, so the two K-module structures agree. We leave the proof for $n>0$ to the reader (use V.7.1).

Now let Λ and Σ be two K-algebras (still ungraded). A Λ-Σ-bimodule ${}_\Lambda A_\Sigma$ is then a bimodule over the rings Λ, Σ such that the two induced K-module structures agree. They induce identical K-module structures on $\mathrm{Hom}_{\Lambda-\Sigma}(C,A)$. The corresponding K-modules $\mathrm{Ext}^n_{\Lambda-\Sigma}(C,A)$ for $n>0$ could be defined as the congruence classes of exact sequences of bimodules leading from A to C through n intermediate steps, just as before. Equivalently, turn A and C into left $\Lambda\otimes\Sigma^{\mathrm{op}}$-modules, and define $\mathrm{Ext}^n_{\Lambda-\Sigma}$ as $\mathrm{Ext}^n_{(\Lambda\otimes\Sigma^{\mathrm{op}})}$. Similarly, bimodules ${}_\Sigma B_\Lambda$, ${}_\Lambda C_\Sigma$ are one-sided modules $B_{(\Lambda\otimes\Sigma^{\mathrm{op}})}$, ${}_{(\Lambda\otimes\Sigma^{\mathrm{op}})}C$ and so have a tensor product $B\otimes_{(\Lambda\otimes\Sigma^{\mathrm{op}})}C$ and torsion products $\mathrm{Tor}^{(\Lambda\otimes\Sigma^{\mathrm{op}})}(B,C)$ which are K-modules. We also write these products as $\mathrm{Tor}^{\Lambda-\Sigma}(B,C)$. We next show that Ext for left Λ-modules sometimes reduces to a Λ-bimodule Ext.

Theorem 3.3. *Let* Λ *be a* K*-algebra and* C *and* A *left* Λ*-modules. Assume that* Λ *and* C *are projective* K*-modules (for instance, this automatically holds if* K *is a field). Then adjoint associativity induces a natural isomorphism*

$$\eta_*\colon \mathrm{Ext}^n_\Lambda(C,A)\cong\mathrm{Ext}^n_{\Lambda-\Lambda}(\Lambda,\mathrm{Hom}_{\mathsf{K}}(C,A)),\qquad n=0,1,\ldots,\tag{3.2}$$

of K*-modules. For* $n=0$, $\mathrm{Ext}^0_\Lambda(C,A)=\mathrm{Hom}_\Lambda(C,A)=\mathrm{Hom}_\Lambda(\Lambda\otimes_\Lambda C,A)$, *and* η *is the ordinary adjoint associativity.*

In (3.2), $\mathrm{Hom}_{\mathsf{K}}(C,A)$ is a left Λ-module via the left Λ-module structure of A and a right Λ-module via the left Λ-module structure of the contravariant argument C.

Proof. Take a free resolution $\varepsilon\colon X\to\Lambda$ of the Λ-Λ-bimodule Λ. As a free bimodule, each X_n has the form $X_n=\Lambda\otimes F_n\otimes\Lambda$ for some free K-module F_n. Now a projective module is a direct summand of a free module, so the tensor product of two projective K-modules is a projective K-module. Since we have assumed Λ and C projective as K-modules, $\Lambda\otimes F_n$ and $F_n\otimes C$ are projective K-modules, so, by Prop. VI.8.1, $X_n=(\Lambda\otimes F_n)\otimes\Lambda$ is a projective right Λ-module and $X_n\otimes_\Lambda C\cong\Lambda\otimes(F_n\otimes C)$ is a projective left Λ-module.

Adjoint associativity is natural, so yields an isomorphism of complexes

$$\eta\colon \mathrm{Hom}_\Lambda(X\otimes_\Lambda C,A)=\mathrm{Hom}_{\Lambda-\Lambda}(X,\mathrm{Hom}_{\mathsf{K}}(C,A)).\tag{3.3}$$

The cohomology groups of the right hand complex are

$$\mathrm{Ext}_{\Lambda-\Lambda}(\Lambda,\mathrm{Hom}_{\mathsf{K}}(C,A)).$$

Examine those of the left hand complex. Since $\varepsilon: X \to \Lambda$ is a projective resolution of Λ as a right Λ-module, the homology of the complex $X \otimes_\Lambda C$ is $\mathrm{Tor}^\Lambda(\Lambda, C)$. But Λ itself is a free right Λ-module, so all $\mathrm{Tor}_n^\Lambda(\Lambda, C) = 0$ for $n > 0$, so the complex $X \otimes_\Lambda C$ with $\varepsilon \otimes 1: X_0 \otimes_\Lambda C \to \Lambda \otimes_\Lambda C = C$ constitutes a projective resolution of ${}_\Lambda C$. Therefore its cohomology over Λ, as on the left side of (3.3), is $\mathrm{Ext}_\Lambda(C, A)$. Thus η induces an isomorphism of these cohomology groups, as asserted.

This isomorphism can be described as follows in terms of long exact sequences.

Corollary 3.4. *For any long exact sequence* $S \in\in \mathrm{Ext}_\Lambda^n(C, A)$ *with* $n > 0$ *the isomorphism* η *of* (3.2) *carries the class of* S *into the class of*

$$[\mathrm{Hom}_K(C, S)]\, \eta(1_C) \in\in \mathrm{Ext}^n_{\Lambda-\Lambda}(\Lambda, \mathrm{Hom}_K(C, A)).$$

Proof. First analyze $[\mathrm{Hom}(C, S)]\, \eta(1_C)$. Since $1_C \in \mathrm{Hom}_\Lambda(C, C)$ and $\eta: \mathrm{Hom}_\Lambda(C, C) = \mathrm{Hom}_\Lambda(\Lambda \otimes_\Lambda C, C) \cong \mathrm{Hom}_{\Lambda-\Lambda}(\Lambda, \mathrm{Hom}(C, C))$, $\eta(1_C)$ is a map $u: \Lambda \to \mathrm{Hom}(C, C)$ (actually, with $(u\lambda)c = \lambda c$). If

$$S: 0 \to A \to B_{n-1} \to \cdots \to B_0 \to C \to 0$$

is exact, and Hom is short for Hom_K, then $\mathrm{Hom}(C, S)$ is the sequence

$$0 \to \mathrm{Hom}(C, A) \to \mathrm{Hom}(C, B_{n-1}) \to \cdots \to \mathrm{Hom}(C, B_0) \to \mathrm{Hom}(C, C) \to 0;$$

since C is K-projective, it is an exact sequence of Λ-Λ-bimodules. Acting on the right of this sequence with $\eta(1_C)$, we get a long exact sequence of bimodules from $\mathrm{Hom}(C, A)$ to Λ, as in the conclusion of the corollary.

To apply the canonical isomorphism $\zeta: \mathrm{Ext}_\Lambda^n(C, A) \cong H^n(X \otimes_\Lambda C, A)$ of (III.6.3), we regard S as a resolution of C, lift 1_C to $f: X \otimes_\Lambda C \to S$, and obtain $\zeta(\mathrm{cls}\, S)$ as the class of the cocycle f_n. But apply adjoint associativity; $\eta f: X \to \mathrm{Hom}(C, S)$ lifts $\eta(1_C): \Lambda \to \mathrm{Hom}(C, C)$, so ηf factors through a chain transformation $g: X \to [\mathrm{Hom}(C, S)]\eta(1_C)$ lifting 1_Λ with $\eta f_n = g_n$. Thus $\eta_* \,\mathrm{cls}\, f_n = \mathrm{cls}\, g_n$, $\zeta(\mathrm{cls}\, S) = \mathrm{cls}\, f_n$, and (again by the definition of ζ) $\zeta\, \mathrm{cls}[\mathrm{Hom}(C, S)\, \eta(1_C)] = \mathrm{cls}\, g_n$, whence the conclusion.

Exercises

1. If Λ is an algebra over a field, P a projective Λ-Λ-bimodule, and B a left Λ-module, show that $P \otimes_\Lambda B$ is a projective left Λ-module.

2. If $T \in\in \mathrm{Ext}^n_{\Lambda-\Lambda}(\Lambda, \mathrm{Hom}(C, A))$, as in Cor. 3.4, is the exact sequence

$$T: \mathrm{Hom}(C, A) \rightarrowtail B_{n-1} \to X_{n-2} \to \cdots \to X_0 \twoheadrightarrow \Lambda$$

with all X_i projective show $\eta^{-1} \mathrm{cls}\, T = \mathrm{cls}(e(T \otimes_\Lambda C))$, where e is the evaluation map $e: \mathrm{Hom}(C, A) \otimes C \to A$.

3. For Λ an algebra over a field F, $\Omega = \Lambda \otimes \Lambda^{\mathrm{op}}$, and modules C_Λ, ${}_\Lambda A$, prove $\mathrm{Tor}_n^\Lambda(C, A) \cong \mathrm{Tor}_n^\Omega(\Lambda, A \otimes_K C)$.

4. For K-algebras Λ and Σ, modules G_Λ and ${}_\Lambda A_\Sigma$ and an injective right Σ-module J use Ex. III.7.3 to establish the isomorphism ("duality"; CARTAN-EILENBERG VI.5)

$$\operatorname{Ext}^n_\Lambda(G, \operatorname{Hom}_\Sigma(A,J)) \cong \operatorname{Hom}_\Sigma(\operatorname{Tor}^\Lambda_n(G,A),J).$$

4. Global Dimensions of Polynomial Rings

We can now compute the global dimensions of polynomial rings over a field.

Proposition 4.1. *If the modules C and A over the commutative ring* K *are regarded as modules over the polynomial ring $P=\mathrm{K}[x]$, by pull-back along ε: $P\to\mathrm{K}$ with $\varepsilon(x)=0$, then* $\operatorname{Hom}_P(C,A)=\operatorname{Hom}_\mathrm{K}(C,A)$ *and, for $n>0$, there is an isomorphism of P-modules*

$$\operatorname{Ext}^n_P(C,A)\cong\operatorname{Ext}^n_\mathrm{K}(C,A)\oplus\operatorname{Ext}^{n-1}_\mathrm{K}(C,A). \tag{4.1}$$

Here the $\operatorname{Ext}_\mathrm{K}$ on the right are K-modules, hence P-modules by pull-back.

Proof. Take a K-projective resolution $\eta\colon X\to C$. The exterior algebra $E=E_P[u]$ provides a resolution $\varepsilon\colon E\to\mathrm{K}$ of K by free P-modules $E_0\cong E_1\cong P$ and the boundary $\partial\colon E_1\to E_0$ is given by multiplication by x. Now P is a free K-module, hence so are E_1, E_0, $H(E)$, and the cycles of E. The Künneth tensor formula (Thm. V.10.1) asserts that $H(E\otimes X)\cong H(E)\otimes H(X)$, so that $H_n(E\otimes X)=0$ for $n>0$ and $\varepsilon\otimes\eta\colon H_0(E\otimes X)\cong\mathrm{K}\otimes C=C$. Thus $\varepsilon\otimes\eta\colon E\otimes X\to C$ is a resolution of C by projective P-modules. Hence $\operatorname{Ext}_P(C,A)$ is the cohomology of the complex $\operatorname{Hom}_P(E\otimes X,A)$.

Now $(E\otimes X)_n=E_0\otimes X_n\oplus E_1\otimes X_{n-1}\cong P\otimes X_n\oplus P\otimes X_{n-1}$, so by adjoint associativity

$$\begin{aligned}\operatorname{Hom}_P((E\otimes X)_n,A)&\cong\operatorname{Hom}_P(P,\operatorname{Hom}(X_n,A))\oplus\operatorname{Hom}_P(P,\operatorname{Hom}(X_{n-1},A)),\\&\cong\operatorname{Hom}(X_n,A)\oplus\operatorname{Hom}(X_{n-1},A).\end{aligned}$$

Since the boundary $\partial\colon E_1\to E_0$ is multiplication by x and since A and X_n are P-modules via ε with $\varepsilon(x)=0$, these isomorphisms carry the coboundary on the left into the coboundary on the right (induced by ∂ in X). This isomorphism of cochain complexes gives the asserted isomorphism (4.1).

Theorem 4.2. *If the commutative ring* K *has global dimension $r\leq\infty$, then the polynomial ring $P=\mathrm{K}[x]$ has global dimension $r+1$ (or ∞, if $r=\infty$).*

Since K and P are commutative, we can omit "left" in l. gl. dim.

Proof. Let G be any P-module. The first r terms of a free resolution of G as a P-module give an exact sequence $S\colon G_r\rightarrowtail Y_{r-1}\to\cdots\to Y_0\twoheadrightarrow G$.

Now P itself and hence each Y_i is also a free K-module, so $\mathrm{h.dim}_K G \leqq \mathrm{gl.dim}\, K = r$ implies that the K-module G_r is projective. For any P-module H we have the isomorphisms

$$\mathrm{Ext}_P^{r+2}(G, H) \cong \mathrm{Ext}_P^2(G_r, H) \cong \mathrm{Ext}_{P-P}^2(P, \mathrm{Hom}(G_r, H)),$$

the first by the iterated connecting homomorphism of the sequence S and the second by adjoint associativity (Thm. 3.3). On the right regard the P-bimodules as $P \otimes P^{\mathrm{op}}$-left modules. Then $P^{\mathrm{op}} \cong P$, so $P \otimes P^{\mathrm{op}} \cong P \otimes K[y]$ is isomorphic to a polynomial ring $P[y]$ in one indeterminate y over P. In particular, the P-P-bimodule P becomes a $P[y]$-module, and the injection $I: P \to P[y]$ satisfies $(Ip)p' = pp'$. Hence Thm. 2.2 (with K there replaced by P and P by $P[y]$) gives $\mathrm{h.dim}_{P[y]} P = 1$, which asserts that $\mathrm{Ext}_{P-P}^2(P, -)$ above vanishes, hence that $\mathrm{Ext}_P^{r+2} = 0$, so $\mathrm{gl.dim}\, P \leqq r+1$. On the other hand, $\mathrm{gl.dim}\, K = r$ means that there are K-modules C and A with $\mathrm{Ext}_K^r(C, A) \neq 0$. By Prop. 4.1, this gives $\mathrm{Ext}_P^{r+1}(C, A) \cong \mathrm{Ext}_K^r(C, A) \neq 0$, so $\mathrm{gl.dim}\, P$ is at least $r+1$. This latter argument also gives the result stated for $r = \infty$.

Corollary 4.3. *The global dimension of $Z[x_1, \ldots, x_n]$ is $n+1$.*

Corollary 4.4. *The global dimension of the polynomial ring $P = P_F[x_1, \ldots, x_n]$ in n indeterminates over a field F is n. If J is any ideal in P, $\mathrm{h.dim}_P J \leqq n-1$.*

Only the assertion as to the ideal J requires proof. Any projective resolution of J yields an exact sequence

$$0 \to C_{n-1} \to X_{n-2} \to \cdots \to X_0 \to J \to 0$$

of P-modules with the X_i projective. Compose this sequence with $J \rightarrowtail P \twoheadrightarrow P/J$ to give an exact sequence with n intermediate projective modules, ending in P/J. Since n is the global dimension of P, $\mathrm{h.dim}_P P/J \leqq n$, so by the characterization of homological dimension (Thm. 1.1), C_{n-1} is projective. This proves $\mathrm{h.dim}_P J \leqq n-1$.

5. Separable Algebras

We now consider applications to the classical theory of (ungraded) algebras Λ. Recall that 1_Λ denotes the identity element of Λ.

Proposition 5.1. *The following conditions on an algebra Λ are equivalent:*

(i) $\mathrm{h.dim}_{(\Lambda \otimes \Lambda^{\mathrm{op}})} \Lambda = 0$.

(ii) *Λ is a projective Λ-bimodule.*

(iii) *The product map $\pi: \Lambda \otimes \Lambda \to \Lambda$ has a bimodule right inverse.*

(iv) *There is an element e in $\Lambda \otimes \Lambda$ with $\pi e = 1_\Lambda$ and $\lambda e = e\lambda$ for all λ.*

In this section we denote these equivalent properties by writing bidim $\Lambda=0$ (read: "The homological dimension of Λ as a Λ-bimodule is zero").

Proof. Properties (i) and (ii) are equivalent by the definition of homological dimension. In (iii), the product map $\pi(\lambda\otimes\mu)=\lambda\mu$ is an epimorphism of Λ-bimodules. If Λ is projective, this product map splits by a bimodule homomorphism $\alpha: \Lambda\to\Lambda\otimes\Lambda$ with $\pi\alpha=1$; this proves (ii)$\Rightarrow$(iii). Conversely, if $\pi\alpha=1$, then Λ is a bimodule direct summand of the free bimodule $\Lambda\otimes\Lambda$, hence is projective. If $\pi\alpha=1$, then $\alpha 1_\Lambda=e\in\Lambda\otimes\Lambda$ has $\pi e=1_\Lambda$; since α is a bimodule homomorphism, $\alpha\lambda=\lambda e=e\lambda$. Conversely, an element e with these properties determines such an α.

We now investigate the preservation of the property bidim $\Lambda=0$ under three standard constructions for algebras: Direct products, ground ring extension, and formation of total matrix algebras.

The *direct product* of two K-algebras Γ and Σ is a K-algebra $\Lambda=\Gamma\times\Sigma$; as a K-module it is the direct sum $\Gamma\oplus\Sigma$ with elements all pairs (γ,σ); its multiplication is given by

$$(\gamma,\sigma)(\gamma',\sigma')=(\gamma\gamma',\sigma\sigma'); \tag{5.1}$$

its identity is thus $(1_\Gamma, 1_\Sigma)$. The projections $\pi_1(\gamma,\sigma)=\gamma$, $\pi_2(\gamma,\sigma)=\sigma$ are algebra homomorphisms

$$\Gamma\xleftarrow{\pi_1}\Gamma\times\Sigma\xrightarrow{\pi_2}\Sigma, \tag{5.2}$$

(the injections ι_1, ι_2 are not; they do not map identity to identity). With these maps the algebra $\Gamma\times\Sigma$ is couniversal for Γ and Σ in the category of algebras. This is why we call $\Gamma\times\Sigma$ the direct "product", even though it is often called the direct "sum" of Γ and Σ.

Any Γ-bimodule becomes a $(\Gamma\times\Sigma)$-bimodule by pull-back along π_1 (on both left and right sides); similarly any Σ-bimodule or any $(\Gamma$-$\Sigma)$-bimodule becomes a $(\Gamma\times\Sigma)$-bimodule. In particular, the definition (5.1) shows that $\Lambda=\Gamma\times\Sigma$, regarded as a Λ-bimodule, is the direct sum $\Gamma\oplus\Sigma$ of the Λ-bimodules Γ and Σ. Since the tensor product is additive, $\Lambda\otimes\Lambda=(\Gamma\oplus\Sigma)\otimes(\Gamma\oplus\Sigma)$ is the direct sum of four Λ-bimodules

$$\Lambda\otimes\Lambda\cong(\Gamma\otimes\Gamma)\oplus(\Gamma\otimes\Sigma)\oplus(\Sigma\otimes\Gamma)\oplus(\Sigma\otimes\Sigma). \tag{5.3}$$

Proposition 5.2. *For algebras* Γ *and* Σ, bidim $\Gamma=0=$ bidim Σ *implies* bidim $(\Gamma\times\Sigma)=0$.

Proof. By hypothesis, Prop. 5.1, part (iii) gives bimodule maps α_Γ: $\Gamma\to\Gamma\otimes\Gamma$ and α_Σ: $\Sigma\to\Sigma\otimes\Sigma$ with $\pi\alpha_\Gamma=1$ and $\pi\alpha_\Sigma=1$. They are also maps of Λ-bimodules, hence combine as $\alpha_\Gamma\oplus\alpha_\Sigma$: $\Gamma\oplus\Sigma\to(\Gamma\otimes\Gamma)\oplus(\Sigma\otimes\Sigma)$ which, followed by the injection into (5.3), yields a Λ-bimodule map $\alpha: \Lambda \to \Lambda \otimes \Lambda$. Since the injections $\Gamma \otimes \Gamma \to \Lambda \otimes \Lambda$ and $\Sigma \otimes \Sigma \to \Lambda \otimes \Lambda$ preserve the product, $\pi\alpha = 1$, as required for bidim $\Lambda = 0$.

A ***ground-ring extension*** is the process of passing from algebras over the commutative ground-ring K to algebras over a new ground-ring R, where R is now assumed to be a *commutative* algebra over K. If Λ is a K-algebra, then $R \otimes \Lambda$ is a ring (as the tensor product of rings) and an R-module (via its left factor); since R is commutative, it is also an algebra over R. As an algebra over R we denote $R \otimes \Lambda$ by Λ^R (the standard notation is Λ_R; this would conflict with our previous notation for R-modules).

Proposition 5.3. *If* bidim $\Lambda = 0$, then bidim $\Lambda^R = 0$.

Proof. The Λ^R-bimodule $\Lambda^R \otimes_R \Lambda^R = (R \otimes \Lambda) \otimes_R (R \otimes \Lambda)$ is isomorphic to $R \otimes \Lambda \otimes \Lambda$ under the correspondence $(r \otimes \lambda) \otimes (s \otimes \mu) \to rs \otimes \lambda \otimes \mu$. If $e = \sum \mu_i \otimes \nu_i \in \Lambda \otimes \Lambda$ has the property (iv) of Prop. 5.1 for Λ, one checks that $e' = \sum 1 \otimes \mu_i \otimes \nu_i$ has the corresponding properties for Λ^R.

The ground-ring extension is useful in the classical case of algebras Λ of finite dimension (as vector spaces) over a field F. Any field $L \supset F$ may be regarded as a commutative algebra over F, so that Λ^L is an algebra over L. If Λ has F-basis $u_1, \ldots, u_n$, the product in Λ is determined via $u_i u_j = \sum_k f_k^{ij} u_k$ by n^3 constants $f_k^{ij} \in F$. The extended algebra Λ^L is the vector space over L with basis $1 \otimes u_i$, $i = 1, \ldots, n$ and the same multiplication constants f_k^{ij}. In this case we have a converse of the last proposition.

Proposition 5.4. *If Λ is an algebra over a field F and R a commutative algebra over F, then* bidim $\Lambda^R = 0$ *implies* bidim $\Lambda = 0$.

Proof. For $\otimes = \otimes_F$, the product map for Λ^R is equivalent to the epimorphism $(1 \otimes \pi): R \otimes \Lambda \otimes \Lambda \to R \otimes \Lambda$ of Λ^R bimodules; by hypothesis it has a right inverse α which is a map of Λ^R-bimodules. Since an F-algebra homomorphism $j: \Lambda \to \Lambda^R$ is defined by $j(\lambda) = 1 \otimes \lambda$, each Λ^R-bimodule pulls back along j to become a Λ-bimodule; in particular, we may regard $\alpha: R \otimes \Lambda \to R \otimes \Lambda \otimes \Lambda$ as a map of Λ-bimodules. Now R is a vector space over the field F; choose a basis with first element 1_R. If η maps 1_R to 1_F and the remaining basis elements to zero, $\eta: R \to F$ is an F-module homomorphism whose composite with the injection $\iota: F \to R$ is the identity. Now form the diagram

$$\begin{array}{ccccccccc} F \otimes \Lambda & \xrightarrow{\iota \otimes 1} & R \otimes \Lambda & \xrightarrow{\alpha} & R \otimes \Lambda \otimes \Lambda & \xrightarrow{\eta \otimes 1 \otimes 1} & F \otimes \Lambda \otimes \Lambda & = & \Lambda \otimes \Lambda \\ \| & & & & \downarrow{\scriptstyle 1 \otimes \pi} & & \downarrow{\scriptstyle 1 \otimes \pi} & & \downarrow{\scriptstyle \pi} \\ \Lambda & & & & R \otimes \Lambda & \xrightarrow{\eta \otimes 1} & F \otimes \Lambda & = & \Lambda. \end{array}$$

The squares are commutative; the composite of the top row is a composite of Λ-bimodule maps, hence is a bimodule map $\alpha': \Lambda \to \Lambda \otimes \Lambda$.

Since $(1\otimes\pi)\alpha=1$ and $\eta\iota=1$, the diagram shows $\pi\alpha'=1$, so bidim $\Lambda=0$ by (iii) of Prop. 5.1.

The process of ground-ring extensions also includes the process of "reduction modulo a prime p". Indeed, the ring Z_p of integers modulo p may be regarded as a commutative algebra R over Z. For any Z-algebra Λ, Λ^{Z_p} is then the algebra Λ "reduced modulo p".

The *total matrix algebra* $M_n(F)$ over a field F consists of all $n\times n$ matrices of elements from F with the usual product; as a vector space over F it has a basis consisting of the matrices e_{ij} for $i, j=1, \ldots, n$. Here e_{ij} is the matrix with entry 1 in the i-th row and the j-th column and zeros elsewhere. The multiplication is given by $e_{ij}e_{jk}=e_{ik}$ and $e_{ir}e_{sk}=0$ for $r\neq s$. If $L\supset F$ is a larger field, $[M_n(F)]^L\cong M_n(L)$.

Proposition 5.5. *For any field* F, bidim $M_n(F)=0$.

Proof. The element $e=\sum e_{i1}\otimes e_{1i}$ in $M_n(F)\otimes M_n(F)$ has $\pi e=\sum e_{ii}=1_M$ and $e_{rs}e=e_{r1}\otimes e_{1s}=ee_{rs}$, so that it satisfies the conditions (iv) of Prop. 5.1.

An algebra Λ over a field F is semi-simple (cf. §1) if every left Λ-module is projective. If bidim $\Lambda=0$, Λ is semi-simple: For any left Λ-modules C and A, Thm. 3.3 gives an isomorphism

$$\operatorname{Ext}^1_\Lambda(C, A)\cong\operatorname{Ext}^1_{\Lambda-\Lambda}(\Lambda, \operatorname{Hom}(C,A)),$$

so $\operatorname{Ext}^1_\Lambda(C, -)$ vanishes and C is left-Λ projective.

An algebra Λ over a field F is called *separable* if, for every extension field $L\supset F$, the algebra Λ^L is semi-simple. By Prop. 5.5, each total matrix algebra is separable. It is easy to see that the direct product of separable algebras is separable. Conversely, the Wedderburn structure theorem states that for every separable algebra Λ of finite dimension over a field F there is an extension field L of F (actually of finite dimension as a vector space over F) such that Λ^L is a direct product of a finite number of total matrix algebras. Assuming this result we prove

Theorem 5.6. *If the algebra Λ over a field F has finite dimension as a vector space over F, Λ is separable if and only if* bidim $\Lambda=0$.

Proof. First suppose Λ separable. By the structure theorem, there is a L with $\Lambda^L=\Sigma_1\times\cdots\times\Sigma_m$ with each Σ_i a total matrix algebra over L. By Prop. 5.5, bidim $\Sigma_i=0$, hence by Prop. 5.2 bidim $\Lambda^L=0$, whence by Prop. 5.4, bidim $\Lambda=0$.

Conversely, suppose bidim $\Lambda=0$. For each $L\supset F$ we wish to prove every left Λ^L-module C projective. Let B be another left Λ^L-module. By adjoint associativity (Thm. 3.3),

$$\operatorname{Ext}^1_{\Lambda^L}(C, B)\cong\operatorname{Ext}^1_{\Lambda^L-\Lambda^L}(\Lambda^L, \operatorname{Hom}_L(C, B)).$$

But bidim $\Lambda=0$ implies bidim $\Lambda^L=0$ by Prop. 5.3, so that Λ^L is a projective bimodule, and the Ext on the right vanishes. Therefore $\mathrm{Ext}^1(C, B)=0$ for any B, which states that C is projective, as desired.

Note that the proof has been wholly elementary, except for the use of Ext^1, via adjoint associativity, to switch from the bimodule Λ^L to left modules.

The effect of direct product and ground-ring extensions upon the functor $\mathrm{Ext}(\Lambda, -)$ in the more general case when bidim $\Lambda\neq 0$ will be studied in Chap. X.

Exercises

1. Construct the direct product of two DG-algebras (over the same K) so as to be couniversal.

2. For Γ and Σ algebras over K prove $(\Gamma\otimes\Sigma)^R\cong\Gamma^R\otimes_R\Sigma^R$, $(\Gamma\times\Sigma)^R\cong\Gamma^R\times\Sigma^R$, and $(\Gamma\times\Sigma)^{\mathrm{op}}\cong\Gamma^{\mathrm{op}}\times\Sigma^{\mathrm{op}}$.

3. (Coefficient extensions need not remain semi-simple.) For p a rational prime, Z_p the field of integers mod p, and $L=Z_p(x)$ the field of all rational functions over Z_p in one indeterminate x, let F be the subfield $Z_p(x^p)$. Then L is a commutative algebra over F; let Λ be an isomorphic F-algebra under $x\to u\in\Lambda$. Show Λ but not Λ^L semi-simple. (If M is the ideal in Λ^L generated by $u-x$, the epimorphism $\Lambda^L\to M$ with $1\to u-x$ does not split.)

6. Graded Syzygies

Let $P=F[x_1, \ldots, x_n]$ be the polynomial algebra over a field F in n indeterminates x_i, each of degree 1. Cor. 4.4 shows that any P-module A has a projective resolution

$$0\leftarrow A\leftarrow X_0\leftarrow\cdots\leftarrow X_n\leftarrow 0$$

which stops with the term X_n. The Hilbert syzygy theorem asserts that a graded P-module A has such a resolution with X_k *free graded* modules stopping at the same point. Though closely related, we cannot deduce this syzygy theorem from our previous result, because we do not know that a projective module must be free.

In this section we regard P as an *internally* graded algebra over F; the homogeneous elements of degree m are thus the ordinary homogeneous polynomials of that degree. We work in the category of all internally graded P-modules with morphisms all P-module homomorphisms of degree 0; the kernels and cokernels of such morphisms are again internally graded P-modules. Each internally graded P-module $A=\sum A_n$ is also an ungraded module over the ungraded algebra P. If G is a second such module, we use $G\otimes_P A$ and $\mathrm{Tor}_1^P(G, A)$ to denote the ordinary tensor and torsion products, constructed without regard to the grading. This use of internal grading has the advantage of suiting the classical notion of a polynomial ring and the technical advantage of using the ordinary torsion product. A grading of the torsion product will be introduced in X.8 where it is appropriate.

The coefficient field F is a (trivially) graded P-module under the usual action $x_i f = 0$ for $f \in F$.

Lemma 6.1. *If A is a graded P-module with $A \otimes_P F = 0$, then $A = 0$.*

Proof. Let $J = (x_1, \ldots, x_n)$ be the ideal of all polynomials in P with constant term 0. The exact sequence $J \rightarrowtail P \twoheadrightarrow F$ of P-modules gives $A \otimes_P J \to A \otimes_P P \twoheadrightarrow A \otimes_P F = 0$ exact, so that $A \otimes_P J \twoheadrightarrow A \otimes_P P = A$. This states that each $a \in A$ lies in AJ. If $A \neq 0$, take a non-zero element a of lowest possible degree k. Every product in $AJ = A$ then has degree at least one higher, in contradiction to the assumption $A \neq 0$.

Note that this proof does not work for Z-graded modules, where there could be elements of arbitrary negative degree.

Lemma 6.2. *A graded P-module A with* $\operatorname{Tor}_1^P(A, F) = 0$ *is free.*

Proof. Since A is graded, $A \otimes_P F$ is a graded vector space over F, spanned by homogeneous elements $a \otimes 1$. Take a set S of homogeneous elements such that the $s \otimes 1$ form a basis of this vector space and form the free graded P-module M on the set S. The identity $S \to S \subset A$ gives a homomorphism $\eta: M \to A$ of degree zero; by the choice of S,

$$\eta \otimes 1: M \otimes_P F \cong A \otimes_P F \tag{6.1}$$

is an isomorphism. The kernel B and the cokernel C of η give an exact sequence of graded P-modules

$$0 \to B \to M \xrightarrow{\eta} A \to C \to 0$$

(with homogeneous homomorphisms of degree 0, though we do not need this fact). Applying $\otimes_P F$ to the right hand portion produces an exact sequence

$$M \otimes_P F \to A \otimes_P F \twoheadrightarrow C \otimes_P F.$$

By (6.1), $C \otimes_P F = 0$, so $C = 0$ by the previous lemma. To the remaining short exact sequence $B \rightarrowtail M \twoheadrightarrow A$ apply the fundamental exact sequence for the torsion product (with F) to get the exact sequence

$$0 \to \operatorname{Tor}_1^P(A, F) \to B \otimes_P F \to M \otimes_P F \xrightarrow{\eta \otimes 1} A \otimes_P F \to 0,$$

where the left hand zero stands for $\operatorname{Tor}_1(M, F)$, which vanishes since M is free. By (6.1) again, $B \otimes_P F \cong \operatorname{Tor}_1^P(A, F)$, which vanishes by assumption. Hence $B \otimes_P F = 0$, so $B = 0$ by another application of the previous lemma. Our exact sequence has collapsed to $0 \to M \to A \to 0$, showing A isomorphic to the free module M.

Proposition 6.3. *For each graded P-module A there is a free graded P-module M and an epimorphism $\eta: M \to A$ of degree 0, such that to each epimorphism $\varepsilon: X_0 \to A$ with X_0 free, every $\beta: M \to X_0$ with $\varepsilon\beta = \eta$ has β a monomorphism. The kernel of η is contained in JM; with this property, the pair (M, η) is unique up to isomorphism.*

Proof. Construct η with $\eta \otimes 1$ an isomorphism as in (6.1); the first part of the proof above shows $\eta(M) = A$. The usual comparison gives a homomorphism β; let C be its cokernel. Since $\operatorname{Tor}_1(X_0, F) = 0$, this yields an exact sequence $0 \to \operatorname{Tor}_1(C, F) \to M \otimes_P F \to X_0 \otimes_P F$. But $(\varepsilon \otimes 1)(\beta \otimes 1) = \eta \otimes 1$ is an isomorphism, so $\beta \otimes 1: M \otimes_P F \to X_0 \otimes_P F$ is a monomorphism; therefore $\operatorname{Tor}_1(C, F) = 0$ and C is free. Hence $X_0 \twoheadrightarrow C$ splits, so the image of $\beta: M \to X_0$ is free; it suffices to let X_0 be this image. Construct the diagram, with B the kernel of β,

$$\begin{array}{ccccccc} 0 \to B \otimes_P F \to & M \otimes_P F & \to & X_0 \otimes_P F \\ & & \searrow & \downarrow \\ & & & A \otimes_P F \end{array}$$

where the left-hand zero stands for $\operatorname{Tor}_1^P(X_0, F)$, zero because X_0 is free. The row is exact and the dotted composite is the isomorphism (6.1), hence $B \otimes_P F = 0$, so $B = 0$ by Lemma 6.1. The uniqueness is similar.

The kernel A_1 of $\eta: M \to A$ can be again written as an image $M_1 \to A_1$; iteration yields a unique free resolution $\cdots \to M_2 \to M_1 \to M \to A \to 0$ of A, called a *minimal resolution*. For applications see ADAMS [1960, p. 28]; for a general discussion, EILENBERG [1956].

Theorem 6.4. *(The* HILBERT *Theorem on Syzygies.) If A is a graded module over the graded polynomial ring $P = F[x_1, \ldots, x_n]$ in n indeterminates of degree 1 over a field F, then any exact sequence*

$$T: 0 \leftarrow A \leftarrow X_0 \leftarrow \cdots \leftarrow X_{n-1} \leftarrow A_n \leftarrow 0$$

of graded P-modules with the X_i free has its n-th term A_n free.

Such a sequence can always be constructed, by choosing X_0 free on a set of homogeneous generators of A, X_1 similarly for generators of $\operatorname{Ker}[X_0 \to A]$, and so on. The theorem implies that $\operatorname{h.dim}_P A \leqq n$.

Proof. Since the X_i are free, the connecting homomorphism of the given exact sequence T provides an isomorphism $\operatorname{Tor}_{n+1}^P(A, F) \cong \operatorname{Tor}_1^P(A_n, F)$. But the Koszul resolution for F showed $\operatorname{h.dim}_P F \leqq n$, so $\operatorname{Tor}_{n+1}^P(A, F) = 0$. Then by Lemma 6.2 A_n is free, as asserted.

Any ideal J of P is a submodule of P; as in VI.3 it is called a *homogeneous ideal* if it is a graded submodule; that is, if J is generated by its homogeneous elements.

Corollary 6.5. *If J is a homogeneous ideal in P, any exact sequence $0 \leftarrow J \leftarrow X_0 \leftarrow \cdots \leftarrow X_{n-2} \leftarrow A_{n-1} \leftarrow 0$ of graded P-modules with all X_i free has A_{n-1} free.*

Proof. This implies our previous result that $\text{h.dim}_P J \leqq n-1$. As in that case, we prove it by composing the given sequence with the short exact sequence $P/J \twoheadleftarrow P \leftarrowtail J$ and applying the Syzygy Theorem to the graded quotient module P/J.

Note. HILBERT's Theorem was proved [HILBERT 1890] with a view to invariant theory, especially to the modules of forms invariant under a group of linear transformations; his paper (on pp. 504–508) contains a calculation equivalent to the KOSZUL resolution of F. His proof was simplified by GRÖBNER [1949]; our proof follows CARTAN [1952], who first applied homological methods and established a much more general theorem, valid also for local rings (see § 7, below).

Exercises

1. For $P = F[x, y, z]$ construct an ungraded P-module which has no internal grading consistent with this P-module structure.

2. Show that the HILBERT Syzygy Theorem holds with P replaced by any internally graded ring G for which G_0 is a field.

3. (General KOSZUL resolution.) If A is a right R-module, an element $x \neq 0$ of R is called a *zero-divisor* for A if $ax = 0$ for some $a \neq 0$ in A. Thus x is not a zero-divisor for A exactly when the map $a \rightarrow ax$ is a monomorphism $A \rightarrowtail A$. For $x_1, \ldots, x_n \in R$ let J_k be the right ideal of R generated by $x_1, \ldots, x_k$. If for each $k = 1, \ldots, n$, x_k is not a zero divisor for A/AJ_{k-1}, prove that $A \otimes_R E_R[u_1, \ldots, u_n]$ with differential $\partial u_i = x_i$ and ε: $A \otimes E_0 = A \otimes R \rightarrow A/AJ_n$ given by $\varepsilon(a \otimes r) = ar + aJ_n$ provides a resolution of length n for the R-module A/AJ_n. (Hint: Use induction on n and apply the exact homology sequence to the quotient of $A \otimes E$ by the corresponding complex without u_n.)

Note. This result with $A = R = F[x_1, \ldots, x_n]$ gives the previous Koszul resolution of F as a P-module. The more general case is useful in ideal theory, where the sequence $x_1, \ldots, x_n$ with x_k no zero divisor for A/AJ_{k-1} and $A/AJ_n \neq 0$ is called an *A-sequence* for A [AUSLANDER-BUCHSBAUM 1957, with E in place of our A] while the least upper bound of all n for such A-sequences is the *codimension* of A.

7. Local Rings

In this section we summarize without proofs some of the accomplishments of homological algebra for the study of local rings. All rings will be commutative.

A *prime ideal* P in a ring K is an ideal such that $rs \in P$ implies $r \in P$ or $s \in P$; it is equivalent to require that the quotient ring K$/P$ has no divisors of zero. Any ring K has as ideals the set (0) consisting of 0 alone and the set K; a *proper ideal* J of K is an ideal with $(0) \neq J \neq$ K. A *unit* u *of* K is an element with an inverse v ($vu = 1$) in K. Clearly no proper ideal can contain an unit.

A *local ring* L is a commutative ring in which the non-units form an ideal M; then M must contain all proper ideals of L. If L is not a field, M is the maximal proper ideal of L. In any event, M is a prime ideal. Moreover L/M is a field, the ***residue field*** of L. For a rational prime p, the ring of p-adic integers is a local ring; the residue field is the field of integers mod p. Another local ring is the set of all formal power series in non-negative powers of n indeterminates $x_1, \ldots, x_n$ and with coefficients in a field F; a power series has a (formal) inverse if and only if its constant term is not zero, so the maximal ideal is the set of all formal power series with vanishing constant term, and the residue field is F.

If P is a prime ideal in the integral domain D, the *ring of quotients* D_P is the set of all formal quotients a/b for $a, b \in D$ and b not in P, with the usual equality $a/b = a'/b'$ if and only if $ab' = a'b$. These quotients form a ring under the usual operations $a/b + a'/b' = (ab' + a'b)/bb'$, $(a/b)(a'/b') = aa'/bb'$. Such a quotient a/b has an inverse b/a in D_P if and only if $a \notin P$, hence D_P is a local ring with maximal ideal all a/b with $a \in P$; if we regard D_P as a D-module, this maximal ideal may be written as the product PD_P. For example, if D is the ring of all polynomials in n indeterminates over an algebraically closed field C, the set of all zeros of P — that is, of all points $(c_1, \ldots, c_n)$ with $f(c_1, \ldots, c_n) = 0$ for each $f \in P$ — is an irreducible (affine) algebraic manifold V. The corresponding local ring D_P is then known as the ring of rational functions on the manifold V; indeed, for each formal quotient f/g in D_P we can define the value of the quotient f/g at each point $(c_1, \ldots, c_n)$ of V as $f(c_1, \ldots, c_n)/g(c_1, \ldots, c_n)$. Similarly, a point on the manifold V is associated with a prime ideal containing P, and the ring of rational functions at this point is a local ring. This example explains the terminology "local".

A K-module C is *noetherian* if every submodule of C is of finite type; it is equivalent to require that C satisfy the ascending chain condition for submodules: For any sequence $C_1 \subset \cdots \subset C_k \subset C_{k+1} \subset \cdots$ of submodules of C there is an index n with $C_n = C_{n+1} = \cdots$. The ring K itself is noetherian if it is a noetherian K-module. Hilbert's basis theorem asserts that the ring of polynomials in n indeterminates over a field is noetherian. Also any module of finite type over a noetherian ring is itself noetherian.

Over a noetherian ring it is natural to consider the category of all noetherian modules; every submodule or quotient module of such is again noetherian. With this agreement, the Hilbert Theorem on Syzygies holds for noetherian local rings: In the statement of Theorem 6.4, replace the polynomial ring by a local ring L, the field of coefficients by the residue field L/M, and read "finitely generated module" for "graded module". The crux of the proof lies in the analogue of Lemma 6.1, with the ideal J replaced by M: When $A = AM$, then $A = AM^n$ for each n, and the intersection of the M^n is zero. An instructive presentation of this argument may be found in EILENBERG [1956].

In a noetherian ring K the *Krull dimension* k is the largest integer for which there is a properly ascending sequence of prime ideals $P_0 < P_1 < \cdots < P_k < \mathsf{K}$; if K is finitely generated, this dimension is finite. In a local ring L with maximal ideal M, the quotient M/M^2 is a vector space over the residue field L/M; since M is a finitely generated ideal, the vector space dimension $n = \dim_{L/M} M/M^2$ is finite. It can be shown that the Krull dimension of L is at most n. The local ring is said to be *regular* if its Krull dimension is exactly $n = \dim_{L/M} M/M^2$. These are the local rings of greatest geometric interest.

Using homological methods, SERRE [1956] and later AUSLANDER-BUCHSBAUM [1956] have proved (see also ASSMUS [1959]):

Theorem. *A local ring L with maximal ideal M is regular if and only if* $\text{h.dim}_L L/M < \infty$, *or equivalently, if and only if* $\text{gl.dim}\, L < \infty$.

In particular this characterization of regularity allows an easy proof that if P is a prime ideal in a regular local ring L, then the (local) ring of quotients L_P is also regular. Before the use of homological methods this result had been known only for certain geometrically important cases.

More recently AUSLANDER-BUCHSBAUM [1959] have proved Krull's conjecture:

Theorem. *Any regular local ring is a unique factorization domain.*

The proof made essential use of NAGATA'S [1958] reduction of this conjecture to the case of homological dimension 3. This theorem includes, for example, the classical result of the unique factorization for power series rings.

Note. The torsion product in local rings yields an efficient treatment of intersection multiplicity of submanifolds of an algebraic manifold [SERRE 1958]. Among the many recent studies of homological dimension in noetherian rings we note TATE [1957], AUSLANDER-BUCHSBAUM [1958], MATLIS [1960], JANS [1961]. One of the earliest uses of homological dimension was HOCHSCHILD'S [1945, 1946] discovery of the connection (§ 5) between the bidimension of Λ and separability. The homology theory of Frobenius algebras is analogous to that of groups [NAKAYAMA 1957; NAKAYAMA-TSUZUKU 1960, 1961; KASCH 1961], ROSENBERG-ZELINSKY [1956] show that Theorem 5.6 holds even if $\dim_F \Lambda$ is infinite.

Chapter eight

Products

1. Homology Products

Throughout the study of products there is an interplay between "external" and "internal" products. This relation may be illustrated in the case of homology products. If X_R and $_RY$ are chain complexes of

R-modules the *external homology product* is the homomorphism of abelian groups

$$p\colon H_k(X) \otimes_R H_m(Y) \to H_{k+m}(X \otimes_R Y), \tag{1.1}$$

defined on cycles u of X and v of Y by

$$p(\operatorname{cls} u \otimes \operatorname{cls} v) = \operatorname{cls}(u \otimes v).$$

This mapping p is natural in X and Y; it has already appeared in the Künneth Formula. This product is associative: For rings R and S and complexes X_R, ${}_RY_S$, and ${}_SW$ the composites

$$p(1 \otimes p) = p(p \otimes 1)\colon H_k(X) \otimes_R H_l(Y) \otimes_S H_m(W) \to H_n(X \otimes_R Y \otimes_S W),$$

with $n = k+l+m$, are equal.

On the other hand suppose U a DG-algebra over a commutative ring K. Then $H(U)$ is a graded K-algebra under the product

$$\pi\colon H(U) \otimes H(U) \to H(U)$$

already defined (VI.7) as

$$\pi(\operatorname{cls} u \otimes \operatorname{cls} v) = \operatorname{cls}(uv);$$

we call this the *internal homology product*. The internal product may be obtained from the external product via the product map $\pi_U\colon U \otimes U \to U$, as the composite

$$\pi = (\pi_U)_* p\colon H(U) \otimes H(U) \to H(U \otimes U) \to H(U).$$

The external homology product can be defined with coefficient modules. Take (ungraded) K-algebras Λ and Λ', complexes ${}_\Lambda X$ and ${}_{\Lambda'}X'$ of K-modules, and right modules G_Λ, $G'_{\Lambda'}$, and set $\Omega = \Lambda \otimes \Lambda'$. The *external homology product* is the composite map $p_H = \tau p$ in the diagram

$$\begin{array}{ccc} H_k(G \otimes_\Lambda X) \otimes H_m(G' \otimes_{\Lambda'} X') & \xrightarrow{p} & H_{k+m}((G \otimes_\Lambda X) \otimes (G' \otimes_{\Lambda'} X')) \\ & \searrow^{p_H} & \downarrow \tau \\ & & H_{k+m}((G \otimes G') \otimes_\Omega (X \otimes X')) \end{array} \tag{1.2}$$

where p is the homology product of (1.1) with $R = \mathsf{K}$, while τ is short for $H_{k+m}(\tau)$; that is, for the homology map induced by the middle four interchange of (VI.8.4). This product p_H is natural and associative — the latter meaning that the diagram

$$\begin{array}{ccc} H(X) \otimes H(X') \otimes H(X'') & \xrightarrow{p_H \otimes 1} & H(X \otimes X') \otimes H(X'') \\ \downarrow 1 \otimes p_H & & \downarrow p_H \\ H(X) \otimes H(X' \otimes X'') & \xrightarrow{p_H} & H(X \otimes X' \otimes X''), \end{array}$$

with G's and Λ's everywhere omitted, is commutative.

Theorem 1.1. *For algebras Λ and Λ' over a field, the homology product is an isomorphism*

$$p_H: \sum_{k+m=n} H_k(G \otimes_\Lambda X) \otimes H_m(G' \otimes_{\Lambda'} X') \cong H_n((G \otimes G') \otimes_\Omega (X \otimes X')).$$

Proof. All modules over a field are free, so the Künneth tensor formula makes p an isomorphism, while τ is always an isomorphism.

For left modules ${}_\Lambda A$ and ${}_{\Lambda'}A'$ the *external cohomology product* is the composite map $p^H = \zeta p$ in the diagram

$$\begin{array}{ccc} H^k(\operatorname{Hom}_\Lambda(X,A)) \otimes H^m(\operatorname{Hom}_{\Lambda'}(X',A')) & \xrightarrow{p} & H^{k+m}(\operatorname{Hom}_\Lambda(X,A) \otimes \operatorname{Hom}_{\Lambda'}(X',A')) \\ & \searrow^{p^H} & \downarrow \zeta \\ & & H^{k+m}(\operatorname{Hom}_\Omega(X \otimes X', A \otimes A')) . \end{array} \quad (1.$$

Here p is the homology product of (1.1), written with upper indices, while ζ is the chain transformation determined by the Hom-$\otimes$ interchange of (VI.8.10). This product is natural and associative. Its definition may be rewritten in terms of cochains $h: X_k \to A$, $h': X'_m \to A'$. Regard h and h' as homomorphisms of graded modules. By definition, $\zeta(h \otimes h')$ is the homomorphism

$$h \otimes h': (X \otimes X')_n = \sum_{p+q=n} X_p \otimes X'_q \to A \otimes A'$$

defined for $n = k + m$, $x \in X_p$, $x' \in X'_q$ as

$$\left. \begin{aligned} (h \otimes h')(x \otimes x') &= hx \otimes h'x', \quad p = k, q = m \\ &= 0, \qquad\qquad\qquad p \neq k. \end{aligned} \right\} \qquad (1.4)$$

Then $\delta(h \otimes h') = \delta h \otimes h' + (-1)^k h \otimes \delta h'$ and p^H is given on cohomology classes as $p^H(\operatorname{cls} h \otimes \operatorname{cls} h') = \operatorname{cls}(h \otimes h')$.

Theorem 1.2. *For algebras Λ and Λ' over a field and positive complexes X and X' with each X_k and each X'_m a free Λ- or Λ'-module of finite type, the cohomology product is an isomorphism*

$$p^H: \sum_{k+m=n} H^k(\operatorname{Hom}_\Lambda(X,A)) \otimes H^m(\operatorname{Hom}_{\Lambda'}(X', A'))$$
$$\cong H^{k+m}(\operatorname{Hom}_\Omega(X \otimes X', A \otimes A')) .$$

Proof. Since X and X' are positive, each $(X \otimes X')_n$ is a finite direct sum $\sum X_p \otimes X'_q$, and $\operatorname{Hom}(X, -)$ is additive for finite direct sums (= direct products). The finite type assumption, as in Prop. VI.8.3, insures that the Hom-$\otimes$ interchange is an isomorphism of complexes, while p is an isomorphism by the Künneth tensor formula over a field.

Theorem 1.3. *Connecting homomorphisms, when defined, commute with the homology product p.*

Proof. In (1.1), replace X by a short exact sequence

$$E\colon 0 \to K \to L \to M \to 0$$

of complexes of right R-modules. The homology connecting homomorphisms are

$$\partial_E = E_*\colon H_{k+1}(M) \to H_k(K).$$

The sequence of tensor product complexes

$$E \otimes_R Y\colon 0 \to K \otimes_R Y \to L \otimes_R Y \to M \otimes_R Y \to 0, \tag{1.5}$$

if it is exact, also defines connecting homomorphisms, as in the diagram

$$\begin{array}{ccc} H_{k+1}(M) \otimes_R H_m(Y) & \xrightarrow{p} & H_{k+m+1}(M \otimes_R Y) \\ \downarrow{\scriptstyle E_* \otimes 1} & & \downarrow{\scriptstyle (E \otimes_R Y)_*} \\ H_k(K) \otimes_R H_m(Y) & \xrightarrow{p} & H_{k+m}(K \otimes_R Y). \end{array} \tag{1.6}$$

Our theorem asserts that this diagram is commutative; the proof is a direct application of the "switchback" description of the connecting homomorphisms. A corresponding result holds if Y is replaced by a short exact sequence of complexes.

This result applies whenever $E \otimes_R Y$ is exact. It may not be; to get exactness we should replace the left hand zero in (1.5) by $\sum \operatorname{Tor}_1^R(M_p, Y_q)$. It will be exact in any one of the following cases:

Case 1: Each Y_n is a flat left R-module;

Case 2: Each M_n is a flat right R-module;

Case 3: E is split as a sequence of right R-modules.

The third condition means that each sequence $K_n \rightarrowtail L_n \twoheadrightarrow M_n$ is split.

Corollary 1.4. *Connecting homomorphisms, when defined, commute with the homology and cohomology products p_H and p^H.*

Proof. The result is immediate, since $p_H = \tau p$ and $p^H = \zeta p$ and the natural maps τ and ζ commute with connecting homomorphisms. The statement includes the cases when any one of the arguments G, X, G' or X' for p_H is replaced by an appropriate short exact sequence. For example, replace G by a short exact sequence E of right Λ-modules. Suppose

(i) X is a complex of flat left Λ-modules X_n;

(ii) E is split as a sequence of K-modules;

(iii) X' is a complex of flat left Λ'-modules X'_n.

(These are plausible hypotheses; they hold if X and X' are projective resolutions and K is a field.) In succession, they insure that $E \otimes_\Lambda X$ is a short exact sequence of complexes of K-modules, that $E \otimes_{\mathrm{K}} G'$ is a short exact sequence of $\Omega = \Lambda \otimes \Lambda'$-modules, and that the product

$(E \otimes G') \otimes_{\Omega} (X \otimes X')$ is a short exact sequence of complexes of K-modules. Thus all connecting homomorphisms are defined, and the diagram like (1.6) is commutative.

2. The Torsion Product of Algebras

When X and X' are resolutions, the (co)homology products p_H and p^H will give corresponding products for Tor and Ext.

For K-modules B, A, B', A' the middle four interchange

$$\tau: (B \otimes A) \otimes (B' \otimes A') \cong (B \otimes B') \otimes (A \otimes A') \qquad (2.1)$$

of (VI.8.4) may be regarded as an external product for the functor $\otimes$. Recall that Thm. V.7.3 gives an isomorphism η: $\mathrm{Tor}_0(B, A) \cong B \otimes A$, where the elements of Tor_0 are written as triples $t = (\mu, F, \nu)$ for F a finitely generated free module with dual F^* and homomorphisms $\mu: F \to B$, $\nu: F^* \to A$. Using this isomorphism η, the middle four interchange takes the form

$$\tau[(\mu, F, \nu) \otimes (\mu', F', \nu')] = (\mu \otimes \mu', F \otimes F', \nu \otimes \nu'). \qquad (2.2)$$

Here $\nu \otimes \nu'$: $F^* \otimes F'^* \to A \otimes A'$, but we may regard $\nu \otimes \nu'$ as a map defined on $(F \otimes F')^*$ by the identification $F^* \otimes F'^* = (F \otimes F')^*$, given by the isomorphism of Prop. V.4.3 (incidentally, this identification is consistent with the identification $(F \otimes F') \otimes F'' = F \otimes (F' \otimes F'')$). This formula (2.2) will be extended to higher torsion products.

An element of $\mathrm{Tor}_k(B, A)$ was written as a triple $t = (\mu, L, \nu)$ with L a finitely generated free complex of length k and μ: $L \to B$, ν: $L^* \to A$ chain transformations. Given a second such $t' \in \mathrm{Tor}_m(B', A')$, define a product

$$(\mu, L, \nu)(\mu', L', \nu') = (\mu \otimes \mu', L \otimes L', \nu \otimes \nu'). \qquad (2.3)$$

Here $L \otimes L'$ is a finitely generated free complex of length $k+m$, and $\nu \otimes \nu'$ a chain transformation $L^* \otimes L'^* = (L \otimes L')^* \to A \otimes A'$. This product is well defined with respect to the equality used for the elements of Tor_n and is natural in the four modules concerned. This product tt' is bilinear; we avoid the direct proof, via the addition defined in Tor, by the following use of resolutions.

Theorem 2.1. *For four* K*-modules* B, A, B', A', *the product* (2.3) *is a homomorphism*

$$p_T: \mathrm{Tor}_k(B, A) \otimes \mathrm{Tor}_m(B', A') \to \mathrm{Tor}_{k+m}(B \otimes B', A \otimes A'). \qquad (2.4)$$

It may be computed from projective resolutions $\varepsilon: X \to B$, $\varepsilon': X' \to B'$, *and* ε'': $Y \to B \otimes B'$ *as the composite*

$$H(X \otimes A) \otimes H(X' \otimes A') \xrightarrow{p_H} H(X \otimes X' \otimes A \otimes A') \xrightarrow{f_*} H(Y \otimes A \otimes A')$$

where p_H is the external homology product of (1.2), *with the roles of G and X interchanged, while $f: X\otimes X' \to Y$ is a chain transformation lifting $1_{B\otimes B'}$.*

Proof. The tensor product of free or projective K-modules is free or projective, as the case may be, so $\varepsilon\otimes\varepsilon': X\otimes X' \to B\otimes B'$ is a projective complex over $B\otimes B'$. By the comparison theorem, the map f lifting 1 exists and its homology map f_* is unique. The calculation of $\operatorname{Tor}(B,A)$ from the resolution X is expressed by the isomorphism $\omega: \operatorname{Tor}(B,A)\cong H(X\otimes A)$ of Thm. V.8.1. Let ω' and ω'' be analogous. The statement that p_T can be computed as the composite $f_* p_H$ is

$$\omega''(tt') = f_* p_H(\omega t\otimes\omega' t'). \tag{2.5}$$

Since $\omega'': \operatorname{Tor}(B\otimes B', A\otimes A')\cong H(Y\otimes A\otimes A')$ is an isomorphism, this equation also shows the product tt' bilinear, hence will prove that p_T is a homomorphism as in (2.4).

To prove (2.5), recall that ω was defined by regarding $t=(\mu, L, \nu)$ as a free complex $\mu: L\to B$ of length k over B plus a cycle $(1, L_k, \nu)\in L_k\otimes A$, by lifting 1_B to a chain transformation $h: L\to X$, and by setting $\omega t = (h\otimes 1)_* \operatorname{cls}(1, L_k, \nu)$. But tt' is correspondingly written as the free complex $\mu\otimes\mu': L\otimes L' \to B\otimes B'$ plus the cycle $(1, L_k\otimes L'_m, \nu\otimes\nu')$. This cycle is the homology product $\tau p[(1, L_k, \nu)\otimes(1, L'_m, \nu')]$ while $f(h\otimes h'): L\otimes L'\to Y$ lifts $1_{B\otimes B'}$. Therefore

$$\omega''(tt') = f_*(h\otimes h'\otimes 1\otimes 1)_* p_H\{\operatorname{cls}(1, L_k, \nu)\otimes\operatorname{cls}(1, L'_m, \nu')\},$$

so that (2.5) is a consequence of the naturality of the homology product p_H under the chain transformations h and h'.

Let Λ and Γ be two K-algebras, $\pi: \Lambda\otimes\Lambda\to\Lambda$ and $\varrho: \Gamma\otimes\Gamma\to\Gamma$ their multiplication maps. The composite

$$(\Lambda\otimes\Gamma)\otimes(\Lambda\otimes\Gamma)\xrightarrow{\tau}(\Lambda\otimes\Lambda)\otimes(\Gamma\otimes\Gamma)\xrightarrow{\pi\otimes\varrho}\Lambda\otimes\Gamma$$

gives the product in the algebra $\Lambda\otimes\Gamma$. In other words, the internal product in the tensor product algebra $\Lambda\otimes\Gamma$ is obtained from the external product τ of the modules.

This internal product will now be defined for $\operatorname{Tor}(\Lambda, \Gamma)$.

Theorem 2.2. *For K-algebras Λ and Γ, the family $\{\operatorname{Tor}_n^{\mathsf{K}}(\Lambda, \Gamma)\}$ is a graded K-algebra $\operatorname{Tor}^{\mathsf{K}}(\Lambda, \Gamma)$ in which the elements of degree zero constitute the tensor product algebra $\Lambda\otimes\Gamma$. The product of two elements $t=(\mu, L, \nu)$ and $t'=(\mu', L', \nu')$ is defined by*

$$(\mu, L, \nu)(\mu', L', \nu') = (\pi(\mu\otimes\mu'), L\otimes L', \varrho(\nu\otimes\nu')), \tag{2.6}$$

for π and ϱ the product maps of Λ and Γ.

Proof. The internal product (2.6) is the composite $[\mathrm{Tor}(\pi, \varrho)]p_T$, with p_T the external product. By Thm. 2.1, this product tt' is bilinear; it is manifestly associative. The identity elements of the algebras Λ and Γ are represented by K-module homomorphisms $I: \mathrm{K} \to \Lambda$, $I': \mathrm{K} \to \Gamma$, and the identity element $1_\Lambda \otimes 1_\Gamma$ of $\Lambda \otimes \Gamma$ appears, via $\eta: \mathrm{Tor}_0(\Lambda, \Gamma) \cong \Lambda \otimes \Gamma$, as the triple $1_T = (I, \mathrm{K}, I')$ of Tor_0, where K is regarded as a free K-module on one generator. Then formula (2.6) shows that $1_T t = t = t 1_T$. Hence $\mathrm{Tor}(\Lambda, \Gamma)$ is a graded algebra as asserted.

We record how this product may be computed from a suitable resolution of Λ.

Corollary 2.3. *If U is a DG-algebra and $\varepsilon: U \to \Lambda$ a homomorphism of DG-algebras such that U, regarded as a complex, is a projective resolution of the* K*-module Λ, then the canonical module isomorphism ω:* $\mathrm{Tor}(\Lambda, \Gamma) \cong$ *$H(U \otimes \Gamma)$ which expresses the torsion products by this resolution is also an isomorphism of graded algebras.*

Proof. Here $U \otimes \Gamma$, as tensor product of a DG-algebra U and a trivial DG-algebra Γ, is a DG-algebra, so that $H(U \otimes \Gamma)$ is indeed a graded algebra. In Thm. 2.1 above, we take $B = B' = \Lambda$, so we may choose both X and X' to be the resolution U, while Y is any projective resolution of $\Lambda \otimes \Lambda$. Lift 1 and π to chain transformations f and g, as in

$$\begin{array}{ccccc} U \otimes U & \xrightarrow{f} & Y & \xrightarrow{g} & U \\ \downarrow & & \downarrow & & \downarrow \\ \Lambda \otimes \Lambda & = & \Lambda \otimes \Lambda & \xrightarrow{\pi} & \Lambda. \end{array}$$

Then $\mathrm{Tor}(\pi, \varrho)$ is the homology map induced by $g \otimes \varrho$: $Y \otimes \Gamma \otimes \Gamma \to U \otimes \Gamma$. The product in $\mathrm{Tor}(\Lambda, \Gamma)$ is thus $(g \otimes \varrho)_* f_* p_H$, as in the diagram

$$\begin{array}{ccccc} H(U \otimes \Gamma) \otimes H(U \otimes \Gamma) & \xrightarrow{p_H} & H(U \otimes U \otimes \Gamma \otimes \Gamma) & \xrightarrow{f_*} & H(Y \otimes \Gamma \otimes \Gamma) \\ & & \downarrow{\scriptstyle (\pi_U)_*} & & \downarrow{\scriptstyle (g \otimes \varrho)_*} \\ & & H(U \otimes \Gamma \otimes \Gamma) & \xrightarrow{(1 \otimes \varrho)_*} & H(U \otimes \Gamma). \end{array}$$

But the product π_U: $U \otimes U \to U$ and gf: $U \otimes U \to U$ are both chain transformations of resolutions lifting π: $\Lambda \otimes \Lambda \to \Lambda$, hence are homotopic by the comparison theorem. Therefore the homology diagram above is commutative, so the product in $\mathrm{Tor}(\Lambda, \Gamma)$ is given by $(\pi_U \otimes \varrho)_* p_H$. This is exactly the internal product in the graded algebra $H(U \otimes \Gamma)$.

For the polynomial algebra P we have already noted in Thm. VII.2.2 that the graded algebra $\mathrm{Tor}^P(\mathrm{K}, \mathrm{K})$ is an exterior algebra over P; the proof used the fact that the Koszul resolution of K is a DG-algebra. Indeed, any algebra Λ has a projective resolution which is a DG-algebra U (Ex. 2).

Our product definition (2.3) is new, but the external product p_T which it defines is exactly the product $\pitchfork$ defined by Cartan-Eilenberg (Chap. XI.4). Their definition uses resolutions of A and A', but this is irrelevant (Ex. 3).

Exercises

1. If U is a DG-algebra, A a K-module, and $\varphi: A \to U_n$ a module homomorphism with $\partial_n \varphi = 0$, show that the graded algebra $U \otimes T(A)$ has a unique DG-structure with $\partial | U = \partial_U$, $\partial | A = \varphi$, and A of degree $n+1$.

2. For any K-algebra Λ construct a DG-algebra U and a homomorphism $\varepsilon: U \to \Lambda$ of graded algebras, so that, as in Cor. 2.3, U is a projective resolution of Λ as a K-module. Hint: Use Ex. 1 to construct a DG-algebra $U^{(n)}$ by recursion on n so that it is a projective resolution up to dimension n.

3. Describe the external product in $\operatorname{Tor}(B, A)$ using resolutions of both B and A, or of A only.

4. For K-algebras Λ and Λ' and modules B_Λ, $B'_{\Lambda'}$, ${}_\Lambda A$, ${}_{\Lambda'}A'$ show that the formula (2.3) provides an external product

$$\operatorname{Tor}^\Lambda(B, A) \otimes \operatorname{Tor}^{\Lambda'}(B', A') \to \operatorname{Tor}^{\Lambda \otimes \Lambda'}(B \otimes B', A \otimes A'),$$

describe its properties, and show that it commutes with all four connecting homomorphisms. This is the product $\top$ of Cartan-Eilenberg XI.1.

3. A Diagram Lemma

In the next section we need the following anticommutative rule on the splicing of exact sequences.

Lemma 3.1. *(The 3×3 splice.) If a commutative 3×3 diagram of modules has columns the short exact sequences E', E, and E'', rows the short exact sequences E_A, E_B, and E_C, then*

$$E_A \circ E'' \equiv -E' \circ E_C.$$

Proof. The given 3×3 diagram has the form

$$\begin{array}{rcccc}
E_A\colon & A' & \to & A & \to & A'' \\
& \downarrow & & \downarrow & & \downarrow \\
E_B\colon & B' & \to & B & \xrightarrow{\beta} & B'' \\
& \downarrow & & \downarrow{\scriptstyle\sigma} & & \downarrow{\scriptstyle\tau} \\
E_C\colon & C' & \to & C & \xrightarrow{\gamma} & C''
\end{array} \tag{3.1}$$

(zeros on the edges not shown). Construct the diagram

$$\begin{array}{ccccccccccc}
0 & \to & A' & \to & A & \to & B'' & \to & C'' & \to & 0 \\
& & \| & & \downarrow & & \downarrow{\scriptstyle\iota_2} & & \downarrow{\scriptstyle -1} & & \\
0 & \to & A' & \to & B & \xrightarrow{\varphi} & C \oplus B'' & \xrightarrow{\psi} & C'' & \to & 0 \\
& & \| & & \uparrow & & \uparrow{\scriptstyle\iota_1} & & \| & & \\
0 & \to & A' & \to & B' & \to & C & \to & C'' & \to & 0
\end{array} \tag{3.2}$$

with $\varphi b=(\sigma b, \beta b)$ and $\psi(c, b'')=\gamma c-\tau b''$ (note the sign), while the other unlabelled arrows are maps or composites from (3.1). The diagram is commutative; a diagram chase shows the middle row exact. The top row is the composite $E_A \circ E''$; by the vertical maps with $-1_{C''}: C'' \to C''$ at the right, it is congruent to the negative of the middle row, which in turn is congruent to the bottom row $E' \circ E_C$. This is the desired result.

A related and frequently used result is

Lemma 3.2. *For right R-modules $A \subset B$, left R-modules $A' \subset B'$,*

$$(B/A) \otimes_R (B'/A') \cong [B \otimes_R B'] / [\operatorname{im}(A \otimes_R B') \cup \operatorname{im}(B \otimes_R A')]. \quad (3.3)$$

Proof. The first image here is that of $A \otimes B' \to B \otimes B'$. This and the symmetric map yield the exact sequence

$$A \otimes_R B' \oplus B \otimes_R A' \to B \otimes_R B' \to (B/A) \otimes_R (B'/A') \to 0.$$

This sequence can also be derived (cf. Ex. 2 below) from a diagram like (3.1) with first row $A \otimes A'$, $A \otimes B'$, $A \otimes (B'/A')$.

Exercises

1. In (3.1) assume only that the rows and columns are right exact, with the third row and the third column short exact. Prove that (3.2), with the left hand zeros omitted, is commutative with exact rows.

2. Prove Lemma 3.1 by a diagram like (3.2) with vertical arrows reversed and middle row $A' \rightarrowtail B' \oplus A \to B \twoheadrightarrow C''$.

4. External Products for Ext

The composition of long exact sequences yields an external product in Ext. For a single Λ-module A, composition is a homomorphism

$$\operatorname{Ext}^k_\Lambda(A, A) \otimes \operatorname{Ext}^m_\Lambda(A, A) \to \operatorname{Ext}^{k+m}_\Lambda(A, A).$$

By Thm. III.5.3, this makes $\operatorname{Ext}_\Lambda(A, A)$ a graded ring; indeed (by VII.3.1) a graded K-algebra. In this algebra, the elements of degree zero form the K-algebra of Λ-module endomorphisms of A. We now describe how this product can sometimes be obtained from the cohomology product for resolutions.

Let Λ and Λ' be algebras over a commutative ring K, while C and A are left Λ-modules, C' and A' are left Λ'-modules. Write Ω for $\Lambda \otimes \Lambda'$, where $\otimes$ is short for $\otimes_K$, and note that $C \otimes C'$ and $A \otimes A'$ are left Ω-modules. We wish to define a K-module homomorphism

$$\vee: \operatorname{Ext}^k_\Lambda(C, A) \otimes \operatorname{Ext}^m_{\Lambda'}(C', A') \to \operatorname{Ext}^{k+m}_\Omega(C \otimes C', A \otimes A') \quad (4.1)$$

called the external or *wedge* product; for $\sigma \in \operatorname{Ext}_\Lambda$ and $\sigma' \in \operatorname{Ext}_{\Lambda'}$ we will write $\vee(\sigma \otimes \sigma')$ as $\sigma \vee \sigma'$. Take free resolutions $\varepsilon: X \to C$ and $\varepsilon': X' \to C'$

by Λ- and Λ'-modules, respectively. The cohomology product (1.3) is

$$p^H\colon H^k(\operatorname{Hom}_\Lambda(X, A)) \otimes H^m(\operatorname{Hom}_{\Lambda'}(X', A')) \to H^{k+m}(\operatorname{Hom}_\Omega(X \otimes X', A \otimes A')).$$

With the canonical isomorphisms $\operatorname{Ext}_\Lambda^k(C, A) \cong H^k(\operatorname{Hom}_\Lambda(X, A))$, this will define the desired wedge product (4.1) *provided* $(\varepsilon \otimes \varepsilon')\colon X \otimes X' \to C \otimes C'$ is a free Ω-module resolution, for standard comparison arguments show that the result is independent of the resolutions used. In any event, each $X_k \otimes X'_m$ is a free left Ω-module. The proviso that $X \otimes X'$ is a resolution holds in two cases.

Case 1. K is a field. By the Künneth tensor formula, valid over a field K, $H_n(X \otimes X') = 0$ for $n > 0$ and $\varepsilon \otimes \varepsilon'\colon H_0(X \otimes X') \cong C \otimes C'$, so $X \otimes X'$ is a resolution.

Case 2. Λ and Λ' are free as K-modules and C is a flat K-module. For, each free Λ-module X_n is a direct sum of copies of the free K-module Λ, so X_n is a free K-module. Then $X \to C$ is also a free K-module resolution of C, so $\operatorname{Tor}_n^{\mathsf{K}}(C, C')$ may be calculated (Thm. V.9.3) from X and X' as $H_n(X \otimes X')$. But C is flat, so $\operatorname{Tor}_n(C, -) = 0$ for $n > 0$, hence $X \otimes X' \to C \otimes C'$ is a resolution.

Other cases will occur in the exercises and in our subsequent discussion of relative Ext functors (Chap. X). From the definition, it follows that the wedge product commutes with connecting homomorphisms, and is associative; for $k = m = 0$, it reduces to the Hom-$\otimes$ interchange. In Case 1, the wedge product may be expressed by the Yoneda composition product.

Theorem 4.1. [Yoneda 1958.] *For algebras Λ and Λ' over a field and $\sigma \in \operatorname{Ext}_\Lambda^k(C, A)$, $\sigma' \in \operatorname{Ext}_{\Lambda'}^m(C', A')$ the wedge product is given by*

$$\sigma \vee \sigma' = (\sigma \otimes A') \circ (C \otimes \sigma') = (-1)^{km}[(A \otimes \sigma') \circ (\sigma \otimes C')]. \tag{4.2}$$

Here $\sigma \otimes A'$ has an evident meaning, as follows. If $k = 0$, σ is a homomorphism $C \to A$; let $\sigma \otimes A'$ mean $\sigma \otimes 1_{A'}\colon C \otimes A' \to A \otimes A'$. If $k > 0$ and $m > 0$, σ and σ' are the congruence classes of long exact sequences

$$S\colon 0 \to A \to B_{k-1} \to \cdots \to B_0 \to C \to 0,$$
$$S'\colon 0 \to A' \to B'_{m-1} \to \cdots \to B'_0 \to C' \to 0.$$

Since K is a field, $\otimes_{\mathsf{K}}$ preserves exactness, so gives long exact sequences

$$S \otimes A'\colon 0 \to A \otimes A' \to B_{k-1} \otimes A' \to \cdots \to C \otimes A' \to 0,$$
$$C \otimes S'\colon 0 \to C \otimes A' \to C \otimes B'_{m-1} \to \cdots \to C \otimes C' \to 0.$$

Take $\sigma \otimes A' = \operatorname{cls}(S \otimes A')$ and $C \otimes \sigma' = \operatorname{cls}(C \otimes S')$, so the Yoneda composite $(\sigma \otimes A') \circ (C \otimes \sigma')$ is defined; for k or m zero it is the usual composite of a homomorphism with a long exact sequence.

Proof. First assume $k>0$ and $m>0$. Regard S as a resolution of C; the comparison theorem lifts 1_C to a chain transformation $f\colon X\to S$. Similarly, $1_{C'}$ lifts to $f'\colon X'\to S'$; in particular, $f'_m\colon X'_m\to A'$ is a cocycle of X' and its class represents $\operatorname{cls} S'$ in the isomorphism $H^m(X', A')\cong \operatorname{Ext}^m(C', A')$ of Thm. III.6.4. The complex $X\otimes X'$ is the first row of the diagram

$$\begin{array}{ccccccc}
(X\otimes X')_{m+1} & \xrightarrow{\partial} & (X\otimes X')_m & \xrightarrow{\partial} & (X\otimes X')_{m-1} & \xrightarrow{\partial} & (X\otimes X')_{m-2}\\
\downarrow & & \downarrow & & \downarrow & & \downarrow\\
X_1\otimes X'_m & \to & X_0\otimes X'_m & \to & X_0\otimes X'_{m-1} & \to & X_0\otimes X'_{m-2}\\
\downarrow{\scriptstyle f_1\otimes f'_m} & & \downarrow{\scriptstyle f_0\otimes f'_m} & & \downarrow{\scriptstyle \varepsilon\otimes f'_{m-1}} & & \downarrow{\scriptstyle \varepsilon\otimes f'_{m-2}}\\
B_1\otimes A' & \to & B_0\otimes A' & \cdots\to & C\otimes B'_{m-1} & \to & C\otimes B'_{m-2}\\
 & & & \searrow \; C\otimes A' \; \nearrow & & &
\end{array}$$

which extends in the same fashion left and right, ending with a column $C\otimes C'$ on the right. The first row of vertical maps projects each $(X\otimes X')_n$ to the indicated one of its direct summands. The bottom row is the composite long sequence $T=(S\otimes A')\circ(C\otimes S')$, with the splice at $C\otimes A'$ displayed. The top squares do not commute, but erase the middle row; the resulting diagram is commutative, even at the splice. Hence the composite vertical map is a chain transformation $h\colon X\otimes X' \to T$ which lifts the identity on $C\otimes C'$. To read off the cohomology class of $X\otimes X'$ corresponding to T, take h on dimension $k+m$. But h there is just

$$(X\otimes X')_{k+m}\to X_k\otimes X'_m \xrightarrow{f_k\otimes f'_m} A\otimes A';$$

the cohomology class of this cocycle is exactly the one obtained from $\operatorname{cls} f_k\otimes \operatorname{cls} f'_m$ by the cohomology product p^H. Since $\operatorname{cls} f_k$ and $\operatorname{cls} f'_m$ represent S and S', respectively, this proves the first equation of the theorem for $k>0$ and $m>0$. The proof for $k=0$ (or $m=0$) uses a similar diagram, with splicing of sequences replaced by the action of a homomorphism $\sigma\colon C\to A$ on a sequence.

The second equality in (4.2) is an (anti-) commutation rule. It is immediate from the definition if $k=0$ or $m=0$. Since any long exact sequence is a composite of short ones, it suffices to give a proof in the case $k=m=1$, for short exact sequences E and E'. Here the commutative square diagram

$$\begin{array}{lccccc}
A\otimes E'\colon & A\otimes A' & \to & A\otimes B' & \to & A\otimes C'\\
 & \downarrow & & \downarrow & & \downarrow\\
B\otimes E'\colon & B\otimes A' & \to & B\otimes B' & \to & B\otimes C'\\
 & \downarrow & & \downarrow & & \downarrow\\
C\otimes E'\colon & C\otimes A' & \to & C\otimes B' & \to & C\otimes C'
\end{array} \qquad (4.3)$$

and Lemma 3.1 prove $(A\otimes E')\circ(E\otimes C')\equiv -(E\otimes A')\circ(C\otimes E')$, as required.

From this theorem it again follows that the $\vee$ product is associative.

Theorem 4.2. *If the K-algebras Λ and Λ' are free as K-modules, while C, A, C', A' are all flat as K-modules, the wedge product* (4.1) *is defined. It may be expressed by the composition product as in* (4.2).

Proof. This falls under Case 2 above. The previous argument applies, since $X\otimes X'$ is a resolution and the tensor product $S\otimes_K A'$ of a long exact sequence S with a K-flat module A' is still exact.

Corollary 4.3. *If Λ and Λ' are augmented K-algebras which are free as K-modules, the wedge product of $\sigma\in \mathrm{Ext}^k_\Lambda(K, K)$ and $\sigma'\in\mathrm{Ext}^m_{\Lambda'}(K, K)$ is given by*

$$\sigma\vee\sigma' = {}_{\varepsilon'}\sigma\circ{}_{\varepsilon}\sigma' = (-1)^{km}\,{}_{\varepsilon}\sigma'\circ{}_{\varepsilon'}\sigma \in \mathrm{Ext}^{k+m}_{\Omega}(K, K). \qquad (4.4)$$

Here K is to be regarded as a Λ- or Λ'-module by pull-back along the augmentations $\varepsilon\colon \Lambda\to K$ and ε', while ${}_{\varepsilon'}\sigma$ is short for $(1\otimes\varepsilon')^*\sigma$; i.e., for the exact sequence in σ pulled back along $1\otimes\varepsilon'\colon \Lambda\otimes\Lambda'\to \Lambda\otimes K=\Lambda$.

Now let V be a Hopf algebra with counit $\varepsilon\colon V\to K$ and diagonal map $\psi\colon V\to V\otimes V$. Pull-back along ψ turns $(V\otimes V)$-modules into V-modules, exact sequences into exact sequences, and so gives a change of rings map $\psi^\#\colon \mathrm{Ext}_{V\otimes V}\to\mathrm{Ext}_V$. If C, A, C', and A' are left V-modules, so are ${}_\psi(C\otimes C')$ and ${}_\psi(A\otimes A')$, and the composite $\psi^\#\vee$ of wedge product and pull-back is a K-module homomorphism

$$\psi^\#\vee\colon \mathrm{Ext}^k_V(C, A)\otimes\mathrm{Ext}^m_V(C', A')\to\mathrm{Ext}^{k+m}_V({}_\psi(C\otimes C'), {}_\psi(A\otimes A')) \qquad (4.5)$$

called the *Hopf wedge product*. It is defined when K is a field, or when C is K-flat and V is free as a K-module, and the analogues of Thms. 4.1 and 4.2 hold. Since ψ is associative, so is this product.

By pull-back, each K-module becomes a V-module ${}_\varepsilon M$.

Lemma 4.4. *For a K-module M and a module C over the Hopf algebra V*

$${}_\psi({}_\varepsilon M\otimes C)\cong M\otimes C\,, \quad {}_\psi(C\otimes{}_\varepsilon M)\cong C\otimes M \qquad (4.6)$$

are isomorphisms of V-modules, with the V-module structure on the right induced by that of C.

Proof. The Hopf algebra, as a coalgebra, satisfies the identity $(\varepsilon\otimes 1)\psi=1$ of (VI.9.1). Pull-back yields

$${}_\psi[{}_\varepsilon M\otimes C] = {}_\psi[{}_{(\varepsilon\otimes 1)}(M\otimes C)] = {}_{(\varepsilon\otimes 1)\psi}(M\otimes C) = M\otimes C\,,$$

and similarly on the other side. Hence a curious result:

Proposition 4.5. *If V is a Hopf algebra over a field* K *and M, N are* K*-modules, C, A V-modules, the Hopf wedge products*

$$\operatorname{Ext}_V({}_\varepsilon M, A)\otimes \operatorname{Ext}_V(C', {}_\varepsilon N')\to \operatorname{Ext}_V(M\otimes C', A\otimes N'),$$

$$\operatorname{Ext}_V(C, {}_\varepsilon N)\otimes \operatorname{Ext}_V({}_\varepsilon M', A')\to \operatorname{Ext}_V(C\otimes M', N\otimes A')$$

are independent of the diagonal map ψ; that is, depend only on V as an augmented algebra $\varepsilon: V\to$ K.

Proof. These wedge products are still given in terms of composition of long exact sequences by the formulas (4.2), where the modules in these long exact sequences are pulled back to V-modules by ψ. The Lemma asserts that the resulting V-module structure is independent of ψ.

In particular, let all modules in sight be ${}_\varepsilon$K; then K$\otimes$K$=$K, and the external wedge product becomes an *internal product*

$$\operatorname{Ext}_V(\mathsf{K},\mathsf{K})\otimes \operatorname{Ext}_V(\mathsf{K},\mathsf{K})\to \operatorname{Ext}_V(\mathsf{K},\mathsf{K}) \tag{4.7}$$

which makes $\operatorname{Ext}_V(\mathsf{K},\mathsf{K})$ a graded K-algebra. Since $\sigma\otimes\mathsf{K}=\sigma$, the formula (4.4) shows this algebra commutative.

Note. The external product for Tor arises from the middle four interchange and agrees with that map for $\operatorname{Tor}_0=\otimes$; it may be obtained, as in (2.5), by replacing suitable arguments by resolutions, and composing with the homology product and a comparison of resolutions. The external product for Ext arises similarly from the Hom-$\otimes$ interchange. Various other "products" involving Tor and Ext arise by the same mechanism from identities on Hom and $\otimes$; for example, there is one arising from the mixed adjoint associativity

$$\operatorname{Hom}(A\otimes A', \operatorname{Hom}(C, C'))\to \operatorname{Hom}(C\otimes A, \operatorname{Hom}(A', C')).$$

These are given in detail, via resolutions, in Cartan-Eilenberg Chap. XI. Description in terms of the invariant definition of Tor and Ext would be of interest. Other types of products will appear in Chap. X below.

Exercises

1. Describe how the external product in Ext commutes with connecting homomorphisms.

In the following exercises, K is a commutative ring, not necessarily a field.

2. If P and P' are projective Λ- and Λ'-modules, respectively, show $P\otimes P'$ a projective $(\Lambda\otimes\Lambda')$-module. If Λ and Λ' are projective as K-modules, show also that $P\otimes P'$ is a projective K-module.

3. Show that the wedge product for K a ring can still be defined, using projective resolutions, provided Λ and Λ' are projective as K-modules and $\operatorname{Tor}_n^{\mathsf{K}}(C, C')=0$ for $n>0$. If, in addition, A and A' are K-flat, show that Thm. 4.1 still holds.

5. Simplicial Objects

The cohomology $H(X, Z)$ of a topological space X with coefficients Z is a graded ring under a product known as the *cup product*. This product can be defined not only for spaces but for other complexes with a "simplicial" structure. Hence we now analyze the combinatorial structure of a simplex; more exactly of a p-dimensional simplex Δ^p with ordered vertices.

For each non-negative integer p, let $[p]$ denote the set $\{0, 1, \ldots, p\}$ of integers in their usual order. A (weakly) *monotonic* map $\mu: [q] \to [p]$ is a function on $[q]$ to $[p]$ such that $i \leqq j$ implies $\mu i \leqq \mu j$. The objects $[p]$ with morphisms all weakly monotonic maps μ constitute a category $\mathcal{M}$ (for monotonic). Note that a monotonic μ is determined by the sequence of $q+1$ integers $\mu_0 \leqq \mu_1 \leqq \cdots \leqq \mu_q$ in $[p]$ where $\mu_0 = \mu 0, \ldots$; hence we regard μ as the affine simplex $(\mu_0, \ldots, \mu_q)$ determined by the vertices μ_i on the standard p-simplex Δ^p.

Let $\mathcal{C}$ be any category. A contravariant functor $S: \mathcal{M} \to \mathcal{C}$ will be called a *simplicial object* in $\mathcal{C}$. Specifically, S assigns to each non-negative integer q (to each object of $\mathcal{M}$) an object S_q of $\mathcal{C}$, and to each monotonic μ: $[q] \to [p]$ a morphism $\mu^* = S(\mu): S_p \to S_q$ of $\mathcal{C}$, with $S(1) = 1$ and $S(\mu\nu) = S(\nu) S(\mu)$. By a *simplicial set* is meant a simplicial object in the category of sets; by a *simplicial Λ-module* is meant a simplicial object in the category of all Λ-modules.

If $F: \mathcal{C} \to \mathcal{D}$ is a covariant functor, each simplicial object S in $\mathcal{C}$ determines a simplicial object FS in $\mathcal{D}$, with $(FS)_q = F(S_q)$, $FS(\mu) = F(S\mu)$. In particular, if Λ is an algebra, and F_Λ the functor which assigns to each set Y the free (left) Λ-module with generators Y, then each simplicial set S determines a simplicial Λ-module $F_\Lambda S$.

The singular simplices (II.7) of a topological space X constitute a simplicial set $\tilde{S}(X)$. In detail, let $\tilde{S}_p(X)$ be the set of all singular p-simplices T of X; each T is a continuous map $T: \Delta^p \to X$ defined on the standard affine p-simplex Δ^p. Now each monotonic μ: $[q] \to [p]$ determines a unique affine map $\mu: \Delta^q \to \Delta^p$ carrying vertex i of Δ^q onto vertex μ_i of Δ^p; the composite $\mu^* T = T\mu: \Delta^q \to X$ defines a map $\mu^* = \tilde{S}(\mu): \tilde{S}_p(X) \to \tilde{S}_q(X)$ which makes $\tilde{S}$ a functor on $\mathcal{M}$ and hence a simplicial set. For Z the ring of integers, $S' = F_Z \tilde{S}$ is a simplicial abelian group with S'_p the free abelian group generated by all singular p-simplices of X. In other words, S'_p is just the usual group of singular p-chains of the space X. We shall soon see that the usual boundary of a singular p-chain is also determined by the simplicial structure of $S'(X)$.

It is convenient to use two special families of monotonic maps

$$\varepsilon^i = \varepsilon^i_q: [q-1] \to [q], \qquad \eta^i = \eta^i_q: [q+1] \to [q] \tag{5.1}$$

defined for $i=0,\ldots,q$ (and for $q>0$ in the case of ε^i) by

$$\begin{aligned}\varepsilon^i(j)&=j \quad \text{for } j<i, & \eta^i(j)&=j \quad \text{for } j\leqq i,\\ &=j+1 \quad \text{for } j\geqq i, & &=j-1 \quad \text{for } j>i.\end{aligned}$$

In other words, ε^i may be described as the $(q-1)$-face of Δ^q with vertices $(0,1,\ldots,\hat{i},\ldots,q)$ — omit index i — and η^i is the $(q+1)$-face with vertices $(0,1,\ldots,i,i,\ldots,q)$ — double the vertex i. From this description one verifies the identities

$$\varepsilon^j_{q+1}\varepsilon^i_q = \varepsilon^i_{q+1}\varepsilon^{j-1}_q, \quad i<j, \tag{5.2}$$

$$\eta^j_q\,\eta^i_{q+1} = \eta^i_q\,\eta^{j+1}_{q+1}, \quad i\leqq j, \tag{5.3}$$

$$\left.\begin{aligned}\eta^j_{q-1}\varepsilon^i_q &= \varepsilon^i_{q-1}\eta^{j-1}_{q-2}, && i<j,\\ &= 1, && i=j,\quad i=j+1,\\ &= \varepsilon^{i-1}_{q-1}\eta^j_{q-2}, && i>j+1.\end{aligned}\right\} \tag{5.4}$$

We normally omit the subscripts q on ε and η.

Lemma 5.1. *Any monotonic μ: $[q]\to[p]$ has a unique factorization*

$$\mu=\varepsilon^{i_1}\ldots\varepsilon^{i_s}\eta^{j_1}\ldots\eta^{j_t}, \tag{5.5}$$

with $p\geqq i_1>\cdots>i_s\geqq 0$, $0\leqq j_1<\cdots<j_t<q$, and $q-t+s=p$.

Proof. Let the elements of $[p]$ not in $\mu[q]$ be $i_1,\ldots,i_s$ in reverse order, while those elements j of $[q]$ with $\mu(j)=\mu(j+1)$ are $j_1,\ldots,j_t$ in order. Then (5.5) holds, and presents μ as the composite of a monotonic epimorphism (the product of the η's) with a monotonic monomorphism (the product of the ε's).

This lemma allows an alternative definition of a simplicial object.

Theorem 5.2. *A simplicial object S in a category $\mathscr{C}$ is a family $\{S_q\}$ of objects of $\mathscr{C}$ together with two families of morphisms of $\mathscr{C}$,*

$$d_i\colon S_q\to S_{q-1}, \quad s_i\colon S_q\to S_{q+1}, \quad i=0,\ldots,q,$$

(and with $q>0$ in the case of d_i) which satisfy the identities

$$d_i d_j = d_{j-1} d_i, \quad i<j, \tag{5.6}$$

$$s_i s_j = s_{j+1} s_i, \quad i\leqq j, \tag{5.7}$$

$$\left.\begin{aligned}d_i s_j &= s_{j-1} d_i, && i<j,\\ &=1, && i=j,\quad i=j+1,\\ &= s_j d_{i-1}, && i>j+1.\end{aligned}\right\} \tag{5.8}$$

Proof. Since S is contravariant, the morphisms $d_i=S(\varepsilon^i)$, $s_i=S(\eta^i)$ satisfy the identities (5.6)—(5.8), which are the duals of (5.2)—(5.4).

Conversely, given the d_i and the s_i, write any monotonic μ in the unique form (5.5) and define

$$S(\mu)=s_{j_k}\dots s_{j_1} d_{i_h}\dots d_{i_1}\colon S_p\to S_q.$$

The identities (5.6)—(5.8) suffice to commute any two of d_i, s_j, hence to calculate the factorization of a composite $\mu\nu$ from that of μ and of ν, hence to prove that $S(\mu\nu)=S(\nu)S(\mu)$. This makes $S\colon \mathcal{M}\to\mathcal{C}$ contravariant.

We call d_i the i-th *face operator* and s_j the j-th *degeneracy operator* of S. Note that (5.6) and (5.7) imply

$$d_i d_j = d_j d_{i+1}, \quad i \geqq j, \tag{5.9}$$

$$s_i s_j = s_j s_{i-1}, \quad i > j. \tag{5.10}$$

For example, let V be any partly ordered set (I.8); call an ordered $(q+1)$-tuple $(v_0, \dots, v_q)$ with elements $v_0 \leqq \dots \leqq v_q$ in the given partial order of V a *q-simplex* of V. Let $S_q(V)$ be the set of all q-simplices of V. Then $S(V)$ is a simplicial set under the face and degeneracy operators defined by

$$d_i(v_0, \dots, v_q) = (v_0, \dots, \hat{v}_i, \dots, v_q) \quad \text{(omit } v_i), \tag{5.11}$$

$$s_i(v_0, \dots, v_q) = (v_0, \dots, v_i, v_i, \dots, v_q) \quad \text{(double } v_i). \tag{5.12}$$

Geometrically, V may be regarded as a schematic description of a polyhedron with partly ordered vertices v_i.

If S and S' are simplicial objects in a category $\mathcal{C}$, a *simplicial map* $\sigma\colon S\to S'$ is a natural transformation of the contravariant functors $S, S'\colon \mathcal{M}\to\mathcal{C}$. In other words, a simplicial map σ is a family of morphisms $\sigma_q\colon S_q\to S'_q$ of $\mathcal{C}$ such that $\sigma_q S(\mu)=S'(\mu)\sigma_p$ for each monotonic $\mu\colon [q]\to[p]$, or, equivalently, such that $\sigma d_i = d_i\sigma$ and $\sigma s_i = s_i\sigma$ for every i. The simplicial objects in $\mathcal{C}$ form a category with morphisms the simplicial maps.

Each simplicial module S determines a (positive) chain complex $K=K(S)$ with $K_q=S_q$ and with boundary homomorphism $\partial\colon K_q\to K_{q-1}$ the alternating sum of the face homomorphisms:

$$\partial = d_0 - d_1 + \dots + (-1)^q d_q\colon K_q\to K_{q-1}. \tag{5.13}$$

The identities (5.6) for $d_i d_j$ imply that $\partial\partial=0$. This allows us to speak of the homology or cohomology modules of a simplicial module S, meaning those of the associated chain complex $K(S)$. For a topological space X, (5.13) gives the usual boundary operator ∂ in the singular complex $S(X)$. More formally, X determines the simplicial set $\tilde{S}(X)$ described above, hence the simplicial abelian group $F_Z\tilde{S}(X)$, hence the chain complex $KF_Z\tilde{S}(X)$; with boundary ∂, this complex is the usual singular complex $S(X)$.

A simplicial module S over the ring R is *augmented* if there is a module homomorphism $\varepsilon\colon S_0 \to R$ with $\varepsilon d_0 = \varepsilon d_1\colon S_1 \to R$; the associated chain complex is then augmented by ε.

Notes. Simplicial sets, under the name *complete semisimplicial complexes*, arose in the study by EILENBERG-ZILBER [1950, 1953] of the singular homology of spaces and their cartesian products. Simplicial abelian groups, under the name FD-complexes (F for face, D for degeneracy) arose simultaneously in the analysis by EILENBERG-MACLANE [1953] of the spaces $K(\Pi, n)$ with one non-vanishing homotopy group Π in dimension n. Simplicial sets satisfying the additional "Kan condition" and simplicial (multiplicative) groups subsequently proved to provide the suitable algebraic formulation of homotopy theory; see KAN [1958b]. The normalization theorem of the next section and its proof are due to EILENBERG-MACLANE [1947]. Each simplicial module is determined by its normalized chain complex; this gives an equivalence between the categories of simplicial modules and (positive) chain complexes of modules, DOLD [1958].

6. Normalization

Let S be a simplicial module. In each dimension n, define $(DS)_n$ to be the submodule of S_n generated by all degenerate elements; that is, set $(DS)_0 = 0$ and

$$(DS)_n = s_0 S_{n-1} \cup \cdots \cup s_{n-1} S_{n-1}, \qquad n > 0.$$

By the identities (5.8) for $d_i s_j$, DS is closed under ∂, so is a subcomplex of the associated chain complex KS of S. The quotient $KS/DS = K_N S$ is known as the *normalized* chain complex of the simplicial module S.

Theorem 6.1. *(Normalization Theorem.) For each simplicial module S the canonical projection $\pi\colon KS \to K_N S = KS/DS$ is a chain equivalence.*

For the proof, we interpret the degeneracies s_i as homotopies. For each non negative k, let $D_k S$ be the graded submodule of S generated by all degenerate elements $s_i a$ with $i \leqq k$; that is, set

$$\begin{aligned}(D_k S)_n &= s_0 S_{n-1} \cup \cdots \cup s_{n-1} S_{n-1}, && n-1 \leqq k,\\ &= s_0 S_{n-1} \cup \cdots \cup s_k S_{n-1}, && n-1 > k.\end{aligned}$$

By (5.8), each $D_k S$ is a subcomplex, while DS is the union of all $D_k S$. Define $t_k\colon S \to S$ of degree 1 by

$$\begin{aligned} t_k a &= (-1)^k s_k a, && k \leqq \dim a,\ a \in S,\\ &= 0, && k > \dim a,\ a \in S,\end{aligned}$$

and set $h_k = 1 - \partial t_k - t_k \partial$. This makes $h_k\colon K(S) \to K(S)$ a chain transformation and $t_k\colon 1 \simeq h_k$ a chain homotopy. Since $t_k S \subset D_k$ and $\partial D_k \subset D_k$

$$h_k a \equiv a \ (\mathrm{mod}\ DS), \qquad a \in S. \tag{6.1}$$

Moreover we claim that

$$h_k D_k S \subset D_{k-1} S, \quad h_k D_j S \subset D_j S, \quad j<k. \tag{6.2}$$

Since $s_k s_j = s_j s_{k-1}$ by (5.10), the second inclusion is immediate. As for the first, the identities (5.8) for $k \leqq \dim a$, $a \in S$, give

$$\begin{aligned} d_i t_k s_k a &= (-1)^k s_{k-1} s_{k-1} d_i a, & i<k, \\ &= (-1)^k s_k a, & i=k,\ k+1,\ k+2, \\ &= (-1)^k s_k s_k d_{i-2} a, & i>k+2, \end{aligned}$$

while, for $k \leqq \dim a$, (5.8) and (5.10) give

$$\begin{aligned} t_k d_i s_k a &= (-1)^k s_{k-1} s_{k-1} d_i a, & i<k, \\ &= (-1)^k s_k a, & i=k,\ k+1, \\ &= (-1)^k s_k s_k d_{i-1} a, & i>k+1. \end{aligned}$$

With $\partial = \sum (-1)^i d_i$, these combine to give $(\partial t_k + t_k \partial) s_k a \equiv s_k a \pmod{D_{k-1} S}$ for $k \leqq \dim a$ and hence the first inclusion of (6.2). In particular, $h_0 D_0 S = 0$.

Now set $h = h_0 h_1 \dots h_k \dots$. Since $h_k a = a$ for $k > \dim a$, this composite is finite in each dimension, and defines a chain transformation h: $KS \to KS$. By (6.1), $h_k DS \subset DS$, so an iteration of (6.1) gives

$$ha \equiv a \pmod{DS}. \tag{6.3}$$

By (6.2), $hDS = 0$. Since each h_k is chain homotopic to 1, there is a composite homotopy t: $1 \simeq h$. Because $hD = 0$, $g(a + DS) = ha$ defines a chain transformation g: $KS/DS \to KS$; by (6.3), $\pi g = 1$, where π is the projection $KS \to KS/DS$. Moreover, $g\pi = h$: $KS \to KS$ is chain homotopic to 1, by construction, so π is a chain equivalence, as asserted.

7. Acyclic Models

The treatment of products of simplicial modules in the next section will require the use of acyclic models; here we state the preliminaries, for simplicial modules over some fixed ring R.

For each non-negative integer n a simplicial R-module M^n is defined by taking M_p^n to be the free module with generators all monotonic maps λ: $[p] \to [n]$, while $\mu^* = M^n(\mu)$: $M_p^n \to M_q^n$ is defined for each monotonic μ: $[q] \to [p]$ as $\mu^* \lambda = \lambda \mu$. This makes M^n a contravariant functor. Observe that the generators λ of M_p^n are all the p-dimensional faces $(\lambda_0, \dots, \lambda_p)$, degenerate or not, on the usual n-simplex, and that M^n is augmented by $\varepsilon(\lambda_0) = 1$; often M^n is denoted as Δ^n. We call M^n the n-dimensional *model* simplicial module and the identity map $\varkappa^n = 1$: $[n] \to [n]$ the *basic cell* on this model; thus $\varkappa^n \in M_n^n$.

As in the case of spaces (II.7), an augmented chain complex $\varepsilon: K \to R$ is *acyclic* if $H_n(K)=0$ for $n>0$ and $\varepsilon: H_0(K) \cong R$.

Proposition 7.1. *For each non-negative integer n, $K(M^n)$ is acyclic.*

Proof. It will suffice to construct a contracting homotopy. Define a homomorphism $s: M_p^n \to M_{p+1}^n$ by $s(\lambda_0, \ldots, \lambda_p) = (0, \lambda_0, \ldots, \lambda_p)$. By (5.11) and (5.12),

$$d_0 s = 1, \quad d_i s = s d_{i-1}, \quad i>0 \tag{7.1}$$

and $s s_i = s_{i+1} s$. Hence, in the associated chain complex, s induces a chain homotopy $s: 1 \simeq f\varepsilon$, where $f: R \to S$ is defined by $f 1_R = (0)$.

Proposition 7.2. *For each simplicial module S and each $a \in S_n$ there is a unique simplicial map $\alpha: M^n \to S$ with $\alpha \varkappa^n = a$.*

Proof. Each free generator λ of M_p^n can be written uniquely in terms of the basic cell $\varkappa^n$ as $\lambda^* \varkappa^n = \varkappa^n \lambda$. Hence $\alpha(\lambda^* \varkappa^n) = \lambda^* a$ defines a simplicial map $\alpha: M^n \to S$; it is clearly the only such with $\alpha \varkappa^n = a$.

To summarize: the models are acyclic and represent each $a \in S_n$. Similarly, in the proof (II.8) of the homotopy axiom for the singular complex $S(X)$ of a topological space, the models $S(\Delta^n)$ and $S(\Delta^n \times I)$ are acyclic and represent each singular simplex T via $T: \Delta^n \to X$. This situation recurs in many connections as a means of constructing chain transformations and chain homotopies. It can be described in categorical terms (EILENBERG-MACLANE [1953], GUGENHEIM-MOORE [1957]); it is more efficient to apply it directly in each case, as in the argument to follow in the next section.

Exercise

1. If V is any set with the partial order defined by $v \leqq v'$ for every $v, v' \in V$, $K(F_Z SV)$ is acyclic.

8. The Eilenberg-Zilber Theorem

If U and V are simplicial sets, their *cartesian product* $U \times V$ is the simplicial set with $(U \times V)_n = U_n \times V_n$ the cartesian product of sets and

$$d_i(u, v) = (d_i u, d_i v), \quad s_i(u, v) = (s_i u, s_i v), \quad i = 0, \ldots, n, \tag{8.1}$$

for $u \in U_n$, $v \in V_n$, and $n>0$ in the case of d_i. This definition is suggested by the case of topological spaces. If $X \times Y$ is the cartesian product of two spaces X and Y, with projections π_1 and π_2 on X and Y, respectively, each singular simplex $T: \Delta^n \to X \times Y$ is determined by its projections $\pi_1 T$ and $\pi_2 T$, while $d_i \pi_j T = \pi_j d_i T$, $s_i \pi_j T = \pi_j s_i T$. Hence $T \to (\pi_1 T, \pi_2 T)$ provides an isomorphism $S(X \times Y) \cong S(X) \times S(Y)$ of simplicial sets. The computation of the singular homology of $X \times Y$ is thus reduced to the computation of the homology of a cartesian product of simplicial sets.

There is a parallel product for simplicial modules A and B over a commutative ring. The *cartesian product* $A \times B$ is defined to be the simplicial module with $(A \times B)_n = A_n \otimes B_n$ and

$$d_i(a \otimes b) = d_i a \otimes d_i b, \quad s_i(a \otimes b) = s_i a \otimes s_i b, \quad i = 0, \ldots, n, \tag{8.2}$$

for $a \in A_n$, $b \in B_n$, and $n > 0$ in the case of d_i. To avoid confusion with the tensor product of complexes we shall write $a \times b$ for the element $a \otimes b$ of $A_n \otimes B_n$. For simplicial sets U and V, this definition insures that there is a natural isomorphism of simplicial modules

$$F(U \times V) \cong FU \times FV; \tag{8.3}$$

for $F(U \times V)$ in dimension n is the free module generated by the set $U_n \times V_n$, and this free module is naturally isomorphic to the tensor product $(FU_n) \otimes (FV_n)$.

The associated chain complex $K(A \times B)$ now reduces to the tensor product of the chain complexes $K(A)$ and $K(B)$.

Theorem 8.1. (Eilenberg-Zilber.) *For simplicial modules A and B over a commutative ring there is a natural chain equivalence*

$$K(A \times B) \underset{g}{\overset{f}{\rightleftarrows}} K(A) \otimes K(B). \tag{8.4}$$

In view of the normalization theorem, $K(A) \to K_N(A)$ is a chain equivalence, so there is also a natural chain equivalence

$$K_N(A \times B) \rightleftarrows K_N(A) \otimes K_N(B). \tag{8.5}$$

The proof, as recorded in the following lemmas, will use the method of acyclic models. Note that $K_0(A \times B) = A_0 \otimes B_0 = K_0(A) \otimes K_0(B)$; hence we can choose maps f and g in (8.4) to be the identity in dimension zero.

Lemma 8.2. *For simplicial modules A and B there exists a natural chain transformation f: $K(A \times B) \to K(A) \otimes K(B)$ which is the identity in dimension zero. Any two such natural maps f are chain homotopic via a homotopy which is natural.*

Proof. Since f_0 is given, suppose by induction on n that f_q is already defined for all $q < n$ and natural on $K_q(A \times B)$, with $\partial f_q = f_{q-1} \partial$. We wish to define f_n with $\partial f_n = f_{n-1} \partial$; we do this first for the product $\varkappa^n \times \varkappa^n$ of the two basic cells in the model $A = M^n = B$. We require that $\partial f_n(\varkappa^n \times \varkappa^n) = f_{n-1} \partial(\varkappa^n \times \varkappa^n)$. The right hand side e is already defined and has $\partial e = 0$ (or $\varepsilon e = 0$, if $n = 1$); it is thus a cycle in the complex $K(M^n) \otimes K(M^n)$, which is acyclic as the tensor product of two acyclic complexes (Prop. 7.1). Hence there is in this complex a chain c of dimension n with $\partial c = e$. We set $f_n(\varkappa^n \times \varkappa^n) = c$, so that

$$\partial f_n(\varkappa^n \times \varkappa^n) = \partial c = f_{n-1}\partial(\varkappa^n \times \varkappa^n). \tag{8.6}$$

Now consider $a \in A_n$, $b \in B_n$. By Prop. 7.2, there are simplicial maps $\alpha\colon M^n \to A$, $\beta\colon M^n \to B$ with $\alpha\varkappa^n = a$, $\beta\varkappa^n = b$. Then $K(\alpha)\colon K(M^n) \to K(A)$ is a chain transformation which we again denote as α, and $\alpha \otimes \beta\colon K(M^n) \otimes K(M^n) \to K(A) \otimes K(B)$ is a chain transformation. Set $f_n(a \times b) = (\alpha \otimes \beta)\, c$, for c as in (8.6); since the simplicial maps α and β are unique, the right hand side is bilinear in a and b, so defines $f_n\colon K_n(A \times B) \to [K(A) \otimes K(B)]_n$. Moreover,

$$\partial f_n(a \times b) = \partial(\alpha \otimes \beta)\, c = (\alpha \otimes \beta)\, \partial c = (\alpha \otimes \beta)\, f_{n-1}\partial(\varkappa^n \times \varkappa^n).$$

Now f_{n-1} is natural, so

$$\partial f_n(a \times b) = f_{n-1}\partial(\alpha\varkappa^n \times \beta\varkappa^n) = f_{n-1}\partial(a \times b).$$

Thus f is indeed a chain transformation up to dimension n.

To prepare for the next induction step it remains to show that f_n is natural. Let $\eta\colon A \to A'$, $\zeta\colon B \to B'$ be any simplicial maps, with $\eta a = a'$, $\zeta b = b'$. Then $\eta\alpha\colon M^n \to A'$ has $\eta\alpha\varkappa^n = \eta a = a'$, so is the unique simplicial map carrying $\varkappa^n$ to a'. Hence

$$(\eta \otimes \zeta)\, f_n(a \times b) = (\eta \otimes \zeta)(\alpha \otimes \beta)\, c = (\eta\alpha \otimes \zeta\beta)\, c = f_n(a' \times b'),$$

and f_n is natural.

Now let f and f' be two such chain transformations. By induction on n we may assume that the $t_q\colon K_q(A \times B) \to (K(A) \otimes K(B))_{q+1}$ are maps defined for $q = 0, \ldots, n-1$ with $\partial t + t\partial = f - f'$ in dimensions $q < n$. (For $q = 0$, $f_0 = f_0'$; so choose $t_0 = 0$.) Again we define t_n first on $\varkappa^n \times \varkappa^n$. We require

$$\partial t_n(\varkappa^n \times \varkappa^n) = f(\varkappa^n \times \varkappa^n) - f'(\varkappa^n \times \varkappa^n) - t_{n-1}\partial(\varkappa^n \times \varkappa^n).$$

By the induction assumption, $\partial(f - f' - t\partial) = 0$, so the right hand side is a cycle in an acyclic complex, hence is the boundary of some chain d. Set $t_n(\varkappa^n \times \varkappa^n) = d$, $t_n(a \times b) = (\alpha \otimes \beta)\, d$ for α, β with $\alpha\varkappa^n = a$, $\beta\varkappa^n = b$. The previous type of argument then shows t_n natural and $\partial t_n + t_{n-1}\partial = f - f'$ for all $a \times b$.

Lemma 8.3. *For simplicial modules A and B there is a natural chain transformation $g\colon K(A) \otimes K(B) \to K(A \times B)$ which is the identity in dimension zero. Any two such g are homotopic by a chain homotopy natural in A and B.*

The proof is analogous. A typical chain of $K(A) \otimes K(B)$ in dimension n has the form $a \otimes b$, with $a \in K_p(A)$, $b \in K_q(B)$, and $p + q = n$. Use the models M^p and M^q and maps $\alpha: M^p \to A$, $\beta: M^q \to B$ with $\alpha\,\varkappa^p = a$, $\beta\,\varkappa^q = b$. Now the complex $K(M^p \times M^q)$ is acyclic, for the homotopies s of (7.1) for M^p and M^q yield a contracting homotopy $s(a \times b) = s\,a \times s\,b$ on $K(M^p \times M^q)$. Using this acyclicity, the construction of g proceeds as before.

We now have the chain transformations f and g of the theorem; it remains to establish homotopies $1 \simeq fg$, $1 \simeq gf$. These are done by exactly the same method; for instance the homotopy $1 \simeq gf$ in $K(A \times B)$ is obtained, using the acyclicity of $K(M^p \times M^p)$, by comparing $h = 1$ with $h' = gf$ as follows.

Lemma 8.4. *If $h, h': K(A \times B) \to K(A \times B)$ are two natural chain transformations, both the identity in dimension zero, there is a natural chain homotopy $t: h \simeq h'$.*

These proofs are actually constructive; explicit formulas for f and g can be found by calculating the chain c used at each stage of the induction (e.g., in (8.6)) from the explicit contracting homotopies given in the proof of Prop. 7.1 for the models. We do not need the explicit homotopies $1 \simeq fg$, $1 \simeq gf$, but the explicit formulas so obtained for f and g are useful. To write them out, denote the "last" face in a simplicial object S by $\tilde{d}$; that is, for a in S_n set $\tilde{d}a = d_n a$. Thus, for any exponent $n-i$, $\tilde{d}^{n-i} a = d_{i+1} \dots d_n a$.

Theorem 8.5. *For any simplicial modules A and B, a natural chain transformation $f: K(A \times B) \to K(A) \otimes K(B)$ for the* Eilenberg-Zilber *theorem is given by*

$$f(a \times b) = \sum_{i=0}^{n} \tilde{d}^{n-i} a \otimes d_0^i b, \qquad a \in A_n,\ b \in B_n. \tag{8.7}$$

Proof. Since f is defined by face operators, it is natural. It reduces to the identity in dimension $n=0$. It remains to prove that $\partial f(a \times b) = f \partial (a \times b)$; in view of naturality, it suffices to prove this for $a = \varkappa^n = b$ in the model M^n. Now $\varkappa^n = (0, 1, \dots, n)$, $\tilde{d}^{n-i} \varkappa^n$ is the simplex $(0, 1, \dots, i)$ and

$$f(\varkappa^n \times \varkappa^n) = \sum_{i=0}^{n} (0, \dots, i) \otimes (i, i+1, \dots, n). \tag{8.8}$$

In $\partial f(\varkappa^n \times \varkappa^n)$ the last face of each first factor cancels with the term arising from the initial face of the second factor, and the remaining terms assemble to give $f \partial (\varkappa^n \times \varkappa^n)$, as required.

The chain transformation f of (8.7) is known as the *Alexander-Whitney* map, since it appeared in the simultaneous and independent definition of the cup product in topology by these authors. The explicit map f calculated from our contracting homotopy differs from the

Alexander-Whitney map, but only by terms which are degenerate. Moreover,

Corollary 8.6. *The Alexander-Whitney map f induces a chain transformation on the associated normalized chain complexes,*

$$f_N\colon K_N(A\times B)\to K_N A\otimes K_N B.$$

Proof. By definition, $K_N A = KA/DA$; by (3.3) regard $K_N A\otimes K_N B$ as $KA\otimes KB$ modulo the subcomplex spanned by the images of both $DA\otimes KB$ and $KA\otimes DB$. In (8.7) suppose $a\times b\in K(A\times B)$ degenerate, so of the form $s_k a'\times s_k b'$ for some k. In each term on the right of (8.7) one of the factors is degenerate. Specifically, if $i\leq k$, (5.8) shows $d_0^i s_k b'$ degenerate, while if $i>k$, $\tilde{d}^{n-i} s_k a'$ is degenerate, whence the desired result.

Geometrically, f is an "approximation to the diagonal". Consider for instance the cartesian product $\Delta^1\times\Delta^1$ of two 1-simplices (= intervals); it is a square with four vertices. Algebraically, Δ^1 is represented by $K(M^1)$; in $K_N(M^1)\otimes K_N(M^1)$ the group of 1-chains is a free group on four generators, corresponding to the four edges of the square. The diagonal of the square does not appear directly as a chain. However,

$$f(\varkappa^1\times\varkappa^1)=(0)\otimes(0\ 1)+(0\ 1)\otimes(1)$$

is the chain represented by left hand edge plus top edge of the square. This chain is "homotopic" to the diagonal, hence an "approximation" to the diagonal. Observe that the bottom edge plus the right hand edge would give a different approximation, which could be developed algebraically by interchanging the roles of initial and final faces in the formula (8.7). Comparison of these two different approximations to the diagonal leads to the Steenrod squaring operations (Steenrod [1953], Milnor [1958], Dold [1961], Steenrod-Epstein [1962]).

For three simplicial modules A, B, and C, any natural Eilenberg-Zilber map f may be iterated, as in

$$K(A\times B\times C)\xrightarrow{f} K(A)\otimes K(B\times C)\xrightarrow{1\otimes f} K(A)\otimes K(B)\otimes K(C).$$

Proposition 8.7. *Any natural f is associative up to homotopy, in the sense that there is a natural chain homotopy $(1\otimes f)f\simeq(f\otimes 1)f$. The Alexander-Whitney map is associative.*

Proof. Since $(1\otimes f)f$ and $(f\otimes 1)f$ are each the identity in dimension 0, a natural homotopy between them may be constructed by the method of acyclic models. The associativity (no homotopy necessary) of the Alexander-Whitney map can be computed directly, say by (8.8).

To describe the second map g of the Eilenberg-Zilber theorem we introduce certain "shuffles". If p and q are non-negative integers, a *(p, q)-shuffle* (μ, ν) is a partition of the set $[p+q-1]$ of integers into two disjoint subsets $\mu_1 < \cdots < \mu_p$ and $\nu_1 < \cdots < \nu_q$ of p and q integers, respectively. Such a partition describes a possible way of shuffling a deck of p cards through a deck of q cards, placing the cards of the first deck in order in the position $\mu_1, \ldots, \mu_p$ and those of the second deck in order in the positions $\nu_1, \ldots, \nu_q$. A shuffle may be pictured as a sequence of moves in the lattice of points (m, n) in the plane with integral coordinates: Start at $(0, 0)$ at time 0; at time k move to the right if k is one of $\mu_1, \ldots, \mu_p$ and up if k is one of $\nu_1, \ldots, \nu_q$; the result is a "staircase" from $(0, 0)$ to (p, q). A (p, q)-shuffle can also be defined to be a permutation t of the set of integers $\{1, \ldots, p+q\}$ such that $t(i) < t(j)$ whenever $i < j \leqq p$ or $p < i < j$; for, each such permutation t determines the μ_i as $t(i) - 1$, the ν_j as $t(p+j) - 1$, and conversely. The *signature* $\varepsilon(\mu)$ of the shuffle (μ, ν) is the integer $\varepsilon(\mu) = \sum_{i=1}^{p} \mu_i - (i-1)$; then $(-1)^{\varepsilon(\mu)}$ is the sign of the associated permutation t.

Theorem 8.8. *For any simplicial modules A and B a natural chain transformation g for the Eilenberg-Zilber theorem is given, for $a \in A_p$, $b \in B_q$, by*

$$g(a \otimes b) = \sum_{(\mu, \nu)} (-1)^{\varepsilon(\mu)} (s_{\nu_q} \ldots s_{\nu_1} a \times s_{\mu_p} \ldots s_{\mu_1} b), \tag{8.9}$$

where the sum is taken over all (p, q)-shuffles (μ, ν).

Clearly g is natural, $a \otimes b$ has dimension $p+q$, and so do $s_{\nu_q} \ldots s_{\nu_1} a$ and $s_{\mu_p} \ldots s_{\mu_1} b$. The proof that g is a chain transformation is a straightforward verification which we omit (details in EILENBERG-MACLANE [1953b], § 5, where the shuffles were first introduced).

Geometrically this function g provides a "triangulation" of the cartesian product $\Delta^p \times \Delta^q$ of two simplices. Specifically, take $a = \varkappa^p \in M^p$ and $b = \varkappa^q \in M^q$, so $\varkappa^p$ has vertices $(0, 1, \ldots, p)$. In this vertex notation,

$$s_{\nu_q} \ldots s_{\nu_1} \varkappa^p = (i_0, i_1, \ldots, i_{p+q}),$$

with $0 = i_0 \leqq i_1 \leqq \cdots \leqq i_{p+q} = p$, and $i_k = i_{k+1}$ precisely when k is one of $\nu_1, \ldots, \nu_q$. Similarly $s_{\mu_p} \ldots s_{\mu_1} \varkappa^q = (j_0, \ldots, j_{p+q})$, with $j_k = j_{k+1}$ precisely when k is one of $\mu_1, \ldots, \mu_p$. The simplex displayed on the right of (8.9) then has the form

$$(i_0, \ldots, i_{p+q}) \times (j_0, \ldots, j_{p+q}),$$

where the first factor is degenerate at those indices k for which the second factor is not degenerate. This symbol may be read as the $(p+q)$-dimensional affine simplex with vertices (i_k, j_k) in the product $\Delta^p \times \Delta^q$.

These simplices, for all (p, q)-shuffles, provide a simplicial subdivision of $\Delta^p \times \Delta^q$. For example, if $p=2$, $q=1$, $\Delta^2 \times \Delta^1$ is a triangular prism and the three possible (2, 1) shuffles triangulate this prism into three simplices

$$(0\,1\,2\,2) \times (0\,0\,0\,1), \quad (0\,1\,1\,2) \times (0\,0\,1\,1), \quad (0\,0\,1\,2) \times (0\,1\,1\,1),$$

each of dimension three. (Draw a figure!)

This description also shows that if either factor a or b is degenerate, so is each term on the right in (8.9). Hence

Corollary 8.9. *The shuffle map g of* (8.9) *induces a chain transformation on the normalized chain complexes*

$$g_N: K_N(A) \otimes K_N(B) \to K_N(A \times B).$$

For these normalized complexes, the composite $f_N g_N$ can be shown to be the identity (no homotopy $1 \simeq f_N g_N$ is needed).

Exercises

1. Exhibit a second explicit formula for f, with first and last faces interchanged in (8.7).

2. Establish associativity for the shuffle map g.

3. Prove the normalization theorem of § 6 by the method of acyclic models.

4. Show that the Eilenberg-Zilber theorem holds for A a simplicial right R-module, B a simplicial left R-module, and R any ring.

5. Calculate the integral homology of a torus $S^1 \times S^1$ from that of a circle S^1 (Eilenberg-Zilber plus Künneth).

9. Cup Products

For any simplicial set U, $\Delta u = u \times u$ defines a simplicial map $\Delta: U \to U \times U$ called the *simplicial diagonal map*. Now U determines the simplicial abelian group $F_Z U$ and hence the chain complex $K(F_Z U)$ which we write simply as $K(U)$; each $K_n(U)$ is the free abelian group generated by the set U_n, with $\partial = \sum (-1)^i d_i$. The diagonal induces a chain transformation $K(U) \to K(U \times U)$, also denoted by Δ. If f is any one of the natural maps from the Eilenberg-Zilber theorem the composite

$$\omega = f\Delta: K(U) \to K(U \times U) \to K(U) \otimes K(U) \tag{9.1}$$

is called a *diagonal map* in $K(U)$. Since f is unique up to a (natural) chain homotopy, so is ω. Since Δ is associative — $(\Delta \times 1)\Delta = (1 \times \Delta)\Delta$ — and f is associative up to homotopy (Prop. 8.7), there is a homotopy $(\omega \otimes 1)\omega \simeq (1 \otimes \omega)\omega$. The complex $K(U)$ is augmented by $\varepsilon(u) = 1$ for $u \in U_0$. We assert that there are homotopies

$$(\varepsilon \otimes 1)\,\omega \simeq 1 \simeq (1 \otimes \varepsilon)\,\omega: \ K(U) \to K(U). \tag{9.2}$$

Indeed, each of $(\varepsilon \otimes 1)\omega$ and $(1 \otimes \varepsilon)\omega$ is natural and is the identity in dimension zero, so natural homotopies may be constructed by using the acyclic models M^n (taken this time as simplicial sets U). Now equalities in (9.2) and in associativity are exactly the conditions (VI.9.1) required to make ω a coproduct with counit ε, so that we might say that $K(U)$ with diagonal ω is a differential graded coalgebra "up to homotopy".

If we choose for f the Alexander-Whitney map, then ω is associative and it is easy to check that $(\varepsilon \otimes 1)\omega = (1 \otimes \varepsilon)\omega$. Hence with this choice $K(U)$ is a differential graded coalgebra, and so is the normalized complex $K_N(U)$.

Now let A and A' be abelian groups, and write $H^k(U, A)$ for the cohomology group $H^k(\operatorname{Hom}(K(U), A))$. The composite $\smile = \omega^* p^H$,

$$\begin{array}{ccc} H^k(U,A)\otimes H^m(U,A') & & \\ \downarrow p^H & \searrow \smile & \\ H^{k+m}(K(U)\otimes K(U), A\otimes A') & \xrightarrow{\omega^*} & H^{k+m}(U, A\otimes A'), \end{array} \tag{9.3}$$

where p^H is the cohomology product of (1.3), is called the (external) *simplicial cup product*. With cochains h and h', the definition reads $(\operatorname{cls} h) \smile (\operatorname{cls} h') = \operatorname{cls}(h \smile h')$, where

$$(h \smile h')u = (h \otimes h') f \Delta u, \tag{9.4}$$

for $h \otimes h'$ as in (1.4). In particular, if $U = S(V)$ is the simplicial set associated with a partly ordered set V of vertices and f is the Alexander-Whitney map, while $h \in H^k$, $h' \in H^{n-k}$, then

$$(h \smile h')(v_0, \ldots, v_n) = h(v_0, \ldots, v_k) \otimes h'(v_k, \ldots, v_n). \tag{9.5}$$

If $A = A'$ is the additive group of a commutative ring R with product $\pi\colon R \otimes R \to R$, the composite $\pi_* \smile$ is a map

$$H^k(U, R) \otimes H^m(U, R) \to H^{k+m}(U, R) \tag{9.6}$$

called the *internal* simplicial cup product.

Theorem 9.1. *For each simplicial set U and each coefficient ring R the cohomology modules $H^k(\operatorname{Hom}(K(U), R)) = H^k(U, R)$ constitute a graded ring under the internal simplicial cup product. If R is commutative, so is this cohomology ring.*

Proof. The associativity of the product is known. The augmentation $\varepsilon\colon K(U) \to Z$ composed with $I\colon Z \to R$ gives a zero dimensional cocycle $I\varepsilon$ of $K(U)$. Then $(h \smile I\varepsilon)\, u = \pi(h \otimes I)(1 \otimes \varepsilon)\,\omega\, u$, where $\pi(h \otimes I)$: $K \otimes Z \to R$ is h when $K \otimes Z$ is identified with K, while $(1 \otimes \varepsilon)\,\omega \simeq 1$. Hence the cohomology class e of the cocycle $I\varepsilon$ acts as the identity for the cup product. Similarly, to show that the cup product is commutative, it suffices to establish a chain homotopy $f \simeq \tau f$ for the usual interchange $\tau\colon K_k \otimes K_m \cong K_m \otimes K_k$. Both f and τf are the identity in dimension 0, and this homotopy is given by using acyclic models.

When f is the Alexander-Whitney map, the cochains themselves form a graded ring, but the ring is not commutative: Commutativity holds only for cohomology classes.

This theorem shows that the singular cohomology of a topological space X with coefficients Z is a commutative ring under the cup product.

The simplicial cup product also applies to the cohomology of a group Π. By a *Π-set* S is meant a set S together with an action of Π on S; more formally, this action is given by $\varphi\colon \Pi \to \operatorname{Aut}(S)$, a homomorphism of Π into the group of 1-1 maps of S onto S. The Π-sets form a category. For example, take $\tilde{B}_n(\Pi)$ to be the set of all $(n+1)$-tuples $(x_0, \ldots, x_n)$ with the action of Π given by $x(x_0, \ldots, x_n) = (x x_0, \ldots, x x_n)$. The usual face and degeneracy operators

$$d_i(x_0, \ldots, x_n) = (x_0, \ldots, \hat{x}_i, \ldots, x_n), \qquad 0 \leqq i \leqq n,\ n > 0,$$

$$s_i(x_0, \ldots, x_n) = (x_0, \ldots, x_i, x_i, \ldots, x_n), \qquad 0 \leqq i \leqq n,$$

are Π-maps, so $\tilde{B}(\Pi)$ is a simplicial Π-set. The associated simplicial abelian group $F_Z(\tilde{B}(\Pi))$ is a simplicial Π-module, while $K = KF_Z(\tilde{B}(\Pi))$ is a complex of Π-modules, with K_n the free abelian group generated by the $(x_0, \ldots, x_n)$ and with boundary

$$\partial(x_0, \ldots, x_n) = \sum_{i=0}^{n} (-1)^i (x_0, \ldots, \hat{x}_i, \ldots, x_n).$$

We have recovered the homogeneous description (IV.5.13) of the unnormalized bar resolution $\beta(\Pi) = KF_Z(\tilde{B}(\Pi))$, while $K_N F_Z(\tilde{B}(\Pi))$ is the normalized bar resolution $B(\Pi)$.

Now recall that the group ring $Z(\Pi)$ is a Hopf algebra with coproduct

$$\psi\colon Z(\Pi) \to Z(\Pi) \otimes Z(\Pi), \qquad \psi(x) = x \otimes x.$$

By pull-back along the corresponding diagonal map $\Pi \to \Pi \times \Pi$, the cartesian product $\tilde{B}(\Pi) \times \tilde{B}(\Pi)$ of two Π-sets is a Π-set. The diagonal map ω for $\beta(\Pi) = K\tilde{B}(\Pi)$ is the composite

$$\omega\colon \beta(\Pi) \xrightarrow{\Delta} K(\tilde{B}(\Pi) \times \tilde{B}(\Pi)) \xrightarrow{f} {}_{\psi}[\beta(\Pi) \otimes \beta(\Pi)];$$

here Δ is a Π-map, f is natural, so commutes with the action of Π and is also a Π-map. Therefore ω is a chain transformation for complexes of Π-modules. This implies that the simplicial cup product is defined, for two Π-modules A and A', as a homomorphism

$$\smile\colon H^k(\Pi, A) \otimes H^m(\Pi, A') \to H^{k+m}(\Pi, {}_{\psi}(A \otimes A')). \qquad (9.7)$$

This product is associative.

A homomorphism $\alpha: {}_\psi(A \otimes A') \to A''$ of Π-modules is called a *pairing* of A and A' to A''. The cup product followed by the homomorphism induced by the pairing α yields an "internal" cup product which is a homomorphism $H(\Pi, A) \otimes H(\Pi, A') \to H(\Pi, A'')$.

The discussion of Π-sets in the definition of this cup product could be short cut by simply giving the direct description of the cup product by cochains. If h and h' are cochains of dimensions k and m, respectively, regarded as functions on the homogeneous generators $(x_0, \ldots, x_k)$ and $(x_0, \ldots, x_m)$ of $\beta(\Pi)$, their cup product is the cochain defined, via Alexander-Whitney, by

$$(h \smile h')(x_0, \ldots, x_n) = h(x_0, \ldots, x_k) \otimes h'(x_k, \ldots, x_n), \qquad n = k + m. \tag{9.8}$$

This $h \smile h'$ is clearly a Π-module homomorphism into $A \otimes A'$ with diagonal operators, that is, into ${}_\psi(A \otimes A')$.

In particular, if A and A' are both the ring Z with trivial operators $(Z = {}_\varepsilon Z)$, then ${}_\psi({}_\varepsilon Z \otimes {}_\varepsilon Z)$ is ${}_\varepsilon Z$. It follows that $H^k(\Pi, {}_\varepsilon Z)$ is a commutative graded ring under the simplicial cup product.

Theorem 9.2. *Under the isomorphism $H^n(\Pi, A) \cong \operatorname{Ext}^n_{Z(\Pi)}(Z, A)$, for any Π-module A, the simplicial cup product is mapped onto the Hopf wedge product defined in* Ext.

The crux of the proof is the observation that the diagonal map

$$\omega: \beta(\Pi) \to {}_\psi[\beta(\Pi) \otimes \beta(\Pi)]$$

of complexes of Π-modules commutes with the augmentation, hence is a comparison of the resolution $\varepsilon: \beta(\Pi) \to Z$ to the resolution given by ${}_\psi[\beta(\Pi) \otimes \beta(\Pi)] \to Z$. Both H^n and Ext^n are $H^n(\beta(\Pi), A)$. The Hopf wedge product of (4.5) is $\psi^\# p^H$, where p^H is the cohomology product and $\psi^\#$ the change of rings defined by $\psi: Z(\Pi) \to Z(\Pi) \otimes Z(\Pi)$. Now Thm. III.6.7 asserts that this change of rings can be calculated as $\psi^\# = f^* \psi^*$, where ψ^* maps $\operatorname{Hom}_{Z(\Pi) \otimes Z(\Pi)}$ to $\operatorname{Hom}_{Z(\Pi)}$, while the map $f: \beta(\Pi) \to {}_\psi[\beta(\Pi) \otimes \beta(\Pi)]$ is a comparison. Choose f to be the comparison ω; then $\psi^\# p^H$ becomes $\omega^* \psi^* p^H$, which is the simplicial cup product.

The cup product in the cohomology ring $H^*(\Pi, Z)$ can thus be defined in three equivalent ways:

(i) As the simplicial cup product;

(ii) As the wedge product induced by the diagonal map ψ;

(iii) As the Yoneda product, by composites of long exact sequences.

Still a fourth definition will appear in Chap. XII and will facilitate the computation of examples.

One application is the "cup product reduction theorem". Suppose $\Pi = F/R$ where F is a free multiplicative group. Let $[R, R]$ be the commutator subgroup of R and set $F_0 = F/[R, R]$, $R_0 = R/[R, R]$. Then R_0

is abelian and $\Pi \cong F_0/R_0$, so F_0 is an extension of R_0 by Π with factor set a 2-cocycle f_0 of Π in the Π-module R_0. For any Π-module A, $\mathrm{Hom}_Z(R_0, A)$ is a Π-module with operators $x\alpha$ defined for $\alpha: R_0 \to A$ by $(x\alpha)\,r = x[\alpha(x^{-1}r)]$, while $\alpha \otimes r \to \alpha\,r$ is a pairing $\mathrm{Hom}(R_0, A) \otimes R_0 \to A$. The internal cup product of an n-cocycle with f_0 then determines a homomorphism

$$H^n(\Pi, \mathrm{Hom}(R_0, A)) \to H^{n+2}(\Pi, A).$$

The cup product reduction theorem asserts that this is an isomorphism for $n > 0$. The theorem is due to EILENBERG-MACLANE [1947]; an elegant proof, using relative cohomology and the characteristic class (IV.6) of an extension is given by SWAN [1960] and also below in IX.7, Ex. 7—10.

The cohomology groups $H^n(\Pi, Z)$ were shown in IV.11 to be the singular homology groups of the space X/Π when Π operates properly on the acyclic space X. The comparison made there evidently preserves the simplicial structure, hence the cup product, so $H(\Pi, Z) \cong H(X/\Pi, Z)$ is an isomorphism of cohomology rings.

Exercise

1. Show that $\beta(\Pi)$ with non-homogeneous generators (IV.5.11) has degeneracies and faces given by

$$\begin{aligned} s_i(x[x_1, \ldots, x_n]) &= x[x_1, \ldots, x_i, 1, x_{i+1}, \ldots, x_n], & 0 \leq i \leq n,\\ d_i(x[x_1, \ldots, x_n]) &= x\,x_1\,[x_2, \ldots, x_n], & i = 0,\\ &= x[x_1, \ldots, x_i x_{i+1}, \ldots, x_n], & 0 < i < n,\\ &= x[x_1, \ldots, x_{n-1}], & i = n \end{aligned}$$

and that the map ω, for f Alexander-Whitney, is

$$\omega(x[x_1| \ldots |x_n]) = \sum_{i=0}^{n} x[x_1| \ldots |x_i] \otimes x\,x_1 \ldots x_i[x_{i+1}| \ldots |x_n].$$

Notes. For topological discussion of the cup product (in contrary terminology) see HILTON-WYLIE [1960]. For the cup product for groups see EILENBERG-MAC LANE [1947], ECKMANN [1945—1946], [1954]. A fiber space may be regarded as a sort of "twisted" cartesian product; there is a corresponding twisted version of the EILENBERG-ZILBER theorem (BROWN [1959], GUGENHEIM [1960], SZCZARBA [1961]). Simplicial fiber bundles are treated in BARRATT-GUGENHEIM-MOORE [1959].

Chapter nine

Relative Homological Algebra

Introduction. When we described the elements of $\mathrm{Ext}^n(C, A)$ as long exact sequences from A to C we supposed that A and C were left modules over a ring. We could equally well have supposed that they were right modules, bimodules, or graded modules. An efficient formulation of this situation is to assume that A and C are objects in a category with suitable properties: One where morphisms can be added and kernels and cokernels constructed. The first three sections of this chapter are devoted to the description of such "abelian" categories.

If Π is a group, each Π-module is also an abelian group; this gives a homomorphism of the category of all Π-modules to that of all abelian groups. If Λ is an algebra over the ground ring K, each Λ-module is also a K-module, while each Λ-bimodule is also a right Λ-module. If $R \supset S$ are rings, each R-module is an S-module. In each such case we have a homomorphism of one abelian category to a second which leads naturally to the definition of "relative" functors Ext and Tor; for further introductory explanation, see § 8 below. The general method is described in this chapter and will be applied in the next chapter to study the cohomology of various types of algebraic systems.

1. Additive Categories

First examine the categories in which suitable pairs of morphisms can be added. An *additive category* $\mathscr{C}$ is a class of objects $A, B, C, \ldots$ together with

(i) A family of disjoint abelian groups $\hom(A, B)$, one for each ordered pair of objects. We write $\alpha: A \to B$ for $\alpha \in \hom(A, B)$ and call α a *morphism* of $\mathscr{C}$.

(ii) To each ordered triple of objects A, B, C a homomorphism

$$\hom(B, C) \otimes \hom(A, B) \to \hom(A, C) \tag{1.1}$$

of abelian groups. The image of $\beta \otimes \alpha$ under composition is written $\beta\alpha$, and called the *composite* of β and α.

(iii) To each object A a morphism $1_A: A \to A$, called the *identity* of A.

These data are subject to the following four axioms:

Associativity: If $\alpha: A \to B$, $\beta: B \to C$, and $\gamma: C \to D$, then

$$\gamma(\beta\alpha) = (\gamma\beta)\alpha; \tag{1.2}$$

Identities: If $\alpha: A \to B$, then

$$\alpha 1_A = \alpha = 1_B \alpha; \tag{1.3}$$

Zero: There is an object $0'$ such that $\hom(0', 0')$ is the zero group.

Finite Direct Sums: To each pair of objects A_1, A_2 there exists an object B and four morphisms forming a diagram

$$A_1 \underset{\pi_1}{\overset{\iota_1}{\rightleftarrows}} B \underset{\pi_2}{\overset{\iota_2}{\leftrightarrows}} A_2$$

with

$$\pi_1\iota_1 = 1_{A_1}, \quad \pi_2\iota_2 = 1_{A_2}, \quad \iota_1\pi_1 + \iota_2\pi_2 = 1_B. \tag{1.4}$$

To avoid foundational difficulties, two further axioms of a set-theoretic character are required; they will be stated at the end of this section.

These axioms are like those for a category (I.7). Indeed, an additive category may be defined as a category with zero and direct sums, as above, in which each set $\hom(A, B)$ of morphisms has the structure of an abelian group such that the distributive laws

$$\beta(\alpha_1 + \alpha_2) = \beta\,\alpha_1 + \beta\,\alpha_2, \quad (\beta_1 + \beta_2)\,\alpha = \beta_1\,\alpha + \beta_2\,\alpha \tag{1.5}$$

are valid whenever both sides are defined. (This insures that composition is bilinear, as required by (1.1).)

If the existence of direct sums is not required, we speak of a *preadditive category*. As in the case of categories, we can omit the objects and work only with the morphisms, using the identity morphisms 1_A in place of objects. The axioms are then like the axioms for a ring in which the compositions $\alpha_1 + \alpha_2$ and $\beta\,\alpha$ are not always defined but, whenever defined, satisfy the usual ring axioms such as (1.2), (1.3), and (1.5). Thus HILTON-LEDERMANN [1958] call a preadditive category a *ringoid*, following the terminology of BARRATT [1954].

By 0 we denote (ambiguously) the zero element of any group $\hom(A, B)$; then $0\,\alpha = 0 = \beta\,0$ whenever defined (proof: $0\,\alpha = (0+0)\,\alpha$; use the distributive law). An object $0'$ with $\hom(0', 0')$ the zero group is called a *zero object*. Then $1_{0'} = 0$, hence $\hom(A, 0')$ and $\hom(0', B)$ are the zero groups whatever the objects A and B, and any two zero objects are equivalent.

Examine next the consequences of the finite direct sum axiom. By (1.4),

$$\pi_1\iota_2 = \pi_1(\iota_1\pi_1 + \iota_2\pi_2)\,\iota_2 = 1\,\pi_1\iota_2 + \pi_1\iota_2 1 = \pi_1\iota_2 + \pi_1\iota_2,$$

hence $\pi_1\iota_2 = 0$ and $\pi_2\iota_1 = 0$, as usual. Props. 4.1 and 4.3—4.5 of Chap. I follow; in particular, the diagram (1.4) determines the object B up to equivalence, and we usually write such a B as $A_1 \oplus A_2$. Each morphism $\gamma\colon A_1 \oplus A_2 \to C$ determines a pair of morphisms $\gamma_j = \gamma\,\iota_j\colon A_j \to C$; the correspondence $\varphi(\gamma) = (\gamma_1, \gamma_2)$ is an isomorphism

$$\varphi\colon \hom(A_1 \oplus A_2, C) \cong \hom(A_1, C) \oplus \hom(A_2, C)$$

of abelian groups. The inverse is given by $\varphi^{-1}(\gamma_1, \gamma_2) = \gamma_1\pi_1 + \gamma_2\pi_2\colon A_1 \oplus A_2 \to C$. Thus $\gamma = \gamma_1\pi_1 + \gamma_2\pi_2$ is the unique morphism $A_1 \oplus A_2 \to C$

with $\gamma \iota_j = \gamma_j$, $j=1, 2$, so the injections $\iota_j\colon A_j \to A_1 \oplus A_2$ of the direct sum constitute a universal diagram. Here a diagram $\{\alpha_t\colon A_t \to B \mid t \in T\}$ of coterminal morphisms α_t, with T any set of indices, is ***universal*** if to each diagram $\{\gamma_t\colon A_t \to C \mid t \in T\}$ there is a unique morphism $\gamma\colon B \to C$ with $\gamma \alpha_t = \gamma_t$ for each t. Dually, there is an isomorphism

$$\psi\colon \hom(C, A_1 \oplus A_2) \cong \hom(C, A_1) \oplus \hom(C, A_2)$$

with $\psi \gamma = (\pi_1 \gamma, \pi_2 \gamma)$ and $\psi^{-1}(\gamma_1, \gamma_2) = \iota_1 \gamma_1 + \iota_2 \gamma_2$. Consequently, the diagram $\{\pi_j\colon A_1 \oplus A_2 \to A_j \mid j = 1, 2\}$ is couniversal. The usual diagonal and codiagonal morphisms

$$\Delta_A = \iota_1 + \iota_2\colon A \to A \oplus A, \qquad \nabla_A = \pi_1 + \pi_2\colon A \oplus A \to A, \tag{1.6}$$

are characterized by the respective properties

$$\pi_1 \Delta_A = 1_A = \pi_2 \Delta_A, \quad \nabla_A \iota_1 = 1_A = \nabla_A \iota_2. \tag{1.7}$$

Given two direct sums $A_1 \oplus A_2$ and $A_1' \oplus A_2'$ and morphisms $\alpha_j\colon A_j \to A_j'$ there is a unique morphism $\alpha_1 \oplus \alpha_2\colon A_1 \oplus A_2 \to A_1' \oplus A_2'$ with

$$\pi_1(\alpha_1 \oplus \alpha_2) = \alpha_1 \pi_1, \quad \pi_2(\alpha_1 \oplus \alpha_2) = \alpha_2 \pi_2. \tag{1.8}$$

The same morphism is characterized by the dual properties

$$(\alpha_1 \oplus \alpha_2)\iota_1 = \iota_1 \alpha_1, \quad (\alpha_1 \oplus \alpha_2)\iota_2 = \iota_2 \alpha_2. \tag{1.9}$$

The iterated direct sum $A_1 \oplus (A_2 \oplus \cdots \oplus A_n)$ with the corresponding injections is a universal diagram, and any universal diagram on $A_1, \ldots, A_n$ is equivalent to this iterated direct sum. Dually, the projections π_j of an (iterated) direct sum provide a couniversal diagram. The axiom requiring the existence of finite direct sums may be replaced either by the assumption that there exists a universal diagram for any two objects A_1 and A_2, or by the dual requirement. In any event, the axioms for an additive category are self-dual.

In an additive category $\mathscr{C}$, $\hom(A, B)$ is a bifunctor on the category $\mathscr{C}$ to the category of abelian groups.

To prepare the way for the study of kernels, we formulate definitions of "monic" and "epic" in categories to agree in the standard examples with monomorphisms and epimorphisms. In the category of sets, a function f on X to Y is *surjective* if $f(X) = Y$ (f is onto Y) and *injective* if $f(x) = f(x')$ always implies $x = x'$ (f is 1-1 into Y). In any category, a morphism $\varkappa\colon A \to B$ is said to be *monic* if each induced map $\varkappa_*\colon \hom(C, A) \to \hom(C, B)$ is injective. Thus $\varkappa$ monic means that $\varkappa \alpha = \varkappa \alpha'$ implies $\alpha = \alpha'$ for all $\alpha, \alpha'\colon C \to A$, hence that $\varkappa$ is *left cancellable*. In an additive category, $\varkappa$ is monic if and only if $\varkappa \alpha = 0$ implies $\alpha = 0$ whenever $\varkappa \alpha$ is defined. Dually, a morphism $\sigma\colon B \to C$ in any category is said to be *epic* if each induced map σ^*: $\hom(C, G) \to \hom(B, G)$ is injective. Thus σ epic means that $\alpha\sigma = \alpha'\sigma$ always implies $\alpha = \alpha'$, hence that σ is *right cancellable*. In an additive category, σ is epic if and only if $\alpha\sigma = 0$ always implies $\alpha = 0$. In this chapter we systematically denote morphisms which are monic by $\varkappa$, λ, μ, ν and those which are epic by ϱ, σ, τ. If $\varkappa$ and λ are monic, so is $\varkappa\lambda$ whenever it is defined, and dually. *Warning:* In certain additive categories of modules, "monic" may not agree with "monomorphism" (see Ex. 5) though the agreement does hold in the category of all modules with morphisms all homomorphisms.

An *equivalence* is a morphism θ with a two-sided inverse ψ ($\psi\theta = 1$, $\theta\psi = 1$). Two morphisms α: $S \to A$ and α': $S' \to A$ with the same range are called *right equivalent* if there is an equivalence θ: $S \to S'$ with $\alpha'\theta = \alpha$; this relation is reflexive, symmetric, and transitive, so allows the formulation of right equivalence classes of morphisms with range A. If $\varkappa$ is monic, so is each right equivalent of $\varkappa$. In the additive category of all modules, two monomorphisms with range A are right equivalent if and only if their images are identical, as submodules of A. Hence in any additive category we say that the right equivalence class of a monic $\varkappa$: $S \to A$ is a *subobject* of A. It is convenient to say that $\varkappa$ itself is a subobject of A — meaning thereby the right equivalence class, cls $\varkappa$, of $\varkappa$. Observe that a "subobject" so defined is *not* an object of the category; for example, we cannot regard A as a subobject of itself but we must use instead cls 1_A, which is the class of all equivalences with range A.

The dual definitions are: α: $A \to T$ and α': $A \to T'$ are *left equivalent* if $\theta\alpha' = \alpha$ for some equivalence θ. The left equivalence class of an epic σ: $A \to T$ consists of epic morphisms and is called a *quotient object* of A.

For modules, the kernel K of a homomorphism α: $A \to B$ is the largest submodule of A mapped by α into 0 and is characterized by the property that each morphism β with $\alpha\beta = 0$ factors uniquely through the injection $\varkappa$: $K \to A$ as $\beta = \varkappa\beta'$. This can be paraphrased in any additive category $\mathscr{C}$: A *kernel* of α: $A \to B$ is a monic $\varkappa$ with range A such that

$$\alpha\varkappa = 0, \quad \text{while} \quad \alpha\beta = 0 \quad \text{implies} \quad \beta = \varkappa\beta' \tag{1.10}$$

for some β', necessarily unique. In other words, the right annihilators of α are exactly the right multiples of its kernel $\varkappa$. Hence any two kernels $\varkappa$ and $\varkappa'$ of α are right equivalent, so the class of all kernels of α, if not vacuous, is a subobject of A which we write as $\ker\alpha$. Dually, a *cokernel* of α: $A \to B$ is an epic σ with domain B such that

$$\sigma\alpha = 0, \quad \text{while} \quad \gamma\alpha = 0 \quad \text{implies} \quad \gamma = \gamma'\sigma \tag{1.11}$$

for some γ', necessarily unique. The left annihilators of α are thus the left multiples of a cokernel σ of α. Any two cokernels of α are left equivalent; if α has a cokernel, the class of all cokernels of α is a quotient object of B, so $\sigma \in \operatorname{coker} \alpha$ states that σ is one of the cokernels of α. In the category of modules, the projection $B \to B/\alpha A$ is a cokernel of α.

An immediate consequence of the respective definitions is

Lemma 1.1. *If the composites $\alpha\beta$, $\varkappa\alpha$, and $\alpha\sigma$ are defined, the following implications hold:*

$$\alpha\beta \text{ monic} \Rightarrow \beta \text{ monic}, \qquad \alpha\beta \text{ epic} \Rightarrow \alpha \text{ epic},$$

$$\varkappa \text{ monic} \Rightarrow \ker(\varkappa\alpha) = \ker\alpha, \qquad \sigma \text{ epic} \Rightarrow \operatorname{coker}(\alpha\sigma) = \operatorname{coker}\alpha.$$

Also $\ker 1_A = 0$, $\operatorname{coker} 1_A = 0$, *and, for* $0\colon A \to B$, $1_A \in \ker 0$ *and* $1_B \in \operatorname{coker} 0$. (*Here* $\ker 1 = 0$ *is short for* $0 \in \ker 1$.)

Finally we introduce a notation for short exact sequences, defining

$$\alpha \| \beta \quad \Leftrightarrow \quad \alpha \in \ker\beta \;\&\; \beta \in \operatorname{coker}\alpha. \tag{1.12}$$

This implies α monic and β epic, so we may read "$\varkappa\|\sigma$" as "$\varkappa$ and σ are the morphisms of a short exact sequence".

To keep the foundations in order we wish the collection of all subobjects of an object A and the collection of all extensions of A by C both to be sets and not classes. Hence, for an additive category we assume two additional axioms:

Sets of sub- and quotient objects. For each object A there is a set of morphisms $\varkappa$, each monic with range A, which contains a representative of every subobject of A and dually, for quotient objects of A.

Set of extensions. For each pair of objects C, A and each $n \geqq 1$ there is a set of n-fold exact sequences from A to C containing a representative of every congruence class of such sequences (with "congruence" defined as in III.5).

Both axioms hold in all the relevant examples.

Exercises

1. Prove: If $0\colon A \to B$ is monic, then A is a zero object, and conversely.
2. In the isomorphism φ above, show that $\varphi^{-1}(\gamma_1, \gamma_2) = \nabla_C(\gamma_1 \oplus \gamma_2)$.
3. For $\alpha, \beta\colon A \to B$ prove that $\alpha + \beta = \nabla_B(\alpha \oplus \beta)\Delta_A$.
4. Show that the direct sum of two short exact sequences is exact.
5. Construct an additive category of (some) abelian groups in which a morphism which is monic need not be a monomorphism. (Hint: Omit lots of subgroups.)
6. Construct an additive category of some abelian groups in which some of the morphisms do not have kernels or cokernels.
7. In the categories of sets, of modules, and of (not necessarily abelian) groups, show that a morphism is monic if and only if it is injective, and epic if and only if it is surjective (as a function on sets).

2. Abelian Categories

To use effectively the notions of kernel and cokernel just introduced, we need conditions to insure that these classes are not empty. Furthermore, each monomorphism should be the kernel of its cokernel, and conversely. For modules, the image of a homomorphism $\alpha: A \to B$ appears in its factorization $A \to \alpha A \to B$, with first factor $A \to \alpha A$ an epimorphism and second factor the injection $\alpha A \to B$, which is a monomorphism. Corresponding properties hold in other familiar categories: The category of all complexes of modules over a fixed ring, with morphisms the chain transformations; the category of all modules over a given graded algebra, with morphisms of degree zero; the category of all modules over a given DG-algebra. Thus an *abelian category* is to be an additive category $\mathscr{A}$ satisfying the following further axioms:

(Abel-1). For every morphism α of $\mathscr{A}$ there exists a $\varkappa \in \ker \alpha$ and a $\sigma \in \operatorname{coker} \alpha$.

(Abel-2). For $\varkappa$ monic and σ epic, $\varkappa \in \ker \sigma$ if and only if $\sigma \in \operatorname{coker} \varkappa$.

(Abel-3). Every morphism of $\mathscr{A}$ can be factored as $\alpha = \lambda \sigma$ with λ monic and σ epic.

(Abel-2) may be restated thus: σ epic and $\varkappa \in \ker \sigma$ imply $\varkappa \| \sigma$, and dually. The three axioms together are subsumed in

Theorem 2.1. *To each morphism α there exist morphisms $\varkappa$, σ, λ, τ forming the following diagram with the indicated properties*

$$\bullet \xrightarrow{\varkappa} \bullet \underset{\alpha}{\xrightarrow{\sigma} \bullet \xrightarrow{\lambda}} \bullet \xrightarrow{\tau} \bullet, \qquad \alpha = \lambda \sigma, \qquad \varkappa \| \sigma, \qquad \lambda \| \tau. \tag{2.1}$$

Here and below the *dots* designate unnamed objects.

Proof. By (Abel-3), write $\alpha = \lambda \sigma$; by (Abel-1), $\varkappa \in \ker \alpha = \ker \sigma$ and $\tau \in \operatorname{coker} \alpha = \operatorname{coker} \lambda$ exist; by (Abel-2), $\varkappa \| \sigma$ and $\lambda \| \tau$. The converse proof that this theorem implies the three axioms is left to the reader.

The diagram (2.1) is called an *analysis* of α, and $\alpha = \lambda \sigma$ is a *standard factorization* of α.

Proposition 2.2. *The analysis of a morphism $\alpha \in \mathscr{A}$ is a functor.*

Here we regard the analysis (2.1) of α as a functor on the category $\mathscr{M} = \operatorname{Morph}(\mathscr{A})$ of morphisms of $\mathscr{A}$; the objects of $\mathscr{M}$ are the morphisms $\alpha: A \to B$ of $\mathscr{A}$; the maps $\Xi: \alpha \to \alpha'$ of $\mathscr{M}$ are the pairs $\Xi = (\xi_1, \xi_2)$ of morphisms of $\mathscr{A}$ with $\alpha' \xi_1 = \xi_2 \alpha$. The values of the "analysis" functor lie in a similar category of diagrams from $\mathscr{A}$. As the analysis is not uniquely determined, we assert more exactly that any choice of analyses, one for each α, provides such a functor.

Thus, given $\Xi=(\xi_1,\xi_2)\colon\alpha\to\alpha'$ and analyses of α and α', we assert that there are unique morphisms η_1, η_2, and η_3 of $\mathscr{A}$ which render the diagram

$$\begin{array}{ccccccccccl} \bullet & \xrightarrow{\varkappa} & \bullet & \xrightarrow{\sigma} & \bullet & \xrightarrow{\lambda} & \bullet & \xrightarrow{\tau} & \bullet, & & \alpha=\lambda\sigma, \\ \downarrow{\scriptstyle\eta_1} & & \downarrow{\scriptstyle\xi_1} & & \downarrow{\scriptstyle\eta_2} & & \downarrow{\scriptstyle\xi_2} & & \downarrow{\scriptstyle\eta_3} & & \\ \bullet & \xrightarrow{\varkappa'} & \bullet & \xrightarrow{\sigma'} & \bullet & \xrightarrow{\lambda'} & \bullet & \xrightarrow{\tau'} & \bullet, & & \alpha'=\lambda'\sigma', \end{array} \tag{2.2}$$

commutative. (In ordinary parlance, η_1 is the map *induced* by ξ_1 on the kernels, etc.) Indeed, $\alpha'(\xi_1\varkappa)=\xi_2\alpha\varkappa=0$ implies that $\xi_1\varkappa$ factors through $\varkappa'\in\ker\alpha'$ as $\xi_1\varkappa=\varkappa'\eta_1$; since $\varkappa'$ is left cancellable, η_1 is unique. Dually, $\tau'\xi_2=\eta_3\tau$ for a unique η_3. Also $\lambda'\sigma'\xi_1\varkappa=\xi_2\alpha\varkappa=0$; since λ' is left cancellable, $\sigma'\xi_1\varkappa=0$ and $\sigma'\xi_1$ factors through $\sigma\in\operatorname{coker}\varkappa$ as $\sigma'\xi_1=\eta_2\sigma$, with η_2 unique. Then $\xi_2\lambda\sigma=\lambda'\sigma'\xi_1=\lambda'\eta_2\sigma$; cancelling σ, $\xi_2\lambda=\lambda'\eta_2$. This proves the diagram commutative and unique. Applied to $1\colon\alpha\to\alpha$, with two different analyses of α, this argument gives equivalences η_1, η_2, η_3; thus

Corollary 2.3. *An analysis* (2.1) *of* α *is uniquely determined up to equivalences of the three objects domain* $\varkappa$, *range* $\sigma=$ *domain* λ, *range* τ.

In the analysis (2.1) the unique right equivalence class of λ is the *image* of α and the unique left equivalence class of σ the *coimage* of α. The image of α is a subobject of the range of α, the coimage a quotient of the domain. An analysis of $\alpha\colon A\to B$ has the form of a commutative diagram

$$\begin{array}{ccccc} \bullet & \xrightarrow{\ker\alpha} & A & \xrightarrow{\operatorname{coim}\alpha} & \bullet \\ & & & \searrow^{\alpha} & \downarrow{\scriptstyle\operatorname{im}\alpha} \\ & & & & B \\ & & & & \downarrow{\scriptstyle\operatorname{coker}\alpha} \\ & & & & \bullet \end{array} \tag{2.3}$$

with row and column short exact sequences. Here "ker α" of course stands for any morphism in the class $\ker\alpha$. With the same convention we may read off the relations

$$(\ker\alpha)\,\|\,(\operatorname{coim}\alpha), \qquad (\operatorname{im}\alpha)\,\|\,(\operatorname{coker}\alpha), \tag{2.4}$$

$$\operatorname{coim}\alpha=\operatorname{coker}(\ker\alpha), \qquad \operatorname{im}\alpha=\ker(\operatorname{coker}\alpha), \tag{2.5}$$

$$\ker\alpha=\ker(\operatorname{coim}\alpha), \qquad \operatorname{coker}\alpha=\operatorname{coker}(\operatorname{im}\alpha). \tag{2.6}$$

Hence also $\ker(\operatorname{coker}(\ker\alpha))=\ker\alpha$, and dually.

Proposition 2.4. *A morphism* α *is monic if and only if* $\ker\alpha=0$, *epic if and only if* $\operatorname{coker}\alpha=0$, *and an equivalence if and only if both* $\ker\alpha$ *and* $\operatorname{coker}\alpha$ *are zero. In particular, a morphism which is both monic and epic is an equivalence.*

Here $\ker\alpha = 0$ is short for $0 \in \ker\alpha$; it means that every element of the class $\ker\alpha$ is a zero morphism.

Proof. The definition states that a monic α has only zeros as right annihilators, so that necessarily $\ker\alpha = 0$. Conversely, if $0 \in \ker\alpha$, then any right annihilator of α factors through zero, hence is zero, so that α is monic by definition. The proof for α epic is dual; both proofs use only the axioms of an additive category. Finally, an equivalence α is both monic and epic, so that $\ker\alpha = 0 = \operatorname{coker}\alpha$. Conversely, if $\ker\alpha = 0 = \operatorname{coker}\alpha$, then $1 \in \ker 0 = \ker(\operatorname{coker}\alpha) = \operatorname{im}\alpha$ by (2.5), so $\operatorname{im}\alpha$ is equivalent to 1, hence an equivalence. Dually, $\operatorname{coim}\alpha$ is an equivalence, and thus so is $\alpha = (\operatorname{im}\alpha)(\operatorname{coim}\alpha)$.

Exact sequences operate as usual and can be defined in two (dual) ways.

Proposition 2.5. *If the composite* $\beta\alpha$ *is defined, then* $\operatorname{im}\alpha = \ker\beta$ *if and only if* $\operatorname{coim}\beta = \operatorname{coker}\alpha$.

When this is the case, we say that (α, β) is *exact*. In particular, $\varkappa \| \sigma$ implies $(\varkappa, \sigma)$ exact.

Proof. If $\operatorname{im}\alpha = \ker\beta$, then $\operatorname{coim}\beta = \operatorname{coker}(\ker\beta) = \operatorname{coker}(\operatorname{im}\alpha) = \operatorname{coker}\alpha$ by (2.5) and (2.6), and dually.

Proposition 2.6. *The short five lemma holds in any abelian category.*

Proof. Given a commutative diagram

$$\begin{array}{ccccc} \bullet & \xrightarrow{\varkappa} & \bullet & \xrightarrow{\sigma} & \bullet \\ \downarrow{\scriptstyle\alpha} & & \downarrow{\scriptstyle\beta} & & \downarrow{\scriptstyle\gamma} \\ \bullet & \xrightarrow{\varkappa'} & \bullet & \xrightarrow{\sigma'} & \bullet \end{array}$$

with $\varkappa \| \sigma$ and $\varkappa' \| \sigma'$, we wish to prove that α, γ monic imply β monic, and dually. But take $\mu \in \ker\beta$. Then $\beta\mu = 0$ gives $0 = \sigma'\beta\mu = \gamma\sigma\mu$; since γ is monic, $\sigma\mu = 0$. This implies that μ factors through $\varkappa \in \ker\sigma$ as $\mu = \varkappa\nu$, for a ν which is necessarily monic. Then $\varkappa'\alpha\nu = \beta\varkappa\nu = \beta\mu = 0$. But $\varkappa'$ and α are monic, so $\nu = 0$ and thus $\ker\beta = \mu = \varkappa\nu = 0$, so β is monic, as desired.

The Five Lemma, the Four Lemma, and the 3×3 Lemma also hold in an abelian category. The proofs, which depend on certain additional techniques, will be given in Chap. XII.

Call an abelian category *selective* if

(Select 1). There is a function assigning to each pair of objects A_1, A_2, a direct sum diagram, of the form specified in (1.4).

(Select 2). There is a function selecting a unique representative $\varkappa$ for each subobject and a unique representative σ for each quotient object.

In a selective abelian category $\mathscr{A}$ we can assign an object K as Kernel for each morphism α: Take K to be the domain of the selected representative $\varkappa: K \to A$ of the right equivalence class $\ker \alpha$. (Observe that now "Kernel" in capitals means an object, in lower case, a morphism.) Similarly, we can assign Cokernels and form Quotients of subobjects: In this regard we operate as if we were in the category of all R-modules. The various cited examples of abelian categories are all selective; by the axiom of choice, any small abelian category is selective.

Note on Terminology. The possibility of doing homological algebra abstractly in a suitable category was first demonstrated by MacLane [1950], working in an "abelian bicategory" which was substantially an abelian category with a canonical selection of representatives of subobjects and quotient objects. This canonical selection proved cumbersome and was dropped at the price of the present arrangement in which a subobject is not an object. The formulation of Buchsbaum [1955] uses exact categories, which are our abelian categories minus the direct sum axiom, while Grothendieck's extensive study [1957] introduced the term "abelian category" in the sense used here. Other authors have used "abelian category" in other meanings. Atiyah [1956] established the Krull-Schmidt theorem stating the uniqueness of a direct sum decomposition into indecomposable objects for an abelian category satisfying a "bichain" condition. Set-theoretical questions about abelian categories are considered in MacLane [1961 b]. Various other types of categories may be constructed by imposing additional structure on the sets hom(A, B). Thus a graded category (XII.4 below) has each hom(A, B) a graded group; a differential category (Eilenberg-Moore, unpublished) has each hom(A, B) a positive complex of K-modules. One might wish categories with a tensor product functor satisfying suitable axioms, as in our treatment (Chap. VI) of types of algebras. Noetherian Categories have been studied by Gabriel [1962].

Exercises

1. Given (Abel-2) and (Abel-3), show that (Abel-1) may be replaced by the weaker statement that each epic has a kernel and each monic a cokernel.

2. In (2.2), show ξ_1 monic implies η_1 monic and ξ_2 monic implies η_2 monic.

3. Categories of Diagrams

Let $\mathscr{A}$ be an additive category and $\mathscr{C}$ a category which is small (i.e., the class of objects in $\mathscr{C}$ is a set). By Dgram$(\mathscr{C}, \mathscr{A})$ we denote the category with objects the covariant functors $T: \mathscr{C} \to \mathscr{A}$ and morphisms the natural transformations $f: T \to S$ of functors. The sum of two natural transformations f and g: $T \to S$ is defined for each object $C \in \mathscr{C}$ by $(f+g)(C) = f(C) + g(C)$. The axioms for an additive category hold in Dgram$(\mathscr{C}, \mathscr{A})$; in particular, the direct sum of two diagrams T_1 and T_2 is $(T_1 \oplus T_2)(C) = T_1(C) \oplus T_2(C)$: Take the direct sum at each vertex. Here, as in I.8, we can regard each $T: \mathscr{C} \to \mathscr{A}$ as a "diagram" in $\mathscr{A}$ with "pattern" $\mathscr{C}$. For example, if $\mathscr{C}_0$ is the category with two objects

C, C' and three morphisms 1_C, $1_{C'}$, and $\gamma: C \to C'$, then each $T: \mathscr{C}_0 \to \mathscr{A}$ is determined by a morphism $T(\gamma)$ in $\mathscr{A}$, so that $\mathrm{Dgram}(\mathscr{C}_0, \mathscr{A})$ is the category $\mathrm{Morph}(\mathscr{A})$ of § 2, with objects the morphisms of $\mathscr{A}$.

Proposition 3.1. (Grothendieck [1957].) *If the category $\mathscr{C}$ is small and $\mathscr{A}$ is abelian, $\mathscr{D} = \mathrm{Dgram}(\mathscr{C}, \mathscr{A})$ is an abelian category. If f and g are morphisms of $\mathscr{D}$, $f \| g$ in $\mathscr{D}$ if and only if, for each C, $f(C) \| g(C)$ in $\mathscr{A}$.*

Proof. Let $f: T \to S$ be natural. Since $\mathscr{C}$ is small, we can choose for each object C a monic $k(C) \in \ker f(C)$, with domain, say, $K(C)$. Thus $k(C): K(C) \to T(C)$ is a morphism of $\mathscr{A}$. Since f is natural, each $\gamma: C \to C'$ gives a commutative diagram

$$\begin{array}{ccccccc} 0 \to & K(C) & \xrightarrow{k(C)} & T(C) & \xrightarrow{f(C)} & S(C) \\ & \vdots & & \downarrow T(\gamma) & & \downarrow S(\gamma) \\ 0 \to & K(C') & \xrightarrow{k(C')} & T(C') & \xrightarrow{f(C')} & S(C') \end{array}$$

with exact rows. Since $f(C')[T(\gamma)\, k(C)] = 0$, and $k(C')$ is the kernel of $f(C')$, there is a unique $K(\gamma)$ (dotted arrow) with $T(\gamma)\, k(C) = k(C')\, K(\gamma)$. It follows that $K: \mathscr{C} \to \mathscr{A}$ with mapping function $K(\gamma)$ is a functor and $k: K \to T$ natural. As a morphism of $\mathscr{D}$, k is monic, for if $kh = 0$, then $(kh)(C) = k(C)\, h(C) = 0$ for each object C; since $k(C)$ is monic in $\mathscr{A}$, $h(C) = 0$. Furthermore, if $g: R \to T$ is natural with $fg = 0$, each $g(C)$ factors uniquely through $k(C)$ as $g(C) = k(C)\, h(C)$, $h: R \to K$ is natural, and $g = kh$. Therefore $k \in \ker_{\mathscr{D}} f$. This argument with its dual proves (Abel-1) in $\mathscr{D}$ and also gives

$$f \text{ monic in } \mathscr{D} \Leftrightarrow \text{each } f(C) \text{ monic in } \mathscr{A},$$

$$k \in \ker_{\mathscr{D}} f \Leftrightarrow \text{each } k(C) \in \ker_{\mathscr{A}} f(C).$$

These statements with their duals prove (Abel-2).

To get a standard factorization (Abel-3) for $f: T \to S$, choose for each C a standard factorization $f(C) = l(C)\, t(C)$; the range $R(C)$ of $t(C)$ yields a functor $R: \mathscr{C} \to \mathscr{A}$, $t: T \to R$ is epic and $l: R \to S$ monic in $\mathscr{D}$, and $f = lt$. Since $\mathscr{C}$ is small, we can also select for each T a *set* of representatives of the subobjects of T and for each S and T a *set* of representatives for the extensions of S by T, thus proving that $\mathscr{D}$ satisfies the supplementary set-theoretical axioms (§ 1) for an additive category.

Next consider the diagrams which involve zero objects. In any category $\mathscr{C}$ call an object N a *null* object if for each object C of $\mathscr{C}$ there is exactly one morphism $C \to N$ and exactly one morphism $N \to C$; write $0_C: C \to N$ and $0^C: N \to C$ for these morphisms. Any two null objects in $\mathscr{C}$ are equivalent, and any object equivalent to a null object is null.

For given objects C and D, the composite morphism $0^D 0_C$: $C \to N \to D$ is independent of the choice of the intermediate null object N; it may be called the ***null morphism*** 0_C^D: $C \to D$. A (new) null object may be adjoined to any category. In an additive category the null objects are exactly the zero objects, and the null morphism 0: $C \to D$ is the zero of hom(C, D).

If $\mathscr{C}$ and $\mathscr{A}$ have null objects, a ***normalized*** functor T: $\mathscr{C} \to \mathscr{A}$ is one with $T(N)$ null for some (and hence for any) null N of $\mathscr{C}$. It follows that T maps null morphisms to null morphisms. $\text{Dgram}_N(\mathscr{C}, \mathscr{A})$ will denote the category of all such T. Prop. 3.1 again applies.

An example is the category of complexes. To get this, take $\mathscr{C}$ to be the following small category:

$$N; \quad \cdots \leftarrow -2 \xleftarrow{\partial_{-1}} -1 \xleftarrow{\partial_0} 0 \xleftarrow{\partial_1} 1 \leftarrow \cdots;$$

the objects of $\mathscr{C}$ are all integers n plus a null object N; the morphisms are all identities, the null morphisms $n \to N$, $N \to n$, and $n \to m$, and morphisms ∂_n: $n \to (n-1)$. The composite of morphisms is defined by requiring $\partial_{n-1}\partial_n$ to be null. Take any abelian category $\mathscr{A}$. A normalized covariant functor T: $\mathscr{C} \to \mathscr{A}$ is then given by $\cdots \leftarrow T_{n-1} \leftarrow T_n \leftarrow \cdots$, a sequence of objects and morphisms of $\mathscr{A}$, with $\partial_{n-1}\partial_n = 0$, so is just a chain complex of objects from $\mathscr{A}$ (in brief, an $\mathscr{A}$-complex). A natural transformation f: $T \to S$ is a chain transformation. Therefore $\text{Dgram}_N(\mathscr{C}, \mathscr{A})$ is the category of all $\mathscr{A}$-complexes; by Prop. 3.1 it is an abelian category. If $\mathscr{A}$ is selective, the homology objects $H_n(T) = \text{Ker}\,\partial_n/\text{Im}\,\partial_{n+1}$ may be defined as usual; the reader should show that each f: $T \to S$ induces f_*: $H_n(T) \to H_n(S)$, so that H_n is a covariant functor on this category, and that homotopies have the usual properties.

Exercises

1. If $\mathscr{A}$ is abelian, show that the category of graded objects of $\mathscr{A}$ is abelian.

2. Describe an abelian category whose objects include the analyses (2.1).

3. Show that the category of positive complexes of objects from an abelian category $\mathscr{A}$ is abelian.

4. (Mac Lane [1950].) In a category $\mathscr{C}$ with null, assume that to each pair of objects A_1, A_2 there is a diagram $A_1 \rightleftarrows B \leftrightarrows A_2$ in which the two morphisms with range B are universal and the two with domain B couniversal. In each hom(A, C) introduce a binary operation of addition as in Ex. 1.3 and show this addition commutative, associative, and distributive.

5. In Ex. 4, assume also that there is a natural transformation V_A: $A \to A$ with $\nabla_A (V_A \oplus 1_A) \Delta_A = 0$ for all A. Use V to define $-\alpha$ for each morphism α, and prove that $\mathscr{C}$ becomes an additive category.

4. Comparison of Allowable Resolutions

If Λ is an algebra over a fixed commutative ground ring K, many concepts are appropriately taken "relative to K". Each left Λ-module A is also a K-module and each Λ-module homomorphism $\alpha: A \to B$ is also a homomorphism of K-modules, but not conversely. Call such a homomorphism α "allowable" relative to (Λ, K) if there is a K-module homomorphism $t: B \dashrightarrow A$ (backwards!) with $\alpha t \alpha = \alpha$. In particular, a monomorphism α is allowable if there is a t with $t\alpha = 1_A$; that is, if α has a left inverse t which is a K-module homomorphism but not necessarily a homomorphism of Λ-modules. Similarly, a Λ-module epimorphism σ is allowable if and only if it has a K-module right inverse t. Hence a short exact sequence $\varkappa \| \sigma$ of Λ-module homomorphisms is allowable if $\varkappa$ has a left K-inverse and σ has a right K-inverse. These properties state that the sequence $(\varkappa, \sigma)$ becomes a direct sum sequence when regarded just as a sequence of K-modules. More briefly, the sequence of Λ-modules is K-*split* (for some authors, *weakly split*). The use of such a class of "K-split" or "allowable" short exact sequences is typical of relative homological algebra. We shall now show how the comparison theorem for resolutions applies to any such situation.

In any abelian category $\mathscr{A}$, a class $\mathscr{E}$ of short exact sequences of $\mathscr{A}$ will be called *allowable* if $\mathscr{E}$ contains, with any one short exact sequence $(\varkappa, \sigma)$, all isomorphic short exact sequences of $\mathscr{A}$ and if also $\mathscr{E}$ contains the short exact sequences $(0, 1_A)$ and $(1_A, 0)$ for any object A in an allowable s.e.s. Write $\varkappa \mathscr{E} \sigma$ if $(\varkappa, \sigma)$ is one of the short exact sequences of $\mathscr{E}$ and call $(\varkappa, \sigma)$ $\mathscr{E}$-allowable. Call a monic $\varkappa$ of $\mathscr{A}$ *allowable* and write $\varkappa \in \mathscr{E}_m$ if $\varkappa \mathscr{E} \sigma$ for some σ; this is the case if and only if $\varkappa \mathscr{E} (\operatorname{coker} \varkappa)$. Dually, call an epic σ *allowable* and write $\sigma \in \mathscr{E}_e$ if and only if $(\ker \sigma)\, \mathscr{E}\, \sigma$. Since $\varkappa \mathscr{E} \sigma$ if and only if $\varkappa \in \mathscr{E}_m$ and $\varkappa \| \sigma$, the class $\mathscr{E}$ is determined by the class $\mathscr{E}_m$ of allowable monics, or by the class $\mathscr{E}_e$ of allowable epics. Thus $\mathscr{E}_e$ determines $\mathscr{E}_m$; for $\varkappa$ monic, $\varkappa \in \mathscr{E}_m$ if and only if $\operatorname{coker} \varkappa \in \mathscr{E}_e$. If $\varkappa \in \mathscr{E}_m$, any left or right equivalent of $\varkappa$ is also in $\mathscr{E}_m$.

From the properties of an analysis of α we derive at once

Proposition 4.1. *For a given allowable class $\mathscr{E}$, the following conditions on a morphism α are equivalent:*

(i) $\operatorname{im} \alpha \in \mathscr{E}_m$ *and* $\operatorname{coim} \alpha \in \mathscr{E}_e$;

(ii) $\ker \alpha \in \mathscr{E}_m$ *and* $\operatorname{coker} \alpha \in \mathscr{E}_e$;

(iii) *In a standard factorization $\alpha = \lambda \sigma$, $\lambda \in \mathscr{E}_m$ and $\sigma \in \mathscr{E}_e$;*

(iv) *Each analysis of α consists of allowable monics and epics.*

The morphism $\alpha: A \to B$ is called *allowable* when it satisfies these conditions. If α happens to be monic, then $\operatorname{coim} \alpha = 1_A$, so α is allowable and monic if and only if $\alpha \in \mathscr{E}_m$. Likewise, the allowable morphisms which are epic are the elements of $\mathscr{E}_e$. The composite of allowable morphisms need not be allowable.

For example, in the category of all left Λ-modules, the K-split short exact sequences form an allowable class, and the corresponding allowable morphisms can be shown to be those α with $\alpha t \alpha = \alpha$ for some t, as above. Additional properties which hold in this case will be studied in Chap. XII.

Let $\mathscr{E}$ be any allowable class in $\mathscr{A}$. An $\mathscr{E}$-projective object P (or, an *allowable projective object*) is any object P of $\mathscr{A}$ such that, for every allowable epic $\sigma: B \to C$, each morphism $\gamma: P \to C$ of $\mathscr{A}$ can be factored through σ as $\gamma = \sigma \gamma'$ for some $\gamma': P \to B$. As before, this condition can be formulated in several equivalent ways:

Proposition 4.2. *For a given allowable class $\mathscr{E}$ of short exact sequences the following conditions on an object P are equivalent:*

(i) *P is an allowable projective;*

(ii) *Each $\sigma: B \to C$ in $\mathscr{E}_e$ induces an epimorphism* $\hom(P, B) \twoheadrightarrow \hom(P, C)$;

(iii) *For each allowable short exact sequence $A \rightarrowtail B \twoheadrightarrow C$ the induced sequence* $\hom(P, A) \rightarrowtail \hom(P, B) \twoheadrightarrow \hom(P, C)$ *of abelian groups is short exact.*

We say that there are *enough allowable projectives* if to each object C of $\mathscr{A}$ there is at least one morphism $\varrho: P \to C$ which is an allowable epic with an allowable projective domain P. The dual notion is that there are *enough allowable injectives.*

Any long exact sequence in an abelian category can be written as a Yoneda composite of short exact sequences; we call the long exact sequence allowable if and only if each of these short exact sequences is allowable.

Consider a complex $\cdots \to X_n \to \cdots \to X_1 \to X_0 \to C \to 0$ over an object C of $\mathscr{A}$. Call it an *allowable resolution* if it is an allowable long exact sequence, and an *allowable projective complex* over C if each X_n is an allowable projective. If both conditions hold, it is an *allowable projective resolution* of C.

Theorem 4.3. *(Comparison Theorem.) Let $\mathscr{E}$ be an allowable class of short exact sequences in the abelian category $\mathscr{A}$. If $\gamma: C \to C'$ is a morphism of $\mathscr{A}$, $\varepsilon: X \to C$ an allowable projective complex over C and $\varepsilon': X' \to C'$ an allowable resolution of C', then there is a chain transformation $f: X \to X'$ of morphisms of $\mathscr{A}$ with $\varepsilon' f = \gamma \varepsilon$. Any two such chain transformations are chain homotopic.*

The proof is substantially a repetition of the previous argument for the case of modules (Thm. III.6.1). Since X_0 is an allowable projective and ε': $X_0' \to C'$ an allowable epic, $\gamma\varepsilon$: $X_0 \to C$ factors as $\varepsilon' f_0$ for some f_0. We next wish to construct f_1 so that the diagram

$$\begin{array}{ccccc} X_1 & \xrightarrow{\partial} & X_0 & \xrightarrow{\varepsilon} & C \\ \downarrow f_1 & & \downarrow f_0 & & \downarrow \gamma \\ X_1' & \xrightarrow[\sigma \quad \lambda]{\partial'} & X_0' & \xrightarrow{\varepsilon'} & C' \to 0 \end{array}$$

will be commutative. Take a standard factorization $\partial' = \lambda\sigma$ as displayed; since X' is allowable, $\lambda \mathscr{E} \varepsilon'$. But $\varepsilon' f_0 \partial = \gamma\varepsilon\partial = 0$, so $f_0\partial$ factors through $\lambda \in \ker \varepsilon'$ as $f_0\partial = \lambda\beta$ for some β. Now σ is an allowable epic and X_1 an allowable projective, so $\beta = \sigma f_1$ for some f_1, and $\partial' f_1 = \lambda\sigma f_1 = \lambda\beta = f_0\partial$, as desired. The construction of $f_2, f_3, \ldots$ and of the homotopy is similar.

Note on injective envelopes. A family $\{a_t\}$ of subobjects of A is *directed* by inclusion if each pair of subobjects a_s, a_t of the family is contained in a third subobject of the family (in the obvious sense of "contained", as defined in Chap. XII below). An abelian category $\mathscr{A}$ satisfies GROTHENDIECK's axiom AB-5 if for each A, each subobject b, and each family a_t directed by inclusion, $b \cap (\bigcup_t a_t) = \bigcup_t (b \cap a_t)$ holds in the lattice of subobjects of $\mathscr{A}$ and if $\mathscr{A}$ has infinite direct sums. An object U is a *generator* of $\mathscr{A}$ if to each non-zero morphism α: $A \to B$ there is a morphism ξ: $U \to A$ with $\alpha\xi \neq 0$. Both conditions hold in the category of all Λ-modules, with Λ a generator. GROTHENDIECK [1957, Thm. 1.10.1] shows that an abelian category with AB-5 and a generator has enough injectives; MITCHELL [1962] constructs the ECKMANN-SCHOPF injective envelope under these hypotheses. In particular this shows that there are enough injectives in the category of sheaves over a fixed topological space (though in this case there are not enough projectives): See GROTHENDIECK [1957], GODEMENT [1958].

Exercise

1. (Characterization of allowable short exact sequences by allowable projectives [HELLER 1958].) If $\mathscr{E}$ is an allowable class of short exact sequences satisfying the condition $\alpha\beta \in \mathscr{E}_e \Rightarrow \alpha \in \mathscr{E}_e$, and if there are enough allowable projectives, show that an epic σ: $B \to C$ is allowable if and only if $\hom(P, B) \to \hom(P, C)$ is an epimorphism for all allowable projectives P.

5. Relative Abelian Categories

Let S be a subring of R with the same identity as R. Some short exact sequences of R-modules will split when regarded as sequences of S-modules. Each R-module A is also an S-module ${}_\iota A$, by pull-back along the injection ι: $S \to R$, and a function which is an R-module homomorphism α: $A \to B$ is also an S-module homomorphism ${}_\iota\alpha$: ${}_\iota A \to {}_\iota B$. Thus $\square(A) = {}_\iota A$, $\square(\alpha) = {}_\iota\alpha$ is a functor $\square$ on the category $\mathscr{A}$ of all left R-modules to the category $\mathscr{M}$ of all left S-modules; it "forgets" or "neglects" part of the structure of an R-module. We have in mind many other examples, such as modules over an algebra Λ and modules over the ground ring K, as explained in the introduction of § 4. In each example there will be a similar functor $\square$. Let us state the appropriate general properties of such a functor.

A *relative abelian category* $\square$ will mean a pair of selective abelian categories $\mathscr{A}$ and $\mathscr{M}$ together with a covariant functor $\square\colon \mathscr{A}\to\mathscr{M}$ (write $\square A = A_\square$, $\square\alpha=\alpha_\square$) which is additive, exact, and faithful.

Additive means that $\alpha, \beta\in\hom_{\mathscr{A}}(A, B)$ implies $(\alpha+\beta)_\square=\alpha_\square+\beta_\square$ in $\hom_{\mathscr{M}}(A_\square, B_\square)$. It follows that $0_\square=0$ and that $(A\oplus B)_\square\cong A_\square\oplus B_\square$.

Exact means that $\alpha\|\beta$ in $\mathscr{A}$ implies $\alpha_\square\|\beta_\square$ in $\mathscr{M}$. It follows that $\varkappa$ monic and σ epic in $\mathscr{A}$ imply $\varkappa_\square$ monic and $\sigma_\square$ epic in $\mathscr{M}$, that $\square$ carries each analysis (2.1) of α in $\mathscr{A}$ into an analysis of $\alpha_\square$ and hence that $\varkappa\in\ker\beta$ implies $\varkappa_\square\in\ker\beta_\square$, and similarly for coker, im, and coim. Moreover, $\square$ carries exact sequences into exact sequences.

Faithful means that $\alpha_\square=0$ implies $\alpha=0$. It follows that $A_\square=0$ implies $A=0'$, but $A_\square=B_\square$ need not imply $A=B$. However $\alpha_\square$ epic (or monic) in $\mathscr{M}$ implies α epic (or monic, respectively) in $\mathscr{A}$.

Write objects of $\mathscr{A}$ as $A, B, C, \ldots$ and morphisms $\alpha\colon A\to B$ in Greek letters with solid arrows. Write objects of $\mathscr{M}$ as $L, M, N, \ldots$ and morphisms $t\colon L \dashrightarrow M$ in lower case Latin letters with broken arrows.

A short exact sequence $\varkappa\|\sigma$ in $\mathscr{A}$ is said to be *relatively split* (or, $\square$-split) if the exact sequence $\varkappa_\square\|\sigma_\square$ splits in $\mathscr{M}$; that is, if $\sigma_\square$ has a right inverse k or (equivalently) $\varkappa_\square$ has a left inverse t in $\mathscr{M}$. This gives two diagrams

$$A\overset{\varkappa}{\rightarrowtail}B\overset{\sigma}{\twoheadrightarrow}C,\qquad A_\square\underset{t}{\overset{\varkappa_\square}{\rightleftarrows}}B_\square\underset{k}{\overset{\sigma_\square}{\rightleftarrows}}C_\square \tag{5.1}$$

the first an exact sequence in $\mathscr{A}$, the second a direct sum diagram in $\mathscr{M}$. For simplicity, we often replace these by a single *schematic diagram*

$$A\underset{t}{\overset{\varkappa}{\rightleftarrows}}B\underset{\sigma}{\overset{k}{\leftrightarrows}}C \tag{5.2}$$

(solid arrows for the $\mathscr{A}$ part, solid and broken arrows for $\mathscr{M}$). Similarly the equations

$$t\varkappa_\square=1_{A_\square},\quad \sigma\varkappa=0,\quad \varkappa_\square t+k\sigma_\square=1_{B_\square},\quad \ldots,$$

valid in the direct sum diagram (5.1), will be written schematically as

$$t\varkappa=1_A,\quad \sigma\varkappa=0,\quad \varkappa t+k\sigma=1_B,\quad \ldots,$$

without the $\square$, so that a composite $t\varkappa$ is short for $t\varkappa_\square$ in $\mathscr{M}$.

The class of $\square$-split short exact sequences of $\mathscr{A}$ is allowable in the sense of § 4; the conditions of Prop. 4.1 then describe certain morphisms α of $\mathscr{A}$ as allowable (say, $\square$-*allowable*). In detail,

Proposition 5.1. *A morphism $\alpha: A \to B$ with standard factorization $\alpha = \lambda\sigma$ is $\square$-allowable in the relative abelian category $\square$ if and only if it satisfies any one of the following equivalent conditions:*

(i) *$\lambda_\square$ has a left inverse and $\sigma_\square$ a right inverse in $\mathcal{M}$;*

(ii) *$(\operatorname{im}\alpha)_\square$ and $(\ker\alpha)_\square$ have left inverses in $\mathcal{M}$;*

(iii) *There is a morphism $u: B \dashrightarrow A$ in $\mathcal{M}$ with $\alpha_\square u \alpha_\square = \alpha_\square$;*

(iv) *There is a morphism $v: B \dashrightarrow A$ in $\mathcal{M}$ with both*

$$\alpha_\square v \alpha_\square = \alpha_\square, \quad v \alpha_\square v = v.$$

Condition (ii) may be read: The image of α is an $\mathcal{M}$-direct summand of $B_\square$ and the kernel of α an $\mathcal{M}$-direct summand of $A_\square$ — or dually.

Proof. The equivalence of (i), (ii), and the allowability of α is immediate, by Prop. 4.1. If $t\lambda = 1$ and $\sigma k = 1$, as in

$$A \underset{k}{\overset{\sigma}{\rightleftarrows}} \bullet \underset{t}{\overset{\lambda}{\rightleftarrows}} B, \quad \alpha = \lambda\sigma,$$

then $v = kt: B \dashrightarrow A$ has $\alpha v \alpha = \lambda\sigma k t \lambda\sigma = \lambda\sigma = \alpha$ and $v\alpha v = kt\lambda\sigma kt = kt = v$; this proves (i) $\Rightarrow$ (iv). Trivially, (iv) $\Rightarrow$ (iii). Finally, to get (iii) $\Rightarrow$ (i), assume $\alpha u \alpha = \alpha$ and set $\alpha = \lambda\sigma$. Thus $\lambda\sigma u \lambda\sigma = \lambda\sigma$ with λ monic (in $\mathcal{M}$!) and σ epic implies $\sigma u \lambda = 1$, so λ has a left inverse σu in $\mathcal{M}$ and σ a right inverse $u\lambda$.

If X is an $\mathcal{A}$-complex in the sense of § 3 (X_n objects and ∂_n morphisms in $\mathcal{A}$) then $\square X$ is an $\mathcal{M}$-complex; since $\square$ is exact, it follows that $\square[H_n(X)] \cong H_n(\square X)$.

Theorem 5.2. *If X is an $\mathcal{A}$-complex (not necessarily positive) then $\square X$ has a contracting homotopy s with $\partial s + s\partial = 1$ (and each $s_n: X_n \dashrightarrow X_{n+1}$ a morphism of $\mathcal{M}$) if and only if all $H_n(X)$ vanish and all boundary homomorphisms ∂ are allowable. When these conditions hold, s may be chosen so that $s^2 = 0$.*

Proof. Given s, we know that all $\square H_n(X) \cong H_n(\square X) = 0$. But $\square$ is faithful, so $H_n(X) = 0$. Moreover, $\partial = \partial s \partial + s \partial \partial = \partial s \partial$, so each ∂ is allowable by part (iii) of the preceding proposition.

Conversely, suppose the sequence $\cdots \to X_{n+1} \to X_n \to X_{n-1} \to \cdots$ exact and all ∂ allowable. Take a standard factorization $\partial = \lambda\sigma$ for each ∂_n. Then X factors into $\square$-split short exact sequences $D_n \rightarrowtail X_n \twoheadrightarrow D_{n-1}$, and each X_n is an $\mathcal{M}$-direct sum via morphisms $t = t_n$, $k = k_n$, as in the schematic diagram

$$X_{n+1} \underset{\sigma}{\overset{k}{\leftrightarrows}} D_n \underset{t}{\overset{\lambda}{\rightleftarrows}} X_n \underset{\sigma}{\overset{k}{\leftrightarrows}} D_{n-1} \underset{t}{\overset{\lambda}{\rightleftarrows}} X_{n-1}$$

with the usual direct sum identities

$$1_{X_n} = \lambda t + k\sigma, \quad t\lambda = 1, \quad \sigma k = 1, \quad tk = 0, \quad \sigma\lambda = 0$$

in $\mathscr{M}$. Now define each s_n: $X_n \dashrightarrow X_{n+1}$ as $s_n = kt$, so that $s^2 = 0$ and

$$\partial s + s\partial = \lambda(\sigma k)t + k(t\lambda)\sigma = \lambda t + k\sigma = 1_{X_n}.$$

Now a complex $\varepsilon: X \to C$ is allowable if both $\varepsilon: X_0 \to C$ and also each $\partial_n: X_n \to X_{n-1}$ are allowable, and a resolution if $\varepsilon: H_0(X) \cong C$ and $H_n(X) = 0$ for $n > 0$.

Corollary 5.3. *A complex* $\varepsilon: X \to C$ *in* $\mathscr{A}$ *over* C *is a* $\square$*-allowable resolution of* C *if and only if the complex* $\varepsilon_\square: X_\square \dashrightarrow C_\square$ *in* $\mathscr{M}$ *over* $C_\square$ *has a contracting homotopy. When this is the case, there is such a homotopy* s *with* $s^2 = 0$.

As usual, s consists of morphisms $s_{-1}: C \dashrightarrow X_0$, $s_n: X_n \dashrightarrow X_{n+1}$ in $\mathscr{M}$ with

$$\varepsilon s_{-1} = 1_C, \quad \partial s_0 + s_{-1}\varepsilon = 1_{X_0}, \quad \partial s + s\partial = 1_{X_n}, \quad n > 0.$$

The condition $s^2 = 0$ means $s_n s_{n-1} = 0$ for all $n = 0, 1, \ldots$. The proof is immediate.

A $\square$-allowable projective object P in $\mathscr{A}$ will also be called a *relative projective* object for $\square$.

Any projective object P in $\mathscr{A}$ is *a fortiori* relatively projective, but this does not show that there are enough relative projectives: If we write an object as the image $P \to A$ of a projective, we do not know $P \to A$ to be an allowable epic.

Exercises

(The first three exercises deal with the absolute case $\mathscr{A} = \mathscr{M}$.)

1. A complex X in an abelian category $\mathscr{A}$ has a contracting homotopy s if and only if $(\operatorname{im} \partial_{n+1}, \operatorname{coim} \partial_n)$: $\bullet \to X_n \to \bullet$ is a direct sum representation of each X_n. When these conditions hold, there is an s with $s^2 = 0$.

2. A complex X of modules has a contracting homotopy if and only if, for each n, the module of n-cycles is a direct summand of X_n.

3. A (not necessarily positive) complex X of free abelian groups has homomorphisms $s: X_n \to X_{n+1}$ with $\partial s + s\partial = 1$ if and only if all $H_n(X)$ vanish.

4. Deduce Thm. 5.2 from the result of Ex. 1.

6. Relative Resolutions

To construct enough relative projectives, we further specialize our relative abelian categories. By a *resolvent pair* $\mathscr{R}$ of categories we mean a relative abelian category $\square: \mathscr{A} \to \mathscr{M}$ together with

(i) A covariant functor $F: \mathscr{M} \to \mathscr{A}$.

(ii) A natural transformation $e: I_{\mathscr{M}} \dashrightarrow \square F$, for $I_{\mathscr{M}}$ the identity functor, such that every morphism $u: M \dashrightarrow A_\square$ in $\mathscr{M}$ has a factorization $u = \alpha_\square e_M$ for a unique morphism $\alpha: F(M) \to A$ of $\mathscr{A}$.

Thus each M determines FM in $\mathscr{A}$ and a morphism $e_M: M \dashrightarrow \Box FM$, and each u lifts uniquely to FM, as in the schematic diagram

$$\begin{array}{ccc} FM & \dashrightarrow & A \\ {\scriptstyle e_M}\uparrow & \nearrow{\scriptstyle u} & \\ M & & \end{array} ;$$

in other words, FM is the "relatively free" object in $\mathscr{A}$ to the given object M in $\mathscr{M}$. The lifting property states that $e^*\alpha = \alpha_\Box e$ defines a natural isomorphism

$$e^*: \hom_{\mathscr{A}}(FM, A) \cong \hom_{\mathscr{M}}(M, \Box A);$$

this last property states that the functor $F: \mathscr{M} \to \mathscr{A}$ is a *left adjoint* [KAN 1958] of the functor $\Box: \mathscr{A} \to \mathscr{M}$ (see note below).

Conversely, the conditions (i) and (ii) for a resolvent pair may be replaced by the requirement that the functor $\Box$ has a left adjoint F. Indeed, this requirement means that there is a natural isomorphism

$$\varphi: \hom_{\mathscr{A}}(FM, A) \cong \hom_{\mathscr{M}}(M, \Box A)$$

(of abelian groups). Take $A = FM$ in this isomorphism; then 1_{FM} in the group on the left gives $\varphi(1_{FM}) = e_M: M \dashrightarrow \Box FM$. That $e: I_{\mathscr{M}} \dashrightarrow \Box F$ is a natural transformation follows by taking any $\mu: M \to M'$ and applying φ to the diagram

$$\hom_{\mathscr{A}}(FM, FM) \xrightarrow{\mu_*} \hom_{\mathscr{A}}(FM, FM') \xleftarrow{\mu^*} \hom_{\mathscr{A}}(FM', FM').$$

Next take any A and any $\alpha: FM \to A$. Since φ is natural, the diagram

$$\begin{array}{ccc} \hom_{\mathscr{A}}(FM, FM) & \xrightarrow{\varphi} & \hom_{\mathscr{M}}(M, \Box FM) \\ \downarrow{\scriptstyle \alpha_*} & & \downarrow{\scriptstyle \alpha_*} \\ \hom_{\mathscr{A}}(FM, A) & \xrightarrow{\varphi} & \hom_{\mathscr{M}}(M, \Box A) \end{array}$$

commutes. Take 1_{FM} in the group at the upper left; it goes to α below and to e_M at the right, so commutativity gives $\varphi\alpha = \alpha_\Box e_M$. Since φ is an isomorphism, this proves that each $u: M \dashrightarrow \Box A$ in the group at the lower right has the form $u = \alpha_\Box e_M$ for a unique α, as required in our previous condition (ii).

For example, two rings $R \supset S$ yield a resolvent pair, denoted $\mathscr{R}(R, S)$ or just (R, S), with $\mathscr{A}$ and $\mathscr{M}$ the categories of R- and S-modules, respectively, $\Box$ the usual "neglect" functor, and

$$F(M) = R \otimes_S M, \quad e_M(m) = 1 \otimes m \in F(M).$$

Again, for each K-algebra Λ there is a resolvent pair with $\mathscr{A}$ the left Λ-modules, $\mathscr{M}$ the K-modules, $F(M) = \Lambda \otimes_K M$ (Prop. VI.8.2). Other examples of resolvent pairs appear in Ex. 2 below.

Theorem 6.1. *In a resolvent pair of categories, each $F(M)$ is relatively projective in $\mathscr{A}$. For each object A, the factorization $1_A = \alpha\, e_{A_\square}$ yields a $\square$-allowable epimorphism $\alpha\colon F(A_\square) \to A$. Hence there are enough relative projectives.*

The proof that $F(M)$ is relatively projective is the familiar argument (Lemma I.5.4) that each free module is projective. Indeed, let $\gamma\colon F(M) \to C$ be a morphism and $\sigma\colon B \to C$ an allowable epic of $\mathscr{A}$, so that $\sigma_\square$ has a right inverse k. Form the schematic diagram

$$k\,\gamma_\square\, e_M = k\,\gamma\, e_M \qquad \begin{array}{ccc} M & \overset{e_M}{\dashrightarrow} & FM \\ \vdots & & \downarrow\gamma \\ B & \underset{k}{\overset{\sigma}{\rightleftarrows}} & C \end{array}, \qquad \sigma_\square k = 1.$$

The composite $k\gamma e_M$ displayed factors uniquely through FM as $k\gamma e_M = \beta e_M$ for some $\beta\colon FM \to B$ in $\mathscr{A}$. Hence $\sigma\beta e_M = \gamma e_M$; but γe_M factors uniquely through e_M, so $\sigma\beta = \gamma$; this states that FM is relatively projective.

The usual comparison theorem maps a projective complex to a resolution. The comparison of a relatively free complex to an allowable resolution can be put in canonical form. A *relatively free complex* $\varepsilon_X\colon X \to A$ over A in $\mathscr{A}$ has each X_n of the form $F(M_n)$ for some object M_n of $\mathscr{M}$; we write e_n for $e_{M_n}\colon M_n \to X_n$. An *allowable resolution* $\varepsilon_Y\colon Y \to B$ has an $\mathscr{M}$-contracting homotopy s with $s^2 = 0$, as in Cor. 5.3 (in particular, $s_{-1}\colon B \dashrightarrow Y_0$).

Theorem 6.2. *Let $\varepsilon_X\colon X \to A$ be a relatively free complex over A in $\mathscr{A}$ and $\varepsilon_Y\colon Y \to B$ an allowable resolution. Each morphism $\alpha\colon A \to B$ in $\mathscr{A}$ lifts to a unique chain transformation $\varphi\colon X \to Y$ of $\mathscr{A}$-complexes such that each $\varphi_n e_n\colon M_n \dashrightarrow Y_n$ factors through s_{n-1}. This φ is determined from the data e_n and s_n by the recursive formulas*

$$\varphi_0 e_0 = s_{-1}\alpha\, \varepsilon_X e_0, \qquad \varphi_{n+1} e_{n+1} = s_n \varphi_n \partial\, e_{n+1}.$$

We call φ the *canonical* comparison for the given representation $X_n = F(M_n)$ and the given homotopy s in Y. In case $\mathscr{M}$ is a category of modules, the condition that each $\varphi_n e_n$ factors through s_{n-1} can be written

$$\varphi_0 e_0 M_0 \subset s_{-1} B, \qquad \varphi_{n+1} e_{n+1} M_{n+1} \subset s_n Y_n. \tag{6.1}$$

We write this more briefly as $\varphi e M \subset s Y$.

Proof. We construct $\varphi_n\colon X_n \to Y_n$ with $\varepsilon_Y \varphi_0 = \alpha\, \varepsilon_X$, $\partial\varphi_{n+1} = \varphi_n \partial$ and show it unique, all by induction on n. If $\varphi_0 e_0$ factors through s_{-1}, then $s^2 = 0$ gives $s_0 \varphi_0 e_0 = 0$ and

$$\varphi_0 e_0 = 1\, \varphi_0 e_0 = (\partial s_0 + s_{-1}\varepsilon_Y)\, \varphi_0 e_0 = s_{-1} \varepsilon_Y \varphi_0 e_0 = s_{-1}\alpha\, \varepsilon_X e_0.$$

By the lifting property, there is a unique such φ_0; this φ_0 does satisfy $\varepsilon_Y \varphi_0 = \alpha\, \varepsilon_X$. Given $\varphi_0, \ldots, \varphi_{n-1}$ unique, any $\varphi_n e_n$ which factors through s_{n-1} has $s_n \varphi_n e_n = 0$; hence

$$\varphi_n e_n = 1\, \varphi_n e_n = (\partial s + s \partial)\, \varphi_n e_n = s \partial \varphi_n e_n = s \varphi_{n-1} \partial e_n .$$

This uniquely determines φ_n so as to satisfy $\partial \varphi_n = \varphi_{n-1} \partial$. The proof is complete.

Next, each object C of $\mathscr{A}$ has a canonical $\square$-split resolution. Write $\tilde{F} C$ for $F \square C \in \mathscr{A}$, $\tilde{F}^n$ for the n-fold iteration of $\tilde{F}$, and construct the objects

$$\beta_n(C) = \beta_n(\mathscr{R}, C) = \tilde{F}^n \tilde{F} C, \qquad n = 0, 1, 2, \ldots$$

of $\mathscr{A}$. Define morphisms s between the corresponding objects

$$\square\, C \xrightarrow{s_{-1}} \square\, \beta_0 C \xrightarrow{s_0} \square\, \beta_1 C \to \cdots \to \square\, \beta_n C \xrightarrow{s_n} \square\, \beta_{n+1} C, \ldots \tag{6.2}$$

in $\mathscr{M}$ as $s_{-1} = e(\square\, C)$ and $s_n = e(\square\, \beta_n C)$.

Theorem 6.3. *There are unique morphisms*

$$\varepsilon\colon \beta_0(C) \to C, \qquad \partial_n\colon \beta_n(C) \to \beta_{n-1}(C), \qquad n = 1, 2, \ldots$$

of $\mathscr{A}$ which make $\beta(\mathscr{R}, C) = \{\beta_n(\mathscr{R}, C)\}$ a relatively free allowable resolution of C with s as contracting homotopy in $\mathscr{M}$. This resolution, with its contracting homotopy, is a covariant functor of C.

We do not claim $s^2 = 0$ — because it usually isn't so.

Proof. We wish to fill in the schematic diagram

$$C \underset{s_{-1}}{\overset{\varepsilon}{\leftrightarrows}} \beta_0 C \underset{s_0}{\overset{\partial_1}{\leftrightarrows}} \beta_1 C \underset{s_1}{\overset{\partial_2}{\leftrightarrows}} \beta_2 C \underset{s_2}{\overset{\partial_3}{\leftrightarrows}} \cdots$$

at the solid arrows (morphisms of $\mathscr{A}$) to get a contracting homotopy. By the properties of e, 1_C factors uniquely as $1_C = \varepsilon\, e_C$; this gives ε uniquely and shows ε allowable. The boundary operators are now defined by recursion so that s will be a contracting homotopy; given ε, $1 - s_{-1} \varepsilon$ factors uniquely as $\partial_1 s_0 = 1 - s_{-1} \varepsilon$ for some $\partial_1\colon \beta_1 \to \beta_0$, and similarly $\partial_{n+1} s_n = 1 - s_{n-1} \partial_n\colon \beta_n \to \beta_n$ determines ∂_{n+1}, given ∂_n. Using this equation,

$$\partial_n \partial_{n+1} s_n = \partial_n - \partial_n s_{n-1} \partial_n = \partial_n - (1 - s_{n-2} \partial_{n-1})\, \partial_n = s_{n-2} \partial_{n-1} \partial_n$$

so, by induction and the uniqueness of the factorization, $\varepsilon \partial = 0$ and $\partial^2 = 0$. Moreover, $\partial_{n+1} s_n \partial_{n+1} = \partial_{n+1}$, so ∂_{n+1} is allowable.

This resolution $\beta(\mathscr{R}, C)$ is clearly functorial; it is called the (unnormalized) *bar resolution*; for a concrete example, see § 8 below.

A "relative" ext bifunctor may now be defined by

$$\operatorname{Ext}^n_{\mathscr{R}}(C, A) = H^n(\hom_{\mathscr{A}}(\beta(\mathscr{R}, C), A)). \qquad (6.3)$$

The comparison theorem shows that we can equally well use any $\square$-split relatively projective resolution $\varepsilon: X \to C$ to calculate $\operatorname{Ext}^n_{\mathscr{R}} = \operatorname{Ext}^n_{\square}$ as

$$\operatorname{Ext}^n_{\mathscr{R}}(C, A) \cong H^n(\hom_{\mathscr{A}}(X, A)); \qquad (6.4)$$

note that, in each dimension n, $\hom_{\mathscr{A}}(X_n, A)$ stands for the group of *all* morphisms $\xi: X_n \to A$ in $\mathscr{A}$ — not just the allowable ones. In particular, $\operatorname{Ext}^0_{\mathscr{R}}(C, A) = \hom_{\mathscr{A}}(C, A)$. Replacement of C by a short exact sequence E will give the usual long exact sequence for Ext^n, as in Thm. III.9.1, provided E is $\square$-split. The analogous result holds if A is replaced by a $\square$-split short exact sequence; the proof uses either an exact sequence of resolutions (Ex. III.9.1) or the assumption that there are enough relative injectives. These long exact sequences are actually valid in any relative abelian category without the assumption that there are enough relative projectives or injectives. The proof, to be given in Chap. XII, depends on the interpretation of $\operatorname{Ext}^n_{\square}(C, A)$ as congruence classes of n-fold, $\square$-split exact sequences from A to C. In particular, $\operatorname{Ext}^1_{\square}$, unlike $\operatorname{Ext}^0_{\square}$, depends on $\square$.

Note on Adjoints. If $\mathscr{C}$ and $\mathscr{A}$ are categories, a functor $T: \mathscr{C} \to \mathscr{A}$ is called a *right adjoint* of $S: \mathscr{A} \to \mathscr{C}$ if there is a natural equivalence

$$\hom_{\mathscr{A}}(A, T(C)) \cong \hom_{\mathscr{C}}(S(A), C);$$

here both sides are bifunctors of A and C with values in the category of sets (or, if $\mathscr{C}$ and $\mathscr{A}$ are additive, in the category of abelian groups). For example, adjoint associativity

$$\operatorname{Hom}(A \otimes B, C) \cong \operatorname{Hom}(A, \operatorname{Hom}(B, C))$$

states for fixed B that $T(C) = \operatorname{Hom}(B, C)$ is a right adjoint to $S(A) = A \otimes B$. There are many other examples [KAN 1958].

Exercises

1. If the relative abelian category $\square$ is a resolvent pair of categories for two functors F and F', show that there is a unique natural isomorphism $\eta: F \to F'$ with $\eta e = e'$.

2. Construct resolvent pairs of categories in the following cases:

(a) For graded rings $R \supset S$; $\mathscr{A}$ and $\mathscr{M}$ as in the text.

(b) For $\varrho: R' \to R$ any ring homomorphism, $\mathscr{A}$ = left R-modules, $\mathscr{M}$ = left R'-modules, $\square A = {}_{\varrho}A$ the R-module A pulled back along ϱ to be an R'-module.

(c) For Λ, Σ both K-algebras, $\mathscr{A} = \Lambda$-Σ-bimodules, $\mathscr{M}$ = K-modules.

3. In case (b) of Ex. 2 show that the allowable exact sequences and the relative Ext functor are identical with those for $\mathscr{R} = (R, S)$ when $S = \varrho R'$.

7. The Categorical Bar Resolution

The (normalized) bar resolution $\varepsilon: B(Z(\Pi)) \to Z$ for a group ring $Z(\Pi)$, as presented in Chap. IV, provides a standard Z-split resolution of the trivial Π-module Z. For each Π-module A the cohomology of A is defined via the bar resolution as $H^n(\mathrm{Hom}_\Pi(B(Z(\Pi)), A))$. Hence

$$H^n(\Pi, A) \cong \mathrm{Ext}^n_{Z(\Pi)}(Z, A) \cong \mathrm{Ext}^n_{(Z(\Pi), Z)}(Z, A).$$

In other words, the cohomology of a group is an instance of both the absolute and relative functor Ext. The same (normalized) bar resolution will be used in the next chapter in many other cases. It may be defined in the context of any resolvent pair

$$\mathscr{R}: \quad \square: \mathscr{A} \to \mathscr{M}, \quad F: \mathscr{M} \to \mathscr{A}, \quad e_M: M \dashrightarrow \square F M$$

of categories. To each object $M \in \mathscr{M}$, select $p_M \in \operatorname{coker} e_M$ and $\overline{F}(M) = \operatorname{Coker} e_M$. Thus

$$M \overset{e_M}{\dashrightarrow} \square F M \overset{p_M}{\dashrightarrow} \overline{F} M \dashrightarrow 0 \tag{7.1}$$

is exact in $\mathscr{M}$, $\overline{F}: \mathscr{M} \to \mathscr{M}$ is a covariant functor, and $p: \square F \to \overline{F}$ is a natural transformation. Apply $\square F$ to $\overline{F} M$ to form the diagram

$$\begin{array}{ccccc} M & \overset{e_M}{\dashrightarrow} & \square F M & \overset{p_M}{\dashrightarrow} & \overline{F} M \dashrightarrow 0 \\ & & & \overset{s_M}{\searrow} & \downarrow e_{\overline{F}M} = e' \\ & & & & \square F \overline{F} M \end{array} ; \tag{7.2}$$

the composite $s_M = e' p$ is a natural transformation $\square F \dashrightarrow \square F \overline{F}$. Its characteristic property is

Lemma 7.1. *The morphisms $e = e_M$ and $s = s_M$ induce for every object A a left exact sequence of abelian groups,*

$$0 \to \hom_{\mathscr{A}}(F\overline{F}M, A) \overset{s^*}{\to} \hom_{\mathscr{M}}(\square F M, \square A) \overset{e^*}{\to} \hom_{\mathscr{M}}(M, \square A).$$

Proof. Each morphism $\alpha: F\overline{F}M \to A$ of $\mathscr{A}$ yields $\alpha_\square: \square F \overline{F} M \dashrightarrow \square A$, and $s^*\alpha$ is the composite $\alpha_\square s: \square F M \dashrightarrow \square A$, a morphism in $\mathscr{M}$. Clearly $e^* s^* \alpha = \alpha_\square s e = \alpha_\square 0 = 0$. If $0 = \alpha_\square s = \alpha_\square e' p$, with p epic, then $\alpha_\square e' = 0$. But the factorization of 0 through e' is unique, so $\alpha = 0$. Next, if some $v: \square F M \dashrightarrow \square A$ has $e^* v = 0$, construct the commutative diagram

$$\begin{array}{ccccccc} M & \overset{e}{\dashrightarrow} & \square F M & \overset{p}{\dashrightarrow} & \overline{F} M & \overset{e'}{\dashrightarrow} & \square F \overline{F} M \\ & & \downarrow v & & \downarrow u & & \downarrow \alpha_\square \\ & & \square A & = & \square A & = & \square A \end{array}$$

as follows. Since $ve = 0$, v factors through $p = \operatorname{coker} e_M$ as $v = up$. By the definition of e', u in turn factors through e' as $u = \alpha_\square e'$ for some α. All told, $v = \alpha_\square e' p = \alpha_\square s$, which gives the asserted exactness.

Each object C of $\mathscr{A}$ yields a sequence of objects $M_n = \bar{F}^n \square C$ of $\mathscr{M}$. The bar resolution consists of the associated relatively free objects

$$B_n(C) = B_n(\mathscr{R}, C) = F\bar{F}^n \square C, \qquad n = 0, 1, 2, \ldots \tag{7.3}$$

of $\mathscr{A}$. Define morphisms s between the corresponding "neglected" objects,

$$\square C \xrightarrow{s_{-1}} \square B_0 C \xrightarrow{s_0} \square B_1 C \xrightarrow{s_1} \square B_2 C \dashrightarrow \cdots, \tag{7.4}$$

of $\mathscr{M}$ by $s_{-1} = e(\square C)$ and

$$s_n = s(M_n) \colon \square F M_n \dashrightarrow \square F\bar{F} M_n = \square B_{n+1} C. \tag{7.5}$$

This construction at once gives $s^2 = 0$.

Theorem 7.2. *There are unique morphisms*

$$\varepsilon \colon B_0(C) \to C, \quad \partial_n \colon B_n(C) \to B_{n-1}(C), \qquad n = 1, 2, \ldots$$

of $\mathscr{A}$ which make $B(\mathscr{R}, C) = \{B_n(\mathscr{R}, C)\}$ an $\mathscr{A}$-complex and a relatively free allowable resolution of C with s as contracting homotopy of square zero in $\mathscr{M}$. This resolution, with its contracting homotopy, is a covariant functor of C.

Proof. We are required to fill in the schematic diagram

$$C \underset{s_{-1}}{\overset{\varepsilon}{\leftrightarrows}} B_0 C \underset{s_0}{\overset{\partial_1}{\leftrightarrows}} B_1 C \underset{s_1}{\overset{\partial_2}{\leftrightarrows}} B_2 C \underset{s_2}{\overset{\partial_3}{\leftrightarrows}} \cdots \tag{7.6}$$

at the solid arrows (morphisms of $\mathscr{A}$) so as to satisfy the conditions

$$\varepsilon s_{-1} = 1, \quad \partial_1 s_0 = 1 - s_{-1}\varepsilon, \quad \partial_{n+1} s_n = 1 - s_{n-1}\partial_n, \quad n > 0, \tag{7.7}$$

for a contracting homotopy. But $1 \colon C \to C$ factors through $e_{\square C} = s_{-1}$ as $1 = \varepsilon s_{-1}$; this gives ε. The morphisms ∂_n are then constructed by recursion. Given $\partial_1, \ldots, \partial_n$ satisfying (7.7),

$$(1 - s_{n-1}\partial_n) s_{n-1} = s_{n-1} - s_{n-1}(1 - s_{n-2}\partial_{n-1}) = 0 + s_{n-1}s_{n-2}\partial_{n-1} = 0;$$

since $s_{n-1} = ep$ with p epic, $(1 - s_{n-1}\partial_n)e = 0$. By Lemma 7.1, $1 - s_{n-1}\partial_n$ factors as $(1 - s_{n-1}\partial_n) = \alpha s_n$, which gives $\partial_{n+1} = \alpha$ satisfying (7.7). These morphisms ε, ∂_n are uniquely determined, again by Lemma 7.1. Moreover, (7.7) gives

$$\partial_n \partial_{n+1} s_n = \partial_n - \partial_n s_{n-1} \partial_n = \partial_n - \partial_n - s_{n-2}\partial_{n-1}\partial_n = -s_{n-2}\partial_{n-1}\partial_n,$$

so an induction using Lemma 7.1 shows $\varepsilon\partial_1 = 0$ and $\partial^2 = 0$. This shows $B(\mathscr{R}, C)$ a complex over C and completes the proof of the theorem.

We call $B(\mathscr{R}, C)$ the *bar resolution* of C. By the comparison theorem, it is chain equivalent to our previous "unnormalized" bar resolution $\beta(\mathscr{R}, C)$ (see Ex. 3).

To show that this description of the bar resolution agrees with the previous usage for a group Π, take $\mathscr{A}$ to be the category of left Π-modules, $\mathscr{M}$ that of abelian groups, $F(M)=Z(\Pi)\otimes M$, and $e_M(m)=1\otimes m$ for each $m\in M$. This gives a resolvent pair of categories. Now the sequence $Z\rightarrowtail Z(\Pi)\twoheadrightarrow Z(\Pi)/Z$ of free abelian groups is exact, hence so is its tensor product with M:

$$0\to M=Z\otimes M\to F(M)=Z(\Pi)\otimes M\to[Z(\Pi)/Z]\otimes M\to 0.$$

Therefore $\overline{F}(M)\cong[Z(\Pi)/Z]\otimes M$. Take for C the trivial Π-module Z. Then

$$\overline{F}^n(Z)=[Z(\Pi)/Z]\otimes\cdots\otimes[Z(\Pi)/Z],\qquad n \text{ factors}.$$

But $Z(\Pi)/Z$ is the free abelian group with generators all $x\neq 1$ in Π. Hence $\overline{F}^n(Z)$ may be identified with the free abelian group generated by all symbols $[x_1|\dots|x_n]$, with no x_i in Π equal to 1. Then $B_n(\mathscr{R},Z)=Z(\Pi)\otimes\overline{F}^n(Z)$ is the free abelian group with generators all $x[x_1|\dots|x_n]$ with $x\in\Pi$, while the map $s=e\not p\colon B_n\dashrightarrow B_{n+1}$ defined above becomes

$$s(x[x_1|\dots|x_n])=[x|x_1|\dots|x_n],$$

zero when $x=1$. This is exactly the contracting homotopy s used for the bar resolution $B(Z(\Pi))$ in (IV.5.2). The boundary operators are uniquely determined by s (in Chap. IV as here), so must agree. In short, we have proved that in this resolvent pair of categories

$$B(\mathscr{R},Z)=B(Z(\Pi)).$$

The next chapter will develop explicit formulas in other cases.

Exercises

1. Show that the long sequence (7.4) is exact in $\mathscr{M}$.
2. Show that the canonical comparison $\beta(\mathscr{R},C)\to B(\mathscr{R},C)$ is epic.
3. For the case of groups, show that β gives the unnormalized bar resolution.

The following three exercises consider the relative ext functor for the rings $Z(\Pi)$ and Z.

4. For left Π-modules A, B, and C, make $B\otimes_Z C$ and $\operatorname{Hom}_Z(C,A)$ left Π-modules with operators $x(b\otimes c)=xb\otimes xc$ and $(x\alpha)c=x[\alpha(x^{-1}c)]$, $\alpha\colon C\to A$, respectively, and establish a natural isomorphism,

$$\operatorname{Hom}_\Pi(B,\operatorname{Hom}_Z(C,A))\cong\operatorname{Hom}_\Pi(B\otimes_Z C,A).$$

5. If A is relatively injective or C relatively projective, show that $\operatorname{Hom}_Z(C,A)$, with operators as in Ex. 4, is relatively injective.

6. Using axioms for the relative ext functor, establish a natural isomorphism

$$\operatorname{Ext}^n_{Z(\Pi),Z}(C,A)\cong\operatorname{Ext}^n_{Z(\Pi),Z}(Z,\operatorname{Hom}_Z(C,A)).$$

With this result the following exercises, suggested to me by J. Schmid, will yield the cup product reduction theorem as stated in VIII.9 (cf. Schmid [1963]).

7. From a presentation $\Pi = F/R$ with F a free group obtain a group extension E: $R_0 \rightarrowtail B' \twoheadrightarrow \Pi$, where $[R, R]$ is the commutator subgroup of R, $R_0 = R/[R, R]$ and $B' = F/[R, R]$.

8. The characteristic class χ of the group extension E, described as in IV.6, is a two-fold Z-split extension of R_0 by Z. Show that the intermediate module M in χ is free; specifically, let F be the free group on generators g, S the free Π-module on corresponding generators g' and show that $g' \to [\text{cls } g] \in M$ is an isomorphism $S \cong M$. (Hint: Use Lemma IV.7.2 to construct an inverse.)

9. Let A be a Π-module. Show that the iterated connecting homomorphism for χ yields an isomorphism $\operatorname{Ext}^n(R_0, A) \cong \operatorname{Ext}^{n+2}(Z, A)$ for the relative ext functor and $n > 0$ and hence, by Ex. 6, isomorphisms

$$H^{n+2}(\Pi, A) \cong H^n(\Pi, \operatorname{Hom}_Z(R_0, A)), \quad n > 0,$$
$$H^2(\Pi, A) \cong \operatorname{Coker}[\operatorname{Hom}_\Pi(M, A) \to \operatorname{Hom}_\Pi(R_0, A)].$$

10. For $n > 0$ and G a right Π-module obtain the "dual" reduction theorem

$$H_{n+2}(\Pi, G) \cong H_n(\Pi, G \otimes_Z R_0).$$

8. Relative Torsion Products

Let S be a subring of the ring R, with the same identity. This gives a resolvent pair $\mathscr{R} = (R, S)$ of categories, with $\mathscr{A}$ the left R-modules, $\mathscr{M}$ the left S-modules, $\square(A)$ the functor which remembers only the S-module structure, $F(M) = R \otimes_S M$, and $e(m) = 1 \otimes m$. A $\square$-split short exact sequence is thus an exact sequence of R-modules which splits when regarded as a sequence of S-modules; call such a sequence *S-split*. Label the corresponding allowable homomorphisms *(R, S)-allowable*, and the relative projectives *(R, S)-projectives*.

Define a complex $\beta(R)$ of R-modules over R,

$$R \xleftarrow{\varepsilon} \beta_0(R) \leftarrow \beta_1(R) \leftarrow \beta_2(R) \leftarrow \cdots$$

by $\beta_n(R) = R \otimes R^n \otimes R = R^{n+2}$, with $n+2$ factors, $\varepsilon(r_0 \otimes r_1) = r_0 r_1$, $\otimes$ short for $\otimes_S$, and

$$\partial(r_0 \otimes \cdots \otimes r_{n+1}) = \sum_{i=0}^{n} (-1)^i r_0 \otimes \cdots \otimes r_i r_{i+1} \otimes \cdots \otimes r_{n+1}. \tag{8.1}$$

Theorem 8.1. *For $R \supset S$, ε: $\beta(R) \to R$ is a complex of R-bimodules over R with a contracting homotopy s: $R^{n+2} \to R^{n+3}$, defined for $n \geqq 1$ by*

$$s(r_0 \otimes \cdots \otimes r_{n+1}) = 1 \otimes r_0 \otimes \cdots \otimes r_{n+1}, \tag{8.2}$$

which is a homomorphism of left-S, right-R bimodules.

Proof. First $\varepsilon s r_0 = \varepsilon(1 \otimes r_0) = r_0$, so $\varepsilon s = 1$. Let $u = r_0 \otimes \cdots \otimes r_{n+1}$ with $n \geqq 0$. The first term of $\partial s u$ is u; the remaining terms are $-s\partial u$; hence $\partial s + s\partial = 1$, as desired. From the definition it follows that $\varepsilon\partial = 0$, $\partial\partial = 0$. By symmetry there is also a contracting homotopy

$$t(r_0 \otimes \cdots \otimes r_{n+1}) = r_0 \otimes \cdots \otimes r_{n+1} \otimes 1,$$

which is an R-S-bimodule homomorphism.

Corollary 8.2 *For each left R-module C, $\beta(R)\otimes_R C$ is the bar resolution $\beta(C)$ for the resolvent pair (R, S). Symmetrically, for each right R-module G, $G\otimes_R\beta(R)$ with contracting homotopy t is the (right) bar resolution $\beta(G)$.*

Proof. Since $R\otimes_R C\cong C$, one forms $\beta(R)\otimes_R C$ simply by replacing the last argument r_{n+1} in (8.1) and (8.2) above by $c\in C$. Then $\beta_n(R)\otimes_R C = F^n F(C)$, the contracting homotopy s of (8.2) is that of (6.2), and ∂ is the unique boundary with s as contracting homotopy, by Thm. 6.3. In particular, $\beta(R)$ itself is just the bar resolution (left or right) of the R-module R.

Observe that the boundary operator (8.1) in $\beta(R)$ is the alternating sum of the face operators $d_i\colon \beta_n\to\beta_{n-1}$ defined by

$$d_i(r_0\otimes\cdots\otimes r_{n+1}) = r_0\otimes\cdots\otimes r_{i-1}\otimes r_i r_{i+1}\otimes r_{i+2}\otimes\cdots\otimes r_{n+1}, \qquad (8.3)$$

$i=0, \ldots, n$. The corresponding degeneracy operators $s_i\colon \beta_n\to\beta_{n+1}$ are

$$s_i(r_0\otimes\cdots\otimes r_{n+1}) = r_0\otimes\cdots\otimes r_i\otimes 1\otimes r_{i+1}\otimes\cdots\otimes r_{n+1}, \qquad (8.4)$$

the usual identities for d_i and s_j hold, and $\beta(R)$ is a simplicial R-bimodule in the sense of VIII.5. The reader may show that the simplicial normalization of $\beta(R)$ yields the normalized bar resolution $B(R)$.

Take R-modules G_R and ${}_RC$. The (absolute) torsion products $\operatorname{Tor}_n^R(G, C)$ are calculated from a projective resolution $\varepsilon\colon X\to C$ as $H_n(G\otimes_R X)$. In the present relative case, $\beta(R)\otimes_R C$ provides a canonical and functorial resolution, so we define the n-th *relative torsion product* as:

$$\operatorname{Tor}_n^{(R,S)}(G, C) = H_n(G\otimes_R\beta(R)\otimes_R C); \qquad (8.5)$$

it is a covariant bifunctor of G and C, and is manifestly symmetric in G and C. Since $G\otimes_R R = G$ and $R\otimes_R C = C$, the group of n-chains of the complex $G\otimes_R\beta(R)\otimes_R C$ is $G\otimes_S R^n\otimes_S C$. The boundary formula is obtained from (8.1) by replacing r_0 by $g\in G$ and r_{n+1} by $c\in C$; the complex may be viewed as a simplicial abelian group.

If $E\colon A\rightarrowtail B\twoheadrightarrow C$ is an S-split short exact sequence of left R-modules, its tensor product (over S) with $G\otimes_S R^n$ is still S-split, hence exact, and so is the sequence of complexes

$$G\otimes_R\beta(R)\otimes_R A\rightarrowtail G\otimes_R\beta(R)\otimes_R B\twoheadrightarrow G\otimes_R\beta(R)\otimes_R C.$$

The resulting connecting homomorphisms

$$E_n\colon \operatorname{Tor}_n^{(R,S)}(G, C)\to\operatorname{Tor}_{n-1}^{(R,S)}(G, A), \qquad n>0,$$

are natural in G and E, and yield the exactness of the corresponding long exact sequence

$$\left.\begin{aligned}\cdots\to\operatorname{Tor}_n^{(R,S)}(G, A)\to\operatorname{Tor}_n^{(R,S)}(G, B)\to\operatorname{Tor}_n^{(R,S)}(G, C)&\\ \xrightarrow{E_n}\operatorname{Tor}_{n-1}^{(R,S)}(G, A)\to\cdots,&\end{aligned}\right\} \qquad (8.6)$$

just as for the absolute torsion product, except that here E must be S-split. If $E'\colon G \rightarrowtail K \twoheadrightarrow L$ is an S-split short exact sequence of right R-modules, the same argument (interchanging left and right) gives natural connecting homomorphisms

$$E'_n\colon \operatorname{Tor}_n^{(R,S)}(L, C) \to \operatorname{Tor}_{n-1}^{(R,S)}(G, C), \qquad n>0,$$

and the corresponding long exact sequence in the first argument.

Theorem 8.3. *For rings $R \supset S$ and modules G_R, ${}_RC$, each* $\operatorname{Tor}_n^{(R,S)}(G, C)$ *is a covariant bifunctor of G and C with*

$$\operatorname{Tor}_0^{(R,S)}(G, C) \cong G \otimes_R C, \qquad (natural), \tag{8.7}$$

$$\operatorname{Tor}_n^{(R,S)}(P', C) = 0 = \operatorname{Tor}_n^{(R,S)}(G, P), \qquad n>0, \tag{8.8}$$

when P' and P are (R, S)-projective right and left modules, respectively. If E' and E are S-split short exact sequences of right and left R-modules, respectively, the corresponding connecting homomorphisms are natural and yield long exact sequences (8.6), *and symmetrically.*

In particular, this Theorem leads to a characterization of the relative torsion products as functors of the second argument, by properties (8.7), (8.8), and (8.6), just as in Thm. V.8.5; for this purpose we may replace "(R, S)-projective" by "(R, S)-free" in (8.8).

We need only prove (8.7) and (8.8). First $\beta_1(R) \to \beta_0(R) \to R \to 0$ is exact; since tensor products carry right exact sequences into right exact sequences, so is

$$G \otimes_R \beta_1(R) \otimes_R C \to G \otimes_R \beta_0(R) \otimes_R C \to G \otimes_R R \otimes_R C \to 0.$$

The last term is $G \otimes_R C$; this gives (8.7). To prove (8.8), use

Lemma 8.4. *For rings $R \supset S$, if P is an (R, S)-projective right R-module and $E\colon A \rightarrowtail B \twoheadrightarrow C$ an S-split short exact sequence of left R-modules, then $0 \to P \otimes_R A \to P \otimes_R B \to P \otimes_R C \to 0$ is an exact sequence of abelian groups.*

Proof. Since there are enough relatively free right modules $M \otimes_S R$, each P is an S-split quotient and hence an R-direct summand of some $M \otimes_S R$. Hence it suffices to prove the Lemma with $P = M \otimes_S R$. Then $P \otimes_R A = M \otimes_S R \otimes_R A = M \otimes_S A$, so the sequence in question is isomorphic to $M \otimes_S A \to M \otimes_S B \to M \otimes_S C$, which S-splits because E does, and hence is exact.

Now we prove (8.8). The complex $\beta(R) \otimes_R C$ over C has a left S-module contracting homotopy s as in (8.2). Hence, by the Lemma, $P' \otimes_R (\beta(R) \otimes_R C)$ is exact over $P' \otimes_R C$, so has homology zero in dimensions $n>0$.

The relative torsion products can also be calculated from other resolutions.

Theorem 8.5. *If* $\varepsilon: Y \to G$ *is an* S*-split resolution of the right* R*-module* G *by* (R, S)*-projective modules* Y_n, *there is a canonical isomorphism*

$$\operatorname{Tor}_n^{(R,S)}(G, C) \cong H_n(Y \otimes_R C), \tag{8.9}$$

natural in C. *If* E *is any* S*-split short exact sequence of left* R*-modules, the connecting homomorphisms* E_n *of* (8.6) *are mapped by the isomorphism* (8.9) *into the homology connecting homomorphism of the exact sequence* $Y \otimes_R A \rightarrowtail Y \otimes_R B \twoheadrightarrow Y \otimes_R C$ *of complexes. Symmetrically,* Tor_n *may be calculated from an* S*-split,* (R, S)*-projective resolution of* C *by left* R*-modules.*

Proof. The relative comparison theorem gives a right R-module chain transformation $\varphi: G \otimes_R \beta(R) \to Y$, unique up to homotopy, which induces the isomorphism (8.9). Since each Y_n is (R, S)-projective, each $Y_n \otimes_R A \rightarrowtail Y_n \otimes_R B \twoheadrightarrow Y_n \otimes_R C$ is exact by Lemma 8.4. The chain transformation φ maps the previous exact sequence of complexes onto this exact sequence; hence by naturality of the connecting homomorphism of a sequence of complexes, this yields the method stated for the calculation of the connecting homomorphisms E_n.

Since an (R, S)-projective P' has the resolution $0 \to P' \to P' \to 0$, this gives an immediate proof of (8.8). We leave the reader to verify the other properties of the relative torsion product: Additivity in each argument, anti-commutation of E_n with E'_{n-1} ($E'_{n-1} E_n = -E_{n-1} E'_n$, as in Thm. V.7.7), and the additivity of E_n in E.

The relative torsion product can be considered as a functor of the pair of rings $R \supset S$. More specifically, consider objects $(R, S; G, C, A)$ consisting of rings $R \supset S$ and modules G_R, ${}_RC$, ${}_RA$. A *change of rings* (+ in G and C, − in A) is a quadruple

$$\chi = (\varrho, \zeta, \gamma, \alpha): (R, S; G, C, A) \to (R', S'; G', C', A') \tag{8.10}$$

where $\varrho: R \to R'$ is a ring homomorphism with $\varrho(S) \subset S'$, while

$$\zeta: G \to G'_\varrho, \quad \gamma: C \to {}_\varrho C', \quad \alpha: {}_\varrho A' \to A$$

are homomorphisms of R-modules (note that the direction of α is opposite that for γ). These objects and morphisms χ, with composition $(\varrho, \zeta, \gamma, \alpha)$ $(\varrho', \zeta', \gamma', \alpha') = (\varrho\varrho', \zeta\zeta', \gamma\gamma', \alpha'\alpha)$ constitute the *change of rings* category $\mathscr{R}^{++-}$; omitting A and α gives a "covariant" change of rings category $\mathscr{R}^{++}$. Each χ induces

$$\zeta \otimes \varrho^n \otimes \gamma: G \otimes_R \beta_n(R) \otimes_R C \to G' \otimes_{R'} \beta_n(R') \otimes_{R'} C',$$

a chain transformation, and thence, by the definition (8.5), a map

$$\chi_*: \operatorname{Tor}_n^{(R,S)}(G, C) \to \operatorname{Tor}_n^{(R',S')}(G', C')$$

which makes the relative torsion product Tor_n a covariant functor on $\mathscr{R}^{++}$ to abelian groups. The homomorphism χ_* can also be calculated from S-split relatively free resolutions $\varepsilon: Y \to G$, $\varepsilon': Y' \to G'$; indeed, by pull-back, $\varepsilon': Y'_\varrho \to G'_\varrho$ is a map of complexes of R-modules with an S-module splitting homotopy, so the comparison theorem (relatively projective complexes to split resolutions) lifts $\zeta: G \to G'_\varrho$ to a chain transformation $\varphi: Y \to Y'_\varrho$. The induced map on homology composed with the pull-back map $Y'_\varrho \otimes_R {}_\varrho C' \to Y' \otimes_{R'} C'$ gives χ_* as the composite of

$$H_n(Y \otimes_R C) \xrightarrow{(\varphi\otimes\gamma)_*} H_n(Y'_\varrho \otimes_R {}_\varrho C') \to H_n(Y' \otimes_{R'} C')$$

with the isomorphisms (8.9).

By analogy with $\mathrm{Tor}_n^{(R,S)}$, we write $\mathrm{Ext}^n_{(R,S)}$ for the corresponding relative ext functor. Thus by (6.3)

$$\begin{aligned}\mathrm{Ext}^n_{(R,S)}(C, A) &= H^n(\mathrm{Hom}_R(\beta(R) \otimes_R C, A))\\ &\cong H^n(\mathrm{Hom}_{R-R}(\beta(R), \mathrm{Hom}_Z(C, A))),\end{aligned}$$

where the isomorphism on the right is by adjoint associativity, and $\mathrm{Hom}_Z(C, A)$ is an R-bimodule. This Ext is a contravariant functor on the change of rings category $\mathscr{R}^{+-}$ (omit G and ζ in (8.10) above). When R is fixed and $\varrho = 1$, this includes the usual description of $\mathrm{Ext}^n_{(R,S)}(C, A)$ as a bifunctor, contravariant in C and covariant in A.

Exercises

The first six of the following exercises are taken from HOCHSCHILD [1956].

1. Every (R, S)-projective P is an R-direct summand of some $R \otimes_S A$.

2. For each ${}_S M$, $\mathrm{Hom}_S(R, M)$ is (R, S)-relative injective.

3. Prove: There are enough (R, S)-relative injectives.

4. If P is (R, S)-projective and $\alpha: A \to B$ a homomorphism of R-modules with $\mathrm{Hom}_S(P, A) \twoheadrightarrow \mathrm{Hom}_S(P, B)$, then $\mathrm{Hom}_R(P, A) \twoheadrightarrow \mathrm{Hom}_R(P, B)$.

5. For P as in Ex. 4 and α a map of right R-modules, $A \otimes_S P \rightarrowtail B \otimes_S P$ monic implies $A \otimes_R P \rightarrowtail B \otimes_R P$ monic.

6. For (R, S)-projective resolutions $X \to C$ and $Y \to G$ which are S-split, prove that $\mathrm{Tor}_n^{(R,S)}(G, C) \cong H_n(Y \otimes_R X)$.

7. Give a description for elements of $\mathrm{Tor}^{(R,S)}(G, A)$ analogous to the elements (μ, L, ν) used in V.7.

8. Show by example that $\mathrm{Ext}^1_{(R,S)} \neq \mathrm{Ext}^1_R$.

9. Show that $\beta(R)$ is the (unnormalized) bar resolution for the resolvent pair $\mathscr{R}'$ with $\mathscr{A} = R$-bimodules, $\mathscr{M} = S$-R-bimodules, $F(M) = R \otimes_S M$ and $e(m) = 1 \otimes m$.

10. For $\mathscr{R}'$ as in Ex. 9, show that $\mathrm{Ext}^n_{(R,S)}(C, A) \cong \mathrm{Ext}^n_{\mathscr{R}'}(R, \mathrm{Hom}_Z(C, A))$ and $\mathrm{Tor}_n^{(R,S)}(G, C) \cong \mathrm{Tor}_n^{\mathscr{R}'}(R, C \otimes_Z G)$. Here $C \otimes_Z G$ is the bimodule with $r(c \otimes g) = rc \otimes g$, $(c \otimes g)r = c \otimes gr$.

9. Direct Products of Rings

The direct product $R = R' \times R''$ of two rings is the ring with the additive group $R' \oplus R''$ and multiplication $(r'_1, r''_1)(r'_2, r''_2) = (r'_1 r'_2, r''_1 r''_2)$. (This is just the direct product of R' and R'' as Z-algebras, as in (VII.5.1)). Each left R'-module A' is an R-module ${}_{\pi_1}A'$ by pull-back along the projection $\pi_1: R' \times R'' \to R'$, and similarly for R''-modules. In particular, R' and R'' are left R-modules, while the termwise definition of the product in R shows that $R \cong R' \oplus R''$ is an isomorphism of R-modules, and hence that R' and R'' are projective R-modules.

Lemma 9.1. *If the R'-modules C' and G' and the R''-module A'' are regarded as $R = R' \times R''$ modules, then*

$$G' \otimes_R A'' = 0, \qquad \mathrm{Hom}_R(C', A'') = 0. \tag{9.1}$$

Proof. Take the element $(1', 0) \in R$. Then

$$g' \otimes a'' = g'(1', 0) \otimes a'' = g' \otimes (1', 0)a'' = g' \otimes 0 = 0.$$

Similarly, if $f: C' \to A''$, then

$$f(c') = f[(1', 0)c'] = (1', 0)f(c') = 0.$$

The correspondence $A \to A' = R' \otimes_R A$, $\alpha \to \alpha' = 1_{R'} \otimes \alpha$ is a covariant functor on R-modules to R'-modules which is exact: $\alpha \| \beta$ implies $\alpha' \| \beta'$. Moreover,

Proposition 9.2. *Each left $(R' \times R'')$-module A has a representation $A \cong ({}_{\pi_1}A') \oplus ({}_{\pi_2}A'')$ as a direct sum of two R-modules, the first obtained by pull-back from an R'-module A' and the second from an R''-module A''. These modules A' and A'' are determined up to isomorphism as $A' \cong R' \otimes_R A$, $A'' \cong R'' \otimes_R A$. Given such decompositions for A and B, each R-module homomorphism $\alpha: A \to B$ has a unique decomposition as $\alpha = \alpha' \oplus \alpha''$, with $\alpha': A' \to B'$ and $\alpha'': A'' \to B''$ respectively R'- and R''-module maps*

Proof. Using $R \cong R' \oplus R''$ we get the decomposition

$$A = R \otimes_R A \cong (R' \oplus R'') \otimes_R A \cong (R' \otimes_R A) \oplus (R'' \otimes_R A).$$

If $A = A' \oplus A''$ is such a decomposition, (9.1) gives $R' \otimes_R A = R' \otimes_R A' = R' \otimes_{R'} A' \cong A'$. Given α, $\alpha' = 1_{R'} \otimes \alpha: R' \otimes_R A \to R' \otimes_R B$ and $\alpha'' = 1_{R''} \otimes \alpha$ have $\alpha \cong \alpha' \oplus \alpha''$.

Corollary 9.3. *For left R-modules A and C and a right R-module G, each decomposed as in Prop. 9.2, there are natural isomorphisms*

$$\mathrm{Hom}_R(C, A) \cong \mathrm{Hom}_{R'}(C', A') \oplus \mathrm{Hom}_{R''}(C'', A''), \tag{9.2}$$

$$G \otimes_R A \cong G' \otimes_{R'} A' \oplus G'' \otimes_{R''} A''. \tag{9.3}$$

Proof. $\mathrm{Hom}_R(C, A)$ is additive in C and A, and $\mathrm{Hom}_R(C', A') \cong \mathrm{Hom}_{R'}(C', A')$, while $\mathrm{Hom}_R(C', A'')=0$ by (9.1).

Isomorphisms like (9.2) and (9.3) hold for the relative Ext and Tor functors. For example, if $S' \subset R'$ and $S'' \subset R''$ are subrings, then $S = S' \times S''$ is a subring of $R' \times R''$, with $\pi_1 S = S'$, $\pi_2 S = S''$. We treat the more general case of any subring of $R' \times R''$.

Theorem 9.4. *If S is a subring of $R' \times R''$, set $S' = \pi_1 S \subset R'$, $S'' = \pi_2 S \subset R''$. For left R-modules A and C and a right R-module G, each decomposed as in Prop. 9.2, there are natural isomorphisms*

$$\mathrm{Ext}^n_{(R,S)}(C, A) \cong \mathrm{Ext}^n_{(R',S')}(C', A') \oplus \mathrm{Ext}^n_{(R'',S'')}(C'', A''), \tag{9.4}$$

$$\mathrm{Tor}^{(R,S)}_n(G, C) \cong \mathrm{Tor}^{(R',S')}_n(G', C') \oplus \mathrm{Tor}^{(R'',S'')}_n(G'', C''), \tag{9.5}$$

valid for all n. The same isomorphisms hold with S, S', and S'' omitted.

Proof. First observe that an (R', S')-free module $R' \otimes_{S'} M'$ is also (R, S)-projective (though not necessarily (R, S)-free). For, the left S'-module M' is a left S-module by pull-back, and, using the pull-back lemma,

$$R \otimes_S M' \cong R' \otimes_S M' \oplus R'' \otimes_S M' \cong R' \otimes_{S'} M' \oplus R'' \otimes_{S''} M'.$$

Since $R \otimes_S M'$ is (R, S)-projective, so is its R-direct summand $R' \otimes_{S'} M'$.

Now choose relatively free split resolutions $\varepsilon': X' \to C'$ and $\varepsilon'': X'' \to C''$ of the components of C. Then $\varepsilon' \oplus \varepsilon'': X' \oplus X'' \to C' \oplus C''$ is a resolution of the R-module $C' \oplus C''$ which is S-split by the direct sum of the S' and S'' contracting homotopies for X' and X''. By the first observation, each term $X'_n \oplus X''_n$ is an (R, S)-projective. By (9.2) and (9.3) for $X = X' \oplus X''$,

$$\mathrm{Hom}_R(X, A) \cong \mathrm{Hom}_{R'}(X', A') \oplus \mathrm{Hom}_{R''}(X'', A''),$$

$$G \otimes_R X \cong G' \otimes_{R'} X' \oplus G'' \otimes_{R''} X''.$$

Taking cohomology and homology groups gives the desired isomorphisms (9.4) and (9.5).

In the isomorphism (9.5), each projection $\mathrm{Tor}_n(G, C) \to \mathrm{Tor}_n(G', C')$ can be described as the map χ_* induced by that change of rings χ: $(R, S; G, C) \to (R', S'; G', C')$ which is obtained by the projections $R = R' \times R'' \to R'$, $G = G' \oplus G'' \to G'$, etc. In fact, to calculate χ_* one lifts $C \to C'$ to a chain transformation $\varphi: X \to X'$; such a φ is the projection $X = X' \oplus X'' \to X'$ used in deducing (9.5).

The proof for the same results with S omitted is easier; when X'_n is a free R'-module, it is a direct summand of copies of R, hence is R-projective.

This theorem will be applied in the next chapter to algebras (§ 6).

Chapter ten

Cohomology of Algebraic Systems

1. Introduction

The homology of algebraic systems is an instance of relative homological algebra.

For a group Π, use exact sequences of Π-modules which split as sequences of abelian groups. The cohomology of Π for coefficients in a module ${}_{\Pi}A$ is then (cf. IX.7)

$$H^n(\Pi, A) = \mathrm{Ext}^n_{Z(\Pi)}(Z, A) \cong \mathrm{Ext}^n_{(Z(\Pi),Z)}(Z, A). \tag{1.1}$$

Correspondingly, the homology of Π for coefficients G_{Π} will be defined as

$$H_n(\Pi, G) = \mathrm{Tor}_n^{Z(\Pi)}(G, Z) \cong \mathrm{Tor}_n^{(Z(\Pi),Z)}(G, Z). \tag{1.2}$$

For a K-algebra Λ, use exact sequences of Λ-bimodules which split as sequences of right Λ-modules, or those which split just as sequences of K-modules. For a Λ-bimodule A, the cohomology and homology of Λ will be

$$H^n(\Lambda, A) = \mathrm{Ext}^n_{(\Lambda-\Lambda,\,\mathsf{K}-\Lambda)}(\Lambda, A) \cong \mathrm{Ext}^n_{(\Lambda-\Lambda,\,\mathsf{K})}(\Lambda, A); \tag{1.3}$$

$$H_n(\Lambda, A) = \mathrm{Tor}_n^{(\Lambda-\Lambda,\,\mathsf{K}-\Lambda)}(A, \Lambda) \cong \mathrm{Tor}_n^{(\Lambda-\Lambda,\,\mathsf{K})}(A, \Lambda). \tag{1.4}$$

These equivalent descriptions are presented in terms of the bar resolution for algebras, which is given explicitly in § 2 — it is a special case of the bar resolution (IX.7) for a resolvent pair of categories. This chapter examines the properties of H_n and H^n and develops similar (co)-homology for graded and for differential graded algebras, as well as for monoids and for abelian groups.

2. The Bar Resolution for Algebras

Let Λ be an algebra over K. The identity element 1_{Λ} gives a K-module map $I\colon \mathsf{K} \dashrightarrow \Lambda$; its cokernel $\Lambda/I(\mathsf{K}) = \Lambda/(\mathsf{K}1_{\Lambda})$ will be denoted (simply but inaccurately) as Λ/K, with elements the cosets $\lambda + \mathsf{K}$. For each left Λ-module C construct the relatively free Λ-module ($\otimes = \otimes_{\mathsf{K}}$)

$$B_n(\Lambda, C) = \Lambda \otimes (\Lambda/\mathsf{K}) \otimes \cdots \otimes (\Lambda/\mathsf{K}) \otimes C, \qquad (n \text{ factors } \Lambda/\mathsf{K}). \tag{2.1}$$

As a K-module, it is spanned by elements which we write, with a vertical bar replacing "$\otimes$", as

$$\lambda[\lambda_1| \ldots |\lambda_n]c = \lambda \otimes [(\lambda_1 + \mathsf{K}) \otimes \cdots \otimes (\lambda_n + \mathsf{K})] \otimes c; \tag{2.2}$$

in particular, elements of B_0 are written as $\lambda[\,]c$. The left factor λ gives the left Λ-module structure of B_n, and $[\lambda_1| \ldots |\lambda_n]c$ without the operator λ will designate the corresponding element of $(\Lambda/\mathrm{K})^n \otimes C$. These elements are *normalized*, in the sense that

$$[\lambda_1|\ldots|\lambda_n]c=0 \tag{2.3}$$

when any one $\lambda_i \in \mathrm{K}$.

Now construct maps as in the diagram

$$C \underset{s_{-1}}{\overset{\varepsilon}{\leftrightarrows}} B_0(\Lambda, C) \underset{s_0}{\overset{\partial}{\leftrightarrows}} B_1(\Lambda, C) \leftrightarrows \cdots.$$

The K-module homomorphisms s_{-1}: $C \dashrightarrow B_0$ and s_n: $B_n \dashrightarrow B_{n+1}$ are defined by setting $s_{-1}c=1[\,]c$ and

$$s_n(\lambda[\lambda_1|\ldots|\lambda_n]c)=1[\lambda|\lambda_1|\ldots|\lambda_n]c, \quad n \geqq 0. \tag{2.4}$$

By the normalization, $s_{n+1}s_n=0$. Define left Λ-module homomorphisms ε: $B_0 \to C$ and ∂_n: $B_n \to B_{n-1}$ for $n>0$ by $\varepsilon(\lambda[\,]c)=\lambda c$, and

$$\left.\begin{aligned} \partial_n(\lambda[\lambda_1|\ldots|\lambda_n]c) &= \lambda\lambda_1[\lambda_2|\ldots|\lambda_n]c \\ &\quad+\sum_{i=1}^{n-1}(-1)^i\lambda[\lambda_1|\ldots|\lambda_i\lambda_{i+1}|\ldots|\lambda_n]c \\ &\quad+(-1)^n\lambda[\lambda_1|\ldots|\lambda_{n-1}](\lambda_n c). \end{aligned}\right\} \tag{2.5}$$

This definition is legitimate because the right side is K-multilinear and normalized: If some $\lambda_i=1$, the terms with indices $i-1$ and i cancel and the remaining terms are zero.

Theorem 2.1. *For each left Λ-module C, ε: $B(\Lambda, C)\to C$ is a resolution of C by (Λ, K)-relatively free left Λ-modules which is* K-*split by the contracting homotopy s with $s^2=0$. Moreover, $B(\Lambda, C)$ is a covariant functor of C.*

This can be proved directly from the formulas above. Alternatively, apply the resolution of IX.7 for the resolvent pair of categories $\mathscr{R}$ with $\mathscr{A}$ = left Λ-modules, $\mathscr{M}$ = K-modules, $F(M)=\Lambda\otimes M$, $e(m)=1\otimes m$. Since $\mathrm{K} \dashrightarrow \Lambda \dashrightarrow \Lambda/\mathrm{K} \dashrightarrow 0$ is a right exact sequence of K-modules, each K-module M yields a right exact sequence

$$M=\mathrm{K}\otimes M \dashrightarrow F(M)=\Lambda\otimes M \dashrightarrow (\Lambda/\mathrm{K})\otimes M \dashrightarrow 0,$$

so $\overline{F}(M)\cong(\Lambda/\mathrm{K})\otimes M$. Also, s_M: $F(M) \dashrightarrow F\overline{F}(M)$ is given by $s(\lambda\otimes m)=1\otimes(\lambda+\mathrm{K})\otimes m$. Hence, with $B(\mathscr{R}, C)$ as in (IX.7.3),

$$B_n(\mathscr{R}, C)=F\overline{F}^n \,\square\, C=\Lambda\otimes(\Lambda/\mathrm{K})^n\otimes C=B_n(\Lambda, C),$$

with s given by (2.4). The formulas for ε and ∂_n provide the unique boundaries for which s is the contracting homotopy. Hence $B(\mathscr{R}, C)=B(\Lambda, C)$.

There are several variants of the bar resolution, as follows.

The *un-normalized bar resolution* $\beta(\mathcal{R}, C)=\beta(\Lambda, C)$ (cf. IX.6) has

$$\beta_n(\Lambda, C)=\tilde{F}\tilde{F}^n C=\Lambda\otimes\Lambda^n\otimes C, \tag{2.6}$$

where $\Lambda^n=\Lambda\otimes\cdots\otimes\Lambda$, with n factors. The contracting homotopy s, ε, and the boundary are given by the formulas (2.4) and (2.5) with each $\lambda[\lambda_1|\ldots|\lambda_n]c$ replaced by $\lambda\otimes\lambda_1\otimes\cdots\otimes\lambda_n\otimes c$. In this case the boundary may be written, much as in the singular complex of a space, in the form $\partial_n=\Sigma(-1)^i d_i$, where $d_i\colon \beta_n\to\beta_{n-1}$ is the Λ-module homomorphism defined by

$$d_i(\lambda_0\otimes\lambda_1\otimes\cdots\otimes\lambda_n\otimes c)=\lambda_0\otimes\cdots\otimes\lambda_i\lambda_{i+1}\otimes\cdots\otimes c, \quad i=0,\ldots,n \tag{2.7}$$

(for $i=n$, the right side is $\lambda_0\otimes\cdots\otimes\lambda_n c$). Thm. 2.1 holds with $B(\Lambda, C)$ replaced by $\beta(\Lambda, C)$, except that s^2 need not be zero in $\beta(\Lambda, C)$.

The Λ-module map $\eta\colon \beta_n\to B_n$ defined by $\eta(\lambda\otimes\lambda_1\otimes\cdots\otimes\lambda_n\otimes c)=\lambda[\lambda_1|\ldots|\lambda_n]c$ is a Λ-module chain transformation lifting $1_C\colon C\to C$; indeed, it is the canonical comparison map of β_n to B_n. Hence, by the comparison theorem,

Corollary 2.2. *(The "Normalization Theorem".) The projection $\eta\colon \beta(\Lambda, C)\to B(\Lambda, C)$ is a chain equivalence of complexes of Λ-modules.*

The kernel of η is the Λ-module generated by the union of the images of the Λ-module maps $s_i^n\colon \beta_n\to\beta_{n+1}$ defined by

$$s_i^n(\lambda\otimes\lambda_1\otimes\cdots\otimes\lambda_n\otimes c)=\lambda\otimes\cdots\otimes\lambda_i\otimes 1\otimes\lambda_{i+1}\otimes\cdots\otimes\lambda_n\otimes c \tag{2.8}$$

for $i=0,\ldots,n$. With these s_i and d_i as in (2.7), $\beta(\Lambda, C)$ is the associated chain complex of a simplicial Λ-module and η is the simplicial normalization of Thm. VIII.6.1.

For the *bimodule bar resolution* $B(\Lambda, \Lambda)$, take C above to be Λ. Each B_n is then a Λ-bimodule; formula (2.5) with c replaced by $\lambda'\in\Lambda$ shows that ε and each ∂_n is a Λ-bimodule homomorphism. Similarly, s of (2.4) becomes a homomorphism of right Λ-modules. Hence

Corollary 2.3. *If Λ is a* K*-algebra, $\varepsilon\colon B(\Lambda, \Lambda)\to\Lambda$ is a right-Λ-split resolution of the bimodule Λ by (Λ-Λ, right Λ)-free bimodules, and a* K*-split resolution of Λ by (Λ-Λ,* K*)-free bimodules.*

The last clause does *not* mean that $B(\Lambda, \Lambda)$ is the categorical resolution for the resolvent pair (Λ-bimodules, K-modules). Note also that $B(\Lambda, C)\cong B(\Lambda, \Lambda)\otimes_\Lambda C$.

The *left bar resolution* applies to an augmented algebra $\varepsilon\colon \Lambda\to\mathrm{K}$, and is $B(\Lambda)=B(\Lambda, {}_\varepsilon\mathrm{K})$, where ${}_\varepsilon\mathrm{K}$ is K regarded as a left Λ-module by pull-back along ε. Thus $B_n(\Lambda)=\Lambda\otimes(\Lambda/\mathrm{K})^n$ is generated by elements $\lambda[\lambda_1|\ldots|\lambda_n]$, while s and ∂ are given by (2.4) and (2.5) with c omitted, and with the "outside" factor $\lambda_n c$ in the last term of (2.5) replaced by

$\varepsilon(\lambda_n)$. Thus $B(\Lambda)\to{}_\varepsilon K$ is a K-split, (Λ, K)-free resolution of the left Λ-module ${}_\varepsilon K$. In particular, when $K=Z$ and $\Lambda=Z(\Pi)$, this is the bar resolution of IV.5.

The ***reduced bar resolution*** for an augmented algebra Λ is the complex $\bar{B}(\Lambda)=K_\varepsilon\otimes_\Lambda B(\Lambda)$, so $\bar{B}_0(\Lambda)=K$ and $\bar{B}_n(\Lambda)=(\Lambda/K)^n$ for $n>0$. The contracting homotopy does not apply to $\bar{B}$, but the formula for the boundary still applies, with c and the left operator λ omitted: In (2.5) replace the operator λ_1 by $\varepsilon(\lambda_1)$ and $\lambda_n c$ by $\varepsilon(\lambda_n)$. The "reduced bar resolution" is *not* a resolution, but is useful for computations. The left and the reduced bar resolution can also be formed without normalization.

Exercises

1. For an augmented algebra Λ, let X be any relatively free K-split resolution of ${}_\varepsilon K$ by left Λ-modules. Show that the canonical comparisons (Thm. IX.6.2) φ: $B(\Lambda)\to X$, ψ: $X\to B(\Lambda)$ over the identity satisfy $\varphi\psi=1$.

2. (CARTAN.) For Λ as in Ex. 1, show that the left bar resolution $B(\Lambda)$ is characterized up to isomorphism as a K-split resolution X of ${}_\varepsilon K$ with a contracting homotopy s such that $s^2=0$ and $X_n\cong\Lambda\otimes sX_{n-1}$.

3. The normalization theorem can be proved directly. Show that a bimodule chain transformation ζ: $B(\Lambda, C)\to\beta(\Lambda, C)$ with $\varepsilon\zeta=\varepsilon$ can be defined recursively with $\zeta_0=1$, $\zeta_n e_n=s\zeta_{n-1}\partial e_n$, where $e_n=e(\bar{F}^n\square C)$. Prove that $\eta\zeta=1$, and by similar means construct a chain homotopy $\zeta\eta\simeq 1$, all for η as in Cor. 2.2.

4. For left Λ-modules C and A, show that the 1-cocycles of the cochain complex $\mathrm{Hom}_\Lambda(B(\Lambda, C), A)$ can be regarded as factor sets for K-split Λ-module extensions of A by G.

3. The Cohomology of an Algebra

The n-th cohomology module of a K-algebra Λ with coefficients in a Λ-bimodule A is the K-module

$$H^n(\Lambda, A)=H^n(\mathrm{Hom}_{\Lambda\text{-}\Lambda}(B(\Lambda,\Lambda), A)),\qquad n=0, 1, \ldots; \qquad (3.1)$$

it is a covariant functor of A. Here $\mathrm{Hom}_{\Lambda\text{-}\Lambda}$ stands for bimodule homomorphisms. According to the normalization theorem we can replace the bimodule bar resolution $B(\Lambda,\Lambda)$ here by the un-normalized bar resolution $\beta(\Lambda,\Lambda)$. Both $B(\Lambda,\Lambda)$ and $\beta(\Lambda,\Lambda)$ are right Λ-split (Λ-Λ, K-Λ) relative projective resolutions of the bimodule Λ, and also are K-split (Λ-Λ, K) relative projective resolutions of Λ, so $H^n(\Lambda, A)$ is the n-th relative Ext functor in either case, as stated in (1.3).

We call $H^n(\Lambda, A)$ the *Hochschild cohomology* modules of A, since they were originally defined by HOCHSCHILD [1945] using exactly the formulas given by the bar resolution with K a field.

The complex $\mathrm{Hom}_{\Lambda\text{-}\Lambda}(B(\Lambda,\Lambda), A)$ used in (3.1) may be described more directly. Consider K-multilinear functions f on the n-fold cartesian product $\Lambda\times\cdots\times\Lambda$ to A; call f *normalized* if $f(\lambda_1, \ldots, \lambda_n)=0$ whenever one λ_i is 1. For example, the function $[\lambda_1|\ldots|\lambda_n]c$ of (2.3) is K-multilinear and normalized. The universal property of the tensor product $B_n(\Lambda,\Lambda)=\Lambda\otimes(\Lambda/\mathrm{K})^n\otimes\Lambda$ states that each normalized K-multilinear f determines a unique bimodule homomorphism $\hat{f}\colon B_n(\Lambda,\Lambda)\to A$ such that always

$$f(1\otimes[\lambda_1|\ldots|\lambda_n]\otimes 1)=f(\lambda_1,\ldots,\lambda_n).$$

Hence $\mathrm{Hom}_{\Lambda-\Lambda}(B_n(\Lambda,\Lambda),A)$ is isomorphic to the K-module of all K-multilinear normalized f on the n-fold product. The coboundary δf is the function given, with the standard sign, as

$$\left.\begin{aligned}\delta f(\lambda_1,\ldots,\lambda_{n+1})=(-1)^{n+1}\{&\lambda_1 f(\lambda_2,\ldots,\lambda_{n+1})\\&+\sum_{i=1}^{n}(-1)^i f(\lambda_1,\ldots,\lambda_i\lambda_{i+1},\ldots,\lambda_{n+1})\\&+(-1)^{n+1}f(\lambda_1,\ldots,\lambda_n)\lambda_{n+1}\}.\end{aligned}\right\}\qquad(3.2)$$

In particular, a zero-cochain is a constant $a\in A$; its coboundary is the function $f\colon\Lambda\to A$ with $f(\lambda)=a\lambda-\lambda a$. Call an element $a\in A$ *invariant* if $\lambda a=a\lambda$ for all λ, and let A^Λ denote the sub-K-module of all such invariant elements of A; thus

$$H^0(\Lambda,A)\cong A^\Lambda=[a|\,\lambda a=a\lambda \text{ for all } \lambda\in\Lambda].\qquad(3.3)$$

Similarly, a 1-cocycle is a K-module homomorphism $f\colon\Lambda\to A$ satisfying the identity

$$f(\lambda_1\lambda_2)=\lambda_1 f(\lambda_2)+f(\lambda_1)\lambda_2,\qquad \lambda_1,\lambda_2\in\Lambda;\qquad(3.4)$$

such a function f is called a *crossed homomorphism* of Λ to A. It is a coboundary if it has the form $f_a(\lambda)=a\lambda-\lambda a$ for some fixed a; call f_a a *principal crossed homomorphism*. Therefore $H^1(\Lambda,A)$ is the K-module of all crossed homomorphisms modulo the principal ones, exactly as in the case of the cohomology of groups (IV.2).

As in the case of groups, $H^2(\Lambda,A)$ can be interpreted in terms of extensions by the algebra Λ. An *extension* by the algebra Λ is an epimorphism $\sigma\colon\Gamma\twoheadrightarrow\Lambda$ of algebras. The kernel J of σ is a two-sided ideal in Γ, hence a Γ-bimodule. For each n, let J^n denote the K-submodule of Γ generated by all products $j_1j_2\ldots j_n$ of n factors $j_i\in J$. Then $J=J^1\supset J^2\supset J^3\supset\cdots$, and each J^n is a two-sided ideal of Γ. An extension σ is said to be *cleft* if σ has an algebra homomorphism $\varphi\colon\Lambda\to\Gamma$ as right inverse ($\sigma\varphi=1_\Lambda$); that is, if Γ contains a subalgebra mapped isomorphically onto Λ by σ. An extension σ is said to be *singular* if $J=\mathrm{Ker}\,\sigma$ satisfies $J^2=0$. In each singular extension the Γ-bimodule J may be regarded as a Λ-bimodule, for $\sigma\gamma=\sigma\gamma'$ implies $(\gamma-\gamma')\in J$, so $J^2=0$ implies $\gamma j=\gamma' j$ for each $j\in J$. This defines the left action of each $\lambda=\sigma(\gamma)$ on j.

Conversely, given any Λ-bimodule A, a singular extension of A by Λ is a short exact sequence $(\varkappa, \sigma): A \rightarrowtail \Gamma \twoheadrightarrow \Lambda$ where Γ is an algebra, σ a homomorphism of algebras, A is regarded as a Γ-bimodule by pull-back along σ, and $\varkappa: A \rightarrowtail \Gamma$ is a monomorphism of Γ-bimodules. For given A and Λ, two such extensions $(\varkappa, \sigma)$ and $(\varkappa', \sigma')$ are *congruent* if there is an algebra homomorphism $\varrho: \Gamma \to \Gamma'$ with $\varkappa' = \varrho\varkappa$, $\sigma = \sigma'\varrho$. This gives the familiar commutative diagram which implies that ϱ is an isomorphism. One example of an extension of A by Λ is the *semi-direct sum*, defined to be the K-module $A \oplus \Lambda$ with product defined by $(a_1, \lambda_1)(a_2, \lambda_2) = (a_1\lambda_2 + \lambda_1 a_2, \lambda_1\lambda_2)$; with $\varkappa a = (a, 0)$, $\sigma(a, \lambda) = \lambda$, it is a singular extension of A by Λ, cleft by φ with $\varphi\lambda = (0, \lambda)$. Any cleft singular extension is congruent to this semi-direct sum.

Consider those singular extensions $(\varkappa, \sigma)$ which K-split, in the sense that there is a K-module homomorphism $u: \Lambda \dashrightarrow \Gamma$ which is a right inverse to σ. (Any cleft extension is K-split; if K is a field, any extension K-splits.) Identify each $a \in A$ with $\varkappa a \in \Gamma$, so that $\varkappa: A \to \Gamma$ is the identity injection. The right inverse u can be chosen to satisfy the "normalization" condition $u(1_\Lambda) = 1_\Gamma$, for if u does not satisfy this condition, $a_0 = u(1_\Lambda) - 1_\Gamma \in A$ and $u'(\lambda) = u(\lambda) - \lambda a_0$ is a new right inverse which is normalized. Moreover, $\sigma[u(\lambda_1\lambda_2)] = \lambda_1\lambda_2 = \sigma[u(\lambda_1)u(\lambda_2)]$, so there are uniquely determined elements $f(\lambda_1, \lambda_2) \in A$ such that

$$u(\lambda_1)\, u(\lambda_2) = f(\lambda_1, \lambda_2) + u(\lambda_1\lambda_2). \tag{3.5}$$

Call f the *factor set* of the extension corresponding to the representatives u.

Theorem 3.1. *If Λ is a K-algebra and A a Λ-bimodule, each factor set of a K-split singular algebra extension of A by Λ is a 2-cocycle of* $\mathrm{Hom}_{\Lambda-\Lambda}(B(\Lambda, \Lambda), A)$. *The assignment to each extension of the cohomology class of any one of its factor sets is a 1-1-correspondence between the set of congruence classes of K-split singular algebra extensions of A by Λ and $H^2(\Lambda, A)$. Under this correspondence the cleft extensions (in particular, the semi-direct sum) correspond to zero.*

Proof. Regard $u(\lambda)$ as a representative of λ in the extension Γ. The description of the Γ-bimodule structure of A can be written in terms of u as

$$u(\lambda)\, a = \lambda a, \quad a u(\lambda) = a\lambda, \tag{3.6}$$

for any $a \in A$, $\lambda \in \Lambda$. Since u is a K-module homomorphism,

$$u(k_1\lambda_1 + k_2\lambda_2) = k_1 u(\lambda_1) + k_2 u(\lambda_2), \quad k_i \in \mathrm{K}. \tag{3.7}$$

With the factor set f for u defined by (3.5), the rule (3.6) gives

$$[u(\lambda_1)\, u(\lambda_2)]\, u(\lambda_3) = f(\lambda_1, \lambda_2)\, \lambda_3 + f(\lambda_1\lambda_2, \lambda_3) + u(\lambda_1\lambda_2\lambda_3),$$

$$u(\lambda_1)\, [u(\lambda_2)\, u(\lambda_3)] = \lambda_1 f(\lambda_2, \lambda_3) + f(\lambda_1, \lambda_2\lambda_3) + u(\lambda_1\lambda_2\lambda_3).$$

As the product in Γ is associative,

$$\lambda_1 f(\lambda_2, \lambda_3) - f(\lambda_1\lambda_2, \lambda_3) + f(\lambda_1, \lambda_2\lambda_3) - f(\lambda_1, \lambda_2)\,\lambda_3 = 0. \tag{3.8}$$

This is exactly the condition $\delta f = 0$ that the factor set be a 2-cocycle; moreover the choice $u(1) = 1$ implies that f is normalized. A change in the choice of u to u' by $u'(\lambda) = g(\lambda) + u(\lambda)$ for any K-linear $g: \Lambda \to A$ with the normalization $g(1) = 0$ gives a new (normalized) factor set $f + \delta g$. Thus the extension uniquely determines the cohomology class of f.

Any element in the algebra Γ can be written uniquely as $a + u(\lambda)$. The K-module structure of Γ and the sum and product of any two such elements are determined by the equations (3.5)–(3.7). Given A, Λ, and any 2-cocycle f, these equations construct the extension Γ; in particular, the condition $\delta f = 0$ suffices to make the product in Γ associative. When $f = 0$, the construction is the semi-direct sum, so the proof is complete.

A two-sided ideal J is said to be *nilpotent* if $J^n = 0$ for some n.

Theorem 3.2. (J. H. C. Whitehead-Hochschild.) *If* K *is a field and if the* K*-algebra* Λ *has* $H^2(\Lambda, A) = 0$ *for every* Λ*-bimodule* A*, then any extension of* Λ *with a nilpotent kernel is cleft.*

Let the extension $\sigma: \Gamma \twoheadrightarrow \Lambda$ have kernel J with $J^n = 0$. The proof will be by induction on n. If $n = 2$, the extension is singular and K-split; since $H^2(\Lambda, J) = 0$, the extension is cleft by Thm. 3.1.

Suppose the result true for kernels with exponent $n - 1$, and take σ with kernel $J \neq 0$, $J^n = 0$. Then J^2 is properly contained in J, since $J^2 = J$ would give $J^n = J \neq 0$. From the quotient algebra Γ/J^2, form the commutative diagram on the left in

$$\begin{array}{ccccc} J & \to & \Gamma & \longrightarrow & \Lambda \\ & & \downarrow{\scriptstyle p} & & \| \\ J/J^2 & \to & \Gamma/J^2 & \underset{\varphi}{\overset{\sigma'}{\rightleftarrows}} & \Lambda, \end{array} \qquad \begin{array}{ccc} \Gamma' & \xrightarrow{\iota} & \Gamma \\ {\scriptstyle \varphi'}\uparrow\downarrow{\scriptstyle p'} & & \downarrow{\scriptstyle p} \\ \varphi\Lambda & < & \Gamma/J^2. \end{array}$$

The projection p has kernel J^2, while σ' has kernel J/J^2, hence is a singular extension of Λ. By the case $n = 2$, σ' is cleft by some φ. Now $p^{-1}(\varphi\Lambda) = \Gamma'$ is a subalgebra of Γ, and p induces $p': \Gamma' \twoheadrightarrow \varphi\Lambda \cong \Lambda$ with kernel J^2. Since $(J^2)^{n-1} < J^n = 0$, the induction assumption shows p' cleft by some φ', so σ is cleft by $\iota\varphi'\varphi$.

This result includes the Principal Theorem of Wedderburn for an algebra Γ of finite dimension (as a vector space) over a field. Each such algebra has a two-sided nilpotent ideal R, called the radical, such that Γ/R is semi-simple. The Wedderburn Theorem asserts that if Γ/R is separable, then the extension $\Gamma \to \Gamma/R$ is cleft. This follows from

Thm. 3.2, for Γ/R separable implies (Thm. VII.5.6) bidim $\Gamma/R=0$, hence bidim $\Gamma/R\leqq 1$, hence $H^2(\Gamma/R, A)=0$ for all (Γ/R)-bimodules A, hence $\Gamma\to\Gamma/R$ cleft.

Note. For algebras of finite dimension over a field, Thm. 3.2 is also valid without the hypothesis that the kernel is nilpotent [HOCHSCHILD 1945, Prop. 6.1]; [ROSENBERG-ZELINSKY 1956]. The obstruction problem for the construction of non-singular K-split extensions with a given kernel [HOCHSCHILD 1947] leads to an interpretation of $H^3(\Lambda, A)$ parallel to that for groups (IV.8). Extensions which are not K-split require a second, additive, factor set in place of the linearity of u in (3.7); we return to this question in § 13.

The cohomology groups of a fixed K-algebra Λ are characterized by axioms like those for Ext, as follows.

Theorem 3.3. *For each $n\geqq 0$, $H^n(\Lambda, A)$ is a covariant functor of the Λ-bimodule A to K-modules. H^0 is given by (3.3). $H^n(\Lambda, A)=0$ when $n>0$ and A is a bimodule with $A=\operatorname{Hom}_K(\Lambda\otimes\Lambda, M)$ for M a K-module. For each K-split short exact sequence E: $A\rightarrowtail B\twoheadrightarrow C$ of bimodules and each $n\geqq 0$ there is a connecting homomorphism $E_*: H^n(\Lambda, C)\to H^{n+1}(\Lambda, A)$, natural in E, such that the long sequence*

$$\cdots\to H^n(\Lambda, A)\to H^n(\Lambda, B)\to H^n(\Lambda, C)\xrightarrow{E_*} H^{n+1}(\Lambda, A)\to\cdots$$

is exact. These properties determine H^n and the connecting homomorphisms E_* up to natural isomorphisms of H^n.

The proof is left to the reader; note that $\operatorname{Hom}_K(\Lambda\otimes\Lambda, M)$ is a "relatively injective" bimodule.

If $\varepsilon:\Lambda\to K$ is an augmented algebra, each left Λ-module D becomes a Λ-bimodule D_ε by pull-back on the right along the augmentation.

Proposition 3.4. *For a left module D over an augmented algebra (Λ, ε) the Hochschild cohomology of the bimodule D_ε can be computed from the left bar resolution by a natural isomorphism*

$$H^n(\Lambda, D_\varepsilon)\cong H^n(\operatorname{Hom}_\Lambda(B(\Lambda), D)). \tag{3.9}$$

Proof. The canonical isomorphism $\operatorname{Hom}(K, D)\cong D$ of left Λ-modules is also an isomorphism $\operatorname{Hom}({}_\varepsilon K, D)\cong D_\varepsilon$ of Λ-bimodules. Thus, for any bimodule B, adjoint associativity yields a natural isomorphism

$$\operatorname{Hom}_\Lambda(B\otimes_\Lambda({}_\varepsilon K), D)\cong\operatorname{Hom}_{\Lambda-\Lambda}(B, \operatorname{Hom}({}_\varepsilon K, D))\cong\operatorname{Hom}_{\Lambda-\Lambda}(B, D_\varepsilon).$$

When B is the two-sided bar resolution, $B\otimes_\Lambda({}_\varepsilon K)$ is the left bar resolution; hence the result (3.9).

Note. Suppose that the K-algebra Λ is projective as a K-module. Then Λ^n is K-projective (Cor. V.3.3), hence $\beta_n(\Lambda, \Lambda)$ is a projective Λ-bimodule (Prop. VI.8.1). Hence ε: $\beta(\Lambda, \Lambda)\to\Lambda$ is a projective bimodule resolution of Λ. In this case H^n of (3.1) is therefore given as an "absolute" functor Ext:

$$H^n(\Lambda, A)\cong\operatorname{Ext}^n_{\Lambda-\Lambda}(\Lambda, A)\qquad(\text{if }\Lambda\text{ is K-projective}).$$

Using B for β, the same result holds for Λ/K projective as a K-module. CARTAN-EILENBERG define the "Hochschild" cohomology by the absolute Ext functor in *all* cases, so their definition does not always agree with ours.

Exercises

1. Show that A^Λ is a sub-Λ-bimodule of A when Λ is commutative.

2. Construct a "Baer sum" of extensions of A by Λ so that the correspondence of Thm. 3.1 maps the Baer sum into the sum in $H^2(\Lambda, A)$.

3. Show that $H^1(\Lambda, A)$ is the group of congruence classes of those bimodule extensions $A \rightarrowtail B \twoheadrightarrow \Lambda$ which K-split.

4. Show explicitly that each short exact sequence $A \rightarrowtail B \twoheadrightarrow \Lambda$ of bimodules which is K-split is also split as a sequence of right Λ-modules.

5. If Λ is an augmented algebra and M a K-module then the cohomology of M, pulled back to be a bimodule, may be calculated from the reduced bar resolution as $H^n(\Lambda, {}_\varepsilon M_\varepsilon) \cong H^n(\mathrm{Hom}(\overline{B}(\Lambda), M))$.

4. The Homology of an Algebra

Two Λ-bimodules A and B have a "bimodule" tensor product $A \otimes_{\Lambda-\Lambda} B$; it is obtained from the tensor product $A \otimes_{\mathrm{K}} B$ by the identifications

$$a\lambda \otimes b = a \otimes \lambda b, \qquad \lambda a \otimes b = a \otimes b \lambda$$

(middle associativity and outside associativity, as in (VI.5.10)). The canonical isomorphism $A \otimes_\Lambda \Lambda \cong A$ has an analogue for bimodules. Indeed, if A is a bimodule and M is a K-module, a natural isomorphism

$$\theta\colon A \otimes_{\Lambda-\Lambda} (\Lambda \otimes M \otimes \Lambda) \cong A \otimes M, \qquad \otimes = \otimes_{\mathrm{K}}, \tag{4.1}$$

may be defined by $\theta[a \otimes (\lambda \otimes m \otimes \lambda')] = \lambda' a \lambda \otimes m$, for the expression on the right is K-multilinear and satisfies the middle and outside associativity rules. The inverse is given by $\theta^{-1}(a \otimes m) = a \otimes (1 \otimes m \otimes 1)$.

The *Hochschild homology* modules of a K-algebra Λ with coefficients in a Λ-bimodule A are defined via the bar resolution to be the K-modules

$$H_n(\Lambda, A) = H_n(A \otimes_{\Lambda-\Lambda} B(\Lambda, \Lambda)), \qquad n = 0, 1, \ldots, . \tag{4.2}$$

As for cohomology, this is an instance (1.4) of the relative torsion functor, for sequences of Λ-bimodules split either as sequences of right Λ-modules or as sequences of K-modules.

In the definition (4.2) we may replace B by the un-normalized bar resolution $\beta(\Lambda, \Lambda)$ with $\beta_n(\Lambda, \Lambda) = \Lambda \otimes \Lambda^n \otimes \Lambda$. By (4.1), $A \otimes_{\Lambda-\Lambda} \beta_n(\Lambda, \Lambda) \cong A \otimes \Lambda^n$. Hence $H_n(\Lambda, A)$ is the n-th homology module of the complex of K-modules $A \otimes \Lambda^n$ with a boundary $\partial = d_0 - d_1 + \cdots + (-1)^n d_n$, where the d_i are "simplicial" faces:

$$\left.\begin{aligned} d_i(a \otimes \lambda_1 \otimes \cdots \otimes \lambda_n) &= a\lambda_1 \otimes \lambda_2 \otimes \cdots \otimes \lambda_n, & i &= 0, \\ &= a \otimes \lambda_1 \otimes \cdots \otimes \lambda_i \lambda_{i+1} \otimes \cdots \otimes \lambda_n, & 0 &< i < n, \\ &= \lambda_n a \otimes \lambda_1 \otimes \cdots \otimes \lambda_{n-1}, & i &= n; \end{aligned}\right\} \tag{4.3}$$

in the last term, λ_n appears in front in virtue of the "outside" associativity rule. In particular $\partial(a\otimes\lambda)=a\,\lambda-\lambda\,a$, so that H_0 is the quotient of A,

$$H_0(\Lambda, A)\cong A/\{\lambda\, a-a\,\lambda \mid \lambda\in\Lambda, a\in A\}, \tag{4.4}$$

by the sub-K-module generated by all differences $\lambda\, a-a\,\lambda$.

Much as in Thm. 3.3 we have

Theorem 4.1. *For a fixed* K*-algebra* Λ*, each* $H_n(\Lambda, A)$ *is a covariant functor of the* Λ*-bimodule* A *to* K*-modules, with* H_0 *given by* (4.4) *and*

$$H_n(\Lambda, \Lambda\otimes L\otimes\Lambda)=0, \qquad n>0,\ L \text{ a K-module.}$$

If $E\colon A\rightarrowtail B\twoheadrightarrow C$ *is a* K*-split short exact sequence of bimodules, there is for each* $n>0$ *a "connecting" homomorphism* $E_n\colon H_n(\Lambda, C)\to H_{n-1}(\Lambda, A)$*, natural in* E*, such that the long sequence*

$$\cdots\to H_{n+1}(\Lambda, C)\xrightarrow{E_{n+1}} H_n(\Lambda, A)\to H_n(\Lambda, B)\to H_n(\Lambda, C)\to\cdots$$

is exact. These properties characterize the H_n *and* E_n *up to natural isomorphism.*

The functorial behavior of the homology of algebras is like that for groups (IV.2.6). Consider quadruples $(\mathrm{K}, \Lambda, A, C)$ where K is a commutative ring, Λ a K-algebra, and A, C are Λ-bimodules. A *change of algebras* (+ in A, − in C) is a quadruple

$$\zeta=(\varkappa, \varrho, \alpha, \gamma)\colon (\mathrm{K}, \Lambda, A, C)\to(\mathrm{K}', \Lambda', A', C') \tag{4.5}$$

where $\varkappa\colon \mathrm{K}\to\mathrm{K}'$ and $\varrho\colon \Lambda\to\Lambda'$ are ring homomorphisms such that always $\varrho(k\lambda)=(\varkappa k)(\varrho\lambda)$ and where $\alpha\colon A\to{}_{\varrho}A'_{\varrho}$ and $\gamma\colon {}_{\varrho}C'_{\varrho}\to C$ (opposite direction!) are homomorphisms of Λ-bimodules; i.e., $\alpha(\lambda a)=(\varrho\lambda)(\alpha a)$ and $\alpha(a\,\lambda)=(\alpha\, a)(\varrho\,\lambda)$. The category with these morphisms ζ is denoted $\mathscr{B}^{+-}$; here the exponent $+-$ indicates that the change is covariant in the first bimodule A and contravariant in C. Omitting C and γ gives the category $\mathscr{B}^{+}$. We also use the category $\mathscr{B}^{+}_{\mathrm{K}}$, with $\mathrm{K}=\mathrm{K}'$ fixed and $\varkappa$ the identity.

The complex $A\otimes_{\Lambda-\Lambda}B(\Lambda, \Lambda)$ of (4.2) and hence $H_n(\Lambda, A)$ is a covariant functor on $\mathscr{B}^{+}$; in particular, this gives the previous result that $H_n(\Lambda, A)$ for Λ and K fixed is covariant in A. Similarly, the cohomology $H^n(\Lambda, C)$ is a contravariant functor on $\mathscr{B}^{-}$. The action of a change ζ (with α omitted) on a normalized cochain f for Λ', of the form (3.2), is defined by $(\zeta^* f)(\lambda_1, \ldots, \lambda_n)=\gamma f(\varrho\lambda_1, \ldots, \varrho\lambda_n)$.

Exercises

1. Show that the isomorphism (1.1) is natural over the category $\mathscr{B}^{-}$.

2. Let $\varepsilon\colon \Lambda\to\mathrm{K}$ be an augmented algebra. For M a right Λ-module and G a K-module, prove that

$$H_n(\Lambda, {}_{\varepsilon}M)=H_n(M\otimes_{\Lambda}B(\Lambda)), \qquad H_n(\Lambda, {}_{\varepsilon}G_{\varepsilon})=H_n(G\otimes\overline{B}(\Lambda)).$$

5. The Homology of Groups and Monoids

The cohomology of a group Π was treated in Chap. IV, using the functor Hom_Π for $Z(\Pi)$-modules. Now that we have at hand the tensor product $\otimes_\Pi = \otimes_{Z(\Pi)}$ we can define and study the homology of a group Π. It is just as easy to do this for a monoid M, though the added generality is not of great moment.

A *monoid* is a set M with a distinguished element $1 = 1_M$ and a function assigning to each pair $x, y \in M$ a "product" $xy \in M$ in such a way that always $(xy)z = x(yz)$ and $1x = x = x1$. The *monoid ring* $Z(M)$, like the group ring, consists of all finite sums $\sum k_i x_i$ with $k_i \in Z$, $x_i \in M$, under the obvious product, and with augmentation the ring homomorphism $\varepsilon: Z(M) \to Z$ defined by $\varepsilon(\sum k_i x_i) = \sum k_i$. This ring $Z(M)$ may be regarded as the free ring on the monoid M in the sense of Prop. IV.1.1. By a left M-module A we mean a left $Z(M)$-module, and we write $\otimes_M$ for $\otimes_{Z(M)}$. If M is a free commutative monoid on n generators, $Z(M)$ is the polynomial ring in n indeterminates.

The homology of M with coefficients in a right module G_M is now defined by the left bar resolution $B(Z(M))$ as

$$H_n(M, G) = H_n(G \otimes_M B(Z(M))), \qquad n = 0, 1, \ldots. \tag{5.1}$$

Since $B(Z(M))$ is a Z-split projective resolution of the left M-module $Z = {}_\varepsilon Z$, we may also write this definition in terms of the relative torsion product as

$$H_n(M, G) = \mathrm{Tor}_n^{(Z(M), Z)}(G, Z) \cong \mathrm{Tor}_n^{Z(M)}(G, Z). \tag{5.2}$$

In particular, $H_0(M, G) = G \otimes_M Z$. We leave to the reader the description of the cohomology of a monoid.

For a free module the higher torsion products vanish, hence

Proposition 5.1. *For Π a group and F a free Π-module*

$$H_0(\Pi, F) \cong F \otimes_\Pi Z, \qquad H_n(\Pi, F) = 0, \qquad n > 0.$$

Note that if F is the free Π-module on generators $\{t\}$, then $F \otimes_\Pi Z$ is the free abelian group on the generators $\{t \otimes 1\}$.

The *commutator subgroup* $[\Pi, \Pi]$ is the subgroup of Π generated by all commutators $xyx^{-1}y^{-1}$ for x, y in Π. It is a normal subgroup of Π; the factor group $\Pi/[\Pi, \Pi]$ is abelian, and any homomorphism of Π into any abelian group has kernel containing $[\Pi, \Pi]$.

Proposition 5.2. *For Π a group and Z the trivial Π-module*

$$H_0(\Pi, Z) \cong Z, \qquad H_1(\Pi, Z) \cong \Pi/[\Pi, \Pi]. \tag{5.3}$$

Proof. The homology of Z is that of the complex $Z \otimes_\Pi B(Z(\Pi))$ which is the reduced bar resolution $\overline{B}(Z(\Pi))$ of § 2, with $\overline{B}_0 = Z$, $\overline{B}_1$

and $\overline{B}_2$ the free abelian groups on generators $[x]$ and $[x|y]$ for $x \neq 1 \neq y$, and with boundaries $\partial[x]=0$, $\partial[x|y]=[y]-[xy]+[x]$. This gives $H_0 \cong Z$ and each $[x]$ a cycle. By the boundary formula, its homology class satisfies $\operatorname{cls}[xy]=\operatorname{cls}[x]+\operatorname{cls}[y]$. Hence $\varphi x = \operatorname{cls}[x]$ gives a homomorphism $\varphi\colon \Pi/[\Pi,\Pi] \to H_1(\Pi, Z)$. Since $\overline{B}_1$ is free abelian, $[x] \to x[\Pi,\Pi]$ defines a homomorphism $\overline{B}_1 \to \Pi/[\Pi,\Pi]$ which annihilates all boundaries. Thus an inverse of φ may be defined as $\varphi^{-1}\operatorname{cls}[x] = x[\Pi,\Pi]$, so φ is an isomorphism, as required for the second equation of (5.3).

The homology of a group (or a monoid) is a special case of the Hochschild homology of its group ring.

Proposition 5.3. *For a right module G over the monoid M there is an isomorphism $H_n(M, G) \cong H_n(Z(M), {}_{\varepsilon}G)$ of the homology of the monoid M to that of the algebra $Z(M)$. This isomorphism is natural in G.*

Proof. Take $\Lambda = Z(M)$, a Z-algebra. For any Λ-bimodule B an isomorphism ${}_{\varepsilon}G \otimes_{\Lambda-\Lambda} B \cong G \otimes_{\Lambda} (B \otimes_{\Lambda\,\varepsilon} Z)$ is given by $g \otimes b \to g \otimes (b \otimes 1)$. Apply this with $B = B(\Lambda, \Lambda)$; it shows the complex used to define the homology of Λ over ${}_{\varepsilon}G$ is isomorphic to the complex used to define the homology of M over G.

A corresponding result for cohomology is

Proposition 5.4. *For left Π-modules A there is a natural isomorphism*

$$H^n(\Pi, A) \cong H^n(Z(\Pi), A_{\varepsilon}).$$

Proof. This is a consequence of Prop. 3.4, for the cohomology of the group Π on the left was defined by $B(Z(\Pi))$, that of the algebra $Z(\Pi)$ by $B(Z(\Pi), Z(\Pi))$.

These propositions reduce the (co)homology of groups to that of algebras. Conversely, the (co)homology of the Z-algebra $Z(\Pi)$ reduces to that of the group Π. This reduction depends on two special properties of the group ring $Z(\Pi)$. First, $\psi x = x \otimes x$ defines a ring homomorphism $\psi\colon Z(\Pi) \to Z(\Pi) \otimes Z(\Pi)$; indeed, ψ is the coproduct which makes $Z(\Pi)$ a Hopf algebra (VI.9). Second, $Z(\Pi)$ is canonically isomorphic to its opposite ring. Indeed, if the opposite ring $Z(\Pi)^{\mathrm{op}}$ consists as usual of elements r^{op} for $r \in Z(\Pi)$ with product $r^{\mathrm{op}} s^{\mathrm{op}} = (s\,r)^{\mathrm{op}}$, then the function $\xi(x) = (x^{-1})^{\mathrm{op}}$ on Π to $Z(\Pi)^{\mathrm{op}}$ has $\xi(1)=1$, $\xi(xy) = (\xi x)(\xi y)$, hence extends (Prop. IV.1.1) to a ring homomorphism $\xi\colon Z(\Pi) \to Z(\Pi)^{\mathrm{op}}$ which is clearly an isomorphism. Composition with the coproduct gives a ring homomorphism

$$\chi\colon Z(\Pi) \xrightarrow{\psi} Z(\Pi) \otimes Z(\Pi) \xrightarrow{1 \otimes \xi} Z(\Pi) \otimes Z(\Pi)^{\mathrm{op}}; \qquad (5.4)$$

it is that ring homomorphism χ which extends the multiplicative map $\chi(x) = x \otimes (x^{-1})^{\mathrm{op}}$.

This map χ allows a reduction of the (bimodule) cohomology of the algebra $Z(\Pi)$ to the cohomology of the group Π. Each bimodule ${}_\Pi C_\Pi$ is a left $Z(\Pi)\otimes Z(\Pi)^{\mathrm{op}}$-module and hence also a left Π-module ${}_\chi C$, by pull-back along χ. These new left operators of Π on C will be denoted as $x\circ c$ for $x\in\Pi$; they are not the original left operators, but are given in terms of the bimodule operators as $x\circ c=x\,c\,x^{-1}$. Similarly, C_χ denotes the right Π-module with operators $c\circ x=x^{-1}c\,x$.

Theorem 5.5. *For a group Π and a Π-bimodule C there are natural isomorphisms*

$$H^n(Z(\Pi),C)\cong H^n(\Pi,{}_\chi C),\qquad H_n(\Pi,C_\chi)\cong H_n(Z(\Pi),C) \tag{5.5}$$

induced by the chain transformation $h\colon B(Z(\Pi))\to B(Z(\Pi),Z(\Pi))$ defined as

$$h_n(x[x_1|\ldots|x_n])=x[x_1|\ldots|x_n](x\,x_1\ldots x_n)^{-1},\qquad x_i\in\Pi.$$

In brief, "two-sided" operators in the cohomology of groups reduce to "one-sided" operators (Eilenberg-Mac Lane [1947], § 5).

For this proof, write $B^L=B(Z(\Pi))$ for the left bar resolution and $B=B(Z(\Pi),Z(\Pi))$ for the bimodule bar resolution. Since B_n^L is the free abelian group on generators $x[x_1|\ldots|x_n]$, the formula given defines h_n as a homomorphism $B_n^L\to B_n$ of abelian groups. For a left operator $y\in\Pi$,

$$h_n(y\,x[x_1|\ldots|x_n])=y\{x[x_1|\ldots|x_n](x\,x_1\ldots x_n)^{-1}\}y^{-1};$$

this shows $h\colon B^L\to{}_\chi B$ a homomorphism of left Π-modules. Now consider the diagram

$$\begin{array}{ccccccc}
Z & \underset{s_{-1}}{\overset{\varepsilon^L}{\leftrightarrows}} & B_0^L & \underset{s_0}{\overset{\partial}{\leftrightarrows}} & B_1^L & \underset{s_1}{\leftrightarrows} & B_2^L\ldots\\
{\scriptstyle I}\downarrow & & \downarrow{\scriptstyle h_0} & & \downarrow{\scriptstyle h_1} & & \downarrow{\scriptstyle h_2}\\
Z(\Pi) & \underset{s_1}{\overset{\varepsilon}{\leftrightarrows}} & B_0 & \overset{\partial}{\leftrightarrows} & B_1 & \leftrightarrows & B_2\ldots
\end{array} \tag{5.6}$$

with $I\colon Z\to Z(\Pi)$ the injection. The contracting homotopies s above and below are both defined by "moving the front argument inside", hence the commutativity $hs=sh$ (with $h_{-1}=I$). Then ε and ∂ above and below are uniquely determined recursively by the fact that s is a contracting homotopy; it follows that $h\partial=\partial h$, $I\varepsilon^L=\varepsilon h_0$. Alternatively, these commutativities may be verified directly; only the initial and final terms in the boundary formulas require attention. Thus $h\colon B^L\to B$ is a chain transformation.

Now let h^* be the induced map on the cochain complexes $\operatorname{Hom}(B,C)$. Composition with the pull-back $\operatorname{Hom}_{\Pi-\Pi}\to\operatorname{Hom}_\Pi$ gives the cochain transformation

$$\varphi\colon \operatorname{Hom}_{\Pi-\Pi}(B,C)\to\operatorname{Hom}_\Pi({}_\chi B,{}_\chi C)\xrightarrow{h^*}\operatorname{Hom}_\Pi(B^L,{}_\chi C).$$

Explicitly, for an n-cochain f on the left, φf is

$$(\varphi f)(x_1, \ldots, x_n) = f(h[x_1| \ldots |x_n]) = [f(x_1, \ldots, x_n)](x_1 \ldots x_n)^{-1}.$$

But for an n-cochain g of B^L an inverse to φ is given by

$$(\varphi^{-1} g)(x_1, \ldots, x_n) = [g(x_1, \ldots, x_n)](x_1 \ldots x_n).$$

Thus φ is an isomorphism on cochains, hence on cohomology. The argument on homology is similar.

The isomorphisms of this theorem may be described in more invariant terms as an instance of change of rings. In homology, regard $H_n(Z(\Pi), C)$ via (1.4) as the relative torsion product $\operatorname{Tor}_n(C, Z(\Pi))$ for the pair of rings $(Z(\Pi) \otimes Z(\Pi)^{\text{op}}, Z)$, and $H_n(\Pi, C_\chi)$ via (5.2) as the relative torsion product $\operatorname{Tor}_n(C_\chi, {}_\varepsilon Z)$ for the pair of rings $Z(\Pi), Z$. Now χ and $I: Z \to Z(\Pi)$ yield a morphism

$$(\chi, 1_C, I): (Z(\Pi), Z; C_\chi, {}_\varepsilon Z) \to (Z(\Pi) \otimes Z(\Pi)^{\text{op}}, Z; C, Z(\Pi))$$

in the "change of rings" category $\mathscr{R}^{++}$ of (IX.8.10). The diagram (5.6) displays h as the chain transformation found in IX.8 from the comparison theorem, so the isomorphism of the present theorem is just the induced map $(\chi, 1_C, I)_*$ for relative torsion products in the change of rings.

Note. Among explicit calculations of the cohomology and homology of groups we cite LYNDON [1950] for groups with one defining relation; GRUENBERG [1960] for a resolution constructed from a free presentation of Π; WALL [1961] for a "twisted product" resolution for a group extension.

Exercises

1. (CARTAN-EILENBERG, p. 201.) For an abelian group G regarded as a trivial Π-module the homology and cohomology can be calculated from the reduced bar resolution. Establish the exact sequences

$$0 \to H_n(\Pi, Z) \otimes G \to H_n(\Pi, G) \to \operatorname{Tor}(H_{n-1}(\Pi, Z), G) \to 0,$$
$$0 \to \operatorname{Ext}(H_{n-1}(\Pi, Z), G) \to H^n(\Pi, G) \to \operatorname{Hom}(H_n(\Pi, Z), G) \to 0.$$

2. For G an abelian group, show $H_1(\Pi, G) \cong G \otimes (\Pi/[\Pi, \Pi])$.

3. Study the effect of conjugation on $H_n(\Pi, G)$ (cf. Prop. IV.5.6).

4. (CARTAN-EILENBERG, Cor. X.4.2.) If the abelian group Π contains a monoid M which generates Π as a group, then each Π-module A or G is also an M-module. Show that the injection $M \to \Pi$ induces isomorphisms

$$H^n(\Pi, A) \cong H^n(M, A), \quad H_n(M, G) \cong H_n(\Pi, G).$$

6. Ground Ring Extensions and Direct Products

This section will study the effect upon Hochschild homology and cohomology of certain standard constructions on algebras: Ground ring extensions and direct products. Tensor products will be treated in § 7.

Consider the ground ring extension from K to a commutative K-algebra R. Each K-algebra Λ yields an R-algebra $\Lambda^R = R \otimes \Lambda$; there are ring homomorphisms j_K: $K \to R$ and j_Λ: $\Lambda \to \Lambda^R$ given by $j_K(k) = k1_R$ and $j_\Lambda(\lambda) = 1_R \otimes \lambda$, so that (j_K, j_Λ): $(K, \Lambda) \to (R, \Lambda^R)$ is a change of algebras. Each Λ^R-module or bimodule pulls back along j_Λ to be a Λ-module or bimodule. There is also a passage in the opposite direction. Each K-module M determines an R-module $M^R = R \otimes M$ and a homomorphism j_M: $M \to M^R$ of K-modules given as $j_M(m) = 1 \otimes m$. Each K-module homomorphism μ: $M \to N$ determines an R-module homomorphism μ^R: $M^R \to N^R$ by $\mu^R(r \otimes m) = r \otimes \mu m$, so that $\mu^R j_M = j_N \mu$. Thus $T^R(M) = M^R$, $T^R(\mu) = \mu^R$ is a covariant functor on K-modules to R-modules. This functor preserves tensor products (with $\otimes$ for $\otimes_K$, as always), since j_M and j_N yield a natural isomorphism

$$\varphi\colon (M \otimes N)^R \cong M^R \otimes_R N^R, \quad \varphi(r \otimes (m \otimes n)) = r\, j_M m \otimes j_N n \tag{6.1}$$

with an inverse given by $\varphi^{-1}[(r \otimes m) \otimes_R (r' \otimes n)] = r r' \otimes m \otimes n$. We regard φ as an identification.

For any R-module U and any K-module M there is a natural isomorphism

$$\psi\colon U \otimes M \cong U \otimes_R M^R, \quad \psi(u \otimes m) = u \otimes_R j_M m \tag{6.2}$$

of R-modules, where $U \otimes M$ on the left is an R-module via the R-module structure of the left factor U. The inverse of ψ is given by $\psi^{-1}(u \otimes_R (r \otimes m)) = ur \otimes m$. There is a similar natural isomorphism of R-modules

$$\chi\colon \operatorname{Hom}(M, U) \cong \operatorname{Hom}_R(M^R, U), \quad (\chi f)(r \otimes m) = r f(m) \tag{6.3}$$

with inverse defined for each R-module homomorphism g: $M^R \to U$ as $(\chi^{-1} g)(m) = g(1 \otimes m)$.

The homology and cohomology of an extended algebra Λ^R with coefficients in any Λ^R-bimodule A is entirely determined by that of Λ with coefficients in A pulled back along j_Λ: $\Lambda \to \Lambda^R$ to be a Λ-bimodule:

Theorem 6.1. *For K-algebras Λ and R, R commutative, and for each Λ^R-bimodule A there are natural isomorphisms*

$$\tau_*\colon H_n(\Lambda, {}_jA_j) \cong H_n(\Lambda^R, A), \quad \sigma^*\colon H^n(\Lambda^R, A) \cong H^n(\Lambda, {}_jA_j)$$

of R-modules, where $H(\Lambda, {}_jA_j)$ is an R-module through the R-module structure of A. Here τ_* is induced by the change of algebras $\tau = (j_K, j_\Lambda, 1_A)$: $(K, \Lambda, {}_jA_j) \to (R, \Lambda^R, A)$ in $\mathscr{B}^+$, and σ^* by $\sigma = (j_K, j_\Lambda, 1_A)$ in $\mathscr{B}^-$ (cf. § 4).

Proof. On the un-normalized complexes for homology, τ_*: $A \otimes \Lambda^n \to A \otimes_R (\Lambda^R)^n$ is just the composite of ψ: $A \otimes \Lambda^n \cong A \otimes_R (\Lambda^n)^R$ with φ: $(\Lambda^n)^R \cong (\Lambda^R)^n$. By (6.1) and (6.2) both are isomorphisms, hence τ_* is an isomorphism for the complexes and hence for their homology $H_n(\Lambda, {}_jA_j)$,

$H_n(\Lambda^R, A)$. The argument for cohomology is analogous, using χ in place of ψ.

The direct product $\Lambda = \Gamma \times \Sigma$ of two K-algebras may be treated as a special case of the direct product of two rings (IX.9). The Hochschild cohomology $H^n(\Lambda, A)$ is $\operatorname{Ext}^n_{(R,S)}(\Lambda, A)$, where $R = \Lambda \otimes \Lambda^{\mathrm{op}}$ is

$$(\Gamma \times \Sigma) \otimes (\Gamma \times \Sigma)^{\mathrm{op}} \cong (\Gamma \otimes \Gamma^{\mathrm{op}}) \times (\Gamma \otimes \Sigma^{\mathrm{op}}) \times (\Sigma \otimes \Gamma^{\mathrm{op}}) \times (\Sigma \otimes \Sigma^{\mathrm{op}}),$$

while S is the image of I: $\mathrm{K} \to \Lambda \otimes \Lambda^{\mathrm{op}}$; the projection of this image on any one of the four direct factors of $\Lambda \otimes \Lambda^{\mathrm{op}}$ above is then the corresponding image of K in that factor. Prop. IX.9.2 asserts that each Λ-bimodule A has a canonical decomposition

$$A = {}'A' \oplus {}'A'' \oplus {}''A' \oplus {}''A'', \qquad {}_{\Gamma}'A'_{\Gamma}, \quad {}_{\Gamma}'A''_{\Sigma}, \quad {}_{\Sigma}''A'_{\Gamma}, \quad {}_{\Sigma}''A''_{\Sigma}, \tag{6.4}$$

into the bimodules shown; explicitly, ${}'A' = \Gamma \otimes_\Lambda A \otimes_\Lambda \Gamma$, etc. In particular, the Λ-bimodule Λ is represented as the direct sum $\Lambda = \Gamma \oplus \Sigma$ of just two non-vanishing components; the Γ-bimodule Γ and the Σ-bimodule Σ. Thm. IX.9.4 for the case of four direct factors now implies

Theorem 6.2. *For each $(\Gamma \times \Sigma)$-bimodule A there are natural isomorphisms*

$$H^n(\Gamma \times \Sigma, A) \cong H^n(\Gamma, \Gamma \otimes_\Lambda A \otimes_\Lambda \Gamma) \oplus H^n(\Sigma, \Sigma \otimes_\Lambda A \otimes_\Lambda \Sigma), \tag{6.5}$$

$$H_n(\Gamma \times \Sigma, A) \cong H_n(\Gamma, \Gamma \otimes_\Lambda A \otimes_\Lambda \Gamma) \oplus H_n(\Sigma, \Sigma \otimes_\Lambda A \otimes_\Lambda \Sigma). \tag{6.6}$$

Specifically, the projections $\Gamma \times \Sigma \to \Gamma$ and $A \to \Gamma \otimes_\Lambda A \otimes_\Lambda \Gamma$ yield a morphism ζ' in the change of algebras category $\mathscr{B}^+_{\mathrm{K}}$ of § 4, hence a map ζ'_*: $H_n(\Gamma \times \Sigma, A) \to H_n(\Gamma, \Gamma \otimes_\Lambda A \otimes_\Lambda \Gamma)$. Replacement of Γ by Σ gives ζ''_*; the isomorphism (6.6) is $h \to (\zeta'_* h, \zeta''_* h)$. Similarly the isomorphism (6.5), in the opposite direction, is induced by the projection $\Gamma \times \Sigma \to \Gamma$ and the injection $\Gamma \otimes_\Lambda A \otimes_\Lambda \Gamma \to A$ in $\mathscr{B}^-_{\mathrm{K}}$.

Exercises

1. If Π is a group and K a commutative ring, give a direct description of the augmented K-algebra $Z(\Pi)^{\mathrm{K}}$. (It is called the *group algebra* of Π over K.)

2. If A is a Λ-bimodule, show that there is a unique Λ^R-bimodule structure on A^R such that $(j_{\mathrm{K}}, j_\Lambda, j_A)$: $(\mathrm{K}, \Lambda, A) \to (R, \Lambda^R, A^R)$ is a change of rings in $\mathscr{B}^+$. Derive a natural homomorphism $H_n(\Lambda, A) \to H_n(\Lambda^R, A^R)$ and show by example that it need not be an isomorphism. Note also that A^R pulled back by j_Λ to be a Λ-bimodule is not identical with A.

7. Homology of Tensor Products

Consider the tensor product $\Lambda \otimes \Lambda'$ of two K-algebras Λ and Λ'. If A and A' are bimodules over Λ and Λ', respectively, then $A \otimes A'$ is a $\Lambda \otimes \Lambda'$-bimodule, with left operators given as $(\lambda \otimes \lambda')(a \otimes a') = \lambda a \otimes \lambda' a'$ and right operators similarly defined. In certain cases we can compute the homology of $A \otimes A'$ from that of A and A'.

Proposition 7.1. *If* $\varepsilon: X \to A$ *and* $\varepsilon': X' \to A'$ *are* K*-split resolutions of left* Λ*- and* Λ'*-modules, respectively, then* $\varepsilon \otimes \varepsilon': X \otimes X' \to A \otimes A'$ *is a* K*-split resolution of the left* $(\Lambda \otimes \Lambda')$*-module* $A \otimes A'$*. If* X *and* X' *are relatively free, so is* $X \otimes X'$.

Proof. The hypothesis that X is K-split means, as in Cor. IX.5.3, that there is a K-module contracting homotopy s of square zero. These homotopies for X and X' combine, as in (V.9.3), to give a K-module contracting homotopy for $\varepsilon \otimes \varepsilon': X \otimes X' \to A \otimes A'$, also of square zero.

If X and X' are relatively free, $X_p = \Lambda \otimes M_p$ and $X'_q = \Lambda' \otimes M'_q$ for K-modules M_p and M'_q, so $(X \otimes X')_n \cong \sum (\Lambda \otimes \Lambda') \otimes (M_p \otimes M'_q)$, with direct sum over $p+q=n$, is also relatively free.

Applied to the bar resolution this gives

Corollary 7.2. *For modules* ${}_\Lambda A$, ${}_{\Lambda'}A'$ *there is a chain equivalence*

$$B(\Lambda, A) \otimes B(\Lambda', A') \overset{g}{\rightleftarrows} B(\Lambda \otimes \Lambda', A \otimes A') \tag{7.1}$$

in which the maps are chain transformations of complexes of left $\Lambda \otimes \Lambda'$*-modules commuting with* ε *and* ε'.

Proof. By Prop. 7.1, both sides are K-split relatively free resolutions of the left $\Lambda \otimes \Lambda'$-module $A \otimes A'$; apply the comparison theorem.

An explicit chain transformation is given by the following natural map

$$\left.\begin{aligned} &f\{\lambda \otimes \lambda' [\lambda_1 \otimes \lambda'_1 | \dots | \lambda_n \otimes \lambda'_n] a \otimes a'\} \\ &\quad = \sum_{i=0}^{n} \lambda [\lambda_1 | \dots | \lambda_i] \lambda_{i+1} \dots \lambda_n a \otimes \lambda' \lambda'_1 \dots \lambda'_i [\lambda'_{i+1} | \dots | \lambda'_n] a'; \end{aligned}\right\} \tag{7.2}$$

indeed, the reader may verify that this is the canonical comparison. Alternatively, f is the Alexander-Whitney map (VIII.8.7) defined on $B(\Lambda, A) = \beta_N(\Lambda, A)$ by the simplicial structure of $\beta(\Lambda, A)$.

For $A = \Lambda$, $A' = \Lambda'$, this corollary yields a chain equivalence

$$B(\Lambda, \Lambda) \otimes B(\Lambda', \Lambda') \overset{g}{\rightleftarrows} B(\Lambda \otimes \Lambda', \Lambda \otimes \Lambda') \tag{7.3}$$

of $\Lambda \otimes \Lambda'$-bimodules; the map f is again given as in (7.2).

Theorem 7.3. *The homology and cohomology products induce homomorphisms*

$$p_\Lambda: H_k(\Lambda, A) \otimes H_m(\Lambda', A') \to H_{k+m}(\Lambda \otimes \Lambda', A \otimes A'), \tag{7.4}$$

$$p^\Lambda: H^k(\Lambda, A) \otimes H^m(\Lambda', A') \to H^{k+m}(\Lambda \otimes \Lambda', A \otimes A') \tag{7.5}$$

of K*-modules, natural in the bimodules* A *and* A' *and commuting with connecting homomorphisms for* K*-split short exact sequences of bimodules* A *or* A'*. For* $k = m = 0$*, these products are induced by the identity map of* $A \otimes A'$*. The products are associative.*

Proof. The homology $H_k(\Lambda, A)$ is defined as $H_k(A \otimes_\Omega B)$, where Ω is short for $\Lambda \otimes \Lambda^{\mathrm{op}}$ and B short for $B(\Lambda, \Lambda)$. The homology product of (VIII.1.2) is the natural map

$$p_H\colon H_k(A \otimes_\Omega B) \otimes H_m(A' \otimes_{\Omega'} B') \to H_{k+m}[(A \otimes A') \otimes_{\Omega \otimes \Omega'} (B \otimes B')].$$

The right hand side is isomorphic to $H_{k+m}(\Lambda \otimes \Lambda', A \otimes A')$ under the equivalence g_* of (7.3), so the product p_Λ of (7.4) is defined as $g_* p_H$; in dimension zero (cf. (4.4)) it carries $\operatorname{cls} a \otimes \operatorname{cls} a'$ to $\operatorname{cls}(a \otimes a')$. If E is a K-split short exact sequence of Λ-bimodules, the tensor product sequence $E \otimes_{\mathrm{K}} A'$ is also short exact, as a sequence of Λ-bimodules, so appropriate connecting homomorphisms are defined. They commute with p_H by Thm. VIII.1.3 and with the natural map g_*, and hence with p_Λ.

In the above definition of this homology product, the bar resolution $B = B(\Lambda, \Lambda)$ may be replaced by any K-split resolution of Λ by relative projective Λ-bimodules.

The cohomology case is analogous. Write $H^k(\Lambda, A)$ as $H^k(\operatorname{Hom}_\Omega(B, A))$, use the cohomology product

$$p^H\colon H^k((\operatorname{Hom}_\Omega(B, A)) \otimes H^m(\operatorname{Hom}_{\Omega'}(B', A'))$$
$$\to H^{k+m}(\operatorname{Hom}_{\Omega \otimes \Omega'}(B \otimes B', A \otimes A'))$$

of (VIII.1.3), and compose with the isomorphism f^* induced by the chain equivalence f of (7.3) to define p^Λ as $f^* p^H$. Since f is the Alexander-Whitney map, p^Λ may be regarded as a simplicial cup product. If $k = m = 0$, $H^0(\Lambda, A)$ is the K-submodule A^Λ of A consisting of the invariant elements of A, as in (3.3). Now $a \in A^\Lambda$ and $a' \in A'^{\Lambda'}$ imply that $a \otimes a' \in (A \otimes A')^{\Lambda \otimes \Lambda'}$ so the identity induces a K-module homomorphism

$$A^\Lambda \otimes A'^{\Lambda'} \to (A \otimes A')^{\Lambda \otimes \Lambda'} \cong H^0(\Lambda \otimes \Lambda', A \otimes A').$$

The formula above for f in dimension zero shows that this map is p^Λ.

Theorem 7.4. *If Λ and Λ' are algebras over the same field, the homology product for bimodules A and A' yields for each n a natural isomorphism*

$$p_\Lambda\colon \sum_{k+m=n} H_k(\Lambda, A) \otimes H_m(\Lambda', A') \cong H_n(\Lambda \otimes \Lambda', A \otimes A').$$

If in addition Λ and Λ' are K-modules of finite type the cohomology product is a natural isomorphism

$$p^\Lambda\colon \sum_{k+m=n} H^k(\Lambda, A) \otimes H^m(\Lambda', A') \cong H^n(\Lambda \otimes \Lambda', A \otimes A').$$

Proof. The first isomorphism is an immediate application of the Künneth tensor formula, as restated in Thm. VIII.1.1. If Λ is of finite type, each $B_n(\Lambda, \Lambda)$ is a free Λ-bimodule of finite type, so the Hom-$\otimes$ interchange is an isomorphism and Thm. VIII.1.2 applies.

This theorem was first proved by ROSE [1952] before the techniques of resolutions were known, so his proof depended essentially upon a direct construction of the chain equivalence (7.3), using shuffles to describe the map g.

For algebras over a field, $H^n(\Lambda, A) = \operatorname{Ext}^n_{\Lambda-\Lambda}(\Lambda, A)$.

Using bidimΛ to denote the homological dimension of Λ as a bimodule, this theorem shows, for algebras of finite type over a field, that $\operatorname{bidim}(\Lambda \otimes \Lambda') \geqq \operatorname{bidim}\Lambda + \operatorname{bidim}\Lambda'$. Similarly, Thm. 6.2 above shows

$$\operatorname{bidim}(\Gamma \rtimes \Sigma) = \operatorname{Max}(\operatorname{bidim}\Gamma, \operatorname{bidim}\Sigma).$$

This yields a fancier proof of the result (Prop. VII.5.2) that $\operatorname{bidim}\Gamma = 0 = \operatorname{bidim}\Sigma$ implies $\operatorname{bidim}(\Gamma \rtimes \Sigma) = 0$.

Exercises

1. For G a right module and A a left module over Λ, the k-th relative torsion product is $H_k(G \otimes_\Lambda B \otimes_\Lambda A)$, with B short for $B(\Lambda, \Lambda)$. The external product for the relative torsion functor is the map

$$p_T\colon \operatorname{Tor}_k^{(\Lambda,\mathsf{K})}(G, A) \otimes \operatorname{Tor}_m^{(\Lambda',\mathsf{K})}(G', A') \to \operatorname{Tor}_{k+m}^{(\Lambda\otimes\Lambda',\mathsf{K})}(G \otimes G', A \otimes A')$$

defined as the composite of the homology product for complexes, the chain transformation

$$(G \otimes_\Lambda B \otimes_\Lambda A) \otimes (G' \otimes_{\Lambda'} B' \otimes_{\Lambda'} A') \cong (G \otimes G') \otimes_{\Lambda\otimes\Lambda'} (B \otimes B') \otimes_{\Lambda\otimes\Lambda'} (A \otimes A')$$

given by two applications of the middle-four interchange, and the chain equivalence g of (7.3). Show that p_T is natural, commutes with connecting homomorphisms in all four arguments, and reduces for $k = m = 0$ to the middle-four interchange.

2. For K a field, show that the relative torsion product of Ex. 1 gives an isomorphism

$$\sum_{k+m=n} \operatorname{Tor}_k(G, A) \otimes \operatorname{Tor}_m(G', A') \cong \operatorname{Tor}_n(G \otimes G', A \otimes A').$$

3. Show that the product p_Λ of the text is (via (1.4)) a special case of the external product for the relative torsion product.

4. Construct the analogous external product for the relative Ext functor.

8. The Case of Graded Algebras

If G_Λ and ${}_\Lambda A$ are modules over a graded K-algebra Λ, their tensor product $G \otimes_\Lambda A$, as described in (VI.5.7), is a graded K-module. Moreover, the functor $G \otimes_\Lambda A$ is right exact: Each K-split short exact sequence $A \rightarrowtail B \twoheadrightarrow C$ of left Λ-modules yields a right exact sequence

$$G \otimes_\Lambda A \to G \otimes_\Lambda B \to G \otimes_\Lambda C \to 0$$

of graded K-modules. To continue this exact sequence to the left requires the (Λ-K)-relative torsion products $\operatorname{Tor}_n(G, C)$, each of which, like $G \otimes_\Lambda A$, must be a graded K-module $\operatorname{Tor}_n = \{\operatorname{Tor}_{n,p} \mid p = 0, 1, \ldots\}$. We now describe how this comes about.

The bar resolution applies to any graded K-algebra Λ, using the general process of IX.7 for the resolvent pair of categories with $\mathscr{A}$ the category of (automatically graded) left Λ-modules C, $\mathscr{M}$ that of graded K-modules M, $F(M)=\Lambda\otimes M$, $e(m)=1\otimes m$, both categories with morphisms of degree 0. Note that $\Lambda=\{\Lambda_p\}$, $C=\{C_p\}$, and $M=\{M_p\}$ are all graded K-modules. The explicit formulas for the bar resolution in § 2 still apply, with the understanding that each Λ-module $B_n(\Lambda, C)$ is graded; indeed, the *degree* of a generator of B_n is given by

$$\deg \lambda[\lambda_1|\ldots|\lambda_n]\, c=\deg \lambda+\deg \lambda_1+\cdots+\deg \lambda_n+\deg c. \tag{8.1}$$

This element has also ***dimension*** n as an element of $B_n(\Lambda, C)$; in other words, $B(\Lambda, C)$ is bigraded by the submodules $B_{n,p}(\Lambda, C)$ of dimension n and degree p in the sense (8.1).

In consequence, the relative torsion functor $\operatorname{Tor}^{(\Lambda,\mathrm{K})}$ is bigraded. Indeed, if G is a right Λ-module, this torsion functor is calculated as the homology of the complex $X=G\otimes_\Lambda B(\Lambda, C)$, where each $X_n=G\otimes_\Lambda B_n$ is a graded K-module. Specifically, X_n is generated by elements $g[\lambda_1|\ldots|\lambda_n]\, c$ with the degree given by (8.1) (with λ there replaced by g). The boundary homomorphism $\partial\colon X_n\to X_{n-1}$ is of degree 0 in this grading. For each dimension the homology $\operatorname{Tor}_n(G, C)=H_n(X)$ is therefore a graded K-module, so may be written as a family $\{H_{n,p}(X)\}$ of K-modules: The relative torsion functor is the bigraded K-module

$$\operatorname{Tor}^{(\Lambda,\mathrm{K})}_{n,p}(G, C)=H_{n,p}(G\otimes_\Lambda B(\Lambda, C)). \tag{8.2}$$

The first degree n is the resolution dimension; the second degree p is the "internal" degree, inherited from the gradings of G and C. The standard long exact sequences for Tor_n have maps which are of degree 0 in the internal grading p, hence may be regarded as a family of exact sequences in $\operatorname{Tor}_{n,p}$, one for each p and variable n.

Similar remarks apply to the relative functor $\operatorname{Ext}_{(\Lambda,\mathrm{K})}$. It is the cohomology of the complex $\operatorname{Hom}_\Lambda(B(\Lambda, C), A)$, which is a complex of Z-graded K-modules: That is, a family of complexes $\{\operatorname{Hom}^p_\Lambda(B, A)\}$, one for each integer p. Therefore

$$\operatorname{Ext}^{n,p}_{(\Lambda,\mathrm{K})}(C, A)=H^n(\operatorname{Hom}^p_\Lambda(B(\Lambda, C), A))$$

is a bigraded K-module, in which the second grading (by p) is a Z-grading.

It suffices to know this functor for all modules C and A and second grading $p=0$. This we prove by shifting degrees. For each graded K-module M we denote by $L(M)$ the same module with all degrees increased by 1; formally, $L(M)_{n+1}=M_n$. The identity then induces an isomorphism $l\colon M\to L(M)$ of graded K-modules, of degree 1, with inverse l^{-1}: $L(M) \to M$. A homomorphism μ: $M \to M'$ of degree d is a family of K-module homomorphisms μ_n: $M_n \to M'_{n+d}$; the corresponding $L(\mu)$: $L(M) \to L(M')$ of the same degree d is defined by $L(\mu)_{n+1} = (-1)^d \mu_n$: $L(M)_{n+1} \to L(M')_{n+d+1}$; in other words,

$$L(\mu)\, l m = (-1)^{\deg \mu} l \mu m, \quad m \in M_n, \quad l m \in L(M)_{n+1}. \tag{8.3}$$

The sign is the usual one for the commutation of morphisms $L(\mu)$ and l of degrees d and 1. Since $L(\mu'\mu) = L(\mu')L(\mu)$, L is a covariant functor on the category of graded K-modules with morphisms of degree 0, while l: $M \to L(M)$ is a natural transformation. A left Λ-module A is a graded K-module with operators $\Lambda \otimes A \to A$, so $L(A)$ is also a left Λ-module with operators

$$\lambda(l a) = (-1)^{\deg \lambda} l(\lambda a), \quad a \in A_n, \quad l a \in L(A)_{n+1}, \tag{8.4}$$

L is a covariant functor on Λ-modules to Λ-modules, l: $A \to L A$ is a homomorphism of Λ-modules of degree 1 and a natural transformation of the identity functor to L. The sign in (8.4) is exactly that required by the rule $l(\lambda a) = (-1)^{\deg l \deg \lambda} \lambda (l a)$ for a homomorphism of degree 1.

Composition with l yields a natural isomorphism

$$\operatorname{Hom}_\Lambda^p(C, A) \cong \operatorname{Hom}_\Lambda^{p-1}(C, L A)$$

and by iteration a natural isomorphism

$$\operatorname{Hom}_\Lambda^p(C, A) \cong \operatorname{Hom}_\Lambda^0(C, L^p A).$$

With C replaced by the complex $B(\Lambda, C)$, this yields the natural isomorphism

$$\operatorname{Ext}_{(\Lambda, \mathsf{K})}^{n,p}(C, A) \cong \operatorname{Ext}_{(\Lambda, \mathsf{K})}^{n,0}(C, L^p A), \tag{8.5}$$

which for $n = 0$ includes the previous isomorphism. Similarly

$$\operatorname{Ext}_{(\Lambda, \mathsf{K})}^{n,-p}(C, A) \cong \operatorname{Ext}_{(\Lambda, \mathsf{K})}^{n,0}(L^p C, A). \tag{8.6}$$

These functors Ext have proved useful for the Steenrod algebra for a fixed prime number p; this is the algebra over the field Z_p of integers modulo p consisting of all primary cohomology operations, modulo p — Adams [1960], Liulevicius [1960].

Exercise

1. For Λ graded, regard the corresponding internally graded algebra $\Lambda_* = \sum \Lambda_n$ simply as an ungraded K-algebra. Similarly Λ-modules G and C yield Λ_*-modules G_* and C_*. Prove that

$$\operatorname{Tor}_n^{(\Lambda_*, \mathsf{K})}(G_*, C_*) \cong \sum_p \operatorname{Tor}_{n,p}^{(\Lambda, \mathsf{K})}(G, C).$$

9. Complexes of Complexes

In any abelian category we may construct complexes; in particular, there are complexes in the category whose objects are themselves complexes and whose morphisms are chain transformations. These will occur in our study of DG-algebras in the next section.

A complex X of complexes may be displayed as a diagram

$$\begin{array}{ccccccc} X_p\colon & \cdots\to & X_{p,q+1} & \xrightarrow{d} & X_{p,q} & \xrightarrow{d} & X_{p,q-1} \to\cdots \\ \downarrow & & \downarrow{\scriptstyle\partial'} & & \downarrow{\scriptstyle\partial'} & & \downarrow{\scriptstyle\partial'} \\ X_{p-1}\colon & \cdots\to & X_{p-1,q+1} & \xrightarrow{d} & X_{p-1,q} & \xrightarrow{d} & X_{p-1,q-1}\to\cdots \end{array}$$

with additional rows below and above. Each row X_p is a complex with boundary d, while the successive rows form a complex under another boundary ∂' which is a chain transformation $\partial'\colon X_p \to X_{p-1}$. Hence $\partial' d = d\partial'$. Adjust the sign of d by setting $\partial'' x_{p,q} = (-1)^p d x_{p,q}$. This gives two families of boundary operators

$$\partial'\colon X_{p,q} \to X_{p-1,q}, \qquad \partial''\colon X_{p,q} \to X_{p,q-1}$$

with $\partial'\partial' = 0$, $\partial''\partial'' = 0$, and $\partial'\partial'' + \partial''\partial' = 0$. These imply formally that $(\partial' + \partial'')(\partial' + \partial'') = 0$. Thus the family $X^\bullet$, $\partial^\bullet$ defined by

$$(X^\bullet)_n = \sum_{p+q=n} X_{p,q}, \qquad \partial^\bullet = \partial' + \partial''$$

is a (single) complex. We say that $X^\bullet$ is obtained from X by *condensation*; its degree is the sum of the two given degrees; its boundary $\partial^\bullet$ the sum of the two given boundaries, with sign adjustment. This sign adjustment may be made plausible by a more systematic presentation.

Let $\mathscr{A}$ be any abelian category. Recall that a (positive) $\mathscr{A}$-complex X is a family $\{X_p\}$ of objects of $\mathscr{A}$ with $X_p = 0$ for $p < 0$, together with morphisms $\partial\colon X_p \to X_{p-1}$ of $\mathscr{A}$ such that $\partial^2 = 0$. These X are the objects of the category $\mathfrak{X}(\mathscr{A})$ of $\mathscr{A}$-complexes. The morphisms of $\mathfrak{X}(\mathscr{A})$ are the chain *transformations* $f\colon X \to Y$; they are families $\{f_p\colon X_p \to Y_p\}$ of $\mathscr{A}$-morphisms with $\partial f_p = f_{p-1}\partial$ for all p. A chain homotopy $s\colon f \simeq f'\colon X \to Y$ is a family $s_p\colon X_p \to Y_{p+1}$ of $\mathscr{A}$-morphisms with $\partial s + s\partial = f - f'$. We also use chain *maps* $h\colon X \to Y$ of degree d; that is, families $\{h_p\colon X_p \to Y_{p+d}\}$ of $\mathscr{A}$-morphisms with $\partial h = (-1)^d h\partial$. We do not explicitly introduce the category with morphisms all such chain "maps" because our discussion of abelian categories is adapted only to the case of morphisms of degree 0.

The lifting functor L of § 8 gives a covariant functor on $\mathfrak{X}(\mathscr{A})$ to $\mathfrak{X}(\mathscr{A})$, which assigns to each complex X the complex $L(X)$ with $L(X)_{n+1} = X_n$ and differential $L(\partial)$. The identity induces a chain map $l\colon X \to L(X)$ of degree 1; as in (8.3), $L(\partial)l = -l\partial$. In brief, L raises all degrees by 1 and changes the sign of the boundary operator.

Theorem 9.1. *Condensation is a covariant functor* $\mathfrak{X}(\mathfrak{X}(\mathscr{A}))\to\mathfrak{X}(\mathscr{A})$

Proof. Let X be a positive complex of positive complexes, in the form

$$0\leftarrow X_0\leftarrow X_1\leftarrow\cdots\leftarrow X_{p-1}\xleftarrow{\partial_p} X_p\leftarrow\cdots.$$

Each X_p is a complex, each ∂_p a chain transformation of complexes. Replace by the diagram X',

$$0\leftarrow X'_0\leftarrow X'_1\leftarrow\cdots\leftarrow X'_{p-1}\xleftarrow{\partial'_p} X'_p\leftarrow\cdots,$$

where each ∂'_p is a chain *map* of degree -1. More formally, set $X'_p=L^p(X_p)$. The chain maps

$$X'_p=LL^{p-1}(X_p)\xrightarrow{l^{-1}} L^{p-1}(X_p)\xrightarrow{L^{p-1}(\partial_p)} L^{p-1}(X_{p-1})=X'_{p-1}$$

define ∂'_p as $l^{-1}L^p(\partial_p)=L^{p-1}(\partial_p)l^{-1}$. Then $\partial'_p\partial'_{p+1}=0$. Each X'_p is an $\mathscr{A}$-complex with a boundary operator which we denote as ∂''. Therefore $X^\bullet=\sum X'_p$ is an $\mathscr{A}$-complex with boundary ∂''. On the other hand, $\partial'_p\colon X'_p\to X'_{p-1}$ has degree -1, hence gives another boundary operator in $X^\bullet$. Now ∂' is a chain map of degree -1 for the boundary ∂'', so $\partial''\partial'=-\partial'\partial''$. Therefore $\partial^\bullet=\partial'+\partial''$ satisfies $\partial^\bullet\partial^\bullet=0$, so $(X^\bullet,\partial^\bullet)$ is an $\mathscr{A}$-complex, called the *condensation* of X. This description of $X^\bullet$ agrees with the initial description, since the boundary ∂'' of X'_p is that of X_p with p sign changes due to p applications of L. Since $X'_{p,n}=0$ for $p>n$, only finite direct sums are involved in the construction of $X^\bullet$.

Now let $f\colon X\to Y$ be a chain transformation. It is a family of chain transformations $\{f_p\colon X_p\to Y_p\}$ and determines $f'\colon X'\to Y'$ as the family $f'_p=L^p(f_p)\colon X'_p\to Y'_p$. Thus $f'_p\partial''=\partial''f'_p$ and $\partial'f'_p=f'_{p-1}\partial'$. Hence $f^\bullet=\sum f'_p$ satisfies $\partial^\bullet f^\bullet=f^\bullet\partial^\bullet$, so is a chain transformation $f^\bullet\colon X^\bullet\to Y^\bullet$. This shows condensation a functor, as stated.

Proposition 9.2. *Each chain homotopy* $s\colon f\simeq g\colon X\to Y$ *in* $\mathfrak{X}(\mathfrak{X}(\mathscr{A}))$ *determines a chain homotopy* $s^\bullet\colon f^\bullet\simeq g^\bullet\colon X^\bullet\to Y^\bullet$ *of the condensed complexes.*

Proof. We are given a family $\{s_p\colon X_p\to Y_{p+1}\}$ of morphisms of $\mathfrak{X}(\mathscr{A})$ with $\partial_{p+1}s_p+s_{p-1}\partial_p=f_p-g_p$. Each s_p is a chain transformation, so determines a chain map $s'_p\colon X'_p\to Y'_{p+1}$ of degree 1. Specifically, $s'_p=L^{p+1}(s_p)l=lL^p(s_p)\colon L^p(X_p)\to L^{p+1}(Y_{p+1})$. Since s'_p has degree 1, $\partial''s_p=-s'_p\partial''$. On the other hand, by lifting, $\partial's'+s'\partial'=f'-g'$. Adding, $s^\bullet=\sum s'_p$ gives $s^\bullet\colon X^\bullet\to Y^\bullet$ of degree 1 with $\partial^\bullet s^\bullet+s^\bullet\partial^\bullet=f^\bullet-g^\bullet$; hence $s^\bullet$ is a chain homotopy, as asserted.

We also consider the effect of condensation upon tensor products of complexes. In the initial category $\mathscr{A}$, assume a tensor product which is a covariant bifunctor on $\mathscr{A}$ to $\mathscr{A}$. A tensor product is introduced in the category $\mathscr{X}(\mathscr{A})$ of $\mathscr{A}$-complexes X, Y by the usual formulas, with $(X\otimes Y)_n = \sum_{p+q=n} X_p \otimes Y_q$ and

$$\partial = (\partial_X \otimes 1) + (-1)^p 1 \otimes \partial_Y \colon X_p \otimes Y_q \to (X \otimes Y)_{p+q-1}. \tag{9.1}$$

In particular, if $\mathscr{M}$ is the category of modules over some commutative ground ring these formulas introduce tensor products in the category $\mathscr{X}(\mathscr{X}(\mathscr{M}))$ of complexes of complexes.

Proposition 9.3. *There is a natural isomorphism* $\psi\colon (X\otimes Y)^\bullet \cong X^\bullet \otimes Y^\bullet$.

Proof. For K and K' single complexes in $\mathscr{X}(\mathscr{A})$, and any p, q there is a chain isomorphism $\psi_{p,q}\colon L^{p+q}(K\otimes K') \cong L^p K \otimes L^q K'$, given by

$$\psi_{p,q} l^{p+q}(k\otimes k') = (-1)^{q \deg k}\, l^p k \otimes l^q k'.$$

Now let X and Y be complexes of complexes (i.e., in $\mathscr{X}(\mathscr{X}(\mathscr{A}))$). The complex of complexes $X\otimes Y$ has $(X\otimes Y)_n = \sum X_p \otimes Y_q$, so the $\psi_{p,q}$ for $p+q=n$ give a chain isomorphism of single complexes

$$\psi_n\colon L^n((X\otimes Y)_n) \cong \sum_{p+q=n} L^p(X_p) \otimes L^q(Y_q).$$

The complex $(X\otimes Y)^\bullet$ is the direct sum of the $L^n((X\otimes Y)_n)$, with boundary $\partial' + \partial''$. The complex $X^\bullet \otimes Y^\bullet$ is $(\sum L^p X_p) \otimes (\sum L^q Y_q)$ with boundary determined by the usual tensor product formula (9.1) from the boundaries $\partial^\bullet = \partial' + \partial''$ in $X^\bullet$ and in $Y^\bullet$. By construction, ψ_n commutes with the ∂'' part of the boundary; a straightforward calculation shows that it commutes with ∂', and hence with the total boundary $\partial^\bullet$.

Note. The notion of a complex of complexes is not usually distinguished from the closely related notion of a "bicomplex", which will be discussed in XI.6. The superficial difference is just one of sign, in the formula $\partial'' x_{p,q} = (-1)^p d x_{p,q}$.

10. Resolutions and Constructions

From algebras Λ we now shift to DGA-algebras U. When a U-module A is resolved, two boundary operators arise: One from that in A, the other from the resolution. Suitable combination of these boundaries make the resolution into a single U-module, called a "construction"; in particular, the canonical resolution of the ground ring yields the "bar construction" $B(U)$. This might be described directly by the string of formulas (10.4)—(10.8) below, which yield the basic properties of $B(U)$, as formulated in Thm. 10.4, as well as its relation to the "reduced" bar construction of Cor. 10.5. Instead, we first describe the bar construction conceptually by condensing the canonical resolution for a suitable relative category.

Let U be a DGA-algebra (*d*ifferential *g*raded *a*ugmented algebra) over the commutative ring K. Each left U-module A (as defined in (VI.7.3)) is by neglect a DG-module (i.e., a positive complex of K-modules). It follows that U determines a resolvent pair of categories

$$\mathscr{A} = \text{all left } U\text{-modules } A, \text{ with morphisms of degree } 0,$$

$$\mathscr{M} = \text{all DG-modules } M, \text{ with morphisms of degree } 0,$$

$F(M) = U \otimes M$, and $e(m) = 1 \otimes m \in F(M)$. Write $\varepsilon\colon U \to \mathrm{K}$ for the augmentation of U; by pull-back, ${}_\varepsilon\mathrm{K}$ is a left U-module. An *augmentation* of A or of M is a morphism

$$\varepsilon_A\colon A \to {}_\varepsilon\mathrm{K}, \quad \varepsilon_M\colon M \dashrightarrow \mathrm{K}.$$

Proposition 10.1. *Each left U-module A determines a DG-module*

$$\bar{A} = \mathrm{K} \otimes_U A \cong A/JA,$$

where J is the kernel of $\varepsilon\colon U \to \mathrm{K}$. If A is augmented, so is $\bar{A}$.

Proof. Recall (VI.7) that the tensor product of U-modules is a DG-module. Since $J \rightarrowtail U \twoheadrightarrow \mathrm{K}$ is an exact sequence of right U-modules,

$$J \otimes_U A \to U \otimes_U A \to \mathrm{K} \otimes_U A \to 0$$

is a right exact sequence of DG-modules. But $U \otimes_U A \cong A$, so the module $\bar{A}$ on the right is isomorphic to the quotient of A by the image JA of $J \otimes_U A$. If A is augmented by ε_A, define an augmentation of $\bar{A}$ by $\varepsilon_{\bar{A}}(k \otimes a) = k\,\varepsilon_A(a)$.

Call $\bar{A}$ the *reduced module* of A and $p\colon A \to \bar{A} = A/JA$ its *projection*. The U-module A is like a "fiber bundle" with "group" U acting on A and "base" $\bar{A}$ obtained by "dividing out" the action of U. The corresponding analogue of an acyclic fiber bundle is a "construction". (Warning: This terminology does not agree with that of CARTAN [1955].)

A *construction* for U is an augmented left U-module $\varepsilon_C\colon C \to {}_\varepsilon\mathrm{K}$ which has a DG-module contracting homotopy of square zero. This homotopy may be written as

$$t_{-1}\colon \mathrm{K} \dashrightarrow C, \quad t_n\colon C_n \dashrightarrow C_{n+1}, \quad n \geqq 0;$$

t_{-1} is a morphism of DG-modules, $t = \{t_n \mid n \geqq 0\}$ is a homomorphism of graded K-modules, of degree 1, and

$$\varepsilon_C t_{-1} = 1, \quad \partial t + t\partial = 1 - t_{-1}\varepsilon_C, \quad t t_{-1} = 0 = tt. \tag{10.1}$$

A construction C is *relatively free* if there is a graded K-module D and an isomorphism $U \otimes D \cong C$ of modules over the *graded* algebra U. The definition of the reduced module $\bar{C}$ then reads

$$\bar{C} \cong \mathrm{K}_\varepsilon \otimes_U (U \otimes D) = (\mathrm{K}_\varepsilon \otimes_U U) \otimes D = \mathrm{K} \otimes D = D;$$

hence D may be identified with $\overline{C}$, so a construction is relatively free if there is an isomorphism

$$\varphi\colon U\otimes\overline{C}\cong C \qquad \text{(of modules over the graded algebra } U).$$

To repeat: φ commutes with the operators by $u\in U$, but not necessarily with the differential. Moreover, the projection $p\colon C\to\overline{C}=C/JC$ of Prop. 10.1 is given by $p\varphi(u\otimes\bar{c})=\varepsilon(u)\bar{c}$. Hence $i(\bar{c})=\varphi(1\otimes\bar{c})$ is a monomorphism $i\colon\overline{C}\to C$ of graded K-modules with pi the identity $\overline{C}\to\overline{C}$. We can and will use i to identify $\overline{C}$, as a graded K-module (*not* as a DG-module) with a submodule of C.

Theorem 10.2. *Condensation is a covariant functor on $\mathcal{M}$-split resolutions X of ${}_\varepsilon\mathsf{K}$ by U-modules to constructions $X^\bullet$ for U. If X is relatively free, so is $X^\bullet$.*

Proof. Let $\varepsilon_X\colon X\to{}_\varepsilon\mathsf{K}$ be a resolution by U-modules X_p. By neglect each U-module X_p is a DG-module; that is, a positive complex. By the same neglect, X is a complex of complexes, so has a condensation $X^\bullet=\sum L^p(X_p)$ which is a DG-module under boundary operators ∂', ∂'', and $\partial^\bullet=\partial'+\partial''$. But if A is a U-module, then $L(A)$ is a U-module with $u(lm)=(-1)^{\deg u}l(um)$. Hence $L^p(X_p)$ is a U-module with differential ∂'', while $\partial'\colon L^p(X_p)\to L^{p-1}(X_{p-1})$ is a map of U-modules of degree -1, so that, writing ∂u for the differential of $u\in U$,

$$\partial''(ux)=(\partial u)x+(-1)^{\deg u}u(\partial''x), \qquad \partial'(ux)=(-1)^{\deg u}u(\partial'x). \tag{10.2}$$

The augmentation ε_X of X condenses to an augmentation $\varepsilon^\bullet\colon X^\bullet\to{}_\varepsilon\mathsf{K}$. The contracting homotopy of X (present because X is $\mathcal{M}$-split) condenses by Prop. 9.2 to a contracting homotopy $s^\bullet$ of square zero in $X^\bullet$. This $s^\bullet$ satisfies the analogue of (10.1); in particular

$$\partial's^\bullet+s^\bullet\partial'=1-s^\bullet_{-1}\varepsilon^\bullet, \qquad \partial''s^\bullet+s^\bullet\partial''=0. \tag{10.3}$$

If X is relatively free, each X_n has the form $U\otimes M_n$ for some DG-module M_n. Thus $L^p(X_p)\cong U\otimes L^p(M_p)$, so $X^\bullet\cong U\otimes\sum L^p(M_p)$ shows $X^\bullet$ relatively free.

Next we condense the canonical comparison (Thm. IX.6.2).

Theorem 10.3. *(Comparison theorem.) If $X\to{}_\varepsilon\mathsf{K}$ is a relatively free resolution and $Y\to{}_\varepsilon\mathsf{K}$ an $\mathcal{M}$-split resolution, both by U-modules, there is a unique homomorphism $\varphi\colon X^\bullet\to Y^\bullet$ of augmented U-modules with*

$$\varphi\overline{X}^\bullet\subset s^\bullet_{-1}\mathsf{K}\cup s^\bullet Y^\bullet,$$

where $s^\bullet$ is the contracting homotopy of $Y^\bullet$.

The proof is by (IX.6.1); the submodule eM of X is here $\overline{X}^\bullet\subset X^\bullet$.

The left bar resolution $B(U)$ is an $\mathcal{M}$-split resolution of ${}_\varepsilon\mathrm{K}$ by relatively free left U-modules, so its condensation $B^\bullet(U)$ is a construction called the *bar construction*. Specifically, $B^\bullet(U)$ is the graded K-module $\sum U \otimes L^p((U/\mathrm{K})^p)$; as a tensor product, it is generated by elements which we write in the usual form as

$$u[u_1|\dots|u_p]^\bullet = u \otimes (u_1+\mathrm{K}) \otimes \cdots \otimes (u_p+\mathrm{K})$$

for u and $u_i \in U$. By normalization, this element is zero if any $u_i \in \mathrm{K}$. The degree of such an element is

$$\deg(u[u_1|\dots|u_p]^\bullet) = p + \deg u + \deg u_1 + \cdots + \deg u_p; \qquad (10.4)$$

an element is multiplied by $u' \in U$ by multiplying its first factor by u'. The augmentation is

$$\varepsilon_B^\bullet(u[\,]^\bullet) = \varepsilon(u), \qquad (10.5)$$

and the contracting homotopy is determined by $s_{-1}(1) = 1[\,]^\bullet$ and

$$s^\bullet(u[u_1|\dots|u_p]^\bullet) = 1[u|u_1|\dots|u_p]^\bullet, \quad p \geqq 0. \qquad (10.6)$$

The normalization insures that $s^\bullet s^\bullet = 0$. The formulas for the two boundary operators ∂' and ∂'' are most easily found from that for $s^\bullet$ by recursion on p, using (10.3) and (10.2); they are

$$\left.\begin{aligned}\partial''(u[u_1|\dots|u_p]^\bullet) &= \partial u[u_1|\dots|u_p]^\bullet \\ &\quad - \sum_{i=1}^{p} (-1)^{e_{i-1}} u[u_1|\dots|\partial u_i|\dots|u_p]^\bullet,\end{aligned}\right\} \qquad (10.7)$$

$$\left.\begin{aligned}\partial'(u[u_1|\dots|u_p]^\bullet) &= (-1)^{e_0} u u_1[u_2|\dots|u_p]^\bullet \\ &\quad + \sum_{i=1}^{p-1} (-1)^{e_i} u[u_1|\dots|u_i u_{i+1}|\dots|u_p]^\bullet \\ &\quad + (-1)^{e_p} u[u_1|\dots|u_{p-1}]^\bullet \varepsilon(u_p),\end{aligned}\right\} \qquad (10.8)$$

with the exponents e_i of the signs given for $i = 0, \dots, p$ by

$$e_i = i + \deg u + \deg u_1 + \cdots + \deg u_i = \deg(u[u_1|\dots|u_i]). \qquad (10.9)$$

Except for sign, ∂'' is the boundary of a tensor product, and ∂' like that of the bar resolution. Incidentally, the signs in (10.7) and (10.8) can be read as cases of our usual sign conventions.

Thus Thm. 10.2 gives

Theorem 10.4. *For each DGA-algebra U the condensed left bar construction $B^\bullet(U) = \sum_p U \otimes (U/\mathrm{K})^p$ is an augmented left U-module with augmentation $\varepsilon_B^\bullet$, grading given by* (10.4), *boundary $\partial^\bullet = \partial' + \partial''$ by* (10.7) *and* (10.8), *and contracting homotopy by* (10.6).

This theorem can also be proved directly from the formulas above, with proofs of (10.2), (10.3), and $\partial'\partial''+\partial''\partial'=0$ en route.

In the sequel we use only the condensed bar construction for a DGA-algebra, so we shall drop the now superfluous dot. The curious reader may note that the signs occurring in this boundary formula are not those arising in the bar resolution of § 2 for an algebra. The change of signs can be deduced from the lifting operation L^p; we have avoided the meticulous control of this change by deriving the signs from (10.2) and (10.3).

As for any U-module, the reduced bar construction $\overline{B}(U)$ has the form $\mathsf{K}_\varepsilon \otimes_U B(U)$, and $\overline{B}(U)$ is regarded as a graded K-submodule of $B(U)$.

Corollary 10.5. *For each DGA-algebra U the reduced bar construction $\overline{B}(U)$ is a DG-module over K with $\overline{B}(U)=\sum L^p((U/\mathsf{K})^p)$. If elements are denoted by $[u_1|\dots|u_p]$ for $u_i \in U$, the degree of these elements is given by (10.4) with u omitted, the boundary $\partial=\partial'+\partial''$ by (10.7) and (10.8) with $u=1$ and with u_1 replaced by $\varepsilon(u_1)$ in the first term on the right of (10.8).*

Note also that the projection $p\colon B(U)\to\overline{B}(U)\cong B(U)/JB(U)$ is given by $p(u[u_1|\dots|u_p])=\varepsilon(u)[u_1|\dots|u_p]$; it is a morphism of DG-modules of degree zero. The isomorphism $\varphi\colon B(U)\cong U\otimes\overline{B}(U)$ is given by $\varphi(u[u_1|\dots|u_p])=u\otimes[u_1|\dots|u_p]$; it is an isomorphism of modules over the graded algebra of U, but does not respect the differential, because $\varphi\partial'\neq\partial'\varphi$.

The bar construction has the convenient property

$$s_{-1}\mathsf{K} \cup sB(U)=\overline{B}(U); \tag{10.10}$$

in words, the image of the contracting homotopy is exactly the reduced bar construction, regarded as a graded submodule of B.

Corollary 10.6. *Both $\overline{B}(U)$ and $B(U)$, the latter with its contracting homotopy, are covariant functors of the DGA-algebra U with values in the category of DG-modules over K. Moreover, $p\colon B\to\overline{B}$ and $i\colon \overline{B}\to B$ are natural transformations of functors.*

Proof. If $\mu\colon U\to V$ is a homomorphism of DGA-algebras, then $B(V)$ pulled back along μ is a U-module, still with a K-module contracting homotopy. Hence the canonical comparison of Thm. 10.3 gives a unique homomorphism

$$B(\mu)\colon\ B(U)\to{}_{\mu}B(V) \tag{10.11}$$

of U-modules with $\varepsilon' B(\mu)=\varepsilon$. Moreover, $JB(U)$ is mapped into $\mu(J)B(V)$, so $B(\mu)$ induces a homomorphism $\overline{B}(\mu)$ such that $pB(\mu)=\overline{B}(\mu)p$. These maps make B and $\overline{B}$ functors, as asserted.

Exercises

1. Describe the bar construction explicitly when $K = Z_p$ is the field of integers modulo p and $U = E(x)$ is the exterior algebra on a generator of odd degree.

2. Obtain a resolution of ${}_\varepsilon K$ when $K = Z_p$, $U = P[x]/(x^p)$, the ring of polynomials in an indeterminate x of even degree, modulo x^p.

3. (Uniqueness of comparison.) If $X \to_\varepsilon K$ is a relatively free resolution, while C is any construction for U with contracting homotopy t, there is at most one homomorphism φ: $X^\bullet \to C$ of augmented U-modules with $\varphi(\overline{X}^\bullet) \subset t_{-1} K \cup t C$.

11. Two-stage Cohomology of DGA-Algebras

The cohomology of a DGA-algebra U with coefficients in a (trivially graded) K-module G can be defined in two ways or "stages". For stage zero, regard U, by neglect, as a complex (= a DG-module); so that $\mathrm{Hom}_K(U, G)$ and $G \otimes_K U$ are complexes with (co)homology the K-modules

$$H^k(U, 0; G) = H^k(\mathrm{Hom}_K(U, G)),$$

$$H_k(U, 0; G) = H_k(G \otimes_K U), \qquad k = 0, 1, \ldots.$$

For stage one, the left bar construction $B(U)$ with its total boundary $\partial^\bullet$ is a left U-module while G is a U-module by pull-back, so $\mathrm{Hom}_U(B(U), {}_\varepsilon G)$ and $G_\varepsilon \otimes_U B(U)$ are DG-modules with (co)homology the K-modules

$$H^k(U, 1; G) = H^k(\mathrm{Hom}_U(B(U), {}_\varepsilon G)), \tag{11.1}$$

$$H_k(U, 1; G) = H_k(G_\varepsilon \otimes_U B(U)), \qquad k = 0, 1, \ldots. \tag{11.2}$$

Since $B(U) \to_\varepsilon K$ arises from a resolution, the definition of $H_k(U, 1; G)$ resembles that of the (U, K)-relative torsion product $\mathrm{Tor}_k(G_\varepsilon, {}_\varepsilon K)$, but it is not a relative torsion product because it uses the total boundary operator $\partial^\bullet$ of $B(U)$ and not just the boundary operator ∂' arising from the resolution.

A homomorphism μ: $(U, \varepsilon) \to (V, \varepsilon')$ of two DGA-algebras over a fixed K is a homomorphism of DG-algebras with $\varepsilon' \mu = \varepsilon$: $U \to K$. Thus $B(V)$ is an augmented U-module by pull-back, and μ induces $B(\mu): B(U) \to_\mu B(V)$, a homomorphism of augmented U-modules which commutes with the contracting homotopy. It follows that $H_k(U, 1; G)$ is a covariant bifunctor of U and G and that $H^k(U, 1; G)$ is a bifunctor covariant in G and contravariant in U. The reduced (condensed) bar construction is also a covariant functor of DGA-algebras to DG-modules.

The (co)homology modules of U may be expressed by the reduced bar construction. Indeed, since G is a K-module, each U-module homomorphism $B(U) \to_\varepsilon G$ must annihilate $JB(U)$, where J is the kernel of the augmentation $\varepsilon\colon U \to \mathsf{K}$, hence induces a K-module homomorphism $\bar{B}(U) = B(U)/JB(U) \to G$. This gives the natural isomorphism

$$H^k(U, 1; G) \cong H^k(\mathrm{Hom}_{\mathsf{K}}(\bar{B}(U), G)). \tag{11.3}$$

Similarly $G_\varepsilon \otimes_U B(U) = (G \otimes \mathsf{K}_\varepsilon) \otimes_U B(U) \cong G \otimes_{\mathsf{K}} \bar{B}(U)$, so

$$H_k(U, 1; G) \cong H_k(G \otimes_{\mathsf{K}} \bar{B}(U)). \tag{11.4}$$

If $X \to_\varepsilon \mathsf{K}$ is any $\mathscr{M}$-split resolution by relatively free U-modules, standard comparison arguments give

$$H^k(U, 1; G) \cong H^k(\mathrm{Hom}_U(X^\bullet, {}_\varepsilon G)) \cong H^k(\mathrm{Hom}_{\mathsf{K}}(\bar{X}^\bullet, G)),$$

and similarly for the functorial behavior and for homology.

"Suspension" maps stage zero homology to that on stage one. Let $S\colon U \to \bar{B}(U)$ be defined by $S(u) = [u]$; note that S is just the contracting homotopy restricted to the subcomplex U of $B(U)$. Thus S is a homomorphism of degree 1 of graded K-modules with $\partial S = -S\partial$, hence induces similar maps $G \otimes U \to G \otimes \bar{B}(U)$ and $\mathrm{Hom}(\bar{B}(U), G) \to \mathrm{Hom}(U, G)$ and thus the homomorphisms

$$S_*\colon H_k(U, 0; G) \to H_{k+1}(U, 1; G), \tag{11.5}$$

$$S^*\colon H^{k+1}(U, 1; G) \to H^k(U, 0; G), \tag{11.6}$$

called *suspension*, and to be used in the next section.

To study the dependence of $H(\bar{B}(U))$ on $H(U)$ we use a filtration of the complex (DG-module) $\bar{B}$. Let $F_p = F_p(\bar{B}(U))$ denote the submodule of $\bar{B}$ spanned by all elements $w = [u_1| \dots |u_k]$ with $k \leqq p$; we say that such an element w has *filtration* at most p.

Proposition 11.1. *For each DGA-algebra U the associated complex $\bar{B}(U)$ has a canonical family of subcomplexes F_p, with $F_0 \subset F_1 \subset \cdots \subset F_p \subset \cdots \subset \bigcup F_p = \bar{B}(U)$. The elements in $\bar{B}(U)$ of total degree n lie in F_n. For $p=0$, $F_0 \cong \mathsf{K}$, with trivial grading and differential, while if $p > 0$, there is a natural isomorphism of chain complexes*

$$F_p/F_{p-1} \cong L(U/\mathsf{K}) \otimes \cdots \otimes L(U/\mathsf{K}) \qquad (p \text{ factors}). \tag{11.7}$$

Only the last statement needs verification. The "internal" boundary operator ∂'' of $\bar{B}$ carries an element of filtration p to one of filtration p, while the "external" boundary operator ∂' maps one of filtration p to one of filtration $p-1$; F_p is indeed closed under the total boundary $\partial = \partial' + \partial''$. Moreover, the formation of the quotient F_p/F_{p-1} drops all the ∂' terms from the total boundary, so the boundary in F_p/F_{p-1} is given by ∂'' as in the formula (10.7), with $u=1$. This is exactly the formula for the boundary in the tensor product of p copies of $L(U/\mathsf{K})$, for the sign exponent e_i is that of the tensor product boundary formula, and the minus sign in front of the summation is that introduced in $L(U/\mathsf{K})$ by the definition $L(\partial) l u = -l \partial u$.

A chain transformation $\mu: X \to Y$ of complexes is called a *homology isomorphism* if, for each dimension n, $H_n(\mu): H_n(X) \cong H_n(Y)$.

Theorem 11.2. (Eilenberg-MacLane [1953b].) *Let $\mu: U \to V$ be a homomorphism of DGA-algebras over K which is a homology isomorphism. Moreover, assume that K is a field or that $\mathsf{K}=Z$ and each U_n and each V_n is a free abelian group (i.e., a free K-module). Then the induced map $\overline{B}(\mu): \overline{B}(U) \to \overline{B}(V)$ is a homology isomorphism, and for each K-module G*

$$\left.\begin{aligned} \mu_*: &\ H_k(U, 1; G) \cong H_k(V, 1; G); \\ \mu^*: &\ H^k(V, 1; G) \cong H^k(U, 1; G). \end{aligned}\right\} \tag{11.8}$$

The proof is an exercise in the use of filtration and the Five Lemma.

First, μ carries 1_U to 1_V, hence induces a chain transformation $U/\mathsf{K} \to V/\mathsf{K}$. We claim that this map is a homology isomorphism. Indeed, the special assumptions (K a field or $\mathsf{K}=Z$, U_0 free) show that $I: \mathsf{K} \to U$ is a monomorphism; hence $\mathsf{K} \rightarrowtail U \twoheadrightarrow U/\mathsf{K}$ is an exact sequence of complexes, which is mapped by μ into the corresponding exact sequence for V. Therefore μ maps the exact homology sequence of the first into that of the second. For $n \geqq 2$, $H_{n-1}(\mathsf{K}) = 0$ and the exact homology sequence reduces to the isomorphism $H_n(U) \cong H_n(U/\mathsf{K})$. For $n=1$ it becomes

$$0 \to H_1(U) \to H_1(U/\mathsf{K}) \to H_0(\mathsf{K}) \to H_0(U) \to H_0(U/\mathsf{K}) \to 0$$

with $H_0(\mathsf{K}) \cong \mathsf{K}$. This is mapped by μ into the corresponding sequence for V. Two applications of the Five Lemma give $H_1(U/\mathsf{K}) \cong H_1(V/\mathsf{K})$, $H_0(U/\mathsf{K}) \cong H_0(V/\mathsf{K})$, so $\mu: U/\mathsf{K} \to V/\mathsf{K}$ is indeed a homology isomorphism.

Next consider the map $\overline{B}(\mu): \overline{B}(U) \to \overline{B}(V)$, given explicitly as

$$\overline{B}(\mu)[u_1| \dots |u_n] = [\mu u_1| \dots |\mu u_n].$$

This map respects the filtration, so carries $F_p = F_p(\overline{B}(U))$ into the corresponding $F'_p = F_p(\overline{B}(V))$. We claim that the induced map $F_p/F_{p-1} \to F'_p/F'_{p-1}$ is a homology isomorphism. Indeed, the quotient F_p/F_{p-1} is just an n-fold tensor product (11.7), and the induced map is $\mu \otimes \dots \otimes \mu$ (n factors). If K is a field, this is a homology isomorphism by the Künneth tensor formula (Thm. V.10.1). If $\mathsf{K}=Z$ and each U_n and each V_n is a free group, this is a homology isomorphism by a consequence of the Künneth formula for this case (Cor. V.11.2).

Finally, we claim that $\mu: F_p \to F'_p$ is a homology isomorphism. The proof is by induction on p. For $p=0$ it is obvious, since $F_0 = \mathsf{K} = F'_0$. For larger p, μ maps the exact sequence $F_{p-1} \rightarrowtail F_p \twoheadrightarrow F_p/F_{p-1}$ of complexes into the corresponding exact sequence for F'_p. The corresponding long exact homology sequences give a commutative diagram with the first row

$$H_{k+1}(F_p/F_{p-1}) \to H_k(F_{p-1}) \to H_k(F_p) \to H_k(F_p/F_{p-1}) \to H_{k-1}(F_{p-1})$$

and vertical maps induced by μ. By the induction assumption and the previous result for F_p/F_{p-1}, the four outside vertical maps are isomorphisms, so the Five Lemma proves $H_k(F_p) \to H_k(F'_p)$ an isomorphism for every k.

Since in each total dimension n, $F_p\overline{B}$ gives all of $\overline{B}$ for p large, it follows now that $\overline{B}(\mu): \overline{B}(U) \to \overline{B}(V)$ is a homology isomorphism. The isomorphisms (11.8) then follow by an application of the appropriate universal coefficient theorem (K a field or K$=Z$ with $\overline{B}$ free).

Exercises

1. (The contraction theorem of EILENBERG-MACLANE [1953b, Thm. 12.1].) If $\mu: U \to V$, $\nu: V \to U$ are homomorphisms of DGA-algebras with $\mu\nu = 1$ and a homotopy t with $\partial t + t\partial = \nu\mu - 1$, $\mu t = 0$, $t\nu = 0$, show that there is a homotopy $\bar{t}$ with $\partial\bar{t} + \bar{t}\partial = \overline{B}(\nu)\,\overline{B}(\mu) - 1$, $\overline{B}(\mu)\bar{t} = 0$, $\bar{t}\,\overline{B}(\nu) = 0$.

2. Obtain the filtration of Prop. 11.1 for an arbitrary $\mathscr{M}$-split relatively free resolution of ${}_\varepsilon$K by U-modules.

12. Cohomology of Commutative DGA-Algebras

Let U and V be two DGA-algebras over K. Their tensor product $U \otimes V$ is also a DGA-algebra, while the tensor product of a U-module by a V-module is a $(U \otimes V)$-module. In particular, the bar constructions $B(U)$ and $B(V)$ yield an augmented $(U \otimes V)$-module $B(U) \otimes B(V)$. Now $B(U) \otimes B(V)$ is a construction, with a contracting homotopy t given in dimension -1 by $s_{-1} \otimes s_{-1}$: K$\to B \otimes B$ and in positive dimensions by the usual formula $t = s \otimes 1 + s_{-1}\varepsilon \otimes s$ for the tensor product of homotopies. Moreover, $B(U) \otimes B(V)$ is relatively free. Indeed, $B(U) \cong U \otimes \overline{B}(U)$ is an isomorphism of modules over the graded algebra of U, so

$$B(U) \otimes B(V) \cong U \otimes \overline{B}(U) \otimes V \otimes \overline{B}(V) \cong U \otimes V \otimes \overline{B}(U) \otimes \overline{B}(V)$$

is an isomorphism of modules over the graded algebra of $U \otimes V$, and $B(U) \otimes B(V)$ is relatively free. One may show that its reduced DG-module is exactly the tensor product $\overline{B}(U) \otimes \overline{B}(V)$ of the DG-modules $\overline{B}(U)$ and $\overline{B}(V)$. Finally, by Prop. 9.3, the construction $B(U) \otimes B(V)$ could also be obtained as a condensation — specifically, as the condensation of the tensor product of the original bar resolutions. Hence we can apply the comparison theorem to obtain homomorphisms of augmented $(U \otimes V)$-modules

$$B(U \otimes V) \underset{g}{\overset{f}{\rightleftarrows}} B(U) \otimes B(V). \qquad (12.1)$$

Let us choose for f and g the canonical comparisons (Thm. 10.3) with

$$f\overline{B}(U\otimes V) \subset t_{-1}\mathsf{K} \cup t(B(U)\otimes B(V)),$$
$$g[\overline{B}(U)\otimes\overline{B}(V)] \subset s_{-1}\mathsf{K} \cup s\,B(U\otimes V).$$

By the comparison theorem again, there is a homotopy $1\simeq gf$. On the other hand, by (10.10), $s_{-1}\mathsf{K}\cup s\,B(U\otimes V)=\overline{B}(U\otimes V)$, so

$$fg[\overline{B}(U)\otimes\overline{B}(V)] \subset t_{-1}\mathsf{K} \cup t(B(U)\otimes B(V)).$$

This shows that fg is the canonical comparison of $B(U)\otimes B(V)$ to itself, so $fg=1$. Since f and g, as canonical comparisons, are unique, they are natural in U and V.

A DGA-algebra U is commutative if $u_p\in U_p$ and $u_q\in U_q$ have $u_p u_q = (-1)^{pq} u_q u_p$; that is, if $\pi\tau=\pi$: $U\otimes U\to U$, where τ: $U\otimes V\to V\otimes U$ is the usual interchange and π the product map for U. Now the tensor product $U\otimes U$ is also a DGA-algebra; a diagram shows that when U is commutative its product mapping π: $U\otimes U\to U$ is a homomorphism of DGA-algebras. Therefore the "external" product g of (12.1) in this case gives an internal product in $B(U)$ as the composite

$$\pi_B\colon\ B(U)\otimes B(U)\xrightarrow{g} B(U\otimes U)\xrightarrow{B(\pi)} B(U). \qquad (12.2)$$

Here $B(U)$ is a $U\otimes U$-module by pull-back along π: $U\otimes U\to U$, while $B(\pi)$ is the canonical map, as in (10.11). Therefore the product π_B of (12.2) can be described as the canonical comparison.

Theorem 12.1. *If U is a commutative DGA-algebra, then $B(U)$ is a commutative DGA-algebra with identity $[\,]$ under the product π_B. Also π_B is a homomorphism of augmented modules over $U\otimes U$. This product induces a product $\overline{B}(U)\otimes\overline{B}(U)\to\overline{B}(U)$ such that $\overline{B}(U)$ is a commutative DGA-algebra, and the projection $B(U)\to\overline{B}(U)$ a homomorphism of DGA-algebras, while inclusion $\overline{B}(U)\to B(U)$ is a homomorphism of graded K-algebras.*

Proof. The identity element of U is represented by the map I: $\mathsf{K}\to U$. With $B(\mathsf{K})=\mathsf{K}$, form the composite map of U-modules

$$B(U)=B(\mathsf{K})\otimes B(U)\xrightarrow{B(I)\otimes 1} B(U)\otimes B(U)\xrightarrow{\pi_B} B(U).$$

Here we regard $B\otimes B$ as a U-module by pull-back along $I\otimes 1$: $\mathsf{K}\otimes U\to U\otimes U$ and then $B(U)$ as a U-module by pull-back along $\pi_U(I\otimes 1)=1$. Hence the composite map is the canonical comparison of $B(U)$ to itself, so is the identity map. This shows that $B(I)1_{\mathsf{K}}=[\,]$ is the identity element of $B(U)$ for the product π_B. Similarly

$$\pi_B(1\otimes\pi_B),\quad \pi_B(\pi_B\otimes 1)\colon\ B(U)\otimes B(U)\otimes B(U)\to B(U)$$

are both canonical, so must be equal. This gives associativity, and makes $B(U)$ a DGA-algebra. There is an analogous proof that the product is commutative.

By definition, $\bar{B}=B/JB$, where J is the kernel of $\varepsilon: U\to \mathsf{K}$; therefore, by Lemma VIII.3.2, the kernel of $p\otimes p: B\otimes B\to \bar{B}\otimes\bar{B}$ is the union of the images of $JB\otimes B$ and $B\otimes JB$. Since π_B is a homomorphism of $(U\otimes U)$-modules, it carries this union into JB and thus induces a unique map $\bar{\pi}$: $\bar{B}\otimes\bar{B}\to\bar{B}$ with $\bar{\pi}(p\otimes p)=p\,\pi_B$. From the uniqueness of this factorization it follows readily that $\bar{B}$ is a DG-algebra under the product $\bar{\pi}$ with augmentation given by $\bar{B}_0\cong\mathsf{K}$, and that p is a homomorphism of augmented algebras.

It remains to show that $i:\bar{B}\to B$ is a homomorphism for the product:

$$\pi_B(i\otimes i)=i\,\bar{\pi}\colon\ \bar{B}(U)\otimes\bar{B}(U)\to B(U).$$

Since π_B is canonical, the image of $\pi_B(i\otimes i)$ lies in $\bar{B}\subset B$; on this submodule ip is the identity, and

$$i p\,\pi_B(i\otimes i)=i\,\bar{\pi}(p\otimes p)(i\otimes i)=i\,\bar{\pi}(p i\otimes p i)=i\bar{\pi},$$

as desired. This completes the proof. Note that the products in $\bar{B}$ and U determine that in B; indeed, since π_B is a homomorphism of $(U\otimes U)$-modules, we have

$$\pi_B[(u_1\otimes\bar{b}_1)\otimes(u_2\otimes\bar{b}_2)]=(-1)^{(\deg u_2)(\deg\bar{b}_1)}\,u_1u_2\,\pi_B(\bar{b}_1\otimes\bar{b}_2)\qquad(12.3)$$

for any two elements $\bar{b}_1,\bar{b}_2\in\bar{B}(U)$.

Since g is canonical, it can be given by an explicit formula; the formula is (except for signs) just the explicit map g of the EILENBERG-ZILBER theorem, as given by the simplicial structure of $B(U)$. As in that case (VIII.8), let t be a (p,q)-shuffle, regarded as a suitable permutation of the integers $\{1,\ldots,p+q\}$. For elements

$$\bar{b}_1=[u_1|\ldots|u_p],\qquad \bar{b}_2=[v_1|\ldots|v_q]\in\bar{B}(V),$$

define a bilinear map (the shuffle product) $*:\bar{B}(U)\otimes\bar{B}(V)\to\bar{B}(U\otimes V)$ by labelling the elements $u_1\otimes 1,\ldots,u_p\otimes 1,1\otimes v_1,\ldots,1\otimes v_q$ of $U\otimes V$ in order as $w_1,\ldots,w_{p+q}$ and setting

$$[u_1|\ldots|u_p]*[v_1|\ldots|v_q]=\sum_t(-1)^{e(t)}[w_{t^{-1}(1)}|\ldots|w_{t^{-1}(p+q)}],\qquad(12.4)$$

where the sum is taken over all (p,q)-shuffles t and the sign exponent $e(t)$ is given in terms of the total degrees as

$$e(t)=\sum(\deg[u_i])(\deg[v_j]),\qquad t(i)>t(p+j)\quad i\leqq p,\ j\leqq q.\qquad(12.5)$$

This sign is exactly that given by the sign convention, since the sum is taken over all those pairs of indices (i,j) for which u_i of degree $\deg[u_i]$ has been shuffled past v_j of degree $\deg[v_j]$.

Theorem 12.2. *The canonical comparison* g *of* (12.1) *is given, for elements* $\bar{b}_1$ *and* $\bar{b}_2$ *of* $\bar{B}(U)$ *and* $\bar{B}(V)$, *respectively, by*

$$g[(u\,\bar{b}_1)\otimes(v\,\bar{b}_2)]=(-1)^{(\deg v)(\deg \bar{b}_1)}(u\otimes v)(\bar{b}_1 * \bar{b}_2),$$

Proof. The formula is suggested by (12.3). It is clearly a homomorphism of modules over the graded algebra of $U\otimes V$, and it carries $\bar{B}(U)\otimes\bar{B}(V)$ into $\bar{B}(U\otimes V)$, so is canonical. The proof is completed by a verification that $\partial g=g\partial$. This is straightforward, using the definition of g and of $\partial=\partial^{\bullet}=\partial'+\partial''$ in the bar construction. We leave the details to the reader, or refer to EILENBERG-MACLANE [1953b], where the proof is formulated in terms of a recursive description of the shuffle product $*$.

Note that the formula (12.4) together with (12.3), in the form

$$(u_1\bar{b}_1)*(u_2\bar{b}_2)=(-1)^{(\deg u_2)(\deg \bar{b}_1)}u_1u_2(\bar{b}_1 * \bar{b}_2),$$

completely determines the product in $B(U)$. For example,

$$[u]*[v]=[u|v]+(-1)^{(1+\deg u)(1+\deg v)}[v|u];$$

again, with an evident "shuffle",

$$[u]*[v|w]=[u|v|w]\pm[v|u|w]\pm[v|w|u].$$

Corollary 12.3. *If* U *is commutative, the algebra* $\bar{B}(U)$ *is strictly commutative.*

Proof. For $\bar{b}=[u_1|\dots|u_p]$, each term in $\bar{b}*\bar{b}$ occurs twice for two shuffles t, t', where

$$e(t)+e(t')=\sum_{i,j}(\deg[u_i])(\deg[u_j])=(\deg\bar{b})^2.$$

When $\deg\bar{b}$ is odd, the signs are opposite, so $\bar{b}*\bar{b}=0$, as required for strict commutativity.

The essential observation is that each *commutative* DGA-algebra U yields a commutative DGA-algebra $\bar{B}(U)$, so allows an iteration to form a commutative DGA-algebra $\bar{B}^n(U)$ for each positive n. This gives an n-th stage cohomology (or homology) of U with coefficients in the K-module G as

$$H^k(U,n;G)=H^k(\operatorname{Hom}(\bar{B}^n(U),G)).$$

This may be applied when $U=Z(\Pi)$ is the group ring of a commutative multiplicative group Π. The n-th stage homology and cohomology groups of this group Π, with coefficients in the abelian group G, are thus

$$H_k(\Pi,n;G)=H_k(G\otimes\bar{B}^n(Z(\Pi))), \tag{12.6}$$

$$H^k(\Pi,n;G)=H^k(\operatorname{Hom}(\bar{B}^n(Z(\Pi)),G)); \tag{12.7}$$

for $n=1$, these are the homology and cohomology of Π as treated in Chap. IV. Note that the suspension $S: \overline{B}^n \to \overline{B}^{n+1}$ of (11.5) gives homomorphisms

$$S_*: H_{n+p}(\Pi, n; G) \to H_{n+1+p}(\Pi, n+1; G), \tag{12.8}$$

$$S^*: H^{n+1+p}(\Pi, n+1; G) \to H^{n+p}(\Pi, n; G). \tag{12.9}$$

The direct limit of $H_{n+p}(\Pi, n; G)$ under S_* gives another set of "stable" homology groups $H_p(\Pi; G)$ for the abelian group Π. They have been studied by EILENBERG-MACLANE [1951, 1955].

For general n, the groups $H^k(\Pi, n; G)$ have a topological interpretation in terms of the so-called Eilenberg-Mac Lane spaces $K(\Pi, n)$. Here $K(\Pi, n)$ is a topological space whose only non-vanishing homotopy group is $\pi_n = \Pi$ in dimension n. It can be proved (EILENBERG-MACLANE [1953b]) that there is a natural isomorphism

$$H^k(K(\Pi, n), G) \cong H^k(\Pi, n; G),$$

with the corresponding result for homology.

Explicit calculations of these groups can be made effectively by using iterated alternative resolutions X, so chosen that $\overline{X}$ has an algebra structure (CARTAN [1955]).

Exercises

1. Show that the image of the contracting homotopy in $B(U) \otimes B(V)$ properly contains $\overline{B}(U) \otimes \overline{B}(V)$.

2. Prove Thm. 12.1 from the explicit formula for the product $*$.

3. Show that $\overline{B}^n(Z(\Pi))$ vanishes in dimensions between 0 and n, and hence that $H^p(\Pi, n; G) = 0 = H_p(\Pi, n; G)$ for $0 < p < n$.

4. Show that $H^n(\Pi, n; G) \cong \operatorname{Hom}(\Pi, G)$ for $n \geqq 1$ and that, for $n \geqq 2$, $H^{n+1}(\Pi, n; G) \cong \operatorname{Ext}^1_Z(\Pi, G)$.

5. (The suspension theorem [EILENBERG-MACLANE 1953b, Thm. 20.4].) For $p < n$, show that S^* and S_* in (12.8) and (12.9) are isomorphisms, while for $p = n$, S^* is a monomorphism and S_* an epimorphism. (Hint: Compare the complexes $\overline{B}^{n+1}(U)$ and $\overline{B}^n(U)$ in the indicated dimensions.)

6. For any K-split relatively free resolution $X \to_\varepsilon K$, written as $X = U \otimes \overline{X}$ as in Thm. 10.2, let $j: U \to X$ be given by $j(u) = u \otimes 1$ (assume $1 \in U = X_0$). Show that the composite psj: $U \to \overline{X}$, with s the contracting homotopy, gives the suspension.

7. For any X as in Ex. 6 find a product $\overline{X} \otimes \overline{X} \to \overline{X}$ associative up to a homotopy.

13. Homology of Algebraic Systems

For groups, monoids, abelian groups, algebras, and graded algebras we have now defined appropriate homology and cohomology groups. A leading idea in each case is that the second cohomology group represents a group of extensions (with given operators) for the type of system in question: See Thm. IV.4.1 for groups, Ex. 12.4 for abelian groups, and Thm. 3.1 for algebras. The third cohomology group has elements which represent the obstructions to corresponding extension problems; see Thm. IV.8.7 for groups and HOCHSCHILD [1947] for algebras. The typical complexes used to construct such homology theories have been described by a notion of "generic acyclicity" (EILENBERG-MAC LANE [1951]). Here we will mention various other algebraic systems for which corresponding homology theories have been developed.

The 2-dimensional cohomology theory for *rings* operates with two factor sets, one for addition and one for multiplication. Let A be an abelian group, regarded as a ring (without identity) in which the product of any two elements is zero. Let R be a ring. A singular extension of A by R is thus a short exact sequence $A \rightarrowtail S \twoheadrightarrow R$ of ring homomorphisms $\varkappa$ and σ, in which S is a ring with identity 1_S and $\sigma 1_S = 1_R$. Regard A as a two-sided ideal in S, with $S/A = R$. To each $x \in R$ choose a representative $u(x) \in S$, with $\sigma u(x) = x$. Then A is an R-bimodule with operators $xa = u(x)a$, $ax = au(x)$, independent of the choice of u. The addition and multiplication in S is determined by two factor sets f and g defined by

$$u(x) + u(y) = f(x, y) + u(x + y), \tag{13.1}$$

$$u(x)u(y) = g(x, y) + u(xy). \tag{13.2}$$

These functions f and g satisfy various identities which reflect the associative, commutative, and distributive laws in S (EVERETT [1942], REDEI [1952], SZENDREI [1952]). One can now construct (MAC LANE [1956]) a cohomology theory for a ring R such that $H^2(R, A)$ has such pairs of functions f, g as cocycles, with cohomology classes representing the extensions of A by R. A part of the corresponding 3-dimensional cohomology group $H^3(R, A)$ then corresponds exactly (MAC LANE [1958]) to the obstructions for the problem of extending a ring T (without identity, but with product not necessarily zero) by the ring R. The results also apply to sheaves of rings (GRAY [1961 a, b]).

SHUKLA [1961] has extended this cohomology theory for rings (Z-algebras) to the case of algebras Λ over an arbitrary commutative ring K. The resulting cohomology of algebras is more refined than the Hochschild cohomology, because the Hochschild cohomology deals systematically with those extensions which are K-split, while in the present case the use of a factor set (13.1) for addition reflects exactly the fact that the extensions concerned do not split additively. SHUKLA's theory is also so arranged that every element of H^3 corresponds to an obstruction. HARRISON [1962] has initiated a cohomology theory for commutative algebras over a field.

A *Lie algebra* L over K is a K-module together with a K-module homomorphism $x\otimes y\to[x,y]$ of $L\otimes L$ into L such that always

$$[x,x]=0, \quad [x,[y,z]]+[y,[z,x]]+[z,[x,y]]=0;$$

a typical example may be constructed by starting with an associative algebra Λ and setting $[x,y]=xy-yx$. Conversely, each Lie algebra L defines an augmented associative algebra L^e as the quotient of the tensor algebra $T(L)$ of the module L by the ideal generated in $T(L)$ by all elements $x\otimes y-y\otimes x-[x,y]$, for $x,y\in L$. The algebra L^e is called the *enveloping* (associative) algebra of L. The homology and cohomology of L are now defined for modules G_{L^e} and ${}_{L^e}C$ as

$$H_n(L,G)=\operatorname{Tor}_n^{L^e}(G,\mathsf{K}), \quad H^n(L,C)=\operatorname{Ext}_{L^e}^n(\mathsf{K},C),$$

though, as in the case of algebras, it may be more appropriate to use the relative Tor and Ext functors for the pair (L^e,K). This theory is developed in CARTAN-EILENBERG, Chap. XIII; cf. JACOBSON [1962]. In case K is a field, the POINCARÉ-BIRKHOFF-WITT Theorem may be used to give an alternative description of these cohomology and homology groups in terms of a standard complex constructed directly from the bracket product in L. Indeed, this is the approach originally used in the first treatment of the cohomology of Lie algebras (CHEVALLEY-EILENBERG [1948], KOSZUL [1950b]). The 2-dimensional cohomology group $H^2(L,C)$ corresponds to K-split extensions for Lie algebras (CARTAN-EILENBERG, XIV.5). In certain cases the elements of the 3-dimensional cohomology group $H^3(L,C)$ are the obstructions to extension problems (HOCHSCHILD [1954]). Analogous results apply to the analytic Lie groups (MACAULEY [1960]), and Lie triple systems (YAMAGUTI [1960], HARRIS [1961]). Shuffle products have been applied to Lie algebras by REE [1958].

Just as the cohomology of rings starts with factor sets for both addition and multiplication, it is possible to construct a cohomology of Lie rings such that H^2 will involve factor sets for both addition and bracket products. Such a theory has been initiated by DIXMIER [1957]; it is to be hoped that subsequent investigation might simplify his formulation.

Topologically, the bar construction starts with a "fiber" U, constructs an acyclic fiber bundle $B(U)$ with the group U and the corresponding base space $\overline{B}(U)$. The converse problem of constructing (the homology of) the fiber from a given base is geometrically important. To this end, J. F. ADAMS has introduced the cobar construction $F(W)$, where W is a graded coalgebra over K. This is a formal dual of the bar construction; for details, see ADAMS [1956], [1960, p. 33].

Notes. The reduced bar construction $\overline{B}(U)$ is due to EILENBERG-MACLANE [1950b]. CARTAN [1954] made the essential observation that $\overline{B}$ could be obtained from the acyclic bar construction B and developed an efficient method of carrying out calculations by "constructions".

Chapter eleven

Spectral Sequences

If Γ is a normal subgroup of the group Π, the homology of Π can be calculated by successive approximations from the homology of Γ and that of Π/Γ. These successive approximations are codified in the notion of a spectral sequence. In this chapter we first formulate the mechanism of these sequences and then proceed to several applications, ending with another general theorem (the comparison theorem). Other applications will appear in the next chapter.

In this chapter, a "module" will mean a left module A over the fixed ring R — though in most cases it could equally well mean a Λ-module or an object of a given abelian category. We deal repeatedly with subquotients S/K of A, where $K \subset S \subset A$: Recall (II.6.3) that each module homomorphism $\alpha: A \to A'$ induces for given subquotients S/K and S'/K' an additive relation $\alpha_{\#}: S/K \to S'/K'$ consisting of all those pairs of cosets $(s+K, \alpha s+K')$ with $s \in S$, $\alpha s \in S'$. If S, T, and U are submodules of A, the *modular* law asserts that $S \cap (T \cup U) = (S \cap T) \cup U$ whenever $S \supset U$. It follows that 1_A induces an isomorphism (the *modular Noether isomorphism*):

$$1_*: S/[U \cup (S \cap T)] \cong (S \cup T)/(U \cup T), \quad S \supset U.$$

Indeed, $S/[U \cup (S \cap T)] = S/[S \cap (T \cup U)]$; by the Noether isomorphism (I.2.5), this is isomorphic to $(S \cup T \cup U)/(T \cup U) = (S \cup T)/(U \cup T)$.

1. Spectral Sequences

A Z-bigraded module is a family $E = \{E_{p,q}\}$ of modules, one for each pair of indices $p, q = 0, \pm 1, \pm 2, \ldots$. A *differential* $d: E \to E$ of bidegree $(-r, r-1)$ is a family of homomorphisms $\{d: E_{p,q} \to E_{p-r,q+r-1}\}$, one for each p, q, with $d^2 = 0$. The homology $H(E) = H(E, d)$ of E under this differential is the bigraded module $\{H_{p,q}(E)\}$ defined in the usual way as

$$H_{p,q}(E) = \operatorname{Ker}[d: E_{p,q} \to E_{p-r,q+r-1}]/dE_{p+r,q-r+1}. \tag{1.1}$$

If E is made into a (singly) Z-graded module $E = \{E_n\}$ with total degree n by the usual process $E_n = \sum_{p+q=n} E_{p,q}$, the differential d induces a differential $d: E_n \to E_{n-1}$ with the usual degree -1, and $H(\{E_n\}, d)$ is the singly graded module obtained from the bigraded module $H_{p,q}(E)$ as $H_n = \sum_{p+q=n} H_{p,q}$.

A *spectral sequence* $E = \{E^r, d^r\}$ is a sequence $E^2, E^3, \ldots$ of Z-bigraded modules, each with a differential

$$d^r\colon E^r_{p,q} \to E^r_{p-r,q+r-1}, \quad r=2,3,\ldots, \tag{1.2}$$

of bidegree $(-r, r-1)$, and with isomorphisms

$$H(E^r, d^r) \cong E^{r+1}, \quad r=2,3,\ldots. \tag{1.3}$$

More briefly, each E^{r+1} is the bigraded homology module of the preceding (E^r, d^r). Thus E^r and d^r determine E^{r+1}, but not necessarily d^{r+1}. The bigraded module E^2 is the *initial term* of the spectral sequence (occasionally it is convenient to start the spectral sequence with $r=1$ and initial term E^1).

If E' is a second spectral sequence, a homomorphism $f\colon E \to E'$ is a family of homomorphisms

$$f^r\colon E^r \to E'^r, \quad r=2,3,\ldots$$

of bigraded modules, each of bidegree $(0,0)$, with $d^r f^r = f^r d^r$ and such that each f^{r+1} is the map induced by f^r on homology (use the isomorphisms (1.3)).

It is instructive to describe a spectral sequence in terms of submodules of E^2 (or of E^1, if this be present). First identify each E^{r+1} with $H(E^r, d^r)$ via the given isomorphism (1.3). This makes $E^3 = H(E^2, d^2)$ a subquotient C^2/B^2 of E^2, where $C^2 = \operatorname{Ker} d^2$ and $B^2 = \operatorname{Im} d^2$. In turn, $E^4 = H(E^3, d^3)$ is a subquotient of C^2/B^2 and so is isomorphic to C^3/B^3, where $C^3/B^2 = \operatorname{Ker} d^3$, $B^3/B^2 = \operatorname{Im} d^3$, and $B^3 \subset C^3$. Upon iteration, the spectral sequence is presented as a tower

$$0 = B^1 \subset B^2 \subset B^3 \subset \cdots \subset \cdots \subset C^3 \subset C^2 \subset C^1 = E^2 \tag{1.4}$$

of bigraded submodules of E^2, with $E^{r+1} = C^r/B^r$, where

$$d^r\colon C^{r-1}/B^{r-1} \to C^{r-1}/B^{r-1}, \; r=2,3,\ldots$$

has kernel C^r/B^{r-1} and image B^r/B^{r-1}. In informal parlance,

C^{r-1} is the module of elements which *live till stage r*,

B^{r-1} is the module of elements which *bound by stage r*.

The module of elements which "live forever" is

C^∞ = intersection of all the submodules C^r, $r=2,3,\ldots$,

while the module of elements which "eventually bound" is

B^∞ = union of all the submodules B^r, $r=2,3,\ldots$.

Then $B^\infty \subset C^\infty$, so the spectral sequence determines a bigraded module

$$E^\infty_{p,q} = C^\infty_{p,q}/B^\infty_{p,q}, \qquad E^\infty = \{E^\infty_{p,q}\}. \tag{1.5}$$

We regard the terms E^r of the spectral sequence as successive approximations (via successive formation of subquotients) to E^∞. In this representation (1.4), a homomorphism $f: E \to E'$ of spectral sequences is a homomorphism $f: E^2 \to E'^2$ of bigraded modules, of bidegree (0, 0), such that $f(C^r) \subset C'^r$, $f(B^r) \subset B'^r$, and such that all the diagrams

$$\begin{array}{ccc} C^{r-1}/B^{r-1} & \xrightarrow{d^r} & C^{r-1}/B^{r-1} \\ \downarrow f_* & & \downarrow f_* \\ C'^{r-1}/B'^{r-1} & \xrightarrow{d^r} & C'^{r-1}/B'^{r-1} \end{array}$$

are commutative. Also $f: E \to E'$ induces $f^\infty: E^\infty \to E'^\infty$.

A *first quadrant* spectral sequence E is one with $E^r_{p,q} = 0$ when $p < 0$ or $q < 0$. (This condition for $r = 2$ implies the same condition for higher r.) It is convenient to display the modules $E^r_{p,q}$ at the lattice points of the first quadrant of the p, q plane:

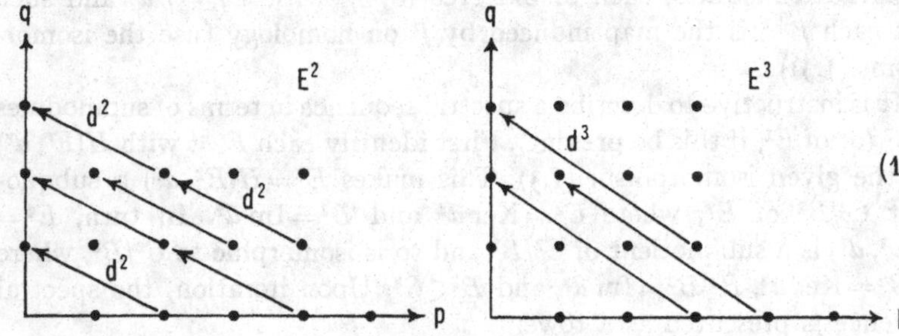

(1.6)

The differential d^r is then indicated by an arrow. The terms of total degree n all lie on the 45° line $p+q=n$; the successive differentials go from a lattice point on this line to one on the next line below, At each lattice point of $E^r_{p,q}$ the next approximation $E^{r+1}_{p,q}$ is formed by taking the kernel of the arrow from that lattice point modulo the image of the arrow which ends there, as in

$$E^r_{p+r,q-r+1} \xrightarrow{d^r} E^r_{p,q} \xrightarrow{d^r} E^r_{p-r,q+r-1}.$$

The outgoing d^r ends outside the quadrant if $r > p$, the incoming d^r starts outside if $r > q+1$, so that

$$E^{r+1}_{p,q} = E^r_{p,q}, \qquad \infty > r > \operatorname{Max}(p, q+1). \tag{1.7}$$

In words: For fixed degrees p and q, $E^r_{p,q}$ is ultimately constant in r.

The terms $E_{p,q}$ on the p-axis are called the *base* terms. Each arrow d^r ending on the base comes from below, hence from 0, so each $E^{r+1}_{p,0}$ is a submodule of $E^r_{p,0}$, namely the kernel of $d^r\colon E^r_{p,0}\to E^r_{p-r,r-1}$. This gives a sequence of monomorphisms

$$E^{\infty}_{p,0}=E^{p+1}_{p,0}\to\cdots\to E^4_{p,0}\to E^3_{p,0}\to E^2_{p,0}. \tag{1.8}$$

The terms $E_{0,q}$ on the q-axis are called the *fiber* terms. Each arrow from a fiber term ends to the left at zero, hence $E^r_{0,q}$ consists of cycles and the next fiber term is a quotient of $E^r_{0,q}$ (the quotient by the image of a d^r). This gives a sequence of epimorphisms

$$E^2_{0,q}\to E^3_{0,q}\to E^4_{0,q}\to\cdots\to E^{q+2}_{0,q}=E^{\infty}_{0,q}. \tag{1.9}$$

These maps (1.8) and (1.9) are known as *edge homomorphisms* (monic on the base, epic on the fiber).

A spectral sequence E is said to be *bounded below* if for each degree n there is an integer $s=s(n)$ such that $E^2_{p,q}=0$ when $p<s$ and $p+q=n$. This amounts to the requirement that on each 45° line ($p+q=n$) the terms are eventually zero as p decreases; thus a first quadrant or a "third quadrant" spectral sequence is bounded below.

Theorem 1.1. *(Mapping Theorem.) If $f\colon E\to E'$ is a homomorphism of spectral sequences, and if $f^t\colon E^t\cong E'^t$ is an isomorphism for some t, then $f^r\colon E^r\cong E'^r$ is an isomorphism for $r\geqq t$. If also E and E' are bounded below, $f^\infty\colon E^\infty\cong E'^\infty$ is an isomorphism.*

Proof. Since f^t is a chain isomorphism and $E^{t+1}=H(E^t,d^t)$, the first assertion follows by induction. When E and E' are bounded below and (p,q) are fixed, $d^r\colon E^r_{p,q}\to E^r_{p-r,q+r-1}$ has image 0 for sufficiently large r. Hence $C^r_{p,q}=C^\infty_{p,q}$ and $C'^r_{p,q}=C'^\infty_{p,q}$ for r large. Thus $a'\in C'^\infty_{p,q}$ lies in $C'^r_{p,q}$, so f^r an epimorphism makes f^∞ an epimorphism. If $a\in C^\infty$ has $fa\in B'^\infty=\bigcup B'^r$, then $fa\in B'^r$ for some r. Hence f^r a monomorphism for all r implies that f^∞ is a monomorphism.

Exercises

1. Show that a tower (1.4) together with a sequence of isomorphisms $\theta^r\colon C^{r-1}/C^r\cong B^r/B^{r-1}$ of bidegrees $(-r,r-1)$ for $r=2,3,\ldots$ determines a spectral sequence with $E^r=C^{r-1}/B^{r-1}$ and d^r the composite $C^{r-1}/B^{r-1}\to C^{r-1}/C^r\to B^r/B^{r-1}\to C^{r-1}/B^{r-1}$, and that every spectral sequence is isomorphic to one so obtained.

2. If E' and E'' are spectral sequences of vector spaces over a field, construct a spectral sequence $E=E'\otimes E''$ with $E^r_{p,q}=\sum E'^r_{p',q'}\otimes E''^r_{p'',q''}$, where the sum is taken over all $p'+p''=p$, $q'+q''=q$, and d^r is given by the usual tensor product differential.

3. If E is a spectral sequence of projective left R-modules, C a left R-module and G a right R-module, construct spectral sequences $\operatorname{Hom}_R(E,C)$ and $G\otimes_R E$ and calculate the terms E^∞.

2. Fiber Spaces

Before studying various algebraic examples of spectral sequences it is illuminating to exhibit some of the formal arguments which can be made directly from the definition of a spectral sequence. For this purpose we cite without proof the important topological example of the spectral sequence of a fibration.

Let I denote the unit interval and P any finite polyhedron; recall that a homotopy is a continuous map $H\colon P\times I\to B$. A continuous map $f\colon E\to B$ of topological spaces with $f(E)=B$ is called a *fiber map* if any commutative diagram of the following form

$$\begin{array}{ccc} P & \xrightarrow{h} & E \\ \downarrow{\scriptstyle i} & \overset{L}{\nearrow} & \downarrow{\scriptstyle f} \\ P\times I & \xrightarrow{H} & B, \end{array} \qquad i(x)=(x,0) \quad \text{for} \quad x\in P,$$

(all maps continuous) can always be filled in at L so as to be commutative. This is the "covering homotopy" property for f: Any homotopy H of P in B whose initial values $H(x,0)$ can be "lifted" to a map $h\colon P\to E$ with $fh(x)=H(x,0)$ can itself be lifted to a homotopy L of P in E with $fL=H$ and $h(x)=L(x,0)$. If b is any point in B, its inverse image $F=f^{-1}b$ is called the *fiber* of f over b. If B is pathwise connected, it can be shown that any two such fibers (over different points b) have isomorphic (singular) homology groups. Hence one may form the singular homology groups $H_p(B, H_q(F))$ of B with coefficients in the homology groups $H_q(F)$ of "the" fiber. Strictly speaking, we should use "local coefficients" which display the action of the fundamental group of B on $H_q(F)$; this we avoid by assuming B simply connected. Since B is pathwise connected, its 0-dimensional singular homology is

$$H_0(B)=Z, \quad H_0(B, H_q(F))\cong H_q(F).$$

The following spectral sequence has been constructed by SERRE [1951] following LERAY's construction [1946, 1950] for the case of cohomology.

Theorem (LERAY-SERRE). *If $f\colon E\to B$ is a fiber map with base B pathwise connected and simply connected and fiber F pathwise connected, there is for each n a nested family of subgroups of the singular homology group $H_n(E)$,*

$$0=H_{-1,n+1}\subset H_{0,n}\subset H_{1,n-1}\subset\cdots\subset H_{n-1,1}\subset H_{n,0}=H_n(E), \tag{2.1}$$

and a first quadrant spectral sequence such that

$$E^2_{p,q}\cong H_p(B, H_q(F)), \quad E^\infty_{p,q}\cong H_{p,q}/H_{p-1,q+1}. \tag{2.2}$$

If e_B is the iterated edge homomorphism on the base, the composite

$$H_p(E) \to H_{p,0}/H_{p-1,1} \cong E^\infty_{p,0} \xrightarrow{e_B} E^2_{p,0} \cong H_p(B, H_0(F)) \cong H_p(B)$$

is the homomorphism induced on homology by the fiber map $f\colon E \to B$. If e_F is the iterated edge homomorphism on the fiber, the composite

$$H_q(F) \cong H_0(B, H_q(F)) \cong E^2_{0,q} \xrightarrow{e_F} E^\infty_{0,q} \to H_q(E)$$

is the homomorphism induced on homology by the inclusion $F \subset E$.

This spectral sequence relates the (singular) homology of the base and fiber, via E^2, to the homology of the "total space" E, with E^∞ giving the successive factor groups in the "filtration" (2.1) of the homology of E.

The universal coefficient theorem (Thm. V.11.1) expresses the first term of (2.2) as an exact sequence

$$0 \to H_p(B) \otimes H_q(F) \to E^2_{p,q} \to \operatorname{Tor}(H_{p-1}(B), H_q(F)) \to 0. \qquad (2.3)$$

In particular, if all $H_{p-1}(B)$ are torsion-free, $E^2_{p,q} = H_p(B) \otimes H_q(F)$.

Assuming this result, we deduce several consequences so as to illustrate how information can be extracted from a spectral sequence.

The LERAY-SERRE theorem holds when all homology groups (of B, F, and E) are interpreted to be homology groups over the field Q of rational numbers. Write $\dim V$ for the dimension of a Q-vector space V over Q. For any space X the n-th *Betti number* $b_n(X)$ and the *Euler characteristic* $\chi(X)$ are defined by

$$b_n(X) = \dim H_n(X, Q), \qquad \chi(X) = \sum_{n=0}^{\infty} (-1)^n b_n(X);$$

more precisely, $\chi(X)$ is defined if each $b_n(X)$ is finite and there is an m such that $b_n(X) = 0$ for $n > m$. If X is a finite polyhedron, $\chi(X)$ is defined.

Corollary 2.1. *If $f\colon E \to B$ is a fiber space with fiber F, with B and F connected as in the Leray-Serre theorem, then if $\chi(B)$ and $\chi(F)$ are defined, so is $\chi(E)$ and $\chi(E) = \chi(B)\chi(F)$.*

Proof. For any bigraded vector space E^r, define a characteristic as $\chi(E^r) = \sum_{p,q} (-1)^{p+q} \dim E^r_{p,q}$. By (2.3) for vector spaces,

$$E^2_{p,q} \cong H_p(B) \otimes H_q(F), \qquad \dim E^2_{p,q} = b_p(B)\, b_q(F) < \infty,$$

and $\chi(E^2) = \chi(B)\chi(F)$. Write $C^r_{p,q}$ and $B^r_{p,q}$ for the cycles and the boundaries of $E^r_{p,q}$ under d^r. The short exact sequences

$$C^r_{p,q} \rightarrowtail E^r_{p,q} \twoheadrightarrow B^r_{p-r,q+r-1}, \qquad B^r_{p,q} \rightarrowtail C^r_{p,q} \twoheadrightarrow E^{r+1}_{p,q}$$

define C^r, B^r, and $E^{r+1}=H(E^r)$. In each sequence the dimension of the middle term is the sum of the dimensions of the end terms, so

$$\dim E^{r+1}_{p,q}=\dim E^r_{p,q}-\dim B^r_{p,q}-\dim B^r_{p-r,q+r-1}.$$

Here the last term has total degree $p-r+q+r-1=(p+q)-1$, so $\chi(E^{r+1})=\chi(E^r)$; by induction $\chi(E^r)=\chi(E^2)$. Since $E^r_{p,q}$ vanishes for p and q large, $E^\infty=E^r$ for r large, and $\chi(E^\infty)=\chi(E^2)$. Now by (2.1) and (2.2),

$$\dim H_n(E)=\sum_{p+q=n}\dim(H_{p,q}/H_{p-1,q+1})=\sum_{p+q=n}\dim E^\infty_{p,q},$$

so $\chi(E)=\chi(E^\infty)=\chi(E^2)=\chi(B)\chi(F)$, as asserted.

Theorem 2.2. *(The* Wang *sequence.) If $f\colon E\to S^k$ is a fiber space with base a k-sphere ($k\geqq 2$) and pathwise connected fiber F, there is an exact sequence*

$$\cdots\to H_n(E)\to H_{n-k}(F)\xrightarrow{d^k}H_{n-1}(F)\to H_{n-1}(E)\to\cdots.$$

Proof. The base S^k is simply connected and has homology $H_0(S^k)\cong Z\cong H_k(S^k)$ and $H_p(S^k)=0$, for $p\neq 0,k$; hence by (2.3)

$$E^2_{k,q}\cong H_q(F),\quad E^2_{0,q}\cong H_q(F),\quad E^2_{p,q}=0,\quad p\neq 0,k.$$

The non-zero terms of $E^2_{p,q}$ all lie on the vertical lines $p=0$ and $p=k$, so the only differential d^r with $r\geqq 2$ which is not zero has $r=k$. Therefore $E^2=E^3=\cdots=E^k$, $E^{k+1}=E^{k+2}=\cdots=E^\infty$. The description of $E^{k+1}=E^\infty$ as the homology of (E^k,d^k) amounts to the exactness of the sequence

$$0\to E^\infty_{k,q}\to E^2_{k,q}\xrightarrow{d^k}E^2_{0,q+k-1}\to E^\infty_{0,q+k-1}\to 0. \tag{2.4}$$

On the other hand, the tower (2.1) has only two non-vanishing quotient modules, so collapses to $0\subset H_{0,n}=H_{k-1,n-k+1}\subset H_{k,n-k}=H_n$. With the isomorphisms for E^∞ in (2.2), this amounts to a short exact sequence

$$0\to E^\infty_{0,n}\to H_n(E)\to E^\infty_{k,n-k}\to 0 \tag{2.5}$$

with $H_n(E)$ in the middle. Now set $q=n-k$ in (2.4), put in the values of E^2 in terms of $H(F)$ and splice the sequences (2.4) and (2.5) together:

$$\begin{array}{ccccccccc}
 & & H_n(E) & & & & 0 & & \\
 & & \downarrow & \searrow & & & \downarrow & & \\
0 & \to & E^\infty_{k,n-k} & \to & H_{n-k}(F)\to H_{n-1}(F) & \to & E^\infty_{0,n-1} & \to & 0\\
 & & \downarrow & & & \searrow & \downarrow & & \\
 & & 0 & & & & H_{n-1}(E)\,. & &
\end{array}$$

The result is the desired long exact sequence. By Leray-Serre, the homomorphism $H_{n-1}(F)\to H_{n-1}(E)$ is that induced by the inclusion $F\subset E$.

Spectral sequences may be used to calculate the homology of certain loop spaces which are used in homotopy theory. Let b_0 be a fixed point in the pathwise connected space B. The space $L(B)$ of paths in B has as points the continuous maps $t\colon I \to B$ with $t(0) = b_0$; here I is the unit interval, and $L(B)$ is given the "compact-open" topology. The map $p\colon L(B) \to B$ with $p(t) = t(1)$ projects each path onto its end point in B; it can be shown to be a fiber map. The fiber $\Omega(B) = p^{-1}(b_0)$ consists of the closed paths t [with $t(0) = b_0 = t(1)$]; it is known as the *loop space* of B.

Corollary 2.3. *The loop space ΩS^k of a k-sphere, $k \geqq 2$, has homology*

$$H_n(\Omega S^k) \cong Z, \quad n \equiv 0 \pmod{k-1},$$
$$= 0, \quad n \not\equiv 0 \pmod{k-1}, \quad n \geqq 0.$$

Proof. Since $k > 1$, S^k is simply connected, so each loop can be contracted to zero; this implies that $\Omega(S^k)$ is pathwise connected, so that $H_0(\Omega S^k) = Z$. The space $E = L(B)$ of paths is contractible, as one may see by "pulling" each path back along itself to the origin. Hence E is acyclic (Ex. II.8.1). Thus every third term $H_n(E)$ in the WANG sequence is zero, except for $H_0(E)$, so the sequence gives isomorphisms $H_{n-k}(\Omega S^k) \cong H_{n-1}(\Omega S^k)$. With the given initial value $H_0 = Z$ this gives the values stated above.

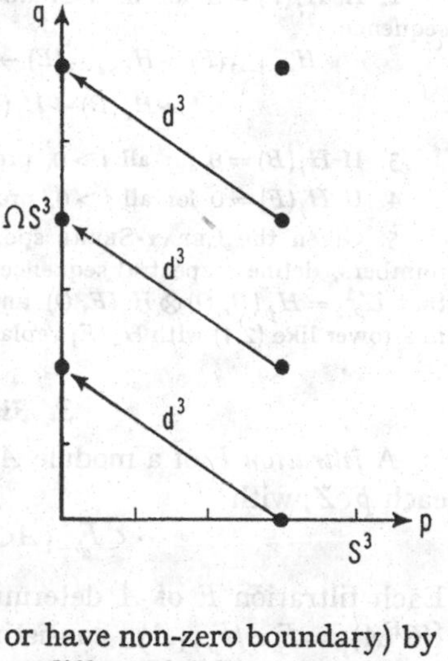

It is instructive to exhibit the diagram of this spectral sequence for $k = 3$. (See the attached diagram.) The heavy dots denote the terms $E_{p,q} \cong Z$, and all others are zero. The only non-zero differential is d^3; these differentials applied to the elements on the line $p = 3$ "kill" the successive elements in the homology of the fiber. This diagram may be constructed directly, without using the WANG sequence. We are given the base with generators $1 \in E^2_{0,0}$ and $x \in E^2_{3,0}$; all elements lie on the vertical lines $p = 0$ and $p = 3$. Since $E^\infty = 0$, every element must be killed (i.e., become a boundary or have non-zero boundary) by some differential. But d^3 is the only non-zero differential. Therefore $d^3 x = y \neq 0$ in $E^2_{0,2}$ on the fiber. The element $x \otimes y \in E^2_{3,2}$ must then also have a non-zero boundary $d^3(x \otimes y) = y'$ in $E^2_{0,4}$ on the fiber, and so on.

Theorem 2.4. *(The* GYSIN *sequence.) If $f\colon E\to B$ is a fiber space with simply connected base B and with fiber F the k-sphere S^k with $k\geqq 1$, there is an exact sequence*

$$\cdots\to H_n(E)\xrightarrow{f_*} H_n(B)\xrightarrow{d^{k+1}} H_{n-k-1}(B)\to H_{n-1}(E)\to\cdots.$$

Proof. Since $H_q(F)=H_q(S^k)=0$ for $q\neq 0, k$, the term E^2 is

$$E^2_{p,q}=H_p(B),\quad q=0,\ q=k;\qquad E^2_{p,q}=0,\quad q\neq 0, k.$$

The spectral sequence then lies on the two horizontal lines $q=0$ and $q=k$; the only non-zero differential is d^{k+1}, and we obtain two exact sequences

$$0\to E^\infty_{n,0}\to E^2_{n,0}\xrightarrow{d^{k+1}} E^2_{n-k-1,k}\to E^\infty_{n-k-1,k}\to 0,$$

$$0\to E^\infty_{n-k-1,k+1}\to H_n(E)\to E^\infty_{n,0}\to 0$$

which splice together to give the sequence of the Theorem.

Exercises

1. If $f\colon S^m\to S^k$ is a fiber map with $k\geqq 2$ and with fiber a sphere S^l prove that one must have $m=2k-1$ and $l=k-1$. (For $k=2$, 4, and 8 there are indeed such fiber maps; they are the *Hopf fibrations*; HOPF [1931, 1935], STEENROD [1951], HU [1959, p. 66].)

In the following three exercises, $f\colon E\to B$ is a fiber space with B pathwise connected and simply connected and fiber F pathwise connected.

2. If $H_j(F)=0$ for $0<j<t$ and $H_i(B)=0$ for $0<i<s$, obtain the exact sequence

$$H_{s+t-1}(F)\to H_{s+t-1}(E)\to H_{s+t-1}(B)\to H_{s+t-2}(F)\to\cdots$$
$$\to H_2(B)\to H_1(F)\to H_1(E)\to H_1(B)\to 0.$$

3. If $H_i(B)=0$ for all $i>0$, prove that $H_n(F)\cong H_n(E)$ for all n.

4. If $H_j(F)=0$ for all $j>0$, prove that $H_n(E)\cong H_n(B)$ for all n.

5. Given the LERAY-SERRE spectral sequence E and Q the field of rational numbers, define a spectral sequence $E'=Q\otimes E$ of vector spaces over Q and show that $E'^2_{p,q}=H_p(B,Q)\otimes H_q(F,Q)$ and $E'^\infty_{p,q}=H'_{p,q}/H'_{p-1,q+1}$, where the H' appear in a tower like (2.1) with $H_n(E)$ replaced by $H_n(E,Q)$.

3. Filtered Modules

A *filtration* F of a module A is a family of submodules F_pA, one for each $p\in Z$, with

$$\cdots\subset F_{p-1}A\subset F_pA\subset F_{p+1}A\subset\cdots. \tag{3.1}$$

Each filtration F of A determines an *associated graded module* $G^FA=\{(G^FA)_p=F_pA/F_{p-1}A\}$, consisting of the successive factor modules in the tower (3.1). If F and F' are filtrations of A and A', respectively, a homomorphism $\alpha\colon A\to A'$ of filtered modules is a module homomorphism with $\alpha(F_pA)\subset F'_pA'$. A filtration F of a differential Z-graded module A is a family of sub-DG_Z-modules F_pA, as in (3.1), with the corresponding definition of a homomorphism. This filtration induces a filtration on the Z-graded homology module $H(A)$, with $F_p(H(A))$ defined as the image of $H(F_pA)$ under the injection $F_pA \to A$. Since A itself is Z-graded by degrees n, the filtration F of A determines a filtration F_pA_n of each A_n, and the differential of A induces homomorphisms $\partial\colon F_pA_n \to F_pA_{n-1}$ for each p and each n. The family $\{F_pA_n\}$ is a Z-bigraded module; it is convenient and customary to write the indices of the grading as (p, q), where p is the *filtration* degree and $q=n-p$ the *complementary degree*; the Z-bigraded module then has the form $\{F_pA_{p+q}\}$. We use "FDG_Z-module" to abbreviate "filtered differential Z-graded module".

A filtration F of a DG_Z-module A is said to be *bounded* if for each degree there are integers $s=s(n)<t=t(n)$ such that $F_sA_n=0$ and $F_tA_n=A_n$. This amounts to the requirement that the filtration of each A_n has "finite length": $0=F_sA_n \subset F_{s+1}A_n \subset \cdots \subset F_tA_n=A_n$.

A spectral sequence $\{E_p^r, d^r\}$ is said to *converge* to a graded module H (in symbols, $E_p^2 \overset{p}{\Rightarrow} H$) if there is a filtration F of H and for each p isomorphisms $E_p^\infty \cong F_pH/F_{p-1}H$ of graded modules. Here, for given r and p, E_p^r denotes the Z-graded module $E_p^r=\{E_{p,q}^r, q=0, \pm 1, \ldots\}$ (graded by the complementary degree q).

The associated spectral sequence of a filtration may now be defined.

Theorem 3.1. *Each filtration F of a differential Z-graded module A determines a spectral sequence (E^r, d^r), $r=1, 2, \ldots$, which is a covariant functor of (F, A), together with natural isomorphisms*

$$E_p^1 \cong H(F_pA/F_{p-1}A); \quad i.e., \quad E_{p,q}^1 \cong H_{p+q}(F_pA/F_{p-1}A). \tag{3.2}$$

If F is bounded, $E_p^2 \Rightarrow H(A)$; more explicitly, there are natural isomorphisms

$$E_p^\infty \cong F_p(HA)/F_{p-1}(HA); \quad i.e., \quad E_{p,q}^\infty \cong F_p(H_{p+q}A)/F_{p-1}(H_{p+q}A). \tag{3.3}$$

For the proof we introduce the submodules

$$Z_p^r = [a \mid a \in F_pA, \ \partial a \in F_{p-r}A], \qquad r=0, 1, \ldots \tag{3.4}$$

of F_pA. An element of Z_p^r may be regarded as an "approximate cycle of level r"; its boundary need not be zero, but lies r stages lower down in the filtration. In particular, $Z_p^0=F_pA$. Each Z_p^r is Z-graded by degrees from A, so we may regard Z^r as the bigraded module with

$$Z_{p,q}^r = [a \mid a \in F_pA_{p+q}, \quad \partial a \in F_{p-r}A_{p+q-1}]. \tag{3.5}$$

Given this notation, the spectral sequence of the filtration F of A is defined by taking

$$E_p^r = (Z_p^r \cup F_{p-1}A)/(\partial Z_{p+r-1}^{r-1} \cup F_{p-1}A), \qquad r=1, 2, \ldots$$

while $d^r: E_p^r \to E_{p-r}^r$ is the homomorphism induced on these subquotients by the differential $\partial: A \to A$. From these definitions the proof of the theorem is fussy but straightforward. In detail:

Set $E_p^0 = F_p A / F_{p-1} A$ and let $\eta_p: F_p A \to E_p^0$ be the canonical projection. Consider the additive relations

$$E_{p+r}^0 \xrightarrow{\partial_1} E_p^0 \xrightarrow{\partial_2} E_{p-r}^0$$

induced on these subquotients by $\partial: A \to A$. Thus ∂_2 consists of the pairs $(\eta_p a, \eta_{p-r} \partial a)$ for $a \in Z_p^r$ (indeed, this is exactly why we need Z_p^r). Moreover $\eta_p a$ lies in the kernel of ∂_2 if $\partial a \in F_{p-r-1} A$, so (where "Def" means the "domain of definition")

$$\operatorname{Def} \partial_2 = \eta_p Z_p^r, \qquad \operatorname{Ker} \partial_2 = \eta_p Z_p^{r+1}.$$

Next, ∂_1 consists of the pairs $(\eta_{p+r} b, \eta_p \partial b)$ for $b \in Z_{p+r}^r$, while $\eta_{p+r} b = 0$ if also $b \in F_{p+r-1} A$; that is, if $b \in Z_{p+r-1}^{r-1}$. Hence (where "Ind" means the "indeterminacy")

$$\operatorname{Im} \partial_1 = \eta_p (\partial Z_{p+r}^r), \qquad \operatorname{Ind} \partial_1 = \eta_p (\partial Z_{p+r-1}^{r-1}).$$

In view of the inclusions $\partial Z_{p+r-1}^{r-1} \subset \partial Z_{p+r}^r \subset Z_p^{r+1} \subset Z_p^r$ we can introduce for each p and r a subquotient of E_p^0 as

$$E_p^r = (\eta_p Z_p^r) / \eta_p (\partial Z_{p+r-1}^{r-1}), \qquad r = 0, 1, 2, \ldots; \tag{3.6}$$

the formulas above show that ∂ induces homomorphisms

$$E_{p+r}^r \xrightarrow{d_1^r} E_p^r \xrightarrow{d_2^r} E_{p-r}^r$$

with image and kernel

$$\operatorname{Im} d_1^r = \eta_p (\partial Z_{p+r}^r) / \eta_p (\partial Z_{p+r-1}^{r-1}),$$
$$\operatorname{Ker} d_2^r = \eta_p (Z_p^{r+1}) / \eta_p (\partial Z_{p+r-1}^{r-1}).$$

Therefore (dropping the subscripts 1 and 2) $d^r d^r = 0$ and

$$H_p(E^r, d^r) \cong \eta_p (Z_p^{r+1}) / \eta_p (\partial Z_{p+r}^r) = E_p^{r+1}.$$

Thus we have a spectral sequence. When $r = 0$, $Z_p^0 = F_p A$ and $d^0: E_p^0 \to E_p^0$ is just the differential of the quotient complex $E_p^0 = F_p A / F_{p-1} A$. This gives (3.2).

This spectral sequence can also be derived from the towers

$$\begin{array}{ccccccccccccc} \partial Z_{p-1}^{-1} & \subset & \partial Z_p^0 & \subset & \partial Z_{p+1}^1 & \subset \cdots \subset & Z_p^2 & \subset & Z_p^1 & \subset & Z_p^0 & = & F_p A \\ \downarrow & & \downarrow & & \downarrow & & \downarrow & & \downarrow & & \downarrow & & \downarrow{\scriptstyle \eta_p} \\ B_p^0 & \subset & B_p^1 & \subset & B_p^2 & \subset \cdots \subset & C_p^2 & \subset & C_p^1 & \subset & C_p^0 & = & E_p^0. \end{array}$$

The tower on the first line, taken modulo $F_{p-1} A$, gives that on the second line, with $B_p^r = \eta_p \partial Z_{p+r-1}^{r-1}$ and $C_p^r = \eta_p Z_p^r$. By II.6 the additive relation $\partial_2 \colon F_pA/F_{p-1}A \to F_{p-r}A/F_{p-r-1}A$ amounts to an isomorphism

$$\operatorname{Def} \partial_2/\operatorname{Ker} \partial_2 \cong \operatorname{Im} \partial_2/\operatorname{Ind} \partial_2 .$$

But this isomorphism is just $C_p^r/C_p^{r+1} \cong B_{p-r}^{r+1}/B_{p-r}^r$. This gives d^r as the composite

$$E_p^r = C_p^r/B_p^r \xrightarrow{\pi} C_p^r/C_p^{r+1} \cong B_{p-r}^{r+1}/B_{p-r}^r \xrightarrow{\iota} C_{p-r}^r/B_{p-r}^r = E_{p-r}^r$$

with π the projection, ι the injection. This yields the spectral sequence, much as in Ex. 1.1 (except that C^r there is C^{r+1} here).

To describe $F_pH/F_{p-1}H$, write $C = \operatorname{Ker} \partial$ and $B = \partial A$ for the cycles and boundaries, respectively, in A. Then F induces on C and B filtrations $F_pC = C \cap F_pA$, $F_pB = B \cap F_pA$. By definition, $F_p(HA) = (F_pC \cup B)/B$. Hence

$$F_p(HA)/F_{p-1}(HA) \cong (F_pC \cup B)/(F_{p-1}C \cup B) \cong F_pC/(F_{p-1}C \cup F_pB),$$

by a modular Noether isomorphism. Another such,

$$F_p(HA)/F_{p-1}(HA) \cong (F_pC \cup F_{p-1}A)/(F_pB \cup F_{p-1}A) \tag{3.7}$$

represents $F_pH/F_{p-1}H$ as a subquotient of $F_pA/F_{p-1}A$.

The numerator of E_p^r in (3.6) is $(Z_p^r \cup F_{p-1}A)/F_{p-1}A \subset F_pA/F_{p-1}A$; the denominator is $(\partial Z_{p+r-1}^{r-1} \cup F_{p-1}A)/F_{p-1}A$, so

$$\left.\begin{aligned} E_p^r &= (Z_p^r \cup F_{p-1}A)/(\partial Z_{p+r-1}^{r-1} \cup F_{p-1}A), \\ E_{p,q}^r &= (Z_{p,q}^r \cup F_{p-1}A_{p+q})/(\partial Z_{p+r-1,q-r+2}^{r-1} \cup F_{p-1}A_{p+q}). \end{aligned}\right\} \tag{3.8}$$

Now suppose F bounded, and consider a fixed (p, q) corresponding to a total degree $n = p+q$. In the numerator of $E_{p,q}^r$, an element $a \in Z_{p,q}^r$ for r large has $\partial a \in F_{p-r}A_{p+q-1} = 0$, hence $a \in F_pC_{p+q}$. Thereafter the numerators are $F_pC_{p+q} \cup F_{p-1}A_{p+q}$. As for the denominator, for r large every element in F_pB_{p+q} is the boundary of an element in $F_{p+r-1}A$; that is, of an element in Z_{p+r-1}^{r-1}. Thereafter the denominators equal $F_pB_{p+q} \cup F_{p-1}A_{p+q}$. But E^∞ is defined as intersection of numerators divided by union of denominators, so

$$E_{p,q}^\infty = (F_pC_{p+q} \cup F_{p-1}A_{p+q})/(F_pB_{p+q} \cup F_{p-1}A_{p+q}), \tag{3.9}$$

which is exactly $F_pH/F_{p-1}H$ as given in (3.7). This proves (3.3).

In the literature, E^∞ is usually *defined* from $H(A)$ by the formula (3.9), so the "convergence" isomorphism (3.3) asserts that this definition agrees with ours.

The convergence (3.3) holds under weaker conditions than boundedness (for a thorough study, see EILENBERG-MOORE [1962]). For example, call a filtration F of the DG_Z-module A *convergent above* if A is the union of all F_pA and *bounded below* if for each degree n there is an integer $s=s(n)$ such that $F_sA_n=0$.

Proposition 3.2. *If F is bounded below and convergent above, then* (3.3) *holds and the spectral sequence of F is bounded below.*

Proof. Since F is bounded below, the intersection of the numerators of E_p^r is $F_pC \cup F_{p-1}A$. Each element of F_pB is a boundary ∂a for some $a \in A = \bigcup F_tA$, hence $a \in F_tA$ for some t. Then $a \in Z_{p+r-1}^{r-1}$ for $r = t-p+1$, so $F_pB \cup F_{p-1}A$ again is the union of the denominators $\partial Z_{p+r-1}^{r-1} \cup F_{p-1}A$, and we have (3.3).

In the formula (3.8) the numerator term Z_p^r gives "approximate" cycles of level r; while ∂Z_{p+r-1}^{r-1} in the denominator is a submodule of the boundaries (boundaries from r levels up). The proof has so chosen these approximations in this quotient that each has the next as its homology. An alternative formula (for the same spectral sequence) appears in Ex. 1.

The filtration F of a DG-module A is *canonically bounded* if $F_{-1}A=0$ and $F_nA_n=A_n$ in each degree n.

Theorem 3.3. *If F is a canonically bounded filtration of a (positively graded) DG-module A, the spectral sequence of F lies in the first quadrant and the induced filtration of HA is finite, of the form*

$$0=F_{-1}H_nA \subset F_0H_nA \subset F_1H_nA \subset \cdots \subset F_nH_nA=H_nA$$

with successive quotients $F_pH_n/F_{p-1}H_n \cong E_{p,n-p}^\infty$, under isomorphisms induced by 1_A. For example, the LERAY-SERRE *theorem arises from a canonically bounded filtration of the singular chains of a fiber space.*

Proof. Since $F_{-1}A=0$, $E_p^1=H(F_pA/F_{p-1}A)=0$ for $p<0$. Since $F_nA_n=A_n$, $q<0$ implies $F_pA_{p+q}=F_{p-1}A_{p+q}$ and hence $E_{p,q}^1=0$ for $q<0$. Therefore all non-zero $E_{p,q}^r$ lie in the first quadrant of the (p,q)-plane, and the induced filtration of $H_n(A)$ is finite as displayed.

For $n=1$ the filtration of H_1 amounts to a description of H_1 as the middle term of a short exact sequence

$$0 \to E_{0,1}^\infty \xrightarrow{\lambda} H_1 \xrightarrow{\sigma} E_{1,0}^\infty \to 0.$$

For each n, the filtration of H_n yields a monomorphism $E_{0,n}^\infty \to H_n(A)$ and an epimorphism $H_n(A) \to E_{n,0}^\infty$. Combined with the edge homomorphisms we get maps

$$E_{0,n}^1 \to H_n(A), \quad H_n(A) \to E_{n,0}^2, \tag{3.10}$$

each induced by 1_A. In general, the spectral sequence of F determines not $H(A)$ but its subquotients $F_pH/F_{p-1}H$, asserting that each is in its turn a subquotient of $E_p^1=H(F_pA/F_{p-1}A)$.

Theorem 3.4. *(The mapping theorem.) Let A, A' be DG_Z-modules with filtrations F and F' bounded below and convergent above. If $\alpha: (F, A) \to (F', A')$ is a homomorphism such that for some t the induced map*

$$\alpha^t: E^t(F, A) \cong E'^t(F', A')$$

is an isomorphism, then α^r is an isomorphism for $\infty \geqq r \geqq t$ and moreover $\alpha_*: H(A) \to H(A')$ is an isomorphism.

Proof. Since both spectral sequences are bounded below, the previous mapping theorem (Thm. 1.1) shows α^r and $\alpha^\infty: E^\infty \to E'^\infty$ isomorphisms. Consider the induced map $\alpha_n: H_n(A) \to H_n(A')$ on homology for a fixed degree n, and the corresponding $\alpha_{p,n}: F_p H_n \to F'_p H'_n$. Since both filtrations are bounded below there is an s with $F_s H_n = 0 = F'_s H'_n$. The convergence isomorphisms (3.3) give the horizontal sequences in the commutative diagram

$$\begin{array}{ccccccccc} 0 \to & F_{p-1}H_n(A) & \to & F_p H_n(A) & \to & E^\infty_{p,n-p} & \to 0 \\ & \downarrow{\scriptstyle \alpha_{p-1,n}} & & \downarrow{\scriptstyle \alpha_{p,n}} & & \downarrow{\scriptstyle \alpha^\infty} & \\ 0 \to & F'_{p-1}H_n(A') & \to & F'_p H_n(A') & \to & E'^\infty_{p,n-p} & \to 0. \end{array}$$

Since α^∞ is an isomorphism, induction on p and the Five Lemma show $\alpha_{p,n}$ an isomorphism. Now the filtration F is convergent above, so $H_n(A) = \cup F_p H_n(A)$; it follows that α_n is an isomorphism, as required.

For $t=1$ the hypothesis of this theorem requires that the induced map $H_n(F_p A / F_{p-1} A) \to H_n(F'_p A' / F'_{p-1} A')$ be an isomorphism for all n and p. This special case of the theorem was proved in Thm. V.9.3 and again in Thm. X.11.2.

Let $\alpha, \beta: (F, A) \to (F', A')$ be homomorphisms of FDG$_Z$-modules. A chain homotopy $s: \alpha \simeq \beta$ is said to have *order* $\leqq t$ if $s(F_p A) \subset F'_{p+t} A'$ for all p.

Proposition 3.5. *If $s: \alpha \simeq \beta$ is a homotopy of order $\leqq t$, then*

$$\alpha^r = \beta^r: E^r(F, A) \to E'^r(F', A')$$

for $r > t$, and $\alpha_* = \beta_*: H(A) \to H(A')$.

Proof. The result $\alpha_* = \beta_*$ follows from the existence of the chain homotopy (irrespective of its "order"). For the rest, it suffices to consider $\gamma = \alpha - \beta$, $s: \gamma \simeq 0$ and prove $\gamma^r = 0$. Write $E^r_{p,q}$ as the subquotient (3.8). If $a \in Z^r_p$, then $\gamma a = \partial s a + s \partial a$, where $\partial a \in F_{p-r} A$, so $s \partial a \in F'_{p-1} A'$ since $t < r$, while $s a \in F'_{p+r-1} A'$, $\partial s a = \gamma a - s \partial a \in F'_p A'$ so $s a \in Z'^{r-1}_{p+r-1}(A')$. Thus $\gamma a \in \partial Z'^{r-1}_{p+r-1} \cup F'_{p-1} A'$ is in the denominator of E'^r_p, so determines zero there.

Exercises

1. Show that $E^r_p = Z^r_p / (\partial Z^{r-1}_{p+r-1} \cup Z^{r-1}_{p-1})$, with $d^r: E^r_p \to E^r_{p-r}$ induced by $\partial: A \to A$, gives a spectral sequence isomorphic to that of Thm. 3.1. (These formulas are often used as the definition.)

2. If the filtration F of the differential graded module A is canonically bounded, show that its spectral sequence yields an exact sequence

$$H_2(A) \xrightarrow{e_B} E^2_{2,0} \xrightarrow{d^2} E^2_{0,1} \xrightarrow{e_F} H_1(A) \xrightarrow{e_B} E^2_{1,0} \to 0.$$

If $E^2_{p,q}=0$ for $0<q<t$ and all p, show that $e_B\colon H_p(A) \cong E^2_{p,0}$ for $0 \leqq p < t$ and establish the exact sequence

$$H_{t+1}(A) \xrightarrow{e_B} E^2_{t+1,0} \xrightarrow{d^{t+1}} E^2_{0,t} \xrightarrow{e_F} H_t(A) \xrightarrow{e_B} E^2_{t,0} \to 0.$$

3. (The exact sequence of "terms of low degree"; cf. Ex. 2.2.) In Ex. 2 suppose

$$E^2_{p,q}=0 \quad \text{when either} \quad 0<q<t \quad \text{or} \quad 0<p<s.$$

Establish an exact sequence, with H_i short for $H_i(A)$,

$$H_{s+t} \xrightarrow{e_B} E^2_{s+t,0} \xrightarrow{d^{s+t}} E^2_{0,s+t-1} \xrightarrow{e_F} H_{s+t-1} \xrightarrow{e_B} E^2_{s+t-1,0} \to \cdots$$

4. (The two-row exact sequence.) In Thm. 3.3 suppose there are two indices $0 \leqq a < b$ such that $E^2_{p,q}=0$ for $q \neq a$, b and all p. Derive the exact sequence

$$\cdots \to E^2_{n-b,b} \to H_n \to E^2_{n-a,a} \xrightarrow{d^r} E^2_{n-b-1,b} \to H_{n-1} \to \cdots$$

with $r=b-a+1$. (Hint: cf. the WANG sequence of Thm. 2.1.)

5. Establish a "two-column" exact sequence analogous to Ex. 4.

6. If A' and A'' are FDG-vector spaces over a field, and if a filtration of $A' \otimes A''$ is defined by $F_p(A' \otimes A'') = \sum F_{p'}(A') \otimes F_{p''}(A'')$ for $p'+p''=p$, prove for the associated spectral sequences that $E(A' \otimes A'') \cong E(A') \otimes E(A'')$.

7. In the spectral sequence of a filtration F of A, show $E^r_{p,q}$ isomorphic to the image of the homomorphism

$$H_{p+q}(F_p A / F_{p-r} A) \to H_{p+q}(F_{p+r-1} A / F_{p-1} A), \qquad r \geqq 1$$

induced by the identity. (This description may be used to define the spectral sequence of a filtration; see FADELL-HUREWICZ [1958, p. 318].)

4. Transgression

In a first quadrant spectral sequence E the last possibly non-zero differential on a term $E^r_{p,0}$ in the base is the differential $d^p\colon E^p_{p,0} \to E^p_{0,p-1}$ which goes from the base all the way to the fiber. With the edge homomorphisms e_B, e_F this yields a diagram

$$\begin{array}{ccccccccc} & & 0 & & E^1_{0,p-1} & & & & \\ & & \downarrow & & \downarrow e_F & & & & \\ 0 \to E^\infty_{p,0} & \to & E^p_{p,0} & \xrightarrow{d^p} & E^p_{0,p-1} & \to & E^\infty_{0,p-1} & \to & 0 \\ & & \downarrow e_B & & \downarrow & & & & \\ & & E^2_{p,0} & & 0 & & & & \end{array} \tag{4.1}$$

with exact row and columns. When (as we have assumed here) the spectral sequence starts with $r=1$, the additive relation

$$\tau = e_F^{-1} d^p e_B^{-1}\colon E^2_{p,0} \to E^1_{0,p-1}, \qquad p=2, 3, \ldots$$

is called the *transgression*. Any additive relation (Prop. II.6.1) is a homomorphism from a submodule of its domain (here called the module of transgressive elements) to a quotient module of its range; in this case, (4.1) represents τ as the homomorphism d^p from the submodule $E^p_{p,0}$ of $E^2_{p,0}$ to the quotient module $E^p_{0,p-1}$ of $E^1_{0,p-1}$. Replacing E^1 by E^2 in this definition of τ gives an additive relation τ': $E^2_{p,0} \to E^2_{0,p-1}$, also called transgression. Each transgression uniquely determines the other via the edge homomorphism e: $E^1_{0,p-1} \to E^2_{0,p-1}$, for $\tau = e^{-1}\tau'$; since e is an epimorphism, $ee^{-1}=1$, so $\tau' = e\tau$.

Proposition 4.1. *The transgression in the spectral sequence E of a canonically bounded filtration of A is the additive relation*

$$\tau\colon E^2_{p,0} \to E^1_{0,p-1}$$

induced by ∂: $A \to A$.

Proof. Here E is a first quadrant spectral sequence. Its edge terms can be written explicitly from (3.8). Since $\partial A_{p+1} \subset A_p = F_p A_p$, $A_{p+1} = Z^{r-1}_{p+r-1,-r+2}$ for any $r \geqq 2$, so, on the base, (3.8) becomes

$$E^r_{p,0} = (Z^r_{p,0} \cup F_{p-1}A_p)/(\partial A_{p+1} \cup F_{p-1}A_p), \qquad r = 2, 3, \ldots . \tag{4.2}$$

The denominator is independent of r; this verifies the fact that the edge homomorphisms e_B are monic. Also $Z^r_{0,q}$ is $F_0 C_q$ when $r \geqq 1$ and C is the kernel of ∂. Hence on the fiber (3.8) is

$$E^r_{0,q} = F_0 C_q / \partial Z^{r-1}_{r-1,q-r+2}, \qquad r = 1, 2, \ldots . \tag{4.3}$$

The numerator is independent of r (edge homomorphisms e_F are epic).

The transgression is the composite relation $\tau = e_F^{-1} d^p e_B^{-1}$, where e_F^{-1} and e_B^{-1} are induced by 1_A and d^p is induced by ∂. The composite τ is then the additive relation induced by ∂, as one sees by calculating τ as the set of all pairs of cosets $(a + D^2_{p,0},\ \partial a + D^1_{0,p-1})$ for $a \in Z^p_{p,0}$ and $D^r_{p,q}$ the denominator of $E^r_{p,q}$, or by applying the composition principle for additive relations (Prop. II.6.3).

The edge maps (3.10) and the transgression can be computed directly from A and two subcomplexes defined by the filtration, without using the whole spectral sequence, but using a generalization of the familiar homology connecting homomorphisms.

If L and M are subcomplexes of a (not necessarily positive) complex K, the *connecting relation*

$$\varrho = \varrho(K; L, M)\colon H_n(K/M) \to H_{n-1}(L) \tag{4.4}$$

is defined to be the additive relation induced by ∂: $K \to K$. Here each homology group is to be regarded as a subquotient of K; for example,

$$H_n(K/M) = C_n(K, M)/(\partial K_{n+1} \cup M_n),$$

where $C_n(K, M)$ is the module of relative cycles (all $k \in K_n$ with $\partial k \in M_{n-1}$). Thus ϱ consists of the pairs of homology classes $(k + (\partial K_{n+1} \cup M_n), \partial k + \partial L_n)$ for all $k \in C_n(K, L \cap M)$. If $M = L$, the connecting relation ϱ is just the usual connecting homomorphism ∂_L for the short exact sequence $L \rightarrowtail K \twoheadrightarrow K/L$ of complexes. More generally,

Proposition 4.2. *If L and M are subcomplexes of the complex K, with $L_{n-1} \subset M_{n-1}$ and $L_n \subset M_n$, then $\varrho = \varrho(K; L, M)$ can be described via connecting homomorphisms as the composite relation*

$$\varrho(K; L, M) = \gamma^{-1} \partial_M = \partial_L \beta^{-1}\colon H_n(K/M) \rightharpoonup H_{n-1}(L)$$

where β and γ are induced by the identity in the commutative diagram

$$\begin{array}{ccccccc} H_n(K) & \to & H_n(K/L) & \xrightarrow{\partial_L} & H_{n-1}(L) & \to & H_{n-1}(K) \\ \| & & \downarrow\beta & & \downarrow\gamma & & \| \\ H_n(K) & \to & H_n(K/M) & \xrightarrow{\partial_M} & H_{n-1}(M) & \to & H_{n-1}(K). \end{array}$$

Proof. The hypotheses $L_{n-1} \subset M_{n-1}$ and $L_n \subset M_n$ show that the identity induces homomorphisms β and γ as displayed. By the equivalence principle (Prop. II.6.2), β^{-1} and γ^{-1} are the additive relations induced by 1. By the composition principle (Prop. II.6.3), each of $\partial_L \beta^{-1}$ and $\gamma^{-1} \partial_M$ turns out to be the additive relation induced by $\partial 1 = 1 \partial$; hence the result.

This result shows that $\operatorname{Def} \varrho = \operatorname{Im} \beta$ and $\operatorname{Ind} \varrho = \operatorname{Ker} \gamma$.

In § 10 we need information as to the effect of a chain equivalence on the connecting relations, as follows.

Lemma 4.3. *Let $f\colon K \to K'$ be a chain transformation which induces homology isomorphisms $f_*\colon H_n(K) \cong H_n(K')$, while L, M are subcomplexes of K and L', M' subcomplexes of K' with $f(L) \subset L'$, $f(M) \subset M'$, so that f induces chain transformations $g\colon L \to L'$, $h\colon K/M \to K'/M'$. Assume that g_* and h_* are homology isomorphisms and that $L_k \subset M_k$, $L'_k \subset M'_k$ for $k = n-1, n$, as in Prop. 4.2. Then the diagram*

$$\begin{array}{rccc} \varrho = \varrho(K; L, M)\colon & H_n(K/M) & \rightharpoonup & H_{n-1}(L) \\ & \downarrow h_* & & \downarrow g_* \\ \varrho' = \varrho(K'; L', M')\colon & H_n(K'/M') & \rightharpoonup & H_{n-1}(L') \end{array}$$

is commutative.

This result computes ϱ' from ϱ as $\varrho' = g_* \varrho h_*^{-1}$, or conversely.

Proof. Since f_* and g_* are homology isomorphisms, the exact homology sequences for L, K, K/L and L', K', K'/L' show that f induces a homology isomorphism $\varphi\colon K/L \to K'/L'$. By Prop. 4.2 we may compute the connecting relations $\varrho = \partial_L \beta^{-1}$ and $\varrho' = \partial_L \beta'^{-1}$ from the rows of the commutative diagram

$$\begin{array}{ccccc} H_n(K/M) & \xleftarrow{\beta} & H_n(K/L) & \xrightarrow{\partial_L} & H_{n-1}(L) \\ \downarrow{h_*} & & \downarrow{\varphi_*} & & \downarrow{g_*} \\ H_n(K'/M') & \xleftarrow{\beta'} & H_n(K'/L') & \xrightarrow{\partial_{L'}} & H_{n-1}(L'). \end{array}$$

Since the diagram commutes, $\beta'\varphi_* = h_*\beta$, or $\beta^{-1}h_*^{-1} = \varphi_*^{-1}\beta'^{-1}$. But h_* and φ_* are isomorphisms, so $\varphi_*\beta^{-1} = \beta'^{-1}h_*$. Now $g_*\varrho = g_*\partial_L\beta^{-1} = \partial_{L'}\varphi_*\beta^{-1} = \partial_{L'}\beta'^{-1}h_* = \varrho' h_*$, as desired.

Theorem 4.4. *If F is a canonically bounded filtration of a DG-module A the "edge effects" in the spectral sequence of F can be computed from A and its subcomplexes $L = F_0A$ and M, where $M_n = F_{n-1}A_n \cup \partial F_n A_{n+1}$. Specifically, the edge homomorphisms*

$$H_n(F_0A) = E^1_{0,n} \to H_n(A), \qquad H_n(A) \to E^2_{n,0} = H_n(A/M)$$

are induced by the injection $F_0A \to A$ and the projection $A \to A/M$, respectively, while the transgression τ is the connecting relation $\varrho(A; F_0A, M)$.

Proof. By (4.3),

$$E^1_{0,n} = F_0C_n/\partial F_0A_{n+1} = H_n(F_0A).$$

By (4.2) and the definition of $H_n(A/M)$ by relative cycles,

$$\begin{aligned} E^2_{n,0} &= (Z^2_{n,0} \cup F_{n-1}A_n)/(\partial A_{n+1} \cup F_{n-1}A_n) \\ &= C_n(A, M)/(\partial A_{n+1} \cup M_n) = H_n(A/M). \end{aligned}$$

But the maps e_B and e_F are induced by the identity, whence the first result. Similarly, each of τ and ϱ is the additive relation $E^2_{n,0} \to E^1_{0,n-1}$ induced by ∂, so $\tau = \varrho$, as desired.

The situation may be visualized in terms of the complexes

$$\begin{array}{ccc} F_0A & \to & A \\ & & \downarrow \\ & & A/M. \end{array}$$

Since $M_n \supset (F_0A)_n$ for $n \geqq 1$, the transgression can also be described in terms of ordinary connecting homomorphisms, as in Prop. 4.2. This theorem shows how additive relations clarify a result of Serre (loc. cit., I.3; his notation $R = F_0A$, $S = A/M$). In the case of a fiber map $f: E \to B$, $H(A) = H(E)$, $H(F_0A)$ is the homology of the fiber, $H_p(A/M) = E^2_{p,0} = H_p(B, Z)$ that of the base. Thus Prop. 4.2 gives for transgression the following "geometric" description (in which it originated): A homology class of the base is transgressive if it can be represented by a cycle z such that $z = fc$, for c a chain of the total space with ∂c in the fiber. An image of cls z under transgression is then the homology class, in the fiber, of any such ∂c.

5. Exact Couples

An alternative description of spectral sequences can be given via "exact couples" (Massey [1952]). Though not necessary to the sequel, they throw some light on the origin and nature of spectral sequences.

An *exact couple* $\mathfrak{C}=\{D, E; i, j, k\}$ is a pair of modules D, E together with three homomorphisms i, j, k,

$$\mathfrak{C} \qquad \begin{array}{ccc} D & \xrightarrow{\;i\;} & D \\ & {}_{k}\nwarrow \quad \swarrow_{j} & \\ & E & \end{array}, \tag{5.1}$$

which form an *exact triangle* in the sense that kernel = image at each vertex. The modules D and E in an exact couple may be graded or Z-bigraded; in the latter case each of i, j, k has some bidegree.

The exactness of $\mathfrak{C}$ shows that the composite $jk\colon E\to E$ has square zero, hence is a differential on E. Form the homology module $H(E, jk)$ for this differential. Construct the triangle

$$\mathfrak{C}' \qquad \begin{array}{ccc} iD & \xrightarrow{\;i'\;} & iD \\ & {}_{k'}\nwarrow \quad \swarrow_{j'} & \\ & H(E, jk) & \end{array}, \tag{5.2}$$

where i' is induced by i and j' and k' are given by

$$j'(id)=jd+jkE, \quad k'(e+jkE)=ke, \quad e\in E, \quad jke=0.$$

Observe that $id=0$ implies $d\in kE$, so $jd\in jkE$ and j' is well defined. Similarly $jke=0$ implies $ke\in iD$, so k' is well defined. Call $\mathfrak{C}'$ the *derived couple* of $\mathfrak{C}$; it is a functor of $\mathfrak{C}$ under the evident definition of homomorphisms for exact couples. A diagram chase proves

Theorem 5.1. *The derived couple of an exact couple is exact.*

There is a whole sequence of derived couples. Iterate i $(r-1)$-times

$$\begin{array}{l} i^{r-1}\colon \; D \xrightarrow{i} D \xrightarrow{i} \cdots \xrightarrow{i} D \\ \qquad\quad \downarrow j \qquad\qquad\quad \| \\ \qquad\quad E \xleftarrow{\quad ji^{1-r}\quad} D. \end{array}$$

Here $i^{1-r}\colon D \rightharpoondown D$ and ji^{1-r} are additive relations, with

$$\operatorname{Ind}(ji^{1-r})=j(\operatorname{Ker} i^{r-1}), \quad \operatorname{Im}(ji^{1-r})=k^{-1}0\subset k^{-1}(i^{r-1}D).$$

Set

$$D^r=i^{r-1}D, \quad E^r=k^{-1}(i^{r-1}D)/j(\operatorname{Ker} i^{r-1}). \tag{5.3}$$

Then i, ji^{1-r}, and k induce homomorphisms i_r, j_r, k_r in the triangle

$$\mathfrak{C}^r \qquad \begin{array}{ccc} D^r & \xrightarrow{\;i_r\;} & D^r \\ & {}_{k_r}\nwarrow \quad \swarrow_{j_r} & \\ & E^r & \end{array}, \qquad r=1, 2, \ldots,$$

called the r-th *derived couple* of $\mathfrak{C}$.

Theorem 5.2. *The r-th derived couple $\mathfrak{C}^r$ is exact with $\mathfrak{C}^1=\mathfrak{C}$, $\mathfrak{C}^2=\mathfrak{C}'$, and $\mathfrak{C}^{r+1}$ is the derived couple of $\mathfrak{C}^r$.*

Proof. For $r=1$, $E^1=E$. For $r=2$, exactness of $\mathfrak{C}$ gives $iD=j^{-1}0$, $\ker i=kE$, hence $E^2=k^{-1}j^{-1}0/jkE=H(E,jk)$ and thus $\mathfrak{C}^2$ the derived couple of $\mathfrak{C}$. For $r>2$, $D^{r+1}=iD^r=i_rD^r$; we need only show that E^{r+1} is the homology of E^r under the differential $j_rk_r\colon E^r\to E^r$. To exhibit this differential, write the definition (5.3) of E^r as

$$E^r=C/B,\quad C=k^{-1}(i^{r-1}D),\quad B=j(\operatorname{Ker} i^{r-1}).$$

An element of E^r is a coset $c+B$, where $kc=i^{r-1}d$ for some d, and

$$j_rk_r(c+B)=j_r(kc)=jd+B,\quad kc=i^{r-1}d. \tag{5.4}$$

It will suffice to prove

$$\operatorname{Ker}(j_rk_r)=k^{-1}(i^rD)/B,\quad \operatorname{Im}(j_rk_r)=j(\operatorname{Ker} i^r)/B.$$

First $j_rk_r(c+B)=0$ gives $jd=ja$ for some $a\in D$ with $i^{r-1}a=0$. By the exactness of $\mathfrak{C}$, $d-a=id'$ for some d', so $kc=i^rd'$ and $c\in k^{-1}(i^rD)$. Conversely, $kc=i^rd'$ gives $j_rk_r(c+B)=0$; the kernel is as stated. Similarly $\operatorname{Im}(j_rk_r)$ consists by (5.4) of elements $jd+B$ with $i^rd=ikc=0$, and conversely $i^rd=0$ implies $i^{r-1}d=kc$ for some c; this gives the stated image. Since $\mathfrak{C}^{r+1}$ is the derived couple of $\mathfrak{C}^r$, it is exact by Thm. 5.1.

Corollary 5.3. *An exact couple of Z-bigraded modules D, E with maps of bidegrees*

$$\deg i=(1,-1),\quad \deg j=(0,0),\quad \deg k=(-1,0) \tag{5.5}$$

determines a spectral sequence (E^r,d^r) with $d^r=j_rk_r$, $r=1,2,\ldots$.

Proof. Given (5.5), the couple $\mathfrak{C}^r$ has maps of the following bidegrees

$$\deg i_r=(1,-1),\quad \deg j_r=(-r+1,r-1),\quad \deg k_r=(-1,0).$$

It follows that $\deg(j_rk_r)=(-r,r-1)$, so each E^{r+1} is the homology of E^r with respect to a differential d^r of the bidegree appropriate to a spectral sequence.

An exact couple $\mathfrak{C}$ with bidegrees (5.5) may be displayed as

$$\begin{array}{ccccccccc}
 & & \vdots & & & & \vdots & & \\
 & & \downarrow i & & & & \downarrow i & & \\
\cdots\to E_{p,q+1} & \xrightarrow{k} & D_{p-1,q+1} & \xrightarrow{j} & E_{p-1,q+1} & \xrightarrow{k} & D_{p-2,q+1} & \xrightarrow{j} & \cdots \\
 & & \downarrow i & & & & \downarrow i & & \\
\cdots\to E_{p+1,q} & \xrightarrow{k} & D_{p,q} & \xrightarrow{j} & E_{p,q} & \xrightarrow{k} & D_{p-1,q} & \xrightarrow{j} & \cdots \\
 & & \downarrow i & & & & \downarrow i & & \\
\cdots\to E_{p+2,q-1} & \xrightarrow{k} & D_{p+1,q-1} & \xrightarrow{j} & E_{p+1,q-1} & \xrightarrow{k} & D_{p,q-1} & \xrightarrow{j} & \cdots \\
 & & \downarrow i & & & & \downarrow i & &
\end{array}$$

Each sequence consisting of a vertical step i, followed by two horizontal steps j and k, followed by a vertical step i, ... is exact; indeed the diagram may be regarded as the intercalation of these various exact sequences, which have the terms D in common. The description of the r-th derived couple at indices (p, q) is visible in the diagram: Form $E^r_{p,q}$ as a subquotient of $E_{p,q}$ with numerator obtained by pulling back (along k) the image of the composite vertical map i^{r-1}, and denominator obtained by pulling forward (along j) the kernel of the corresponding i^{r-1} (see (5.3)).

Each filtration F of a Z-graded differential module A determines an exact couple as follows. The short exact sequence of complexes $F_{p-1} \rightarrowtail F_pA \twoheadrightarrow F_pA/F_{p-1}A$ yields the usual exact homology sequence

$$\cdots \to H_n(F_{p-1}A) \xrightarrow{i} H_n(F_pA) \xrightarrow{j} H_n(F_pA/F_{p-1}A) \xrightarrow{k} H_{n-1}(F_{p-1}A) \to \cdots$$

where i is induced by the injection, j by the projection, and k is the homology connecting homomorphism. These sequences for all p combine to give an exact couple with

$$D_{p,q} = H_{p+q}(F_pA), \quad E_{p,q} = H_{p+q}(F_pA/F_{p-1}A), \tag{5.6}$$

and with degrees of i, j, k as in (5.5). Call this the *exact couple* of the *filtration* F.

Theorem 5.4. *The spectral sequence of F is isomorphic to that of the exact couple of F.*

Proof. The spectral sequence of the exact couple (5.6) of F has

$$E^r = k^{-1}(\operatorname{Im} i^{r-1})/j(\operatorname{Ker} i^{r-1}), \quad i: H(F_{p-1}A) \to H(F_pA).$$

Regard $E_p = E^1_p = H(F_p/F_{p-1})$ and hence also each E^r_p as a subquotient module of F_p/F_{p-1}. Consider the numerator of E^r. Each homology class of E^1_p is represented by a "relative cycle" $c \in F_p$ with $\partial c \in F_{p-1}$, while $k(\operatorname{cls} c) = \operatorname{cls}(\partial c) \in H(F_{p-1})$ lies in $i^{r-1}H(F_{p-r}) \subset H(F_{p-1})$ if $\partial c = a + \partial b$ for some $b \in F_{p-1}$ and some $a \in F_{p-r}$. Then $c-b$ is in the module Z^r_p of (3.5) and $c = (c-b) + b \in Z^r_p \cup F_{p-1}A$. This is the numerator of (3.8).

On the other hand, the denominator of E^r is given by $j(\operatorname{Ker} i^{r-1})$. The kernel of i^{r-1}: $H(F_pA) \to H(F_{p+r-1}A)$ consists of the homology classes of those cycles $c \in F_pA$ with $c = \partial b$ for some $b \in F_{p+r-1}A$, hence for $b \in Z^{r-1}_{p+r-1}$. Then $j(\operatorname{cls} c) = \operatorname{cls}(\partial b)$ has $\partial b \in \partial Z^{r-1}_{p+r-1} \cup F_{p-1}$. This is the denominator of (3.8). All told, E^r_p is given by the formula (3.8) used to define the spectral sequence directly from the filtration. In both cases, d^r is induced by ∂: $A \to A$.

Corollary 5.5. *In the spectral sequence of an FDG-module the first differential d^1 may be described in terms of the maps j and k of the exact* *homology sequence for* $F_pA/F_{p-1}A$ *as the composite* $d^1 = jk$:

$$E^1_{p,q} = H_{p+q}(F_pA/F_{p-1}A) \xrightarrow{jk} H_{p+q-1}(F_{p-1}A/F_{p-2}A) = E^1_{p-1,q}.$$

Note that the sequence of derived couples contains more information than the spectral sequence alone, since it involves not only the E^r, but also the D^r and the maps i_r, j_r, k_r which determine the successive differentials d^r.

An exact couple need not arise from a filtration. An example is the Bockstein exact couple (Browder [1961]; cf. also Ex. II.4.2) for a complex K of torsion-free abelian groups. Let l be a prime number, Z_l the additive group of integers modulo l, and $Z \rightarrowtail Z \twoheadrightarrow Z_l$ the corresponding exact sequence of abelian groups. Since each K_n is torsion free, $K \rightarrowtail K \twoheadrightarrow K \otimes Z_l$ is a short exact sequence of complexes. The usual exact homology sequence is an exact couple

$$\begin{array}{ccc} H(K) & \longrightarrow & H(K) \\ & \nwarrow \quad \swarrow & \\ & H(K \otimes Z_l) & \end{array}$$

of Z-graded (not bigraded) abelian groups.

Another instance arises from tensor products. The tensor product applied to a long exact sequence yields an exact couple and hence a spectral sequence. Indeed, factor the long exact sequence

$$\cdots \to A_{p+1} \to A_p \to A_{p-1} \to A_{p-2} \to \cdots$$

of left R-modules into short exact sequences

$$\ldots, K_p \rightarrowtail A_p \twoheadrightarrow K_{p-1}, \; K_{p-1} \rightarrowtail A_{p-1} \twoheadrightarrow K_{p-2}, \ldots.$$

For a right R-module G and each p we obtain the usual long exact sequence

$$\to \operatorname{Tor}_q(G, K_p) \to \operatorname{Tor}_q(G, A_p) \xrightarrow{k} \operatorname{Tor}_q(G, K_{p-1}) \xrightarrow{i} \operatorname{Tor}_{q-1}(G, K_p) \to \cdots$$

with connecting homomorphisms i. These assemble into an exact couple with

$$D_{p,q} = \operatorname{Tor}_q(G, K_p), \qquad E_{p,q} = \operatorname{Tor}_q(G, A_p)$$

with the degrees of i, j, k as in (5.5); moreover $d = jk$: $\operatorname{Tor}_q(G, A_p) \to \operatorname{Tor}_q(G, A_{p-1})$ is the homomorphism induced by the given mapping $A_p \to A_{p-1}$. Similarly, if C is a left R-module we obtain an exact couple with

$$D_{p,q} = \operatorname{Ext}^{-q}(C, K_p), \qquad E_{p,q} = \operatorname{Ext}^{-q}(C, A_p)$$

and with the degrees of i, j, and k as in (5.5).

Exercises

1. For an exact couple $\mathfrak{C}$ with "first quadrant" term E, show that $D_{p-1,q} = D_{p,q-1}$ for $p<0$ and $q<0$. Describe the upper and lower edges of the corresponding diagram for $\mathfrak{C}$.

2. Show that the exactness of the derived couple $\mathfrak{C}'$ can be deduced from the Ker-coker sequence for the diagram

$$\begin{array}{ccccccc} & & E/jkE & \xrightarrow{k_*} & D & \xrightarrow{i_*} & iD \to 0 \\ & & \downarrow{\scriptstyle (jk)_*} & & \downarrow{\scriptstyle j_*} & & \downarrow \\ 0 & \to & k^{-1}iD & \to & k^{-1}iD & \to & 0\,. \end{array}$$

The following sequence of exercises describe spectral sequences in terms of additive relations and is due to D. Puppe [1962].

3. A *differential relation* d on a module E is an additive relation $d\colon E \rightharpoonup E$ with $\operatorname{Ker} d \supset \operatorname{Im} d$. Define $H(E, d)$.

4. Show that a spectral sequence can be described as a module E together with a sequence of differential relations d_r, $r=2, 3, \ldots$ such that $d_{r+1}0 = d_r E$, $d_{r+1}^{-1}E = d_r^{-1}0$. (Hint: define $E_{r+1} = H(E, d_r)$.)

5. Show that the spectral sequence of an exact couple $\mathfrak{C}$ is E together with the differential relations $d_r = j\,i^{-r+1}k$, $r=2, 3, \ldots$.

6. Show that the spectral sequence of a filtration F is that of the module $E^0 = E_p^0$ with $E_p^0 = F_p/F_{p-1}$ and differentials the additive relations $d^r\colon F_p/F_{p-1} \rightharpoonup F_{p-r}/F_{p-r-1}$ induced by ∂ ($r=0, 1, \ldots$).

6. Bicomplexes

Many useful filtrations arise from bicomplexes. A *bicomplex* (or, a "double complex") K is a family $\{K_{p,q}\}$ of modules with two families

$$\partial'\colon K_{p,q} \to K_{p-1,q}, \qquad \partial''\colon K_{p,q} \to K_{p,q-1} \tag{6.1}$$

of module homomorphisms, defined for all integers p and q and such that

$$\partial'\partial' = 0, \qquad \partial'\partial'' + \partial''\partial' = 0, \qquad \partial''\partial'' = 0. \tag{6.2}$$

Thus K is a Z-bigraded module and ∂', ∂'' are homomorphisms of bidegrees $(-1, 0)$ and $(0, -1)$, respectively. A bicomplex is *positive* if it lies in the first quadrant ($K_{p,q} = 0$ unless $p \geqq 0$, $q \geqq 0$). A homomorphism $f\colon K \to L$ of bicomplexes is a homomorphism of bigraded modules, of degree 0, with $f\partial' = \partial' f$ and $f\partial'' = \partial'' f$. The objects $K_{p,q}$ in a bicomplex may be R-modules, Λ-modules, graded modules, or objects from some abelian category. The *second homology* H'' of K is formed with respect to ∂'' in the usual way as

$$H''_{p,q}(K) = \operatorname{Ker}(\partial''\colon K_{p,q} \to K_{p,q-1})/\partial'' K_{p,q+1}; \tag{6.3}$$

it is a bigraded object with a differential $\partial'\colon H''_{p,q} \to H''_{p-1,q}$ induced by the original ∂'. In turn, its homology

$$H'_p H''_q(K) = \operatorname{Ker}(\partial'\colon H''_{p,q} \to H''_{p-1,q})/\partial' H''_{p+1,q} \tag{6.4}$$

is a bigraded object. The *first homology* $H'(K)$ and the iterated homology $H''H'K$ are defined analogously.

Each bicomplex K determines a single complex $X=\operatorname{Tot}(K)$ as

$$X_n=\sum_{p+q=n} K_{p,q}, \qquad \partial=\partial'+\partial'': X_n \to X_{n-1}. \tag{6.5}$$

The assumptions (6.2) imply that $\partial^2=0$; if K is positive, so is X, and in this case each direct sum in (6.5) is finite. This totalization operator has already been used. Thus, if X and Y are complexes of K-modules with boundary operators ∂' and ∂'', respectively, $X \otimes Y$ is naturally a bicomplex $\{X_p \otimes Y_q\}$ with two boundaries

$$\partial'(x\otimes y)=(\partial' x)\otimes y, \qquad \partial''(x\otimes y)=(-1)^{\deg x} x\otimes \partial'' y$$

which satisfy (6.2); the tensor product of complexes, as defined in Chap. V, is $\operatorname{Tot}(X\otimes Y)$. Similarly $\operatorname{Hom}(X, Y)$ is a bicomplex.

The *first filtration* F' of $X=\operatorname{Tot}(K)$ is defined by the subcomplexes F'_p with

$$(F'_p X)_n=\sum_{h\leqq p} K_{h,n-h}. \tag{6.6}$$

The associated spectral sequence of F is called the *first spectral sequence* E' of the bicomplex.

Theorem 6.1. *For the first spectral sequence E' of a bicomplex K with associated total complex X there are natural isomorphisms*

$$E'^2_{p,q} \cong H'_p H''_q(K). \tag{6.7}$$

If $K_{p,q}=0$ for $p<0$, $E'^2 \Rightarrow H(X)$. If K is positive, E lies in the first quadrant.

In other words, this spectral sequence shows how the iterated homology $H'H''$ approximates the total homology of X.

Proof. Let $E=E'$ be the first spectral sequence. As in (3.2), $E^1_{p,q}=H_{p+q}(F'_p X/F'_{p-1}X)$. But the definition (6.6) of the filtration F' shows that $(F'_p X/F'_{p-1}X)_{p+q} \cong K_{p,q}$. Therefore $E^1_{p,q}=H''_{p,q}(K)$. Moreover, $d^1: E^1 \to E^1$ is induced by $\partial=\partial'+\partial''$, which under the isomorphism $E^1 \cong H''K$ corresponds to ∂'. Therefore $E^2=H(E^1, d^1)=H'H''K$, as asserted in (6.7).

Since each X_n is the union of all $F'_p X_n$, the first filtration is convergent above. When $K_{p,q}=0$ for $p<0$, $F'_{-1}X=0$, and the filtration is bounded below. This gives the convergence $E'^2 \Rightarrow H(X)$. For K positive, (6.7) shows that E lies in the first quadrant.

It is instructive to give a proof of the theorem directly from the definition

$$E^2_{p,q}=(Z^2_{p,q} \cup F_{p-1}X_n)/(\partial Z^1_{p+1,q} \cup F_{p-1}X_n), \qquad n=p+q.$$

An element $a \in F_p X_n$ has the form

$$a = a_{p,q} + a_{p-1,q+1} + a_{p-2,q+2} + \cdots, \qquad a_{p,q} \in K_{p,q},$$
$$\partial a = \partial'' a_{p,q} + (\partial' a_{p,q} + \partial'' a_{p-1,q+1}) + (\partial' a_{p-1,q+1} + \partial'' a_{p-2,q+2}) + \cdots,$$

where we have grouped terms of the same bidegree. Hence $a \in Z^1_{p,q}$ if $\partial'' a_{p,q} = 0$ and $a \in Z^2_{p,q}$ if

$$\partial'' a_{p,q} = 0, \quad \partial' a_{p,q} + \partial'' a_{p-1,q+1} = 0.$$

Therefore $E^2_{p,q} = L_{p,q}/M_{p,q}$, where

$$L_{p,q} = [a_{p,q} \mid \partial'' a_{p,q} = 0 \text{ and } \partial' a_{p,q} = -\partial'' a_{p-1,q+1} \text{ for some } a_{p-1,q+1}],$$
$$M_{p,q} = [\partial' b_{p+1,q} + \partial'' b_{p,q+1} \mid \partial'' b_{p+1,q} = 0].$$

In L the first condition on $a_{p,q}$ makes it a ∂''-cycle, so that it determines $\text{cls}'' a_{p,q} \in H''_{p,q}$; the second condition asserts that this homology class lies in the kernel of $\partial' \colon H''_{p,q} \to H''_{p-1,q}$. The term $\partial'' b_{p,q+1}$ in M can vary $a_{p,q}$ by a ∂''-boundary, leaving $\text{cls}'' a_{p,q}$ unchanged; the term $\partial' b_{p+1,q}$ can vary $\text{cls}'' a_{p,q}$ by $\partial'(\text{cls}'' b_{p+1,q})$. Hence the correspondence given by $a_{p,q} \to \text{cls}'(\text{cls}'' a_{p,q})$ provides the desired isomorphism $E^2_{p,q} \cong H'_p H''_q$.

The second filtration F'' and spectral sequence E'' are defined similarly. To keep p as notation for the filtration degree, write the bicomplex as $K = \{K_{q,p}\}$, so that $\partial' \colon K_{q,p} \to K_{q-1,p}$. Then F'' is defined by $(F''_p X)_n = \sum K_{n-h,h}$ for $h \leqq p$ and has an associated spectral sequence E'' with $E''^2_{p,q} \cong H''_p H'_q(\{K_{q,p}\})$. When $K_{q,p} = 0$ for $p < 0$, this converges to the filtration F'' of $H(X)$. If K is positive, both spectral sequences lie in the first quadrant and converge to different filtrations F' and F'' of the same graded module $H(X)$.

Exercises

1. Let X and Y be complexes of abelian groups, with each X_n a free group. In the first spectral sequence of the bicomplex $K = X \otimes Y$, show that $E^2_{p,q} \cong H_p(X \otimes H_q(Y))$. Use the KÜNNETH formula, with the explicit generators of V.6 for Tor, applied as in Prop. V.10.6, to show that $d^2 = d^3 = \cdots = 0$ and hence that $E^2 = E^\infty$ in this case.
2. Describe $E^3_{p,q}$ by a quotient L/M, as in the second proof of the text.

7. The Spectral Sequence of a Covering

If a group Π operates properly, as in IV.11, on the right on a pathwise connected space X, then X is a "regular covering" of the quotient space (= orbit space) X/Π under the canonical projection

$$f \colon X \to X/\Pi.$$

Each u in Π carries singular simplices of X into such, so the total singular complex $S(X)$ and its homology $H(S(X), C)$ are both right Π-modules.

Theorem 7.1. *If Π operates properly on the pathwise connected space X while C is any abelian group, there is a first quadrant spectral sequence E with*

$$E^2_{p,q} \cong H_p(\Pi, H_q(X, C)) \underset{p}{\Rightarrow} H(X/\Pi, C). \tag{7.1}$$

As always, the convergence means that there is a filtration F of the graded group $H_n(X/\Pi, C)$ and an isomorphism of $E^\infty_{p,q}$ to the associated (bi)-graded group $G^F_p H_{p+q}(X/\Pi, C)$.

For the proof, first recall how the various homologies are computed. The singular homology $H(X, C)$ is that of the complex $C \otimes S(X)$. For any right Π-module A, such as $H(X, C)$, the homology $H_p(\Pi, A)$ is that of $A \otimes_\Pi B(\Pi)$, where $B(\Pi)$ is the bar resolution for Π — any other projective resolution of the trivial Π-module Z would do as well. Finally the homology of the orbit space X/Π is computed from its singular complex $S(X/\Pi)$. There is an isomorphism of complexes

$$\varphi\colon\ S(X) \otimes_\Pi Z \cong S(X/\Pi) \tag{7.2}$$

defined as $\varphi(T' \otimes 1) = f T'$ for each singular simplex T' in X. Indeed, since Z is a trivial Π-module, $T'u \otimes 1 = T' \otimes 1$ for each $u \in \Pi$, so φ is well defined on $\otimes_\Pi$. By Lemma IV.11.3, each singular n-simplex T in X/Π can be lifted to a singular n-simplex T' in X and these T', one for each T, are free Π-module generators of $S(X)$. Thus $S_n(X) \otimes_\Pi Z$ is the free abelian group with generators $T' \otimes_\Pi 1$, $f T' = T$, and φ is an isomorphism. The bicomplex

$$K_{p,q} = (C \otimes S_p(X)) \otimes_\Pi B_q(\Pi)$$

has two filtrations F' and F'' and the corresponding spectral sequences

$$E'^2_{p,q} = H'_p H''_q(K), \quad E''^2_{p,q} = H''_p H'_q(K),$$

each converging to the associated graded group of $H(\operatorname{Tot} K)$ under the corresponding filtration F' or F''.

For the first spectral sequence, $H''_{p,q}(K) = H_q(C \otimes S_p(X) \otimes_\Pi B(\Pi))$ is the homology $H_q(\Pi, C \otimes S_p(X))$ of Π. If $C = Z$, this is just the homology of Π with coefficients in the free Π-module $S_p(X)$, which has been calculated to be $S_p(X) \otimes_\Pi Z$ for $q = 0$ and zero for $q > 0$. Since $S_p \otimes_\Pi B$ is a complex of torsion free abelian groups, the universal coefficient theorem gives

$$\begin{aligned} H''_{p,q}(K) &= C \otimes S_p(X) \otimes_\Pi Z, & q &= 0,\\ &= 0, & q &> 0. \end{aligned}$$

By (7.2), the complex on the right is $C \otimes S(X/\Pi)$. Therefore

$$\begin{aligned} E'^2_{p,q} \cong H'_p H''_q(K) &\cong H_p(X/\Pi, C), & q &= 0,\\ &= 0, & q &> 0. \end{aligned}$$

Hence the spectral sequence "collapses" — it lies on the horizontal axis $q=0$, has all differentials zero, so is equal to its limit with

$$H_n(\mathrm{Tot}\,K)\cong H_n(X/\Pi, C). \tag{7.3}$$

In the second spectral sequence we write the indices of K as $K_{q,p}= C\otimes S_q\otimes_\Pi B_p$, so that p will still denote the filtration degree. The first homology H_q' uses only the boundary in $S_q(X)$; since each B_p is a free Π-module, this gives

$$H'_{p,q}(K)=H'_{p,q}(\{C\otimes S_q\otimes_\Pi B_p\})=H_q(X, C)\otimes_\Pi B_p(\Pi).$$

The second homology H_p'' is then the homology of the group Π with coefficients in $H_q(X, C)$, so that

$$E''^2_{p,q}\cong H_p''H_q'(K)\cong H_p(\Pi, H_q(X, C)). \tag{7.4}$$

This gives the spectral sequence of the theorem. As for any canonically bounded filtration, it converges to $H(\mathrm{Tot}\,K)$ as given in (7.3) by the first spectral sequence. Hence the conclusion (7.1).

This proof is a typical case of two spectral sequences, one of which collapses so as to determine the limit of the second.

Corollary 7.2. *If Π operates properly on the pathwise connected acyclic space X there is a natural isomorphism $H_p(\Pi, C)\cong H_p(X/\Pi, C)$ for each p, where C is any abelian group regarded as a trivial Π-module.*

Proof. Since X is acyclic, $H_q(X, C)=0$ for $q\neq 0$ and is C for $q=0$, so the (second) spectral sequence collapses, so has E^2 isomorphic to the limit, as asserted.

This result is the homology parallel of Thm. IV.11.5 on the cohomology of X/Π. As in that case, this corollary could be proved directly without the use of spectral sequences. Put differently, the spectral sequences allow us to generalize Thm. IV.11.5 to apply to spaces which are not acyclic. For example:

Corollary 7.3. *If the space X has $H_0(X)\cong Z$ and $H_q(X)=0$ for $0<q<t$ and if Π operates properly on X, then*

$$H_n(X/\Pi, C)\cong H_n(\Pi, C),\quad 0\leqq n<t.$$

For $n=t$ there is an exact sequence

$$H_{t+1}(X/\Pi, C)\to H_{t+1}(\Pi, C)\to H_t(X, C)\otimes_\Pi Z\to H_t(X/\Pi, C)\to H_t(\Pi, C)\to 0.$$

Proof. The universal coefficient theorem gives $H_0(X, C)\cong C$ and $H_q(X, C)=0$ for $0<q<t$. The spectral sequence of the theorem then has $E^2_{p,q}=0$ for $0<q<t$, and hence $E^\infty_{t,0}=E^2_{t,0}\cong H_t(\Pi, C)$. The filtration

of $H_t(X/\Pi, C)$ amounts to the exact sequence

$$0 \to E^\infty_{0,t} \to H_t(X/\Pi, C) \to E^\infty_{t,0} \cong H_t(\Pi, C) \to 0,$$

while the description of $E^\infty_{0,t}$ as the homology of $E^2_{0,t}$ under d^{t+1} is the exact sequence

$$H_{t+1}(X/\Pi, C) \xrightarrow{e_B} E^2_{t+1,0} \xrightarrow{d^{t+1}} E^2_{0,t} \to E^\infty_{0,t} \to 0.$$

Replacing $E^2_{t+1,0}$ by its value $H_{t+1}(\Pi, C)$, using (X.5.2) to calculate $E^2_{0,t} \cong H_0(\Pi, H_t(X, C)) \cong H_t(X, C) \otimes_\Pi Z$, and splicing these sequences gives the result. This exact sequence is a particular case of the "exact sequence of terms of low degree" (Ex. 3.3).

This result determines $H_n(X/\Pi)$ for $n<t$ and $H_t(X/\Pi)$ up to a certain group extension. A complete determination of $H_t(X/\Pi)$ in terms of $H(\Pi)$ and $H(X)$ requires an additional invariant, a cohomology class $k \in H^{t+1}(\Pi, H_n(X))$, as introduced by EILENBERG-MAC LANE [1949, 1950].

The spectral sequence of a covering is due to CARTAN-LERAY [1949] and to CARTAN [1948]. For further applications, see CARTAN-EILENBERG, p. 356; HU [1959], p. 287ff.; HILTON-WYLIE [1960], p. 467.

Exercise

1. Show that the use of the first spectral sequence in the proof above may be replaced by proving that $1 \otimes \varepsilon: C \otimes S(X) \otimes_\Pi B(\Pi) \to C \otimes S(X) \otimes_\Pi Z$ is a homology isomorphism, where $\varepsilon: B \to Z$ is the augmentation (use the first filtration and Thm. 3.4).

8. Cohomology Spectral Sequences

For cohomology it is customary and convenient to write a spectral sequence with upper indices and the usual change of signs as $E_r^{p,q} = E^r_{-p,-q}$ (the sign of r is *not* changed). The *same* spectral sequence E then appears as a family E_r of bigraded modules, $r = 2, 3, \ldots$, with differentials

$$d_r: E_r^{p,q} \to E_r^{p+r,q-r+1} \tag{8.1}$$

of bidegree $(r, 1-r)$ and with $H(E_r, d_r) \cong E_{r+1}$. Comparing this with the previous $d^r: E^r_{p+r,q-r+1} \to E^r_{p,q}$, we see that the *formulas* for spectral sequences in the upper indices are obtained from those in the lower indices by reversing all arrows and moving each index up — or down, as the case may be — without a sign change. The limit, E_∞, is defined as before.

A *third quadrant* spectral sequence E is one with $E^r_{p,q} = 0$ when $p>0$ or $q>0$; equivalently, all non-zero terms lie in the first quadrant of the upper indices, and the diagram is simply (1.6) with arrows reversed (differential from fiber toward base, increasing the total degree by 1). The edge homomorphisms on the base are epimorphisms

$$E_2^{p,0} \to E_3^{p,0} \to \cdots \to E_{p+1}^{p,0} = E_\infty^{p,0},$$

on the fiber are monomorphisms

$$E_\infty^{0,q} = E_{q+2}^{0,q} \to E_{q+1}^{0,q} \to \cdots \to E_2^{0,q}.$$

The transgression $\tau\colon E^1_{0,q-1} \rightharpoonup E^2_{q,0}$ is the additive relation (fiber to base) induced by d_q, and defined by (4.1) with all arrows reversed.

Let A be a DG_Z-module, written with upper indices ($A^n = A_{-n}$) and a boundary operation $\delta\colon A^n \to A^{n+1}$. A filtration F of A, written with upper indices $F^p = F_{-p}$, appears as a tower of differential Z-graded submodules

$$\cdots \supset F^{p-1}A \supset F^p A \supset F^{p+1}A \supset \cdots \tag{8.2}$$

— often called a *descending filtration*, though it's really the same filtration in a different notation. Thm. 3.1 applies directly (only the notation is changed): Each such F yields a spectral sequence $\{E_r, d_r\}$ with $E_1^p = H(F^p A/F^{p+1}A)$ and

$$E_r^{p,q} = (Z_r^{p,q} \cup F^{p+1}A^{p+q}) / (\delta Z_{r-1}^{p-r+1,q+r-2} \cup F^{p+1}A^{p+q}),$$

where $Z_r^{p,q} = [a \mid a \in F^p A^{p+q},\ \delta a \in F^{p+r}A^{p+q+1}]$, and d_r is induced by δ. If F is bounded, there are natural isomorphisms $E_\infty^p \cong F^p HA/F^{p+1}HA$, where F^pH denotes the filtration of HA induced by F. These isomorphisms also hold if F is convergent above ($\bigcup F^pA = A$) and bounded below (for each n there is an s with $F^sA^n = 0$). Note that bounded "below" appears as a bound at the right in the descending filtration (8.2).

The filtration F is *canonically cobounded* if $F^0A = A$ and $F^{n+1}A^n = 0$ (note that this is *not* the same as canonically bounded). This implies that the complex A is positive in upper indices ($A^n = 0$ for $n < 0$). An argument like that for Thm. 4.4 proves

Theorem 8.1. *A canonically cobounded filtration of a DG_Z-module A yields a "third quadrant" spectral sequence. The initial edge terms are given in terms of the subcomplexes F^1A and L, where $L^p = Z_1^{p,0}$, as $E_1^{0,n} = H^n(A/F^1A)$ and $E_2^{n,0} = H^n(L)$, and the edge homomorphisms $H^n(A) \to E_0^{0,n}$ and $E_2^{n,0} \to H^n(A)$ are induced by the identity 1_A. The transgression $\tau\colon E_1^{0,n-1} \rightharpoonup E_2^{n,0}$ for $n \geqq 2$ is the additive relation induced by δ, and is also the connecting relation $\varrho = \varrho(A; L, F^1A)$*

$$\varrho(A; L, F^1A)\colon H^{n-1}(A/F^1A) \rightharpoonup H^n(L), \quad n \geqq 2.$$

Explicitly, the edge terms are given for $r \geqq 2$ by

$$E_r^{p,0} \cong F^pC^p/\delta Z_{r-1}^{p-r+1,r-2}, \quad C = \operatorname{Ker}[\delta\colon A \to A], \tag{8.3}$$
$$E_r^{0,q} \cong (Z_r^{0,q} \cup F^1A^q)/(\delta A^{q-1} \cup F^1A^q).$$

Similarly, exact couples and bicomplexes may be written in upper indices. Many cohomology spectral sequences have an (exceedingly useful) product structure, arising from the cup product in cohomology.

Exercises

1. Under the hypotheses of Thm. 7.1, obtain a third quadrant spectral sequence $E_2^{p,q} \cong H^p(\Pi, H^q(X, C)) \underset{p}{\Rightarrow} H^n(X/\Pi, C)$.

2. Prove Thm. 8.1.

3. If $E^r_{p,q}$ is a spectral sequence of vector spaces over some field, and V is a vector space, describe $\operatorname{Hom}(E^r_{p,q}, V)$ as a spectral sequence with upper indices.

9. Restriction, Inflation, and Connection

Our next example of a spectral sequence deals with the cohomology of a group Π with a given normal subgroup Γ. Certain preliminary concepts relating the cohomology groups of Π and Γ are needed.

If Γ is a subgroup of Π and A a left Π-module the injection $\varkappa: \Gamma \to \Pi$ gives a change of groups $(\varkappa, 1_A)$ which induces a homomorphism

$$\operatorname{res}^{\Pi}_{\Gamma}: H^n(\Pi, A) \to H^n(\Gamma, A), \tag{9.1}$$

called *restriction*, which is natural in A. Also $\Delta \subset \Gamma \subset \Pi$ gives $\operatorname{res}^{\Gamma}_{\Delta} \operatorname{res}^{\Pi}_{\Gamma} = \operatorname{res}^{\Pi}_{\Delta}$. Let A^{Γ} denote, as usual, the subgroup of those elements a in A with $ta = a$ for every $t \in \Gamma$. If Γ is a normal subgroup of Π, A^{Γ} is a left (Π/Γ)-module. The projection $\sigma: \Pi \to \Pi/\Gamma$ and the injection $j: A^{\Gamma} \to A$ give a change of groups $(\sigma, j): (\Pi, A) \to (\Pi/\Gamma, A^{\Gamma})$ which induces a homomorphism

$$\inf^{\Pi/\Gamma}_{\Pi}: H^n(\Pi/\Gamma, A^{\Gamma}) \to H^n(\Pi, A) \tag{9.2}$$

called *inflation*, which is natural in A. Moreover, there is an additive relation

$$\varrho^{\Gamma}_{\Pi/\Gamma}: H^n(\Gamma, A) \to H^{n+1}(\Pi/\Gamma, A^{\Gamma}), \quad n > 0 \tag{9.3}$$

called *connection*, and to be defined below.

Recall that $H^n(\Pi, A) = H^n(\operatorname{Hom}_{\Pi}(B(\Pi), A))$, where $B(\Pi) = B(Z(\Pi))$ is the bar resolution. Each $f \in \operatorname{Hom}_{\Pi}(B_n(\Pi), A)$ can be written as a homogeneous cochain; that is, as a function $f(x_0, \ldots, x_n) \in A$ of $n+1$ arguments $x_i \in \Pi$ with $f(x x_0, \ldots, x x_n) = x f(x_0, \ldots, x_n)$, normalized by the condition that $f(x_0, \ldots, x_n) = 0$ if $x_i = x_{i+1}$ for any i. Moreover

$$\delta f(x_0, \ldots, x_{n+1}) = (-1)^{n+1} \sum_{i=0}^{n+1} (-1)^i f(x_0, \ldots, \hat{x}_i, \ldots, x_{n+1}).$$

Then restriction is induced by the chain transformation ψ given by

$$(\psi f)(t_0, \ldots, t_n) = f(t_0, \ldots, t_n), \quad t_i \in \Gamma. \tag{9.4}$$

For $g \in \operatorname{Hom}_{\Pi/\Gamma}(B_n(\Pi/\Gamma), A)$, inflation is induced by the cochain transformation σ^* with

$$(\sigma^* g)(x_0, \ldots, x_n) = g(\sigma x_0, \ldots, \sigma x_n), \quad x_i \in \Pi, \quad \sigma x_i \in \Pi/\Gamma. \tag{9.5}$$

These transformations σ^* and ψ may be recorded in the diagram of chain transformations

$$\begin{array}{ccccc} L & & K & & \\ \| & & \| & & \\ \operatorname{Hom}_\Pi(B(\Pi/\Gamma), A) & \xrightarrow{\sigma^*} & \operatorname{Hom}_\Pi(B(\Pi), A) & & \\ & & \downarrow i & \searrow \psi & \\ & & \operatorname{Hom}_\Gamma(B(\Pi), A) & \xrightarrow{B(\varkappa)^*} & \operatorname{Hom}_\Gamma(B(\Gamma), A) \\ & & \| & & \| \\ & & S' & & S\,, \end{array} \tag{9.6}$$

of complexes denoted as L, K, S', S. Note that the (Π/Γ)-module $B(\Pi/\Gamma)$ is also a Π-module by pull-back along σ, so the complex L at the left is canonically isomorphic to $\operatorname{Hom}_{\Pi/\Gamma}(B(\Pi/\Gamma), A^\Gamma)$ with cohomology $H^n(\Pi/\Gamma, A^\Gamma)$ and σ^* is $B(\sigma)^*$ where $B(\sigma): B(\Pi) \to B(\Pi/\Gamma)$. Each Π-module is also a Γ-module, by pull-back along the injection $\varkappa: \Gamma \to \Pi$, so each Π-module homomorphism is a Γ-module homomorphism. This monomorphism $i: \operatorname{Hom}_\Pi \to \operatorname{Hom}_\Gamma$ gives the vertical chain transformation $i: K \to S'$ in the diagram (9.6), while $\varkappa$ induces $B(\varkappa)^*: S' \to S$ there. Clearly $\psi = B(\varkappa)^* i$.

The chain transformation $B(\varkappa)^*$ is a cohomology isomorphism

$$B(\varkappa)^*: \; H^n(\operatorname{Hom}_\Gamma(B(\Pi), A)) \cong H^n(\operatorname{Hom}_\Gamma(B(\Gamma), A)) = H^n(\Gamma, A). \tag{9.7}$$

Indeed, since Π is a union of cosets Γy of Γ, the free Π-module $Z(\Pi)$ on one generator is the direct sum of the free Γ-modules $Z(\Gamma)y$. Hence any free Π-module is also a free Γ-module, so $\varepsilon: B(\Pi) \to Z$ is also a free Γ-resolution of the trivial Γ-module Z. The map $B(\varkappa): B(\Gamma) \to B(\Pi)$ is a chain transformation lifting the identity 1_Z, hence by the comparison theorem gives an isomorphism (9.7).

Next, if Γ is a normal subgroup of Π and A a Π-module, each $H^n(\Gamma, A)$ is a (Π/Γ)-module. First, for any ${}_\Pi B$, $\operatorname{Hom}_\Gamma(B, A)$ is a (Π/Γ)-module under the definition (Hopf algebra structure!)

$$(x f)(b) = x f(x^{-1} b) \quad \text{for } f\colon \; B \to A, \; x \in \Pi, \; b \in B. \tag{9.8}$$

Indeed, $x f$ so defined is a Γ-module homomorphism when f is, for, with $t \in \Gamma$, $(x f)(t b) = x f(x^{-1} t b) = x(x^{-1} t x) f(x^{-1} b) = t[(x f) b]$ by the normality of Γ. This makes $\operatorname{Hom}_\Gamma$ a Π-module, but since $t f = f$ for $t \in \Gamma$, it may be regarded as a (Π/Γ)-module. This module structure is natural in B, so $\operatorname{Hom}_\Gamma(B(\Pi), A)$ is a (Π/Γ)-module. By the isomorphism $B(\varkappa)^*$ of (9.7), $H^n(\Gamma, A)$ becomes a (Π/Γ)-module, as asserted.

An explicit formula for this (Π/Γ)-module structure in terms of cocycles of $B(\Gamma)$ is given in Ex. 3—5 below.

Lemma 9.1. *For Γ normal in Π, the image of the restriction lies in $H^n(\Gamma, A)^\Pi$.*

Proof. By (9.6), restriction is the composite $\psi = B(\varkappa)^* i$. For each Π-module homomorphism f: $B(\Pi) \to A$, (9.8) gives $x f = f$ for each $x \in \Pi$. Hence, if f is a cocycle, cls f in $H^n(\Gamma, A)$ is invariant under each operator of Π.

In the diagram (9.6), the definitions (9.5) and (9.4) show σ^*: $L \to K$ a monomorphism and ψ: $K \to S$ an epimorphism, with composite $\psi\sigma^*$ zero in dimensions greater than 0. Hence we are in the situation of a complex K with two given subcomplexes $\sigma^* L$ and $M = \operatorname{Ker}\psi$, with $(\sigma^* L)^n \subset M^n$ for $n > 0$ and $S \cong K/M$; in this situation (4.4) defines a homology connecting relation

$$\varrho = \varrho(K; \sigma^* L, \operatorname{Ker}\psi): H^n(S) \rightharpoonup H^{n+1}(L).$$

Take this to be the connection $\varrho^\Gamma_{\Pi/\Gamma}$ of (9.3). Explicitly, ϱ is the additive relation consisting of all pairs of cohomology classes

$$(\operatorname{cls}_S \psi f, \operatorname{cls}_L g), \quad f \in K^n, \quad g \in L^{n+1}, \quad \delta f = \sigma^* g.$$

The last condition implies that $\delta g = 0$ and $\delta\psi f = 0$.

Lemma 9.2. *The module* Def ϱ *for the connection ϱ lies in $H^n(\Gamma, A)^\Pi$.*

Proof. Take $(\operatorname{cls}_S \psi f, \operatorname{cls}_L g) \in \varrho$ as above, and define a cochain $h \in S'^n$ for $x_i \in \Pi$ by

$$h(x_0, \ldots, x_n) = f(x_0, \ldots, x_n) + (-1)^n g(1, \sigma x_0, \ldots, \sigma x_n),$$

where the second term on the right in effect implicitly uses the contracting homotopy in $B(\Pi/\Gamma)$. Since the values of g lie in A^Γ, this function h is indeed a Γ-module homomorphism h: $B_n(\Pi) \to A$. A calculation with the boundary formula in $B(\Pi)$, using $\delta f = \sigma^* g$ and $\delta g = 0$, shows $\delta h = 0$. Moreover, $B(\varkappa)$: $B(\Gamma) \to B(\Pi)$ carries h in S' into ψf in S, so any $\operatorname{cls}_S \psi f$ in Def ϱ is represented by $\operatorname{cls}_{S'} h$ in $H^n(S')$. In this complex S' we can compute the action of any $x \in \Pi$. Let k_x be the cochain with $k_x(x_0, \ldots, x_{n-1}) = g(\sigma x, 1, \sigma x_0, \ldots, \sigma x_{n-1})$. The coboundary formula and the definition (9.8) show that

$$(x h - h - \delta k_x)(x_0, \ldots, x_n) = \delta g(\sigma x, 1, \sigma x_0, \ldots, \sigma x_n) = 0.$$

Hence $x h - h$ is the coboundary of k_x, so the cohomology class of h in S' is invariant under x, as asserted.

By Lemmas 9.1 and 9.2 we may rewrite restriction and connection as

$$\operatorname{res}: H^n(\Pi, A) \to H^n(\Gamma, A)^\Pi \quad \text{and} \quad \varrho: H^n(\Gamma, A)^\Pi \rightharpoonup H^{n+1}(\Pi/\Gamma, A).$$

Two minor observations will be needed in the next section. For modules ${}_{(\Pi/\Gamma)}C$, ${}_{\Pi}B$, and ${}_{\Pi}A$ there is a natural isomorphism

$$\operatorname{Hom}_{\Pi/\Gamma}(C, \operatorname{Hom}_{\Gamma}(B, A)) \cong \operatorname{Hom}_{\Pi}(C \otimes B, A), \tag{9.9}$$

where $\operatorname{Hom}_{\Gamma}$ has operators as in (9.8) and $C \otimes B$ has "diagonal" operators $x(c \otimes b) = (x\,c \otimes x\,b)$. The map in (9.9) is given by adjoint associativity. To check that it respects the operators indicated, consider any group homomorphism $f\colon C \otimes B \to A$. This lies in Hom_{Π} on the right if

$$f(xc \otimes xb) = xf(c \otimes b), \qquad c \in C,\ b \in B,\ x \in \Pi. \tag{9.10}$$

For fixed c, $f(c \otimes -)$ lies in $\operatorname{Hom}_{\Gamma}$ on the left if

$$f(c \otimes tb) = tf(c \otimes b), \qquad t \in \Gamma, \tag{9.11}$$

while the condition that f yield a map in $\operatorname{Hom}_{\Pi/\Gamma}$ is

$$f(xc \otimes b') = xf(c \otimes x^{-1}b'), \qquad b' \in B. \tag{9.12}$$

Now (9.12) with $b' = xb$ is (9.10), while (9.10) with $x = t \in \Gamma$ has $tc = c$, hence gives (9.11). Thus the conditions left and right on f are equivalent.

Lemma 9.3. *For any free Π-module F and any Π-module A,*

$$H^n(\Pi/\Gamma, \operatorname{Hom}_{\Gamma}(F, A)) = 0, \qquad n > 0.$$

Proof. (Cf. Ex. 6.) It suffices to take for F the free Π-module $Z(\Pi)$ on one generator. The cohomology in question is that of the complex

$$\operatorname{Hom}_{\Pi/\Gamma}(B(\Pi/\Gamma), \operatorname{Hom}_{\Gamma}(Z(\Pi), A)) \cong \operatorname{Hom}_{\Pi}(B(\Pi/\Gamma) \otimes Z(\Pi), A).$$

An n-cocycle f of this complex has $f((u_0, \ldots, u_n) \otimes x) \in A$ for $u_i \in \Pi/\Gamma$. Define an $(n-1)$-cochain h, using $\sigma\colon \Pi \to \Pi/\Gamma$, by $h((u_0, \ldots, u_{n-1}) \otimes x) = f((u_0, \ldots, u_{n-1}, \sigma x) \otimes x)$. Then h is a Π-homomorphism and the condition $\delta f((u_0, \ldots, u_n, \sigma x) \otimes x) = 0$, when expanded, gives $f = \delta h$. Hence every cocycle of positive dimension n is a coboundary, q.e.d.

Exercises

1. Show how the restriction homomorphism may be calculated from any free Π-module resolution of Z.

2. If $\Pi = \Gamma \times\!\!\!\!< \Delta$, identify Π/Δ with Γ and show that $\inf_{\Pi}^{\Pi/\Delta} \operatorname{res}_{\Gamma}^{\Pi} = 0$.

3. For a change of groups $\varrho = (\zeta, \alpha)\colon (\Gamma, A, \varphi) \to (\Gamma', A', \varphi')$ show that the homomorphism $\varrho^*\colon H^n(\Gamma', A') \to H^n(\Gamma, A)$ of (IV.5.9) may be calculated from free resolutions $\varepsilon\colon X \to Z$ and $\varepsilon'\colon X' \to Z$ of Z as a trivial Γ- or Γ'-module, respectively, as the composite

$$\varrho^* = f^* \alpha_*\colon H^n(\operatorname{Hom}_{\Gamma'}(X', A')) \to H^n(\operatorname{Hom}_{\Gamma}(X', A)) \to H^n(\operatorname{Hom}_{\Gamma}(X, A)),$$

where $f\colon X \to X'$ is a Γ-module chain transformation lifting 1_Z.

4. For Γ normal in Π, each Π-module A is a Γ-module under the induced $\varphi': \Gamma \to \operatorname{Aut} A$. For each $x \in \Pi$, show that the definitions $\zeta_x t = x^{-1} t x$ for $t \in \Gamma$ and $\alpha_x a = x a$ yield a change of groups

$$\varrho_x = (\zeta_x, \alpha_x): (\Gamma, A, \varphi') \to (\Gamma, A, \varphi')$$

with $\varrho_{xy} = \varrho_y \varrho_x$. For $X \to Z$ a Π-module resolution and f as in Ex. 3 show that $\alpha_* f^*: \operatorname{Hom}_\Gamma(X, A) \to \operatorname{Hom}_\Gamma(X, A)$ is the module operation of x on $\operatorname{Hom}_\Gamma$, as defined in the text.

5. Use Exs. 3, 4 to prove that the module operation of $x \in \Pi$ on $H^n(\Gamma, A)$ is given on a (non-homogeneous) cocycle $h \in \operatorname{Hom}(B_n(\Gamma), A)$ by $\operatorname{cls} h \to \operatorname{cls} h'$, where h' is defined by conjugation as

$$h'(t_1, \ldots, t_n) = x f(x^{-1} t_1 x, \ldots, x^{-1} t_n x), \qquad t_i \in \Gamma, \; x \in \Pi.$$

6. Using (9.9), show that ${}_\Pi F$ free implies $\operatorname{Hom}_\Gamma(F, A)$ relatively injective, for the pair of rings $Z(\Pi/\Gamma), Z$. Hence give a second proof of Lemma 9.3.

10. The Lyndon Spectral Sequence

Theorem 10.1. *For Γ a normal subgroup of Π and A a Π-module there is a third quadrant spectral sequence $\{E_r, d_r\}$, natural in A, with natural isomorphisms*

$$E_2^{p,q} \cong H^p(\Pi/\Gamma, H^q(\Gamma, A)) \underset{p}{\Rightarrow} H^{p+q}(\Pi, A);$$

converging as shown to the cohomology of Π.

Here $H^q(\Gamma, A)$ is a (Π/Γ)-module with operators as described in § 9. This spectral sequence thus relates the cohomology of the subgroup Γ and of the factor group Π/Γ to that of the whole group Π.

Proof. Using the bar resolutions, form the bicomplex K with

$$K^{p,q} = \operatorname{Hom}_{\Pi/\Gamma}(B_p(\Pi/\Gamma), \operatorname{Hom}_\Gamma(B_q(\Pi), A))$$
$$\cong \operatorname{Hom}_\Pi(B_p(\Pi/\Gamma) \otimes B_q(\Pi), A),$$

as by (9.9), and with two differentials given, with the standard signs for a coboundary and a differential in $B_p \otimes B_q$, for $f \in K^{p,q}$ by

$$(\delta' f)(b' \otimes b'') = (-1)^{p+q+1} f(\partial b' \otimes b''), \qquad b' \in B_{p+1}, \; b'' \in B_q,$$
$$(\delta'' f)(b' \otimes b'') = (-1)^{q+1} f(b' \otimes \partial b''), \qquad b' \in B_p, \; b'' \in B_{q+1}.$$

The condition $\delta'\delta'' + \delta''\delta' = 0$ is readily verified. The first and second filtrations of this bicomplex yield corresponding spectral sequences E' and E'', both converging to $H(\operatorname{Tot} K)$.

For the second spectral sequence E'' the filtration index is still to be denoted as p, so we write $K^{q,p} = \operatorname{Hom}_{\Pi/\Gamma}(B_q, \operatorname{Hom}_\Gamma(B_p, A))$ for the terms of K, with second degree labelled as p. As for any bicomplex, $E_2''^{p,q} = H''^p H'^q(K)$. But $H'^q(K)$ is the cohomology of Π/Γ with coefficients in $\operatorname{Hom}_\Gamma(B_p, A)$. By Lemma 9.3, this is zero for $q > 0$; it is

$[\mathrm{Hom}_\Gamma(B_p, A)]^{\Pi/\Gamma} = \mathrm{Hom}_\Pi(B_p, A)$ for $q=0$, this because any $H^0(\Pi, M)$ is the group M^Π of Π-invariant elements of the Π-module M. Next, calculating H''^p of $\mathrm{Hom}_\Pi(B_p, A)$ gives the cohomology of Π, so that

$$E_2''^{p,q} \cong H^p(\Pi, A), \quad q=0,$$
$$= 0, \quad q>0.$$

The non-zero terms all lie in the base $q=0$, so the spectral sequence collapses. For each total degree n there is only one non-zero quotient in the filtration of $H^n(\mathrm{Tot}\, K)$, hence an isomorphism

$$H^n(\Pi, A) \cong H^n(\mathrm{Tot}\, K). \tag{10.1}$$

The proof shows that this isomorphism is induced by the chain transformation

$$\zeta\colon \mathrm{Hom}_\Pi(B(\Pi), A) \to \mathrm{Tot}\, K$$

which assigns to each $f\colon B_n(\Pi) \to A$ the element $\zeta f \in K^{0,n}$ defined by

$$(\zeta f)((u) \otimes b) = f(b), \quad u \in \Pi/\Gamma, \ b \in B_n(\Pi).$$

For the first spectral sequence, $E_2'^{p,q} \cong H'^p H''^q(K)$. Let S' denote the complex $\mathrm{Hom}_\Gamma(B(\Pi), A)$, as in (9.6); the cohomology of S' is $H(\Gamma, A)$. Now $K^p = \mathrm{Hom}_{\Pi/\Gamma}(B_p(\Pi/\Gamma), S')$, with $B_p(\Pi/\Gamma)$ a free (Π/Γ)-module, is exact as a functor of S', so

$$H''^q(K^p) \cong \mathrm{Hom}_{\Pi/\Gamma}(B_p(\Pi/\Gamma), H^q(S')) \cong \mathrm{Hom}_{\Pi/\Gamma}(B_p(\Pi/\Gamma), H^q(\Gamma, A)).$$

Taking H'^p gives the cohomology of Π/Γ, hence an isomorphism

$$\theta\colon E_2'^{p,q} \cong H^p(\Pi/\Gamma, H^q(\Gamma, A)). \tag{10.2}$$

This spectral sequence converges, as for any positive bicomplex, to $H(\mathrm{Tot}\, K)$, which by (10.1) just above is $H^n(\Pi, A)$, q.e.d.

Proposition 10.2. *In the Lyndon spectral sequence $E = E'$ the edge terms are*

$$E_2^{p,0} \cong H^p(\Pi/\Gamma, A^\Gamma), \qquad E_2^{0,q} \cong H^q(\Gamma, A)^{\Pi/\Gamma} = H^q(\Gamma, A)^\Pi \tag{10.3}$$

and $E_1^{0,q} \cong H^q(\Gamma, A)$. The edge homomorphism

$$H^n(\Pi, A) \to E_1^{0,n} \cong H^n(\Gamma, A)$$

on the fiber is the restriction homomorphism res_Γ^Π. *The edge homomorphism*

$$H^n(\Pi/\Gamma, A^\Gamma) \cong E_2^{n,0} \to H^n(\Pi, A)$$

on the base is the inflation $\mathrm{inf}_\Pi^{\Pi/\Gamma}$. *The transgression τ is the connecting relation ϱ of* (9.3),

$$\tau = \varrho_{\Pi/\Gamma}^\Gamma\colon H^{n-1}(\Gamma, A) \to H^n(\Pi/\Gamma, A^\Gamma), \quad n>1. \tag{10.4}$$

The isomorphisms (10.3) are special cases of (10.2). Note that the edge homomorphism on the fiber has its image in $E_2^{0,n}=H^n(\Gamma, A)^\Pi$, exactly as for the restriction map (Lemma 9.1) and that the transgression τ has its domain of definition contained in $H^{n-1}(\Gamma, A)^\Pi$, exactly as for the connecting relation ϱ (Lemma 9.2).

Proof. For the spectral sequence E of the first filtration the edge effects are calculated by Thm. 8.1 from the subcomplexes F^1K and L of Tot K, where $L^p=Z_1^{p,0}$, using the injection ι: $L\to \operatorname{Tot} K$ and the projection π: $\operatorname{Tot} K\to \operatorname{Tot} K/F^1K$. This gives the first line of the following diagram, in which the second line presents the complexes used in § 9 in the calculation of res, inf, and ϱ:

$$\begin{array}{ccccc} L & \xrightarrow{\iota} & \operatorname{Tot} K & \xrightarrow{\pi} & (\operatorname{Tot} K)/F^1K \\ \downarrow{\scriptstyle\lambda} & & \downarrow{\scriptstyle\eta} & & \downarrow{\scriptstyle\varphi} \\ \operatorname{Hom}_\Pi(B(\Pi/\Gamma), A) & \xrightarrow{\sigma^*} & \operatorname{Hom}_\Pi(B(\Pi), A) & \xrightarrow{\psi} & \operatorname{Hom}_\Gamma(B(\Gamma), A)\,. \end{array} \tag{10.5}$$

The maps λ, η, φ comparing these two lines will be defined in terms of the homogeneous generators $(x_0, \dots, x_n)$ of $B(\Pi)$. Specifically, $L^p=Z_1^{p,0}\subset F^pK$ consists of all $g\in K^{p,0}$ with $\delta g\in K^{p+1,0}$; that is, with $\delta''g=0$. Since $B_1(\Pi)$ is the free abelian group on generators (x, y) with $\partial(x, y)=(y)-(x)$,

$$0=\pm\,\delta''g(b'\otimes(x, y))=g(b'\otimes(y))-g(b'\otimes(x)),\quad b'\in B_p(\Pi/\Gamma)\,.$$

Therefore $g(b'\otimes(x))$ is independent of $x\in\Pi$, and $(\lambda g)\,b'=g(b'\otimes(1))$ defines a chain isomorphism λ: $L\cong\operatorname{Hom}_\Pi(B(\Pi/\Gamma), A)$. An element of degree n in Tot K is an $(n+1)$-tuple $h=(h^0, h^1, \dots, h^n)$ with $h^p\in K^{p,n-p}$. It lies in F^1K if $h^0=0$. But $B_0(\Pi/\Gamma)=Z(\Pi/\Gamma)$, so

$$h_0\in\operatorname{Hom}_{\Pi/\Gamma}(B_0(\Pi/\Gamma), \operatorname{Hom}_\Gamma(B_n(\Gamma), A))\cong\operatorname{Hom}_\Gamma(B_n(\Gamma), A)\,.$$

Thus $(\varphi h)(b'')=h^0((1)\otimes b'')$ defines a chain isomorphism φ on the right in (10.5). Finally, a straightforward calculation shows that the definition

$$(\eta h)(x_0, \dots, x_n)=\sum_{p=0}^{n} h^p((\sigma x_0, \dots, \sigma x_p)\otimes(x_p, \dots, x_n))$$

with σ: $\Pi\to\Pi/\Gamma$, $h=(h^0, \dots, h^n)$, gives η: $\operatorname{Tot} K\to\operatorname{Hom}_\Pi(B(\Pi), A)$, a chain transformation which makes the diagram (10.5) commutative. Now ζ: $\operatorname{Hom}_\Pi(B(\Pi), A)\to\operatorname{Tot} K$ as described under (10.1) has $\eta\zeta=1$; since ζ induces a cohomology isomorphism (10.1), so does η.

The vertical maps in (10.5) are thus all cohomology isomorphisms. In the spectral sequence, the edge homomorphisms on base and fiber are (Thm. 8.1) induced by ι and π respectively; under these isomorphisms they correspond to the inflation, as induced by σ^*, and the restriction induced by ψ. Similarly, Lemma 4.3 shows that the transgression, regarded as the connecting relation for the top line, agrees with the group-theoretic transgression computed (as in § 9) from the bottom line.

The terms of low degree in this spectral sequence yield an exact sequence

$$\left.\begin{array}{l} 0\to H^1(\Pi/\Gamma, A^\Gamma)\xrightarrow{\text{inf}} H^1(\Pi, A)\xrightarrow{\text{res}} H^1(\Gamma, A)^\Pi \\ \xrightarrow{\varrho} H^2(\Pi/\Gamma, A^\Gamma)\xrightarrow{\text{inf}} H^2(\Pi, A). \end{array}\right\} \qquad (10.6)$$

In higher degrees the spectral sequence provides a more refined analysis of the kernels and images of the maps inf and res, in terms of a whole sequence of functors $E_r^{p,q}(\Pi, \Gamma, A)$ which may be regarded as "mixed" cohomology groups of the two groups Π and Γ.

As an application, we prove

Corollary 10.3. *If Γ is a normal subgroup of a finite group Π with index $k=[\Pi:\Gamma]$ prime to its order $h=[\Gamma:1]$, then for each Π-module A and each $n>0$, there is a split exact sequence*

$$0\to H^n(\Pi/\Gamma, A^\Gamma)\xrightarrow{\text{inf}} H^n(\Pi, A)\xrightarrow{\text{res}} H^n(\Gamma, A)^\Pi\to 0 \qquad (10.7)$$

which thus gives an isomorphism $H^n(\Pi, A)\cong H^n(\Pi/\Gamma, A^\Gamma)\oplus H^n(\Gamma, A)^\Pi$.

Proof. By Prop. IV.5.3 we know that each element of $H^q(\Gamma, A)$ for $q>0$ has order dividing h, while each element of $H^p(\Pi/\Gamma, M)$, for $p>0$ and M any (Π/Γ)-module, has order dividing k. Therefore $E_2^{p,q}\cong H^p(\Pi/\Gamma, H^q(\Gamma, A))$ for $p>0$ and $q>0$ consists of elements with order dividing both h and k, hence is zero. The non-zero terms of the spectral sequence thus lie on the edges ($p=0$ or $q=0$), and the only non-zero differential is the transgression (fiber to base)

$$d_n\colon H^{n-1}(\Gamma, A)^\Pi = E_n^{0,n-1}\to E_n^{n,0}=H^n(\Pi/\Gamma, A^\Gamma).$$

This is a homomorphism of an abelian group with elements of orders dividing h into one with elements of orders dividing k, where $(h, k)=1$; hence d_n is zero. Thus all differentials in the spectral sequence are zero, $E_2=E_\infty$, and there are only two terms (those on the edges) in each total degree n. The filtration of $H^n(\Pi, A)$ thus amounts to the exact sequence stated. This sequence splits; indeed, a standard argument using the Euclidean algorithm will show that any exact sequence $B\rightarrowtail C\twoheadrightarrow D$ of abelian groups with $kB=0$, $hD=0$, and $(h, k)=1$ must split.

Exercises

(All the exercises refer to the Lyndon spectral sequence)

1. In the filtration of $H^n(\Pi, A)$ show that $F^n H^n$ may be characterized as the image of the inflation map, and $F^1 H^n$ as the kernel of the restriction.

2. Establish the exact sequence

$$0 \to F^1 H^2/F^2 H^2 \to E_2^{1,1} \to E_2^{3,0}.$$

3. (Hochschild-Serre [1953].) Suppose $m \geqq 1$ and $H^n(\Gamma, A) = 0$ for $0 < n < m$. Show that inf: $H^n(\Pi/\Gamma, A^\Gamma) \cong H^n(\Pi, A)$ is an isomorphism for $n < m$, that the transgression τ in dimension m is a homomorphism $\tau: H^m(\Gamma, A)^\Pi \to H^{m+1}(\Pi/\Gamma, A^\Gamma)$, and that the following sequence is exact

$$0 \to H^m(\Pi/\Gamma, A^\Gamma) \xrightarrow{\text{inf}} H^m(\Pi, A) \xrightarrow{\text{res}} H^m(\Gamma, A)^\Pi \xrightarrow{\tau} H^{m+1}(\Pi/\Gamma, A^\Gamma) \xrightarrow{\text{inf}} H^{m+1}(\Pi, A)$$

4. (Hochschild-Serre [1953]; Hattori [1960].) Suppose $m \geqq 1$ and $H^n(\Gamma, A) = 0$ for $1 < n < m$. For $0 < n < m$ establish the exact sequence

$$\begin{aligned} \cdots \to H^n(\Pi/\Gamma, A^\Gamma) \xrightarrow{\text{inf}} H^n(\Pi, A) \to H^{n-1}(\Pi/\Gamma, H^1(\Gamma, A)) \\ \to H^{n+1}(\Pi/\Gamma, A^\Gamma) \to H^{n+1}(\Pi, A) \to \cdots. \end{aligned}$$

5. For C a right Π-module, establish a first quadrant spectral sequence converging to the homology of Π,

$$H_p(\Pi/\Gamma, H_q(\Gamma, C)) \cong E^2_{p,0} \underset{p}{\Rightarrow} H(\Pi, C).$$

11. The Comparison Theorem

In the manipulation of spectral sequences it is useful to be able to conclude from limited data that two spectral sequences are isomorphic. The comparison theorem now to be established does this for first quadrant spectral sequences E of modules over a commutative ring, provided there is a short exact sequence

$$0 \to E^2_{p,0} \otimes E^2_{0,q} \xrightarrow{\varkappa} E^2_{p,q} \xrightarrow{\sigma} \operatorname{Tor}_1(E^2_{p-1,0}, E^2_{0,q}) \to 0 \qquad (11.1)$$

for the term E^2. This hypothesis frequently holds. For example, in the Leray-Serre spectral sequence of a fiber space with simply connected base space, (2.2) gives $E^2_{p,q} \cong H_p(B, H_q(F))$, which by the universal coefficient theorem yields the exact sequence

$$0 \to H_p(B) \otimes H_q(F) \to E^2_{p,q} \to \operatorname{Tor}(H_{p-1}(B), H_q(F)) \to 0.$$

Since B and F are both pathwise connected, $E^2_{p,0} \cong H_p(B, Z) = H_p(B)$ and $E^2_{0,q} \cong H_0(B, H_q(F)) \cong H_q(F)$, and the sequence reduces to (11.1).

Theorem 11.1. *(Comparison Theorem.) Let $f: E \to E'$ be a homomorphism of first quadrant spectral sequences of modules over a commutative* *ring, each of which satisfies* (11.1), *such that f commutes with the maps* $\varkappa, \sigma, \varkappa', \sigma'$ *in* (11.1). *Write* $f^r_{p,q}\colon E^r_{p,q} \to E'^r_{p,q}$. *Then any two of the following conditions imply the third (and hence that f is an isomorphism):*

(i) $f^2_{p,0}\colon E^2_{p,0} \to E'^2_{p,0}$ *is an isomorphism for all* $p \geqq 0$,

(ii) $f^2_{0,q}\colon E^2_{0,q} \to E'^2_{0,q}$ *is an isomorphism for all* $q \geqq 0$,

(iii) $f^\infty_{p,q}\colon E^\infty_{p,q} \to E'^\infty_{p,q}$ *is an isomorphism for all* p, q.

In view of the geometric applications, we read (i) as "f is an isomorphism on the base", while (ii) is "f is an isomorphism on the fiber", and (iii) is "f is an isomorphism on the total space".

Proof. That the first two conditions imply (iii) is elementary. By hypothesis, the diagram

$$\begin{array}{ccccccccc} 0 \to & E^2_{p,0} \otimes E^2_{0,q} & \to & E^2_{p,q} & \to & \operatorname{Tor}_1(E^2_{p-1,0}, E^2_{0,q}) & \to 0 \\ & \downarrow f \otimes f & & \downarrow f^2_{p,q} & & \downarrow \operatorname{Tor}_1(f^2_{p-1,0}, f^2_{0,q}) & \\ 0 \to & E'^2_{p,0} \otimes E'^2_{0,q} & \to & E'^2_{p,q} & \to & \operatorname{Tor}_1(E'^2_{p-1,0}, E'^2_{0,q}) & \to 0 \end{array} \qquad (11.2)$$

has exact rows and is commutative. Conditions (i) and (ii) imply that the outside vertical maps are isomorphisms. By the short Five Lemma, so is the middle vertical map $f^2_{p,q}$. This isomorphism of the complexes (E^2, d^2), (E'^2, d'^2) implies that of their homologies E^3, E'^3, and so on by induction to give (iii), since each $E^r_{p,q}$ is ultimately constant.

The other cases of the proof exploit the fact that a spectral sequence can be regarded as an elaborate congeries of exact sequences in the bigraded modules

$$E^r, \quad C^r = \ker d^r, \quad B^r = \operatorname{im} d^r \quad \text{and} \quad G^r = E^r/B^r.$$

In the application of the Five Lemma (in its refined form, Lemma I.3.3) we shall write down only the first row of commutative diagrams like (11.2).

To prove that (i) and (iii) imply (ii), consider the property

(ii$_m$) $f^2_{0,q}\colon E^2_{0,q} \to E'^2_{0,q}$ is an isomorphism for $0 \leqq q \leqq m$.

Since $E^2_{0,0} = E^\infty_{0,0}$, (iii) implies (ii$_0$). Hence it will suffice to prove by induction on m that (i), (iii), and (ii$_m$) imply (ii$_{m+1}$). Given (ii$_m$), the diagram (11.2) shows that $f^2_{p,q}$ is an isomorphism for $q \leqq m$. By a subsidiary induction on $r \geqq 2$, we prove that

$$f^r_{p,q} \text{ is } \left\{ \begin{array}{l} \text{a monomorphism for } q \leqq m \text{ and all } p, \\ \text{an isomorphism for } q \leqq m - r + 2 \text{ and all } p. \end{array} \right\} \qquad (11.3)$$

This holds for $r=2$; assume it for some r. The Five Lemma for the commutative diagram on the exact sequence

$$0 \to C^r_{p,q} \to E^r_{p,q} \xrightarrow{d^r} E^r_{p-r,q+r-1}$$

which defines the kernel C^r of d^r shows for the map c^r induced by f^r that

$$c^r_{p,q}\colon C^r_{p,q} \to C'^r_{p,q} \text{ is } \left\{\begin{array}{l}\text{a monomorphism for } q \leqq m,\\ \text{an isomorphism for } q \leqq m-r+1.\end{array}\right\} \tag{11.4}$$

Now d^r gives an epimorphism $E^r_{p+r,q-r+1} \twoheadrightarrow B^r_{p,q}$. If $q \leqq m$, $f^r_{p+r,q-r+1}$ is also an epimorphism, hence so is the map b induced by f:

$$b^r_{p,q}\colon B^r_{p,q} \to B'^r_{p,q} \quad \text{is an epimorphism for } q \leqq m. \tag{11.5}$$

Next, E^{r+1} is defined by the short exact sequence

$$0 \to B^r_{p,q} \to C^r_{p,q} \to E^{r+1}_{p,q} \to 0. \tag{11.6}$$

Form the corresponding two-row diagram. For $q \leqq m$ the first vertical map is an epimorphism by (11.5), and the second a monomorphism by (11.4); hence by the Five Lemma the third vertical map $f^{r+1}_{p,q}$ is a monomorphism. If, moreover, $q \leqq m-(r+1)+2 = m-r+1$, the second vertical map is an isomorphism by (11.4), hence so is $f^{r+1}_{p,q}$. This completes the inductive proof of (11.3).

Next we claim that

$$c^r_{p,m-p+2} \quad \text{is an epimorphism for } r \geqq p \geqq 2. \tag{11.7}$$

For $r>p$, $d^r\colon E^r_p \to E^r_{p-r}$ has image zero, so $E^r_p = C^r_p$, $f^r_p = c^r_p$. For r large, $f^r_{p,q} = f^\infty_{p,q}$, so $c^r_{p,q}$ in (11.7) is an isomorphism by the hypothesis (iii). We may then prove (11.7) by descent on r. Assume (11.7) for $r+1$ and take the diagram on (11.6) with $q=m-p+2$. The first vertical map is epic by (11.5); since $E^{r+1}_{p,q} = C^{r+1}_{p,q}$, the third is epic by the case of (11.7) assumed. Hence, by the short Five Lemma, $c^r_{p,q}$ is epic, proving (11.7).

Finally, we prove by descent on r that $f^r_{0,m+1}$ is an isomorphism for $r \geqq 2$. It holds for large r by (iii); assume it for $r+1$ and consider the two-row diagram with first line

$$0 \to C^r_{r,m-r+2} \to E^r_{r,m-r+2} \xrightarrow{d^r} E^r_{0,m+1} \to E^{r+1}_{0,m+1} \to 0.$$

The first vertical map is an epimorphism by (11.7) for $r=p$, the second is an isomorphism by (11.3), and the fourth is an isomorphism by the assumption of descent. Hence the third $f^r_{0,m+1}$ is an isomorphism. For $r=2$, this completes the induction on m in the proof of (ii$_m$).

The proof that (ii) and (iii) imply (i) is analogous.

Notes. Spectral sequences were discovered by LERAY [1946, 1950] for the case of cohomology; their essential features were noted independently by LYNDON [1946, 1948] in the case of the spectral sequence for the cohomology of a group. The algebraic properties of spectral sequences were effectively codified by KOSZUL [1947]. Their utility in calculations for the homotopy groups of spheres was decisively demonstrated by SERRE [1951]. The equivalent formulation by exact couples is due to MASSEY [1952]; for still another formulation see CARTAN-EILENBERG, XV.7. The LERAY-SERRE theorem has been proved by acyclic models [GUGENHEIM-MOORE 1957]; for other proofs see HU [1959, Chap. IX], HILTON-WYLIE [1960, Chap. X] and, with a slightly different notion of fiber space, FADELL-HUREWICZ [1958]. LYNDON's spectral sequence was originally defined by a filtration of Hom$(B(\Pi), A)$; his sequence satisfies Thm. 10.1, but it is at present not known whether it is isomorphic to the spectral sequence we define, which uses a filtration due to HOCHSCHILD-SERRE [1953]. These authors established the edge effects (Prop. 10.2) only for the Lyndon filtration; our proof direct from the Hochschild-Serre filtration depends upon our description of connecting relations, which was concocted for this purpose. The LYNDON spectral sequence has been used by GREEN [1956] to prove for a finite p-group Π of order p^n that $H_2(\Pi, Z)$ has order p^k with $k \leqq n(n-1)/2$. For Π finite, VENKOV [1959] proved topologically that the cohomology ring $H(\Pi, Z)$ is finitely generated as a ring; the algebraic proof of this result by EVENS [1961] uses the product structure of the LYNDON spectral sequence. Among many other applications of spectral sequences, we note BOREL's [1955] proof of the SMITH fixed point theorem and FEDERER's application to function spaces [1956]. In the comparison theorem, due to MOORE [CARTAN seminar 1954—1955], we follow the proof of KUDO and ARAKI [1956]; a closely related proof by ZEEMAN [1957] includes the case where the given isomorphisms are assumed only up to specified dimensions. EILENBERG-MOORE [1962] study convergence and duality properties of spectral sequences in an abelian category.

Chapter twelve

Derived Functors

This chapter will place our previous developments in a more general setting. Fitrs, we have already noted that modules may be replaced by objects in an abelian category; our first three sections develop this technique and show how those ideas of homological algebra which do not involve tensor products can be carried over to any abelian category. Second, the relative and the absolute Ext functors can be treated together, as cases of the general theory of "proper" exact sequences developed here in §§ 4—7. The next sections describe the process of forming "derived" functors: Hom_R leads to the functors Ext_R^n, $\otimes_R$ to the Tor_n^R, and any additive functor T to a sequence of "satellite" functors. Finally, an application of these ideas to the category of complexes yields a generalized KÜNNETH formula in which the usual exact sequence is replaced by a spectral sequence.

1. Squares

Many manipulations in an abelian category depend on a construction of "squares". Let α and β be two coterminal morphisms and consider commutative square diagrams

$$\begin{array}{ccc} D & \xrightarrow{\alpha'} & B \\ {\scriptstyle\beta'}\downarrow & & \downarrow{\scriptstyle\beta} \\ A & \xrightarrow{\alpha} & C, \end{array} \qquad \begin{array}{ccc} D'' & \xrightarrow{\alpha''} & B \\ {\scriptstyle\beta''}\downarrow & & \downarrow{\scriptstyle\beta} \\ A & \xrightarrow{\alpha} & C \end{array} \tag{1.1}$$

formed with the given edges α and β. Call the left hand square *couniversal*, for given α and β, if to each right hand square there exists a unique morphism $\gamma\colon D''\to D$ with $\beta''=\beta'\gamma$, $\alpha''=\alpha'\gamma$. A couniversal square (also called a "pull-back" diagram), if it exists, is unique up to an equivalence of D, so that α and β together determine α' and β' up to a right equivalence. GABRIEL [1962] calls D a fibred product.

Such couniversal squares are familiar in many branches of Mathematics and under more general assumptions (than those made in an abelian category). In the category of sets, if α and β are injections, D is just the intersection of the subsets A and B of C. In the category of topological spaces, if β is a fiber map and $\alpha\colon A\to C$ a continuous map into the base space of β, then β' is the so-called "induced" fiber map. In any abelian category, the couniversal square for $C=0$ is

$$\begin{array}{ccc} A\oplus B & \xrightarrow{\pi_2} & B \\ {\scriptstyle\pi_1}\downarrow & & \downarrow \\ A & \longrightarrow & 0. \end{array}$$

Theorem 1.1. *(Square Construction.) To given coterminal morphisms α, β in an abelian category there exists a couniversal square* (1.1). *In terms of the direct sum $A\oplus B$ with its projections π_1 and π_2, D may be described as the domain of $\nu\in\ker(\alpha\,\pi_1-\beta\,\pi_2)$, with $\alpha'=\pi_2\nu$, $\beta'=\pi_1\nu$.*

Proof. For D, ν, α', and β' as described, consider

$$\begin{array}{ccccccc} & & & & B & & \\ & & & {\scriptstyle\alpha'}\nearrow & \uparrow{\scriptstyle\pi_2} & \searrow{\scriptstyle\beta} & \\ D'' & \overset{\gamma}{\cdots\to} & D & \xrightarrow{\nu} & A\oplus B & & C. \\ & & & {\scriptstyle\beta'}\searrow & \downarrow{\scriptstyle\pi_1} & \nearrow{\scriptstyle\alpha} & \\ & & & & A & & \end{array} \tag{1.2}$$

The two triangles are commutative, by definition of α' and β'. The square (better, the diamond) on D is commutative, for

$$\alpha\,\beta'=\alpha\,\pi_1\nu=(\alpha\,\pi_1-\beta\,\pi_2)\,\nu+\beta\,\pi_2\nu=0+\beta\,\alpha'.$$

Moreover, for any second commutative square on α and β, with upper corner D'' as in (1.1), the couniversality of $A\oplus B$ provides $\xi\colon D''\to A\oplus B$

with $\pi_1 \xi = \beta''$, $\pi_2 \xi = \alpha''$. Therefore $0 = \alpha\,\beta'' - \beta\,\alpha'' = (\alpha\,\pi_1 - \beta\,\pi_2)\,\xi$, so ξ factors through $\nu \in \ker(\alpha\,\pi_1 - \beta\,\pi_2)$ as $\xi = \nu\,\gamma$ for some γ (see (1.2)). Then $\beta'' = \pi_1 \nu\,\gamma = \beta'\gamma$ and $\alpha'' = \alpha'\gamma$. If $\gamma_0 \colon D'' \to D$ is another morphism with $\beta'' = \beta'\gamma_0$, $\alpha'' = \alpha'\gamma_0$, then $\pi_j \nu\,\gamma_0 = \pi_j \nu\,\gamma$ for $j = 1, 2$, so $\nu\,\gamma_0 = \nu\,\gamma$. But ν is monic, so $\gamma_0 = \gamma$, and γ is unique, as required for couniversality.

For modules A, B, C, the corner D might have been described as the module of all pairs (a, b) with $\alpha\,a = \beta\,b$; our argument has shown how to replace the use of the elements a, b by the difference $\alpha\,\pi_1 - \beta\,\pi_2$ and the formation of kernels.

Theorem 1.2. *In a couniversal square, β monic implies β' monic, β epic implies β' epic, and symmetrically for α.*

The proof uses the direct sum $A \oplus B$, with projections π_j and injections ι_j. First take β monic. Suppose $\beta'\omega = 0$ for some ω. Then $\beta\,\pi_2 \nu\,\omega = \beta\,\alpha'\omega = \alpha\,\beta'\omega = 0$; as β is monic, $\pi_2 \nu\,\omega = 0$. But also $\pi_1 \nu\,\omega = \beta'\omega = 0$, so $\nu\,\omega = 0$ and ν monic gives $\omega = 0$. Therefore β' is left cancellable and thus monic.

Next take β epic. Suppose $\omega\,(\alpha\,\pi_1 - \beta\,\pi_2) = 0$ for some ω. Then $0 = \omega\,(\alpha\,\pi_1 - \beta\,\pi_2)\,\iota_2 = -\omega\,\beta\,\pi_2 \iota_2 = -\omega\,\beta$, so $\omega = 0$. Hence $\alpha\,\pi_1 - \beta\,\pi_2$ is epic, thus is the cokernel of its kernel ν. Now suppose that $\xi\,\beta' = 0$ for some ξ. Then $0 = \xi\,\beta' = \xi\,\pi_1 \nu$, so $\xi\,\pi_1$ factors through $\alpha\,\pi_1 - \beta\,\pi_2 \in \operatorname{coker} \nu$ as $\xi\,\pi_1 = \xi'(\alpha\,\pi_1 - \beta\,\pi_2)$. Therefore $0 = \xi\,\pi_1 \iota_2 = -\xi'\beta\,\pi_2 \iota_2 = -\xi'\beta$, so β epic gives $\xi' = 0$, hence $\xi\,\pi_1 = 0$, $\xi = 0$, and β' is epic.

Under duality (reverse arrows, interchange "monic" and "epic", etc.) the axioms of an abelian category are preserved. The dual square construction starts with coinitial morphisms α, β and constructs the commutative square on the left in

$$\begin{array}{ccc} C & \xrightarrow{\alpha} & A \\ \downarrow{\scriptstyle\beta} & & \vdots \\ B & \cdots\rightarrow & D, \end{array} \qquad \begin{array}{ccc} C & \xrightarrow{\alpha} & A \\ \downarrow{\scriptstyle\beta} & & \vdots \\ B & \cdots\rightarrow & D'' \end{array}$$

so as to be universal (or a "push-out" diagram). Here, ***universal*** means that to any other such commutative square with a lower right corner D'' there exists a unique $\gamma \colon D \to D''$ with .... For instance, in the category of groups (*not* an abelian category) with α and β monic, such a universal square exists with corner D the free product of the groups A and B with amalgamated subgroup C (Neumann [1954], Specht [1956]).

Exercises

1. If $\tau\sigma$ is defined with τ, σ epic, then $\tau \in \operatorname{coker}[\sigma(\ker \tau\,\sigma)]$.

2. For $\varkappa$ monic, σ epic, and $\varkappa$, σ coterminal, prove that $\varkappa'$ and σ' in the square construction are determined by the explicit formulas $\varkappa' \in \ker \varrho$, $\sigma' \in \operatorname{coim}(\sigma\,\varkappa')$, with $\varrho = (\operatorname{coker} \varkappa)\,\sigma$. (Use Ex. 1.)

3. If $\varrho(\ker\alpha)=0$ with ϱ epic, show that there is a monic μ and an epic σ with $\mu\varrho=\sigma\alpha$.

4. In a commutative diagram

$$\begin{array}{ccccc} \bullet & \xrightarrow{\xi} & \bullet & \xrightarrow{\eta} & \bullet \\ \downarrow & & \downarrow{\scriptstyle\gamma} & & \downarrow{\scriptstyle\beta}\;\downarrow \\ \bullet & \longrightarrow & \bullet & \longrightarrow & \bullet \end{array}$$

let both squares be couniversal. Show that the square with top and bottom edges $\eta\xi$, $\beta\gamma$ is also couniversal.

5. Construct a couniversal diagram to n given coterminal morphisms.

2. Subobjects and Quotient Objects

A subobject of A is determined by a monic $\varkappa:\bullet\to A$, and is the right equivalence class (all $\varkappa\theta|\theta$ an equivalence) of this $\varkappa$. The class A_s of all subobjects of A may be treated as a set (axiom at end of IX.1).

The ordinary inclusion relation for submodules is matched by the definition that $\operatorname{cls}\varkappa_1\leqq\operatorname{cls}\varkappa_2$ if and only if there is a morphism ω with $\varkappa_1=\varkappa_2\omega$; this ω is necessarily monic. The set A_s is partly ordered by this relation $\leqq$ and has a *zero* 0_A with $0_A\leqq\operatorname{cls}\varkappa$ for each $\varkappa$; namely, 0_A is the class of any zero morphism $0\colon 0'\to A$, where $0'$ is any zero object of the category.

In an abelian category, each morphism α with range A has a standard factorization $\alpha=\lambda\sigma$ (λ monic, σ epic) and $\operatorname{im}\alpha=\operatorname{cls}\lambda\in A_s$. We may thus describe A_s as the set of all images of morphisms α with range A; then equality and inclusion are given by

Proposition 2.1. *In an abelian category, morphisms α_1, α_2 with the same range A have (when "$\Leftrightarrow$" stands for "if and only if")*

$$\operatorname{im}\alpha_1=\operatorname{im}\alpha_2 \;\Leftrightarrow\; \alpha_1\sigma_1=\alpha_2\sigma_2 \qquad \text{for some epics } \sigma_1,\sigma_2;$$

$$\operatorname{im}\alpha_1\leqq\operatorname{im}\alpha_2 \;\Leftrightarrow\; \alpha_1\sigma=\alpha_2\omega \qquad \text{for some epic } \sigma \text{ and some } \omega;$$

$$\operatorname{im}\alpha=0_A \;\Leftrightarrow\; \alpha=0.$$

Proof. The standard factorization of $\alpha_1\sigma_1=\alpha_2\sigma_2$ gives $\operatorname{im}\alpha_1=\operatorname{im}\alpha_1\sigma_1=\operatorname{im}\alpha_2$. Conversely, if α_1 and α_2 both have image $\operatorname{cls}\varkappa$, they have standard factorizations $\alpha_1=\varkappa\varrho_1$, $\alpha_2=\varkappa\varrho_2$ with ϱ_1 and ϱ_2 epic. The square construction on ϱ_1 and ϱ_2 yields, by Thm. 1.2, epics σ_1 and σ_2 with $\varrho_1\sigma_1=\varrho_2\sigma_2$, hence $\alpha_1\sigma_1=\alpha_2\sigma_2$. The rest of the proof is similar.

An element of A_s will be written as $a\in A_s$ or as $\operatorname{im}\alpha$ for some α with range A, according to convenience.

Each morphism $\xi\colon A\to B$ gives a map $\xi_s\colon A_s\to B_s$ of sets, defined by

$$\xi_s(\operatorname{im}\alpha)=\operatorname{im}(\xi\alpha), \qquad \text{range }\alpha=A.$$

The correspondence $A\to A_s$, $\xi\to\xi_s$ provides a "representation" of each abelian category by partly ordered sets with zero. We may also treat

A_s as a "pointed set". By a *pointed set* U is meant a set with a distinguished element, say $0_U \in U$. A map $f\colon U \to V$ of pointed sets is a function on U to V with $f0_U = 0_V$; in particular, $f = 0$ means that $fu = 0_V$ for every $u \in U$. Pointed sets with all these maps f as morphisms constitute a category, in which we can define many familiar notions as follows: For every $f\colon U \to V$:

$$\text{Kernel } f = [\text{all } u \mid u \in U,\ fu = 0_V],$$

$$\text{Image } f = [\text{all } v \mid fu = v \text{ for some } u \in U],$$

$$f \text{ is } \textit{surjective} \text{ if and only if Image } f = V,$$

$$f \text{ is } \textit{injective} \text{ if and only if } fu_1 = fu_2 \text{ implies } u_1 = u_2.$$

If $(f, g)\colon U \to V \to W$, call (f, g) *exact* if Image f = Kernel g. As in abelian categories, (f, g) is exact if and only if $gf = 0$ and Kernel $g \subset$ Image f, where "$\subset$" denotes set-theoretic inclusion.

The fundamental properties of the subobject representation can be formulated in these terms:

Theorem 2.2. *If $\xi\colon A \to B$ is a morphism in an abelian category, then $\xi_s\colon A_s \to B_s$ is a map of partly ordered sets with zero; that is $\xi_s 0_A = 0_B$ and $a \leqq a'$ in A_s implies $\xi_s a \leqq \xi_s a'$. Also*

(i) $\xi = 0 \Leftrightarrow \xi_s = 0$;

(ii) *ξ is epic $\Leftrightarrow$ ξ_s is surjective;*

(iii) *ξ is monic $\Leftrightarrow$ ξ_s is injective $\Leftrightarrow$* Kernel $\xi_s = 0$.

If the composite $\eta\xi$ is defined, $(\eta\xi)_s = \eta_s \xi_s$ and

(iv) *(ξ, η) is exact $\Leftrightarrow$ (ξ_s, η_s) is exact.*

Proof. If $\operatorname{im} \alpha_1 \leqq \operatorname{im} \alpha_2$ in A_s, then by Prop. 2.1 $\alpha_1 \sigma = \alpha_2 \omega$ for some ω and some epic σ, so $\xi \alpha_1 \sigma = \xi \alpha_2 \omega$ and $\operatorname{im}(\xi\alpha_1) \leqq \operatorname{im}(\xi\alpha_2)$. Hence ξ_s respects the partial order. Property (i) is immediate.

If ξ is epic and $\operatorname{im} \beta \in B_s$, the square construction provides ξ' and β' with ξ' epic and $\xi\beta' = \beta\xi'$, whence $\xi_s \operatorname{im} \beta' = \operatorname{im} \beta$ and ξ_s is surjective. Conversely, if ξ_s is surjective there is an α with range A and $\operatorname{im}(\xi\alpha) = \operatorname{im} 1_B$, so $\xi\alpha\sigma_1 = \sigma_2$ for epics σ_1 and σ_2, whence ξ is epic.

If ξ is monic, $\xi_s \operatorname{im} \alpha = \xi_s \operatorname{im} \alpha'$ implies $\xi\alpha\sigma = \xi\alpha'\sigma'$, hence $\alpha\sigma = \alpha'\sigma'$ and $\operatorname{im} \alpha = \operatorname{im} \alpha'$, so ξ_s is injective. If ξ_s is injective, Kernel ξ_s is evidently zero. Finally, if Kernel $\xi_s = 0$, $\xi\alpha = 0$ implies $\operatorname{im}(\xi\alpha) = \xi_s(\operatorname{im} \alpha) = 0$, hence $\operatorname{im} \alpha = 0$ and $\alpha = 0$, so ξ is monic. This proves (iii).

For $\eta\colon B \to C$, the definition of $\ker \eta \in B_s$ shows that

$$\text{Kernel } \eta_s = [b \mid b \in B_s \text{ and } b \leqq \ker \eta]; \tag{2.1}$$

in other words, $\ker \eta$ is the maximal element of the subset $\mathrm{Kernel}\, \eta_s$; note that we write "ker" for a morphism in an abelian category, "Ker" for module homomorphisms, and "Kernel" for pointed sets. Similarly, for $\xi: A \to B$,

$$\mathrm{Image}\, \xi_s = [b \mid b \in B_s \text{ and } b \leqq \mathrm{im}\, \xi]. \tag{2.2}$$

Indeed, if α has range A, $\xi_s \mathrm{im}\, \alpha = \mathrm{im}(\xi\alpha) \leqq \mathrm{im}\, \xi$; conversely $\mathrm{im}\, \beta \leqq \mathrm{im}\, \xi$ implies $\beta\sigma = \xi\alpha$ for some epic σ and some α with range A, so $\xi_s \mathrm{im}\, \alpha = \mathrm{im}(\xi\alpha) = \mathrm{im}\, \beta$. This proves (2.2).

For $\eta\xi$ defined, $(\eta\xi)_s = \eta_s \xi_s$ follows by definition, and (2.1) and (2.2) give part (iv) of the theorem.

Quotients are dual to subobjects. In detail, let B^q denote the set of all quotients of the object B; that is, the set of all left equivalence classes of epics σ with domain B. The set B^q is a partly ordered set with zero; the zero is the class of $0: B \to 0'$; the inclusion $\mathrm{cls}\, \sigma \geqq \mathrm{cls}\, \tau$ is defined to mean $\tau = \beta\sigma$ for some β, necessarily epic. For modules, this inclusion has its expected meaning: If $\sigma: A \to A/S$, $\tau: A \to A/T$, then $\mathrm{cls}\, \sigma \geqq \mathrm{cls}\, \tau$ means $S \subset T$, and hence $A/T \cong (A/S)/(T/S)$.

Each $\xi: A \to B$ induces $\xi^q: B^q \to A^q$ (reverse direction!) by $\xi^q(\mathrm{cls}\, \sigma) = \mathrm{coim}(\sigma\xi)$. By the duality principle we do not need to prove the dual of Thm. 2.2. Recall that the dual of a theorem is formulated by reversing all arrows and leaving unchanged the logical structure of the theorem. Thus "domain" becomes "range", and ξ_s becomes ξ^q. The set-theoretic notions are part of the logical structure of the theorem, so "ξ_s injective" becomes "ξ^q injective".

Theorem 2.3. *If $\xi: A \to B$ is a morphism in an abelian category, then $\xi^q: B^q \to A^q$ is a map of partly ordered sets with zero. Also*

(i) $\xi = 0 \Leftrightarrow \xi^q = 0$;

(ii) *ξ is monic $\Leftrightarrow$ ξ^q is surjective;*

(iii) *ξ is epic $\Leftrightarrow$ ξ^q is injective $\Leftrightarrow$* $\mathrm{Kernel}\, \xi^q = 0$.

If the composite $\xi\eta$ is defined, $(\xi\eta)^q = \eta^q \xi^q$ and

(iv) *(ξ, η) is exact $\Leftrightarrow$ (η^q, ξ^q) is exact.*

These properties have a more familiar form when stated in terms of the "inverse image" of subobjects (Ex. 5, 6).

Exercises

1. Verify directly that each of the assertions of Thm. 2.3 holds in the abelian category of all R-modules.

2. If $\eta\xi$ is defined, show that $\mathrm{im}\, \xi \leqq \ker \eta$ if and only if $\mathrm{coker}\, \xi \geqq \mathrm{coim}\, \eta$ and that $\ker \eta \leqq \mathrm{im}\, \xi$ if and only if $\mathrm{coim}\, \eta \geqq \mathrm{coker}\, \xi$.

3. An *anti-isomorphism* $\varphi: S \to T$ of partly ordered sets S and T is a 1-1 correspondence such that $s \leqq s'$ implies $\varphi s \geqq \varphi s'$. Prove A_s anti-isomorphic to A^q under the correspondence $\text{cls}\, \varkappa \to \text{coker}\, \varkappa$.

4. Prove that A_s is a lattice (I.8), with $(\text{cls}\, \lambda) \cap (\text{cls}\, \mu)$ given in the notation of the square construction by $\text{cls}(\lambda \mu') = \text{cls}(\mu \lambda')$, and with $(\text{cls}\, \lambda) \cup (\text{cls}\, \mu)$ given by duality.

5. For $\xi: A \to B$ define $\xi^s: B_s \to A_s$ by $\xi^s\, \text{im}\, \beta = \ker[\xi^q(\text{coker}\, \beta)]$ (in the notation of Ex. 3, $\xi^s = \varphi^{-1} \xi^q \varphi$). Prove that ξ^s is characterized by the properties $\xi_s(\xi^s\, \text{im}\, \beta) \leqq \text{im}\, \beta$, $\xi_s\, \text{im}\, \alpha \leqq \text{im}\, \beta$ implies $\text{im}\, \alpha \leqq \xi^s\, \text{im}\, \beta$. For modules, conclude that $\xi^s(\text{im}\, \beta)$ is the inverse image of the submodule $\text{im}\, \beta$ under ξ.

6. Restate Thm. 2.3 in terms of the maps ξ^s.

7. Show that $\text{im}\, \alpha$ is the greatest lower bound of the monic left factors of α.

3. Diagram Chasing

Various lemmas about diagrams (Five Lemma, 3×3 Lemma, etc.) hold in abelian categories. The usual proofs by chasing elements can be often carried out by chasing subobjects or quotient objects instead. We give three examples.

Lemma 3.1. *(The Weak Four Lemma.) In any abelian category a commutative* 2×4 *diagram*

$$\begin{array}{ccccccc} A & \to & B & \to & C & \to & D \\ \downarrow{\scriptstyle \xi} & & \downarrow{\scriptstyle \eta} & & \downarrow{\scriptstyle \zeta} & & \downarrow{\scriptstyle \omega} \\ A' & \to & B' & \xrightarrow{\varphi} & C' & \to & D' \end{array}$$

with exact rows (i.e., with rows exact at B, C, B', *and* C'*) satisfies*

(i) ξ *epic,* η *and* ω *monic imply* ζ *monic,*

(ii) ω *monic,* ξ *and* ζ *epic imply* η *epic.*

Proof. Consider the corresponding diagram for the sets of subobjects and write $a \in A_s$, $b' \in B'_s$, etc. To prove (i), consider $c \in C_s$ with $\zeta_s c = 0$ (or, more briefly, take c which goes to 0 in C'_s). Let c go to d in D_s. Then c and hence d go to 0 in D'_s; since ω_s is injective, $d = 0$. By exactness, there is a b which maps to c; this b maps to some $b' \in B'_s$. Both b' and c map to 0 in C'_s, so, by exactness, there is an a' which maps to b'. Since ξ_s is epic, there is an a which maps to a' and thus to b'. Let a map to b_1 in B_s. But b and b_1 in B_s have the same image in B'_s; since η_s is injective, $b = b_1$. Then a maps to b to c, which is zero by exactness of $A \to B \to C$. We have shown that $\text{Kernel}\, \zeta_s$ is 0; by Thm. 2.2, part (iii), ζ is monic.

This proof of (i) is exactly like a chase of elements in a diagram of modules. The dual proof, using quotient objects, gives (ii).

There is a proof of (ii) by subobjects. Given any $b' \in B'_s$, a simple chase gives a $b \in B_s$ with the same image in C'_s as b'; thus $\varphi_s \eta_s b = \varphi_s b'$. With elements, we could subtract, forming $\eta_s b - b'$ in $\operatorname{Ker} \varphi$. Instead, write $b = \operatorname{im} \beta$, $b' = \operatorname{im} \beta'$; then $\operatorname{im}(\varphi \eta \beta) = \operatorname{im}(\varphi \beta')$. By Prop. 2.1, there are epics σ_1, σ_2 with $\varphi \eta \beta \sigma_1 = \varphi \beta' \sigma_2$, and hence $\varphi_s \operatorname{im}(\eta \beta \sigma_1 - \beta' \sigma_2) = 0$. Exactness at B' and ξ epic yield a new element $b_1 = \operatorname{im} \beta_1 \in B_s$ which maps to $\operatorname{im}(\eta \beta \sigma_1 - \beta' \sigma_2)$ in B'_s. Prop. 2.1 again yields epics σ_3, σ_4 with $\eta \beta_1 \sigma_3 = \eta \beta \sigma_1 \sigma_4 - \beta' \sigma_2 \sigma_4$, so

$$b' = \operatorname{im}(\beta' \sigma_2 \sigma_4) = \eta_s(\operatorname{im}(\beta \sigma_1 \sigma_4 - \beta_1 \sigma_3))$$

shows $b' \in \eta_s B_s$, so η is epic. In this fashion, Prop. 2.1 can be used to "subtract" two subobjects with the same image, much as if they were elements of a module.

The weak Four Lemma also gives the Five Lemma (Lemma I.3.3). Recall that $\varkappa \| \sigma$ means that $(\varkappa, \sigma)$ is a short exact sequence.

Lemma 3.2. *(The 3×3 Lemma.) A 3×3 commutative diagram in an abelian category with all three columns and the last two rows short exact sequences has its first row a short exact sequence.*

We prove a little more. Call a sequence $(\alpha, \beta)\colon A \to B \to C$ *left exact* if $0' \to A \to B \to C$ is exact (i.e., exact at A and B). Thus (α, β) left exact means that $\alpha \in \ker \beta$.

Lemma 3.3. *(The sharp 3×3 Lemma.) A 3×3 commutative diagram with all three columns and the last two rows left exact has its first row left exact. If in addition the first column and the middle row are short exact, then the first row is short exact.*

Proof. Consider the diagram (zeros on the top and sides omitted)

$$\begin{array}{ccccc}
A' & \xrightarrow{\alpha} & B' & \xrightarrow{\beta} & C' \\
\downarrow & & \downarrow & & \downarrow \\
A & \longrightarrow & B & \longrightarrow & C \\
\downarrow & & \downarrow & & \downarrow \\
A'' & \longrightarrow & B'' & \longrightarrow & C''.
\end{array}$$

By assumption, $A' \to B$ is monic and has $\alpha\colon A' \to B'$ as right factor; hence α is monic. Since $A' \to C' \to C$ is zero and $C' \to C$ is monic, $\beta \alpha = 0$. To prove exactness at B', take b' in B'_s with image 0 in C'_s, and let b' map to b in B_s. Then b' and b map to 0 in C_s; by left exactness of the row at B, there is an a which maps to b. Then a maps to 0 in B''_s and hence to 0 in A''_s. By left exactness of the first column, there is an a' which maps to a. Then $\alpha_s a'$ and b' have the same image in B_s; since $B' \to B$ is monic, $\alpha_s a' = b'$. This shows the row exact at B'. Again the proof is like a chase of elements.

Now make the added assumptions and use the diagram of the corresponding sets of quotient objects, with all mappings reversed. To prove β epic, by Thm. 2.3 part (iii), consider $c' \in C'^q$ with image 0 in B'^q. By (ii) of the same theorem, there is a c which maps to c'. Let c also map to $b \in B^q$. Since b then maps to 0 in B'^q, exactness of the middle column at B gives a b'' with image b. But b and hence b'' go to 0 in A. By the short exactness of the first column, b'' already goes to 0 in A''. Exactness of the row at B'' gives a c'' with image b''. Let c'' map to c_1 in C^q. Then c and c_1 have the same image in B^q, so $c_1 = c$ by exactness. The original c', as the image of c'', is now zero, so β is epic as desired.

Again, the proof uses quotients to avoid subtraction. For completeness, we adjoin

Lemma 3.4. *(The symmetric 3×3 Lemma.) If a commutative 3×3 diagram has middle row and middle column short exact, then when three of the remaining four rows and columns are short exact, so is the fourth.*

Proof. Use duality and row-column symmetry of Lemma 3.2.

Note. There are several other ways of establishing these and similar lemmas in an abelian category.

The representation theorem (Lubkin [1960]) asserts that for every small abelian category $\mathscr{A}$ there is a covariant additive functor T on $\mathscr{A}$ to the category of abelian groups which is an exact embedding — embedding means that distinct objects or morphisms go to distinct groups or homomorphisms; exact, that a sequence is exact in $\mathscr{A}$ if and only if its image under T is an exact sequence of abelian groups. Freyd's proof [1960] of this theorem studies the category of all functors T and embeds a suitable functor in its injective envelope, as constructed by Mitchell [1962] following the methods of Eckmann-Schopf. Using this important representation theorem, the usual diagram lemmas can be transferred from the category of abelian groups (where they are known) to the small abelian category $\mathscr{A}$.

An *additive relation* $r: A \rightharpoonup B$ in an abelian category can be defined to be a subobject of $A \oplus B$, much as in II.6. Under the natural definition of composition, the additive relations in $\mathscr{A}$ constitute a category with an involution $r \to r^{-1}$. Puppe [1962] has developed an efficient method of proving the diagram lemmas by means of such relations (which he calls *correspondences*); moreover, this provides the natural definition of the connecting homomorphisms for exact sequences of complexes in $\mathscr{A}$. Also, Puppe has achieved a characterization of the category of additive relations in $\mathscr{A}$ by a set of axioms, such that any category satisfying these axioms is the category of additive relations of a uniquely determined abelian category.

Exercises

The first two exercises use the "subtraction" device noted in the proof of the Four Lemma.

1. Prove the strong Four Lemma (Lemma I.3.2) in an abelian category.

2. Prove the middle 3×3 Lemma: If a commutative 3×3 diagram has all three columns and the first and third rows short exact, while the composite of the

two non-zero morphisms in the middle row is zero, then the middle row is short exact (cf. Ex. II.5.2).

Note. Unpublished ideas of R. G. SWAN give a method of chasing diagrams using morphisms $\alpha: P \to A$ with projective domain in place of the elements of A. This method applies to an abelian category which has *barely enough* projectives, in the sense that for each non-zero object A there is an $\alpha: P \to A$ with projective domain P and $\alpha \neq 0$. Let A_p denote the class of all such α (including zero); each $\xi: A \to B$ induces a map $\xi_p: A_p \to B_p$ of pointed sets defined by $\xi_p(\alpha) = \xi\alpha: P \to B$. The method is fomulated in terms of these maps ξ_p, as in Exercises 3–9 below.

3. An epic τ is zero if and only if its range is $0'$, and dually.
4. For $\varkappa$, λ monic, $\gamma\varkappa = 0$ and $\gamma\lambda = 0$ imply $\gamma(\varkappa \cup \lambda) = 0$.

The remaining exercises use Ex. 3 and 4 and chase diagrams in an abelian category $\mathscr{A}$ which is assumed, as in the note, to have *"barely enough"* projectives.

5. Prove: $\xi: A \to B$ is epic if and only if $\xi_p(A_p) = B_p$.
6. Prove: $\xi: A \to B$ is monic if and only if Kernel $\xi_p = 0$.
7. If $\eta\xi = 0$, then $\ker\eta = \operatorname{im}\xi$ if and only if Kernel $\eta_p =$ Image ξ_p.
8. Using the principles of 5–7, prove the weak Four Lemma.
9. By the same methods, prove the 3×3 Lemma.

4. Proper Exact Sequences

In a number of cases we have dealt with a special class of exact sequences in an abelian category and with the corresponding Ext functor; for example, in the category of modules over a K-algebra Λ, $\operatorname{Ext}_{(\Lambda,K)}$ uses those exact sequences of Λ-modules which split as sequences of K-modules.

Another example arises in the category of abelian groups. An abelian group A is said to be a *pure* subgroup of the abelian group B if $a = mb$ for an integer m implies $a = ma'$ for some $a' \in A$; that is, if $mA = mB \cap A$. Equivalently, A is pure in B if and only if each element c of finite order in the quotient group $C = B/A$ has a representative in B of the same order. By $\operatorname{Ext}_f(C, A)$ we denote the set of (congruence classes of) pure extensions of A by C. Topological applications of Ext_f appear in EILENBERG-MAC LANE [1942], algebraic applications in HARRISON [1959], NUNKE [1959], FUCHS [1958], and MAC LANE [1960]. That Ext_f is a bifunctor to abelian groups, entering in suitable exact sequences, will be a consequence of our subsequent theory.

In any abelian category $\mathscr{A}$ let $\mathscr{P}$ be a class of short exact sequences; we write $\varkappa\mathscr{P}\sigma$ to mean that $(\varkappa, \sigma)$ is one of the short exact sequences of $\mathscr{P}$, $\varkappa \in \mathscr{P}_m$ to mean that $\varkappa\mathscr{P}\sigma$ for some σ, and $\sigma \in \mathscr{P}_e$ to mean that $\varkappa\mathscr{P}\sigma$ for some $\varkappa$. Call $\mathscr{P}$ a *proper* class (and any one of its elements a proper short exact sequence) if it satisfies the following self-dual axioms.

(P-1) If $\varkappa \mathscr{P} \sigma$, any isomorphic short exact sequence is in $\mathscr{P}$;

(P-2) For any objects A and C, $A \rightarrowtail A \oplus C \twoheadrightarrow C$ is proper;

(P-3) If $\varkappa\lambda$ is defined with $\varkappa \in \mathscr{P}_m$, $\lambda \in \mathscr{P}_m$, then $\varkappa\lambda \in \mathscr{P}_m$;

(P-3') If $\sigma\tau$ is defined with $\sigma \in \mathscr{P}_e$, $\tau \in \mathscr{P}_e$, then $\sigma\tau \in \mathscr{P}_e$;

(P-4) If $\varkappa$ and λ are monic with $\varkappa\lambda \in \mathscr{P}_m$, then $\lambda \in \mathscr{P}_m$;

(P-4') If σ and τ are epic with $\sigma\tau \in \mathscr{P}_e$, then $\sigma \in \mathscr{P}_e$.

These axioms hold in all of the examples adduced above. They hold if $\mathscr{P}$ is the class of all $\square$-split short exact sequences of a relative abelian category $\square$: $\mathscr{A} \to \mathscr{M}$, or if $\mathscr{P}$ is the class of *all* short exact sequences of the given abelian category.

Note some elementary consequences. The first two axioms imply that $\mathscr{P}$ is an allowable class in the sense of IX.4, so $\mathscr{P}$ is determined by $\mathscr{P}_m$ or $\mathscr{P}_e$. Also, any left or right equivalent of a proper $\varkappa$ is proper; when $\varkappa$ has range A, $\operatorname{cls} \varkappa$ consists of proper monics and is called a *proper subobject* of A. By (P-2), $0' \rightarrowtail 0' \oplus A \twoheadrightarrow A$ is a proper short exact sequence, and dually; hence 1_A and $0: 0' \to A$ are proper monic, 1_A and $0: A \to 0'$ proper epic. A morphism $\alpha: A \to B$ is called *proper* if $\ker \alpha$ and $\operatorname{coker} \alpha$ are proper; as in Prop. IX.4.1, this amounts to the requirement that $\operatorname{im} \alpha$ and $\operatorname{coim} \alpha$ be proper. Any equivalence θ has both $\ker \theta$ and $\operatorname{coker} \theta$ proper, hence is proper and in both $\mathscr{P}_m$ and $\mathscr{P}_e$.

Proposition 4.1. *The direct sum of two proper short exact sequences is proper exact.*

Proof. Morphisms $\alpha_i: A_i \to B_i$ have a direct sum

$$\alpha_1 \oplus \alpha_2 = \iota_1 \alpha_1 \pi_1 + \iota_2 \alpha_2 \pi_2 : A_1 \oplus A_2 \to B_1 \oplus B_2, \tag{4.1}$$

where $\pi_i: A_1 \oplus A_2 \to A_i$ and $\iota_j: B_j \to B_1 \oplus B_2$. If $\varkappa \| \sigma$ and $\lambda \| \tau$, an easy argument shows $(\varkappa \oplus \lambda) \| (\sigma \oplus \tau)$. Hence it is enough to show that $\varkappa, \lambda \in \mathscr{P}_m$ imply $\varkappa \oplus \lambda \in \mathscr{P}_m$. Since $\varkappa \oplus \lambda = (\varkappa \oplus 1)(1 \oplus \lambda)$, it suffices by (P-3) to prove $\varkappa \oplus 1 \in \mathscr{P}_m$. Thus we wish to prove for each D that $(\varkappa, \sigma): A \rightarrowtail B \twoheadrightarrow C$ proper exact implies $(\varkappa \oplus 1, \sigma'): A \oplus D \rightarrowtail B \oplus D \twoheadrightarrow C$ proper exact. Here we have $\sigma' = \sigma\pi$, where $\pi: B \oplus D \to B$ is a projection of the direct sum, hence proper by (P-2). Therefore $\sigma' = \sigma\pi$ is proper by (P-3'), hence $\varkappa \oplus 1 \in \ker \sigma'$ proper, as required.

Two proper short exact sequences $E = (\varkappa, \sigma)$ and $E' = (\varkappa', \sigma')$ from A to C are called *congruent* if there is a morphism θ with $\theta\varkappa = \varkappa'$, $\sigma'\theta = \sigma$. By the short Five Lemma, any such θ is necessarily an equivalence.

Proposition 4.2. *If the proper short exact sequence $E = (\varkappa, \sigma): A \rightarrowtail B \twoheadrightarrow C$ splits by a morphism $\alpha: C \to B$ with $\sigma\alpha = 1_C$, then α is a proper monic and E is congruent to the direct sum. Conversely, any sequence congruent to the direct sum splits.*

Proof. Since $\sigma(1-\alpha\sigma)=0$, $1-\alpha\sigma$ factors through $\varkappa \in \ker\sigma$ as $1-\alpha\sigma=\varkappa\beta$, and $\beta\alpha=0$, $\beta\varkappa=1_A$. The resulting diagram $A \rightleftarrows B \leftrightarrows C$ may be compared with the direct sum diagram by the usual equivalence $\theta\colon A\oplus C\to B$ with $\alpha=\theta\iota_2$ and ι_2 the injection $C\to A\oplus C$. Now θ is an equivalence and hence proper. Also $\alpha=\theta\iota_2$ is the composite of proper monics, hence is a proper monic. The converse proof is easier.

For any objects C and A, $\mathrm{Ext}^1_{\mathscr{P}}(C, A)$ is now defined as the set of all congruence classes of *proper* short exact sequences $E\colon A\rightarrowtail B\twoheadrightarrow C$; by the axiom (IX.1) on sets of extensions, we may take $\mathrm{Ext}^1_{\mathscr{P}}$ to be a set. Now $\mathrm{Ext}^1_{\mathscr{P}}$ has all the formal properties found for Ext^1_R with R a ring:

Theorem 4.3. *For each proper class $\mathscr{P}$ of short exact sequences in an abelian category $\mathscr{A}$, $\mathrm{Ext}^1_{\mathscr{P}}(C, A)$ is a bifunctor on $\mathscr{A}$. The addition $E_1+E_2=\nabla_A(E_1\oplus E_2)\Delta_C$ makes it a bifunctor to abelian groups.*

The proof is like that for R-modules. The essential step is the demonstration that $\mathrm{Ext}^1_{\mathscr{P}}$ is a contravariant functor of C; as in Lemma III.1.2, we must construct to each proper E and each morphism $\gamma\colon C'\to C$ of $\mathscr{A}$ a unique commutative diagram

$$\begin{array}{lccccccccc} E'\colon & 0 & \to & A & \xrightarrow{\varkappa'} & D & \xrightarrow{\sigma'} & C' & \to & 0 \\ & & & \| & & \downarrow{\scriptstyle\beta} & & \downarrow{\scriptstyle\gamma} & & \\ E\colon & 0 & \to & A & \xrightarrow{\varkappa} & B & \xrightarrow{\sigma} & C & \to & 0 \end{array} \qquad (4.2)$$

with first row E' proper exact (here 0 is the zero object $0'$). First build the right-hand square by the square construction of Thm. 1.1. By Thm. 1.2, σ' is epic. Form a second square

$$\begin{array}{ccc} A & \xrightarrow{0} & C' \\ \downarrow{\scriptstyle\varkappa} & & \downarrow{\scriptstyle\gamma} \\ B & \xrightarrow{\sigma} & C. \end{array}$$

The couniversal property of the first square provides $\varkappa'\colon A\to D$ with $\beta\varkappa'=\varkappa$ and $\sigma'\varkappa'=0$. The diagram (4.2) is now constructed and is commutative.

To prove E' exact, consider any ξ with $\sigma'\xi=0$. Thus $\sigma\beta\xi=\gamma\sigma'\xi=0$, so $\beta\xi$ factors through $\varkappa\in\ker\sigma$ as $\beta\xi=\varkappa\alpha=\beta\varkappa'\alpha$ for some α. But also $\sigma'\xi=0=\sigma'\varkappa'\alpha$, so the couniversality of D for the coinitial maps ξ and $\varkappa'\alpha$ with range D gives $\xi=\varkappa'\alpha$. Since any ξ with $\sigma'\xi=0$ factors through $\varkappa'$, and $\sigma'\varkappa'=0$, we have $\varkappa'\in\ker\sigma'$.

The proof that E' is proper uses a direct sum. By the square construction, D, β, and σ' are defined by the left exact sequence

$$0\to D\xrightarrow{\nu} B\oplus C'\xrightarrow{\sigma\pi_1-\gamma\pi_2} C,\quad \pi_1\nu=\beta,\ \pi_2\nu=\sigma'.$$

This ν need not be proper, but

$$\nu\varkappa' = (\iota_1\pi_1 + \iota_2\pi_2)\,\nu\varkappa' = \iota_1\beta\varkappa' + \iota_2\sigma'\varkappa' = \iota_1\varkappa.$$

By axiom P-2, $\iota_1 \in \mathscr{P}_m$; then axiom (P-4) shows $\varkappa' \in \mathscr{P}_m$, and thus E' proper.

From the couniversality of the square on D it now follows that the morphism $(1, \beta, \gamma)\colon E' \to E$ of proper short exact sequences is couniversal for morphisms $(\alpha_1, \beta_1, \gamma)\colon E_1 \to E$, exactly as stated in Lemma III.1.3.

Now define $E\gamma$ to be E': This gives a right operation by γ on E; from the couniversality of E' it follows that $\mathrm{Ext}^1_{\mathscr{P}}$ is a contravariant functor of C. The proof that $\mathrm{Ext}^1_{\mathscr{P}}(C, A)$ is covariant in A is dual, so need not be given; the proof that it is a bifunctor can be repeated verbatim (Lemma III.1.6); a similar repetition, using Prop. 4.1, shows that $\mathrm{Ext}^1_{\mathscr{P}}(C, A)$ is an abelian group.

A long exact sequence is *proper* if each of its morphisms is proper. An n-fold exact sequence S starting at A and ending at C can be written (via standard factorization of its morphisms) as $S = E_n \circ \cdots \circ E_1$, a composite of n short exact sequences. By Prop. IX.4.1, S is proper if and only if each of its factors E_i is a proper short exact sequence. Call two n-fold sequences S and S' from A to C *congruent* if the second can be obtained from the first by a finite number of replacements of an E_i by a congruent E'_i or of two successive factors by the rule $(E\alpha) \circ F \equiv E \circ (\alpha F)$ or $E \circ (\alpha F) \equiv (E\alpha) \circ F$, where E and F are both proper and α is any matching morphism. Now the set $\mathrm{Ext}^n_{\mathscr{P}}(C, A)$ has as elements these congruence classes of such n-fold sequences S, with addition and zero as before. The properties of $\mathrm{Ext}^n_{\mathscr{P}}$ are exactly those summarized in Thm. III.5.3.

These properties may be restated in different language. A *graded additive category* $\mathscr{G}$ is a category in which each $\hom_{\mathscr{G}}(C, A)$ is the set union of a family of abelian groups $\{\hom^n(C, A),\ n = 0, 1, \ldots\}$ in which composition induces a homomorphism $\hom(B, C) \otimes \hom(A, B) \to \hom(A, C)$ of degree 0 of graded abelian groups, and such that $\mathscr{G}$ becomes an additive category when only the morphisms $\hom^0(C, A)$ are considered. In particular, each morphism of a graded additive category has a degree. Now regard a proper n-fold exact sequence S starting at A and ending at C as a morphism of degree n from C to A, while the original morphisms from C to A are taken to have degree 0. The properties of $\mathrm{Ext}_{\mathscr{P}}$ may now be summarized by

Theorem 4.4. *Each proper class $\mathscr{P}$ of short exact sequences in an abelian category $\mathscr{A}$ determines a graded additive category $\mathscr{E}_{\mathscr{P}}(\mathscr{A})$ with objects the objects of $\mathscr{A}$ and* $\hom^n_{\mathscr{E}}(A, B) = \mathrm{Ext}^n_{\mathscr{P}}(A, B)$; *in particular, with*

$\hom^0_{\mathscr{E}}(A, B) = \hom_{\mathscr{A}}(A, B)$. *In* $\mathscr{E}_{\mathscr{P}}$ *composition is given by Yoneda composition of proper long sequences and of homomorphisms with long sequences, while addition is defined by* $\operatorname{cls}(S_1 + S_2) = \operatorname{cls}(\nabla_B (S_1 \oplus S_2) \Delta_A)$.

If $\mathscr{P}$ is any proper class of short exact sequences in the abelian category $\mathscr{A}$, then congruent proper long exact sequences S and S' are also congruent as improper long exact sequences. This gives a natural transformation $\operatorname{Ext}^n_{\mathscr{P}}(C, A) \to \operatorname{Ext}^n_{\mathscr{A}}(C, A)$ of bifunctors. Prop. 4.2 asserts that this transformation is a monomorphism for $n = 1$. This may not be the case when $n > 1$; in any event, in an elementary congruence $(E\alpha) \circ F \equiv E \circ (\alpha F)$ in $\mathscr{A}$, αF is proper does not imply F proper.

Note. The idea of systematically studying exact sequences of R-modules which S-split is due to HOCHSCHILD [1956], with hints in CARTAN-EILENBERG [1956]. Homological aspects of the case of pure extensions of abelian groups were noted by HARRISON ([1959] and in an unpublished manuscript). Possible axioms for proper exact sequences were formulated by BUCHSBAUM [1959, 1960], HELLER [1958], and YONEDA [1960]. Our axioms are equivalent to those of BUCHSBAUM. BUTLER-HORROCKS [1961] consider the interrelations of several proper classes in the same category; instead of the proper class $\mathscr{P}$, they treat the subfunctor $\operatorname{Ext}^1_{\mathscr{P}} \subset \operatorname{Ext}^1$. The functors Ext for the category $\mathscr{M} = \operatorname{Morph}(\mathscr{A})$ of morphisms of $\mathscr{A}$ appear to have a close relation with those for $\mathscr{A}$ [MAC LANE 1960b].

Exercises

1. [BUCHSBAUM.] Show that (P-2) may be replaced by the requirement that $\alpha\beta = 1_A$ implies $\beta \in \mathscr{P}_m$.

2. [HELLER.] If $\varkappa \in \mathscr{P}_m$ and $\varkappa\alpha$ is a proper morphism, α is proper.

3. Construct an example of two pure subgroups in $Z_4 \oplus Z_2$ to show that $\varkappa, \lambda \in \mathscr{P}_m$ need not imply $\varkappa + \lambda \in \mathscr{P}_m$.

4. Construct an example of an impure extension F of abelian groups and an α with αF pure.

5. If $\mathscr{P}$ and $\mathscr{P}'$ are proper classes of short exact sequences, so is $\mathscr{P} \cap \mathscr{P}'$.

6. [HARRISON.] If S is a fixed module, show that the class of all short exact $A \rightarrowtail B \twoheadrightarrow C$ with $\operatorname{Hom}(S, B) \to \operatorname{Hom}(S, C)$ an epimorphism is a proper class.

5. Ext without Projectives

If $\mathscr{A}$ has enough proper projectives for the given proper class $\mathscr{P}$, each object C has a proper projective resolution $\varepsilon\colon X \to C$. Then the natural isomorphism $\operatorname{Ext}^n_{\mathscr{P}}(C, A) \cong H^n(\operatorname{Hom}_{\mathscr{A}}(X, A))$ holds, just as for modules (Thm. III.6.4). As in that case, we can establish the standard long exact sequences for $\operatorname{Ext}^n_{\mathscr{P}}$. Instead, we give a direct proof, using neither projectives nor injectives.

Theorem 5.1. *For $\mathcal{P}$ a proper class of short exact sequences in an abelian category $\mathcal{A}$, $E=(\varkappa,\sigma): A \rightarrowtail B \twoheadrightarrow C$ a proper short exact sequence, and G any object there is an exact sequence of abelian groups*

$$\to \operatorname{Ext}^{n-1}_{\mathcal{P}}(A,G) \xrightarrow{E^{n-1}} \operatorname{Ext}^n_{\mathcal{P}}(C,G) \xrightarrow{\sigma^n} \operatorname{Ext}^n_{\mathcal{P}}(B,G) \xrightarrow{\varkappa^n} \operatorname{Ext}^n_{\mathcal{P}}(A,G) \xrightarrow{E^n} \cdots$$

with maps given by composition; in particular, $E^n(\operatorname{cls} S)=(-1)^n \operatorname{cls}(S\circ E)$.

The dual of this theorem asserts the exactness of the usual long sequence with E placed in the second argument, as in Thm. III.9.1.

Proof. It is immediate that $\sigma^n E^{n-1}=0$, $\varkappa^n\sigma^n=0$, and $E^n\varkappa^n=0$. Write "$\sigma^n|\varkappa^n$" for "$(\sigma^n,\varkappa^n)$" is exact. We must prove

$$E^{n-1}|\sigma^n, \quad \sigma^n|\varkappa^n, \quad \varkappa^n|E^n, \qquad n=0,1,\ldots;\ E^{-1}=0.$$

For $n=0$ and for $E^0|\sigma^1$, the proof is that for modules, with minor variants.

To show $\sigma^1|\varkappa^1$, consider $E'\in\operatorname{Ext}^1_{\mathcal{P}}(B,G)$ with $E'\varkappa\equiv 0$. This states that $E'\varkappa$ splits, so the definition (4.2) of $E'\varkappa$ amounts to a commutative diagram

$$\begin{array}{lccccccc}
E'\varkappa: & 0\to & G & \xrightarrow{\iota_1} & G\oplus A & \xrightarrow{\pi_2} & A & \to 0\\
 & & \| & & \downarrow{\scriptstyle\mu} & & \downarrow{\scriptstyle\varkappa} & \\
E': & 0\to & G & \xrightarrow{\varkappa'} & \bullet & \xrightarrow{\sigma'} & B & \to 0\\
 & & \| & & \vdots{\scriptstyle\varrho} & & \downarrow{\scriptstyle\sigma} & \\
E_0: & & G & \overset{\alpha}{\cdots\to} & \bullet & \overset{\tau}{\cdots\to} & C, &
\end{array}$$

with μ monic by the square construction (Thm. 1.2). Moreover, $\sigma\sigma'\in\operatorname{coker}\mu$, for $\sigma\sigma'\mu=\sigma\varkappa\pi_2=0$, while if $\xi\mu=0$ for some ξ, then $\xi\varkappa'=\xi\mu\iota_1=0$, whence $\xi=\eta\sigma'$ for some η with $0=\eta\sigma'\mu=\eta\varkappa\pi_2$. Since π_2 is epic, $\eta\varkappa=0$, and η thus factors through σ as $\eta=\zeta\sigma$. Hence ξ factors through $\sigma\sigma'$, so $\mu\in\ker(\sigma\sigma')$ is proper by (P-3').

To fill in the dotted portion of the diagram, use the proper injection $\iota_2: A\to G\oplus A$, take $\varrho\in\operatorname{coker}(\mu\iota_2)$ and $\alpha=\varrho\varkappa'$. Since $\sigma\sigma'\mu\iota_2=\sigma\varkappa\pi_2\iota_2=0\,1_A=0$, $\sigma\sigma'$ factors through ϱ as $\sigma\sigma'=\tau\varrho$ with τ proper epic by (P-4'). Now replace both G's in the top row by 0, π_2 by 1_A and μ by $\mu\iota_2$. The resulting 3×3 diagram has proper exact columns and the first two rows exact; by the 3×3 Lemma the third row is exact, and proper as τ is proper. This row is therefore an $E_0\in\operatorname{Ext}^1_{\mathcal{P}}(C,G)$; the diagram states that $E_0\sigma\equiv E'$. Hence $\sigma^1|\varkappa^1$.

Lemma 5.2. *If $\varkappa^n|E^n$ for all proper E, then $E^n|\sigma^{n+1}$ and $\sigma^{n+1}|\varkappa^{n+1}$.*

In the proof, we omit the subscript $\mathcal{P}$ on $\operatorname{Ext}_{\mathcal{P}}$ and write $\varkappa_E$ and σ_E for the two non-zero morphisms of a short exact sequence $E=(\varkappa_E,\sigma_E)$. This gives the convenient congruences (Prop. III.1.7)

$$\varkappa_E E\equiv 0, \qquad E\,\sigma_E\equiv 0.$$

First suppose that $S \in\in \operatorname{Ext}^{n+1}(C, G)$ has $\sigma^{n+1} S \equiv 0$. Write S as a composite $S = T \circ F$ for $T \in\in \operatorname{Ext}^n$. Hence $0 \equiv S\sigma \equiv T(F\sigma)$; the hypothesis (with E replaced by $F\sigma$) gives $U \in \operatorname{Ext}^n$ with $T \equiv U \varkappa_{F\sigma}$, hence $S \equiv U(\varkappa_{F\sigma} F)$. But $(\varkappa_{F\sigma} F)\sigma \equiv \varkappa_{F\sigma}(F\sigma) \equiv 0$, so the assertion $E^0 | \sigma^1$ previously proved gives a morphism α with $\varkappa_{F\sigma} F \equiv \alpha E$. Thus $S \equiv U(\varkappa_{F\sigma} F) \equiv (U\alpha) E \equiv \pm E^n (U\alpha)$, as desired.

Second, we wish to prove that $S \in\in \operatorname{Ext}^{n+1}(B, G)$ with $S\varkappa \equiv 0$ can be written as $S \equiv V\sigma$ for some $V \in\in \operatorname{Ext}^{n+1}$. The proof is similar, using $\sigma^1 | \varkappa^1$ instead of $E^0 | \sigma^1$.

The proof of the theorem is now reduced to showing $\varkappa^n | E^n$ for all $n \geq 1$.

Next consider $\varkappa^1 | E^1$, which asserts that if $F \in\in \operatorname{Ext}^1(A, G)$ has $FE \equiv 0$, then $F \equiv F' \varkappa_E$ for some F'. To deal with this we must enter into the several-step definition of the congruence relation $FE \equiv 0$. We actually prove a little more:

Lemma 5.3. *For $F \in \operatorname{Ext}^1(A, G)$ and $E \in \operatorname{Ext}^1(C, A)$, the following three properties are equivalent:*

(i) *$F \equiv F' \varkappa_E$ for some $F' \in \operatorname{Ext}^1$;*

(ii) *$E \equiv \sigma_F E'$ for some $E' \in \operatorname{Ext}^1$;*

(iii) $FE \equiv 0$.

To prove that (i) implies (ii), write the commutative diagram for the morphism $F \to F'$ defining $F' \varkappa_E$ as

$$\begin{array}{rcccccccc}
F\colon & 0 \to & G & \longrightarrow & \bullet & \xrightarrow{\sigma_F} & A & \to 0 \\
 & & \| & & \downarrow{\scriptstyle \mu} & & \downarrow{\scriptstyle \varkappa_E} & \\
F'\colon & 0 \to & G & \xrightarrow{\varkappa'} & \bullet & \xrightarrow{\sigma'} & B & \to 0 \\
 & & & & \vdots & & \downarrow{\scriptstyle \sigma_E} & \\
 & & & & C & = & C &
\end{array}$$

with last column E. Here μ is monic by the square construction for $F'\varkappa_E$. Insert $\sigma_E \sigma'$ at the dotted arrow. This morphism is proper epic and also in coker μ, by a proof like that for $\sigma\sigma'$ in the previous diagram. The middle column is now a proper short exact sequence E', and the diagram states that $\sigma_F E' = E$, as required. The proof that (ii) implies (i) is dual to this one.

The hypotheses of the lemma insure that $FE \in\in \operatorname{Ext}^2(C, G)$ is defined, and (i) implies that $FE \equiv (F'\varkappa_E) E \equiv F'(\varkappa_E E) \equiv F' 0 = 0$, which is (iii). Dually, (ii) implies (iii). To prove the converse, let $F \# E$ denote the property of F and E given by the equivalent statements (i) and (ii). Now the zero of $\operatorname{Ext}^2(C, G)$ has the factorization $0 = F_0 E_0$, with

$$F_0\colon G \overset{1}{\rightarrowtail} G \twoheadrightarrow 0, \qquad E_0\colon 0 \rightarrowtail C \overset{1}{\twoheadrightarrow} C,$$

and $F_0 \# E_0$, since $F_0 = \varkappa_{E_0} F'$ with $F'\colon G \rightarrowtail G \oplus C \twoheadrightarrow C$. Assume $FE \equiv 0$ as in (iii); this congruence is obtained by a finite number k of applications of the associative law $F'(\gamma E') \equiv (F'\gamma) E'$ to $F_0 E_0 = 0$. We now show that $F \# E$, by induction on the number k of such applications. Since $F_0 \# E_0$, we need only show that $F\gamma \# E'$ implies $F \# \gamma E'$, and conversely by duality. Now, by (ii), $F\gamma \# E'$ states that $E' \equiv \sigma_{F\gamma} E''$ for some E''. The diagram defining $F\gamma$,

$$\begin{array}{rccccc} F\gamma\colon & \bullet & \longrightarrow & \bullet & \xrightarrow{\sigma_{F\gamma}} & \bullet \\ & \| & & \downarrow{\scriptstyle\beta} & & \downarrow{\scriptstyle\gamma} \\ F\colon & \bullet & \longrightarrow & \bullet & \xrightarrow{\sigma_F} & \bullet, \end{array}$$

yields $\gamma \sigma_{F\gamma} = \sigma_F \beta$ for some β. Therefore $\gamma E' \equiv (\gamma \sigma_{F\gamma}) E'' \equiv \sigma_F (\beta E'')$; by (ii), this states that $F \# \gamma E'$. The converse proof uses (i) in place of (ii) for the relation $\#$. We have completed the proof that (iii) implies (i) and (ii).

Lemma 5.4. *Condition* (ii) *of Lemma* 5.3 *is equivalent to*:

(ii') *For some morphism* α *and some* E', $F\alpha \equiv 0$ *and* $E \equiv \alpha E'$.

Proof. Since $F\sigma_F \equiv 0$, (ii) implies (ii'). To prove the converse, write F as $G \rightarrowtail D \twoheadrightarrow A$. For any object L, the dual sequence induced by F begins

$$0 \to \hom(L, G) \to \hom(L, D) \xrightarrow{\sigma_F} \hom(L, A) \xrightarrow{F_*} \operatorname{Ext}^1_{\mathscr{P}}(L, G);$$

we already know this portion to be exact. Therefore $F\alpha \equiv 0$ with $\alpha\colon L \to A$ gives $\alpha = \sigma_F \beta$ for some $\beta\colon L \to D$. Thus, given (ii'), we get $E \equiv \alpha E' \equiv \sigma_F (\beta E')$, which is (ii) of the Lemma.

These lemmas are the first step of an inductive proof of

Lemma 5.5. *For* $n > 0$, $S \in\in \operatorname{Ext}^n(A, G)$, *and* $E \in \operatorname{Ext}^1(C, A)$ *the following three properties are equivalent*

(i) *For some* $S' \in\in \operatorname{Ext}^n$, $S \equiv S' \varkappa_E$;

(ii) *For some morphism* α *and some* E', $S\alpha \equiv 0$ *and* $E \equiv \alpha E'$;

(iii) $SE \equiv 0$.

The implication (iii) $\Rightarrow$ (i) will show $\varkappa^n | E^n$ and complete the proof of the theorem.

To prove that (i) implies (ii), write S' as a composite TF', with $F' \in \operatorname{Ext}^1$. This gives $S \equiv S' \varkappa_E \equiv T(F' \varkappa_E)$. Apply Lemma 5.3 to $F \equiv F' \varkappa_E$ and E; it proves $E \equiv \sigma_F E'$ with $S \sigma_F \equiv T(F\sigma_F) \equiv 0$, which is (ii).

To prove that (ii) implies (i), use the induction assumption. Given $E \equiv \alpha E'$ and $S\alpha \equiv 0$, write S as a composite TF with $T \in\in \operatorname{Ext}^{n-1}$. Now $T(F\alpha) \equiv 0$, so by induction [(iii) implies (i)] there is a $T' \in\in \operatorname{Ext}^{n-1}$

with $T \equiv T' \varkappa_{F\alpha}$. Thus $S = TF \equiv T'(\varkappa_{F\alpha} F)$ and $(\varkappa_{F\alpha} F)\,\alpha \equiv \varkappa_{F\alpha}(F\alpha) \equiv 0$, so $(\varkappa_{F\alpha} F)\,E \equiv 0$. By Lemma 5.3 [(iii) implies (i)], this gives $\varkappa_{F\alpha} F \equiv F' \varkappa_E$ for some F', so $S \equiv (T'F')\,\varkappa_E$, which is (i).

Both (i) and (ii) imply (iii); to get the converse implications, let $S \# E$ again stand for the relation between S and E given by the equivalent statements (i) and (ii). Then $SE \equiv 0$ implies $S \# E$, by induction on the number of steps in the congruence $SE \equiv 0$, just as in the proof of Lemma 5.3.

Note that the condition (ii) of this lemma may be interpreted to say that the congruence $SE \equiv 0$ may be established by one associativity $SE = S(\alpha E') \equiv (S\alpha)\,E'$ involving E, with the remaining associativities all applied within $S\alpha$.

Note. The theorem thus proved was established by Buchsbaum [1959]; the above arrangement of the proof is wholly due to Stephen Schanuel (unpublished).

6. The Category of Short Exact Sequences

Let $\mathscr{P}$ be a proper class of short exact sequences in an abelian category $\mathscr{A}$. Construct the category $\mathrm{Ses}_{\mathscr{P}}(\mathscr{A})$ (brief for *s*hort *e*xact *s*equence of $\mathscr{A}$) with

Objects: All *proper* short exact sequences $E = (\varkappa, \sigma)$ of $\mathscr{A}$,

Morphisms Γ: $E \to E'$: All triples $\Gamma = (\alpha, \beta, \gamma)$ of morphisms of $\mathscr{A}$ which yield a commutative diagram

$$\begin{array}{lccccccccc} E\colon & 0 & \to & A & \xrightarrow{\varkappa} & B & \xrightarrow{\sigma} & C & \to & 0 \\ & & & \downarrow{\scriptstyle\alpha} & & \downarrow{\scriptstyle\beta} & & \downarrow{\scriptstyle\gamma} & & \\ E'\colon & 0 & \to & A' & \xrightarrow{\varkappa'} & B' & \xrightarrow{\sigma'} & C' & \to & 0. \end{array}$$

Under the evident composition and addition of morphisms, $\mathrm{Ses}_{\mathscr{P}}(\mathscr{A})$ is an additive category. However, $\mathrm{Ses}_{\mathscr{P}}(\mathscr{A})$ is never an abelian category. To see this, note that a morphism (α, β, γ) with $\alpha = \beta = 0$ necessarily has $\gamma = 0$, for $\gamma\sigma = \sigma'\beta = \sigma' 0 = 0$ with σ epic implies $\gamma = 0$. The composition rule $(\alpha, \beta, \gamma)(\alpha', \beta', \gamma') = (\alpha\alpha', \beta\beta', \gamma\gamma')$ shows that α and β monic in $\mathscr{A}$ imply (α, β, γ) monic in $\mathrm{Ses}_{\mathscr{P}}(\mathscr{A})$. Dually, β and γ epic in $\mathscr{A}$ imply (α, β, γ) epic in $\mathrm{Ses}_{\mathscr{P}}(\mathscr{A})$. For the zero object $0'$ and any object $G \neq 0'$ in $\mathscr{A}$ construct the morphism $\Gamma = (0, 1, 0)$,

$$\begin{array}{ccccccccc} 0 & \to & 0' & \to & G & \to & G & \to & 0 \\ & & \downarrow{\scriptstyle 0} & & \downarrow{\scriptstyle 1} & & \downarrow{\scriptstyle 0} & & \\ 0 & \to & G & \to & G & \to & 0' & \to & 0, \end{array}$$

of short exact sequences. Since 0 and 1 are monic, Γ is monic; since 1 and 0 are epic, Γ is epic. But Γ is not an equivalence, as it must be in an abelian category.

The cause of this phenomenon is not difficult to see. If we take the "termwise" kernel of this morphism Γ, we get the short sequence $0' \to 0' \to G$ which is not exact; the same applies to the "termwise" cokernel $G \to 0' \to 0'$. Indeed, the Ker-Coker sequence of Lemma II.5.2 indicates that these two sequences must be put together with $1_G: G \to G$ to get an exact sequence. (Using additive relations, one may obtain the ker-coker sequence in any abelian category.)

Now embed $\mathrm{Ses}_{\mathscr{P}}(\mathscr{A})$ in the category $\mathscr{S}(\mathscr{A})$ with

Objects: All diagrams $D: A \to B \to C$ in $\mathscr{A}$ (no exactness required),

Morphisms Γ: $D \to D'$: All triples $\Gamma = (\alpha, \beta, \gamma)$ of morphisms of $\mathscr{A}$ which yield a commutative 2×3 diagram, as above.

Since $\mathscr{S}(\mathscr{A})$ is a category of diagrams in an abelian category, it is abelian; moreover, (α, β, γ) is epic in $\mathscr{S}(\mathscr{A})$ if and only if α, β, and γ are all epic in $\mathscr{A}$, and likewise for monics. A short exact sequence $D' \rightarrowtail D \twoheadrightarrow D''$ in $\mathscr{S}(\mathscr{A})$ then corresponds to a commutative 3×3 diagram

$$\begin{array}{llll} D': & A' & \to B' & \to C' \\ & \downarrow & \downarrow & \downarrow \\ D: & A & \to B & \to C \\ & \downarrow & \downarrow & \downarrow \\ D'': & A'' & \to B'' & \to C'' \end{array}$$

in $\mathscr{A}$, with columns exact in $\mathscr{A}$. Call $D' \rightarrowtail D \twoheadrightarrow D''$ *allowable* in $\mathscr{S}(\mathscr{A})$ if all rows and columns in this diagram are proper short exact sequences of $\mathscr{A}$. This defines an allowable class of short exact sequences in $\mathscr{S}(\mathscr{A})$, in the sense of IX.4, and hence defines allowable morphisms of $\mathscr{S}(\mathscr{A})$.

Proposition 6.1. *A morphism $\Gamma = (\alpha, \beta, \gamma): D \to D''$ of $\mathscr{S}(\mathscr{A})$ is an allowable epic [an allowable monic] of $\mathscr{S}(\mathscr{A})$ if and only if D and D'' are proper short exact sequences of $\mathscr{A}$ and α, β, and γ are proper epics of $\mathscr{A}$ [respectively, proper monics of $\mathscr{A}$].*

Proof. The condition is clearly necessary. Conversely, given α, β, and γ proper epic, form the 3×3 diagram with second and third rows D and D'', first row the kernels of α, β, and γ with morphisms induced by those of D. By the 3×3 lemma, the first row is short exact; by the axiom (P-4), the first row is proper. Hence all rows and columns are proper exact, so Γ is allowable.

Now "proper" projectives are defined as were "allowable" projectives (IX.4). Given a proper class $\mathscr{P}$, an object P of $\mathscr{A}$ is called a *proper projective* for $\mathscr{P}$ if it has the usual lifting properties for the *proper* epics;

that is, if each proper epic $\sigma\colon B \twoheadrightarrow C$ induces an epimorphism $\mathrm{Hom}(P, B) \to \mathrm{Hom}(P, C)$. We say that there are *enough proper projectives* if to each object A there is a proper epic $\tau\colon P \twoheadrightarrow A$ with P proper projective.

Theorem 6.2. *If P and Q are proper projective objects of the abelian category $\mathscr{A}$, then $F\colon P \to P \oplus Q \to Q$ is an allowable projective object in $\mathscr{S}(\mathscr{A})$.*

Proof. Given any commutative diagram in $\mathscr{A}$,

$$\begin{array}{rlccccccc}
 & F\colon & 0 \to P & \xrightarrow{\iota_1} & P \oplus Q & \xrightarrow{\pi_2} & Q \to 0 \\
\mathrm{Z} & \downarrow & \downarrow \xi & & \downarrow \eta & & \downarrow \zeta \\
 & E'\colon & 0 \to A' & \xrightarrow{\varkappa'} & B' & \xrightarrow{\sigma'} & C' \to 0 \\
\Gamma & \uparrow & \uparrow \alpha & & \uparrow \beta & & \uparrow \gamma \\
 & E\colon & 0 \to A & \xrightarrow{\varkappa} & B & \xrightarrow{\sigma} & C \to 0
\end{array}$$

with exact rows and $\Gamma\colon E \to E'$ allowable epic, we are required to find a morphism $\mathrm{Z}'\colon F \to E$ of the first row to the third so that $\Gamma \mathrm{Z}' = \mathrm{Z}\colon F \to E'$. By Prop. 6.1, α, β, and γ are proper epic in $\mathscr{A}$; thus $\gamma\sigma$ is proper epic. Since P and Q are proper projectives in $\mathscr{A}$, ξ can be lifted to $\xi'\colon P \to A$ with $\alpha \xi' = \xi$ and ζ to $\omega\colon Q \to B$ with $\gamma \sigma \omega = \zeta$. Take $\iota_2\colon Q \to P \oplus Q$. Now $\sigma'(\beta \omega - \eta \iota_2) = \gamma \sigma \omega - \zeta \pi_2 \iota_2 = \zeta - \zeta = 0$, so $\beta \omega - \eta \iota_2$ factors through $\varkappa' \in \ker \sigma'$ as $\beta \omega - \eta \iota_2 = \varkappa' \omega'$, for some $\omega'\colon Q \to A'$. Since α is proper epic and Q proper projective in $\mathscr{A}$, ω' lifts to $\psi\colon Q \to A$ with $\alpha \psi = \omega'$, and

$$\beta \omega - \eta \iota_2 = \varkappa' \alpha \psi = \beta \varkappa \psi .$$

Define $\eta'\colon P \oplus Q \to B$ and $\zeta'\colon Q \to C$, using $\pi_1\colon P \oplus Q \to P$, by

$$\eta' = \varkappa \xi' \pi_1 + (\omega - \varkappa \psi) \pi_2, \qquad \zeta' = \sigma \omega .$$

Then $\mathrm{Z}' = (\xi', \eta', \zeta')\colon F \to E$ is the required morphism.

We now show that there are enough allowable projectives, not for all the objects of $\mathscr{S}(\mathscr{A})$, but for the objects in $\mathrm{Ses}_{\mathscr{P}}(\mathscr{A}) \subset \mathscr{S}(\mathscr{A})$.

Theorem 6.3. *If the abelian category $\mathscr{A}$ has enough proper projectives, then to each proper short exact sequence $E\colon A \rightarrowtail B \twoheadrightarrow C$ of $\mathscr{A}$ there is an allowable projective F and an allowable epic $\mathrm{Z} = (\xi, \eta, \zeta)\colon F \to E$ of $\mathscr{S}(\mathscr{A})$.*

We will construct an F of the form given by Thm. 6.2. Since $\mathscr{A}$ has enough proper projectives, we can find proper projectives P and Q and proper epics $\xi\colon P \to A$, $\omega\colon Q \to B$. The composite $\zeta = \sigma \omega\colon Q \to C$ is proper epic, while $\eta = \varkappa \xi \pi_1 + \omega \pi_2\colon P \oplus Q \to B$ provides a morphism $\mathrm{Z} = (\xi, \eta, \zeta)\colon F \to E$. But ξ and ζ epic, by the short Five Lemma, imply η epic. Hence Z is allowable by Prop. 6.1, provided only that η is proper.

But η is determined by $\eta\,\iota_1=\varkappa\,\xi$, $\eta\,\iota_2=\omega$, so may be written as the composite

$$P\oplus Q \xrightarrow{\xi\oplus\omega} A\oplus B \xrightarrow{\varkappa\oplus 1} B\oplus B \xrightarrow{\nabla_B} B.$$

Both factors $\xi\oplus\omega$ and $\nabla_B(\varkappa\oplus 1)$ are proper epics, the latter because it is equivalent to the (proper) projection π_2 of a direct sum, as in the diagram

$$\begin{array}{ccc} A\oplus B & \xrightarrow{\nabla_B(\varkappa\oplus 1)} & B \\ {\scriptstyle\varphi}\uparrow\downarrow{\scriptstyle\psi} & & \| \\ A\oplus B & \xrightarrow{\pi_2} & B \end{array}$$

with φ and ψ automorphisms of $A\oplus B$ defined by

$$\pi_1\varphi\,\iota_1=1,\quad \pi_1\varphi\,\iota_2=0,\quad \pi_2\varphi\,\iota_1=-\varkappa,\quad \pi_2\varphi\,\iota_2=1;$$

$$\pi_1\psi\,\iota_1=1,\quad \pi_1\psi\,\iota_2=0,\quad \pi_2\psi\,\iota_1=\varkappa,\quad \pi_2\psi\,\iota_2=1;$$

(with elements, $\varphi(a,b)$ is $(a, b-\varkappa\,a)$ and $\psi(a,b)=(a, b+\varkappa\,a)$). The proof is complete.

This theorem constructs allowable projective resolutions:

Theorem 6.4. *Let $\mathscr{P}$ be a proper class of short exact sequences in the abelian category $\mathscr{A}$. To each proper short exact sequence E of $\mathscr{A}$ there is an allowable projective resolution $\varepsilon\colon K\to E$ in $\mathscr{S}(\mathscr{A})$, represented by a commutative diagram*

$$\begin{array}{ccccccccccc} \cdots\to & X_n & \to & X_{n-1} & \to\cdots\to & X_0 & \to & A & \to 0 \\ & \downarrow & & \downarrow & & \downarrow & & \downarrow & \\ \cdots\to & W_n & \to & W_{n-1} & \to\cdots\to & W_0 & \to & B & \to 0 \\ & \downarrow & & \downarrow & & \downarrow & & \downarrow & \\ \cdots\to & Y_n & \to & Y_{n-1} & \to\cdots\to & Y_0 & \to & C & \to 0 \end{array} \qquad (6.1)$$

in $\mathscr{A}$, with each row a proper projective resolution in $\mathscr{A}$, each column of K a proper short exact sequence (of proper projective objects) in $\mathscr{A}$, and each $W_n=X_n\oplus Y_n$.

Proof. Thm. 6.3 constructs $\varepsilon\colon K\to E$ by recursion, with each K_n an allowable projective of $\mathscr{S}(\mathscr{A})$ of the form F of Thm. 6.2. Thus K_n is a proper short exact sequence $X_n\rightarrowtail W_n\twoheadrightarrow Y_n$ with X_n, W_n, and Y_n proper projective (and $W_n=X_n\oplus Y_n$). Each $\partial\colon K_n\to K_{n-1}$ and $\varepsilon\colon K_0\to E$ is an allowable morphism of $\mathscr{S}(\mathscr{A})$, so the rows of the diagram above are exact and proper in $\mathscr{A}$. Observe that K may be regarded either as a complex of short exact sequences, or as a short exact sequence $X\rightarrowtail W\twoheadrightarrow Y$ of complexes of $\mathscr{A}$. Observe also that $X\rightarrowtail W\twoheadrightarrow Y$, though split as a sequence of graded objects, need not be split as a sequence of complexes (= graded objects with boundary ∂).

Exercises

1. If $\mathcal{M}$ is the category of all left R-modules, show that every monic in $\mathrm{Ses}(\mathcal{M})$ has a cokernel in $\mathrm{Ses}(\mathcal{M})$, and dually. (Use the Ker-coker sequence.)

2. A morphism $\Gamma=(\alpha, \beta, \gamma)\colon D \to D'$ is allowable in $\mathcal{S}(\mathcal{A})$ if and only if D and D' are proper short exact sequences of $\mathcal{A}$ and the induced map $\ker\beta \to \ker\gamma$ is proper epic in $\mathcal{A}$ (or, dually, the induced map $\operatorname{coker}\alpha \to \operatorname{coker}\beta$ is proper monic in $\mathcal{A}$).

7. Connected Pairs of Additive Functors

The systematic treatment of functors $T\colon \mathcal{A} \to \mathcal{R}$ in the next sections (§§ 7—9) will assume

(i) $\mathcal{A}$ is an abelian category,

(ii) $\mathcal{P}$ is a proper class of short exact sequences in $\mathcal{A}$,

(iii) $\mathcal{R}$ is a selective abelian category (IX.2).

This formulation includes both relative homological algebra (e.g., with $\mathcal{P}$ the class of suitably split exact sequences) and "absolute" homological algebra, with $\mathcal{P}$ all short exact sequences in $\mathcal{A}$. In $\mathcal{R}$ we use the class of *all* short exact sequences. For the applications intended, $\mathcal{R}$ might as well be the category of all modules over some ring or algebra.

An *additive* functor $T\colon \mathcal{A} \to \mathcal{R}$ is a functor (covariant or contravariant) with $T(\alpha+\beta)=T(\alpha)+T(\beta)$ whenever $\alpha+\beta$ is defined. This condition implies $T(0)=0$, $T(-\alpha)=-T(\alpha)$, and $T(A\oplus B)\cong T(A)\oplus T(B)$. Henceforth we assume: *all functors are additive.*

Study the effect of a covariant T upon all the *proper* short exact sequences $(\varkappa, \sigma)\colon A \rightarrowtail B \twoheadrightarrow C$ of $\mathcal{A}$. Call T

$\mathcal{P}$-exact if every $0 \to T(A) \to T(B) \to T(C) \to 0$ is exact in $\mathcal{R}$,

right $\mathcal{P}$-exact if every $T(A) \to T(B) \to T(C) \to 0$ is exact,

left $\mathcal{P}$-exact if every $0 \to T(A) \to T(B) \to T(C)$ is exact,

half $\mathcal{P}$-exact if every $T(A) \to T(B) \to T(C)$ is exact.

If T is $\mathcal{P}$-exact, it carries proper monics to monics, proper epics to epics, and proper long exact sequences to long exact sequences. Moreover, for any proper morphism α, a $\mathcal{P}$-exact functor has

$$\left.\begin{aligned} T(\ker\alpha)&=\ker(T\alpha), & T(\operatorname{im}\alpha)&=\operatorname{im}(T\alpha),\\ T(\operatorname{coker}\alpha)&=\operatorname{coker}(T\alpha), & T(\operatorname{coim}\alpha)&=\operatorname{coim}(T\alpha).\end{aligned}\right\} \tag{7.1}$$

Right exact functors can be described in several equivalent ways. By a proper *right exact* sequence in the category $\mathcal{A}$ we mean a sequence $(\alpha, \sigma)\colon D \to B \to C \to 0$ exact at B and C with α and σ proper.

Lemma 7.1. *A covariant additive functor T is right $\mathscr{P}$-exact if and only if either*

(i) *T carries proper right exact sequences in $\mathscr{A}$ to right exact sequences in $\mathscr{R}$, or*

(ii) $T(\operatorname{coker}\alpha)=\operatorname{coker}(T\alpha)$ *for every proper α in $\mathscr{A}$.*

Proof. Since $\operatorname{coker}\alpha=\sigma$ states that (α,σ) is a right exact sequence, (i) and (ii) are equivalent, and imply T right $\mathscr{P}$-exact. Conversely, let T be right $\mathscr{P}$-exact. Each proper right exact sequence $D\to B\to C\to 0$ in $\mathscr{A}$ yields two proper short exact sequences

$$\begin{array}{c} K \\ \downarrow \\ D \\ \downarrow \\ 0\to A\to B\to C\to 0; \end{array}$$

T carries each to a right exact sequence in $\mathscr{R}$, so $T(D)\to T(B)\to T(C)$ is right-exact.

Similarly, T is left $\mathscr{P}$-exact if and only if $T(\ker\alpha)=\ker(T\alpha)$ for α proper.

If $T\colon \mathscr{A}\to\mathscr{R}$ is a contravariant functor, then, for all proper short exact sequences $A\rightarrowtail B\twoheadrightarrow C$ of $\mathscr{A}$, T is

$\mathscr{P}$-exact if every $0\to T(C)\to T(B)\to T(A)\to 0$ is exact in $\mathscr{R}$,

right $\mathscr{P}$-exact if every $T(C)\to T(B)\to T(A)\to 0$ is exact,

left $\mathscr{P}$-exact if every $0\to T(C)\to T(B)\to T(A)$ is exact,

half $\mathscr{P}$-exact if every $T(C)\to T(B)\to T(A)$ is exact.

The analogue of Lemma 7.1 holds; in particular, T is right $\mathscr{P}$-exact if and only if it carries each proper left exact sequence in $\mathscr{A}$ into a right exact sequence in $\mathscr{R}$.

A *$\mathscr{P}$-connected pair* (S, E_*, T) of covariant functors is a pair of functors $S, T\colon \mathscr{A}\to\mathscr{R}$ together with a function which assigns to each proper exact $E\colon A\rightarrowtail B\twoheadrightarrow C$ in $\mathscr{A}$ a morphism $E_*\colon S(C)\to T(A)$ of $\mathscr{R}$ such that each morphism $(\alpha,\beta,\gamma)\colon E\to E'$ of proper short exact sequences yields a commutative diagram

$$\begin{array}{ccc} S(C) & \xrightarrow{E_*} & T(A) \\ \downarrow{\scriptstyle S(\gamma)} & & \downarrow{\scriptstyle T(\alpha)} \\ S(C') & \xrightarrow{E'_*} & T(A'), \end{array} \qquad \mathscr{R} \tag{7.2}$$

(in the indicated category $\mathscr{R}$). Call E_* the *connecting morphism* of the pair. The condition (7.2) states that E_* is a natural transformation of functors of E. This condition may be replaced by three separate requirements:

$$\text{If } E \text{ is congruent to } E', \qquad E_* = E'_*, \tag{7.2a}$$

$$\text{If } \gamma\colon C' \to C, \text{ then} \qquad (E\gamma)_* = E_* \gamma_*, \quad \gamma_* = S(\gamma), \tag{7.2b}$$

$$\text{If } \alpha\colon A \to A', \text{ then} \qquad (\alpha E)_* = \alpha_* E_*, \quad \alpha_* = T(\alpha). \tag{7.2c}$$

Indeed, (7.2) with $\alpha = 1$ and $\gamma = 1$ gives (a). If $(1, \beta, \gamma)\colon E \to E'$, then $E'\gamma$ is by definition E, so (7.2) gives (b). Dually, (7.2) with $\gamma = 1$ gives (c). Conversely, given (a), (b), and (c) with $(\alpha, \beta, \gamma)\colon E \to E'$, the congruence $\alpha E \equiv E'\gamma$ of Prop. III.1.8 gives (7.2).

If E_0 splits, then $(E_0)_* = 0$. For, if E_0 splits, the morphism $(1_A, \pi_1, 0)$ maps E_0 to the sequence $A \rightarrowtail A \twoheadrightarrow 0$. Since S is additive, $S(0) = 0$, so (7.2) gives $0 = S(0) = T(1)(E_0)_* = (E_0)_*$.

For each proper $E\colon A \rightarrowtail B \twoheadrightarrow C$, the long sequence

$$S(A) \xrightarrow{S(\varkappa)} S(B) \xrightarrow{S(\sigma)} S(C) \xrightarrow{E_*} T(A) \xrightarrow{T(\varkappa)} T(B) \xrightarrow{T(\sigma)} T(C) \tag{7.3}$$

is a complex in $\mathscr{R}$ (the composite of any two successive maps is zero) and a functor of E. Indeed, write $E = (\varkappa, \sigma)$; both $\varkappa E$ and $E\sigma$ split, so $T(\varkappa)E_* = 0$ and $E_* S(\sigma) = 0$, while $S(\sigma)S(\varkappa) = S(\sigma\varkappa) = S(0) = 0$.

For example, if $\mathscr{A}$ is the category of R-modules, with $\mathscr{P}$ all short exact sequences, the functors $S(A) = \operatorname{Tor}_{n+1}(G, A)$ and $T(A) = \operatorname{Tor}_n(G, A)$ for fixed G and n constitute a connected pair with the usual connecting homomorphism.

A *morphism* $(f, g)\colon (S', E_\#, T') \to (S, E_*, T)$ of connected pairs is a pair of natural transformations $f\colon S' \to S$, $g\colon T' \to T$ of functors on $\mathscr{A}$ such that the diagram

$$\begin{array}{ccc} S'(C) & \xrightarrow{E_\#} & T'(A) \\ \downarrow{\scriptstyle f(C)} & & \downarrow{\scriptstyle g(A)} \\ S(C) & \xrightarrow{E_*} & T(A), \end{array} \qquad \mathscr{R}, \tag{7.4}$$

is commutative for each proper E. In other words, a morphism (f, g) assigns to each A morphisms $f(A)\colon S'(A) \to S(A)$ and $g(A)\colon T'(A) \to T(A)$ of $\mathscr{R}$ which taken together form a chain transformation of the complexes (7.3). These conditions on f and g may be summarized as

$$f\alpha_\# = \alpha_* f, \qquad g E_\# = E_* f, \qquad g\alpha_\# = \alpha_* g, \tag{7.4a}$$

where $\alpha_\#$ is short for $S'(\alpha)$ or $T'(\alpha)$, α_* short for $S(\alpha)$ or $T(\alpha)$.

A connected pair (S, E_*, T) is *left $\mathscr{P}$-couniversal* if to each connected pair $(S', E_\#, T')$ and each natural $g\colon T' \to T$ there is a unique $f\colon S' \to S$ such that (f, g) is a morphism of connected pairs. Briefly, (S, E_*, T) left-couniversal means: Given g, (7.4) can be filled in with a unique natural f. Similarly, (S, E_*, T) right $\mathcal{P}$-couniversal means that to a given f there is a unique g; then $T = E_* = 0$. Also, $(S', E_\#, T')$ is *right $\mathcal{P}$-universal* if, given (S, E_*, T) and f, there is a unique g which satisfies (7.4).

Given T, the usual argument shows that there is at most one left couniversal pair (S, E_*, T) up to a natural equivalence of S. This pair — and, by abuse of language, this functor S — is called the *left satellite* of T. Note the curious fact that if (S, E_*, T) is the left satellite, so is $(S, -E_*, T)$ — just change the signs of every E_* and f in (7.4).

Theorem 7.2. *If $\mathcal{A}$ has enough proper projectives, the following conditions on a $\mathcal{P}$-connected pair (S, E_*, T) of covariant functors are equivalent:*

(i) *(S, E_*, T) is left $\mathcal{P}$-couniversal,*

(ii) *For each proper short exact sequence $K \rightarrowtail P \twoheadrightarrow C$ the sequence*

$$0 \to S(C) \to T(K) \to T(P), \qquad \mathcal{R}, \tag{7.5}$$

is left exact whenever P is proper projective.

Since there are enough projectives, there is for each object C of $\mathcal{A}$ a proper epic σ: $P \twoheadrightarrow C$ with P a proper projective; this gives a proper exact sequence

$$E_C\colon\ 0 \to K \xrightarrow{\varkappa} P \to C \to 0, \quad \varkappa = \varkappa_C. \tag{7.6}$$

It is the first step in the construction of a proper projective resolution of C; we call it a *short projective resolution.*

To prove that (ii) implies (i) we must construct f to a given g in (7.4). For $E = E_C$ the commutative diagram:

$$\begin{array}{ccccccc} S'(P) \to S'(C) & \xrightarrow{E_\#} & T'(K) & \xrightarrow{T'(\varkappa)} & T'(P) & & \\ \vdots{\scriptstyle f(C)} & & \downarrow{\scriptstyle g(K)} & & \downarrow{\scriptstyle g(P)} & & \\ 0 \to S(C) & \xrightarrow{E_*} & T(K) & \xrightarrow{T(\varkappa)} & T(P), & & \mathcal{R}, \end{array} \tag{7.7}$$

with top row a complex, has its bottom row exact, by hypothesis. Hence E_* is monic, so that $f(C)$, if it exists, is unique. On the other hand, $T(\varkappa)\, g(K)\, E_\# = g(P)\, T'(\varkappa)\, E_\# = 0$, so $g(K)\, E_\#$ factors through $E_* \in \ker(T(\varkappa))$ as $g(K)\, E_\# = E_* \xi$ for some unique ξ: $S'(C) \to S(C)$. Take $f(C) = \xi$. This fills in the dotted arrow to make the diagram commute.

Now take any proper short exact sequence $E' = (\varkappa', \sigma')$: $A' \rightarrowtail B' \twoheadrightarrow C'$ and any morphism γ: $C \to C'$ of $\mathcal{A}$. The diagram

$$\begin{array}{ll} E\colon & 0 \to K \to P \to C \to 0 \\ & \qquad\qquad\qquad \downarrow{\scriptstyle \gamma} \\ E'\colon & 0 \to A' \to B' \to C' \to 0, \qquad \mathcal{A}. \end{array} \tag{7.8}$$

in $\mathscr{A}$ has P proper projective, so may be filled in (comparison theorem!) to give a morphism $(\alpha, \beta, \gamma)\colon E \to E'$. We claim that

$$E'_* \, S(\gamma) \, f(C) = g(A') \, E'_{\#} S'(\gamma)\colon \; S'(C) \to T(A'), \qquad \mathscr{R}. \tag{7.9}$$

Indeed, $\alpha E \equiv E'\gamma$ and, in the notation (7.4a), $E'_* \gamma_* f = \alpha_* E_* f = \alpha_* g E_{\#} = g \alpha_{\#} E_{\#} = g E'_{\#} \gamma_{\#}$. We specialize this result (7.9) in two ways.

First, let $\gamma\colon C \to C'$ be any morphism of $\mathscr{A}$. Choose for E' the short projective resolution $E_{C'}$ used to define $f(C')$ by $g(K') E'_{\#} = E'_* f(C')$, as in (7.7). Then $A' = K'$ and E'_* is monic, so (7.9) gives $S(\gamma) f(C) = f(C') \, S'(\gamma)$. This asserts that $f\colon S' \to S$ is natural. With $C = C'$ and $\gamma = 1$, it shows that $f(C)$ is independent of the choice of E_C.

Second, let E' be any proper short exact sequence ending in $C' = C$. Take $\gamma = 1$. Then (7.9) becomes $E'_* f(C) = g(A') E'_{\#}$, which states that f and g commute with the connecting homomorphisms and hence, as in (7.4), constitute a morphism $(S', E_{\#}, T') \to (S, E_*, T)$ of pairs

Before proving the converse, note that (7.7) suggests that $S(C)$ might be defined as the kernel of $T(K) \to T(P)$. Regard each proper short exact sequence $E\colon A \rightarrowtail B \twoheadrightarrow C$ as a complex in $\mathscr{A}$, say in dimensions 1, 0, and -1. Then $T(E)\colon T(A) \to T(B) \to T(C)$ is a complex in $\mathscr{R}$; its one dimensional homology $H_1(T(E))$ is the (selected) object of $\mathscr{R}$ which makes

$$0 \to H_1(T(E)) \xrightarrow{\mu} T(A) \to T(B), \qquad \mathscr{R}, \tag{7.10}$$

exact. Each morphism $\Gamma = (\alpha, \beta, \gamma)\colon E \to E'$ of proper short exact sequences in $\mathscr{A}$ gives a chain transformation $T(\Gamma)\colon T(E) \to T(E')$ and hence induces a morphism

$$H_1(\Gamma)\colon \; H_1(T(E)) \to H_1(T(E')), \qquad \mathscr{R},$$

which is characterized by $\mu' H_1(\Gamma) = T(\alpha) \, \mu$. Moreover $H_1(\Gamma)$ depends only on γ, E, and E', and not on α and β. For, let $\Gamma_0 = (\alpha_0, \beta_0, \gamma)\colon E \to E'$ be any other morphism with the same γ. In the diagram

$$\begin{array}{ccccccc} 0 \to & A & \xrightarrow{\varkappa} & B & \longrightarrow & C & \\ & {\scriptstyle \alpha-\alpha_0}\downarrow & \swarrow{\scriptstyle s} & \downarrow{\scriptstyle \beta-\beta_0} & & \downarrow{\scriptstyle 0} & \\ 0 \to & A' & \xrightarrow{\varkappa'} & B' & \xrightarrow{\sigma'} & C', & \qquad \mathscr{A}, \end{array}$$

$\sigma'(\beta - \beta_0) = 0$, so $\beta - \beta_0 = \varkappa' s$ for some $s\colon B \to A'$. Also $\varkappa'(\alpha - \alpha_0) = (\beta - \beta_0) \varkappa = \varkappa' s \varkappa$, so all told $s \varkappa = \alpha - \alpha_0$, $\varkappa' s = \beta - \beta_0$. Thus s is a homotopy $\Gamma \simeq \Gamma_0$. Since T is additive, $T(s)$ is a homotopy $T(\Gamma) \simeq T(\Gamma_0)\colon T(E) \to T(E')$, so $H_1(\Gamma) = H_1(\Gamma_0)$. Now there exists:

To each object C of $\mathscr{A}$ a short projective resolution E_C,

To each $\gamma\colon C \to C'$ in $\mathscr{A}$ a morphism $\Gamma_\gamma = (-, -, \gamma)\colon E_C \to E_{C'}$,

To each proper exact E in $\mathscr{A}$ a morphism $\Lambda_E = (-, -, 1)\colon E_C \to E$.

We now have

Lemma 7.3. *Given* $T: \mathscr{A} \to \mathscr{R}$ *covariant and the data above,*

$$S(C) = H_1(T(E_C)), \quad S(\gamma) = H_1(\Gamma_\gamma): S(C) \to S(C')$$

define a covariant additive functor $S: \mathscr{A} \to \mathscr{R}$, *while, for* μ *as in* (7.10),

$$E_* = \mu\, H_1(\Lambda_E): S(C) \to T(A)$$

defines a natural transformation which makes (S, E_*, T) *a* $\mathscr{P}$*-connected pair satisfying* (ii) *of Thm.* 7.2.

Proof. By the observation on $H_1(\Gamma)$, $S(C)$ is independent of the choice of E_C. Also $S(1) = 1$; by composition, $S(\gamma_1\gamma_2) = S(\gamma_1)\, S(\gamma_2)$. If $\Gamma = (\alpha, \beta, \gamma): E \to E'$ is a morphism of proper short exact sequences, $\Gamma\Lambda_E$ and $\Lambda_{E'}\Gamma_\gamma$: $E_C \to E'$ agree at γ, so E_* is natural. Property (ii) holds by construction.

This proves that (i) implies (ii) in the theorem, for any left couniversal $(S_0, E_{\#}, T)$ must agree with the one so constructed, which does satisfy (ii). This construction also gives an existence theorem:

Theorem 7.4. *If* $\mathscr{A}$ *has enough proper projectives, each covariant additive functor* $T: \mathscr{A} \to \mathscr{R}$ *has a left satellite* (S, E_*, T).

Corollary 7.5. *Let* (S, E_*, T) *be a* $\mathscr{P}$*-connected pair. If to each* $\mathscr{P}$*-connected pair* $(S', E_{\#}, T)$ *with the same* T *there is a unique natural transformation* f: $S' \to S$ *such that* $(f, 1)$: $(S', E_{\#}, T) \to (S, E_*, T)$ *is a morphism of pairs, then* (S, E_*, T) *is left-*$\mathscr{P}$*-couniversal.*

Proof. Use the hypotheses to compare (S, E_*, T) to the left satellite of T, which is known to exist and to be couniversal.

The dual of Thm. 7.2 is

Theorem 7.6. *If* $\mathscr{A}$ *has enough proper injectives, then a* $\mathscr{P}$*-connected pair* (T, E_*, S) *of covariant functors is right* $\mathscr{P}$*-universal if and only if each proper short exact sequence*

$$0 \to C \to J \to K \to 0, \qquad \mathscr{A},$$

with J *proper injective induces a right exact sequence*

$$T(J) \to T(K) \to S(C) \to 0, \qquad \mathscr{R}.$$

Moreover, given T, the S with this property is uniquely determined; it is called the *right satellite* of T. Each T thus has a left satellite (couniversal) and a right satellite (universal).

Proof. Dualization reverses all arrows, both in $\mathscr{A}$ and in $\mathscr{R}$, replaces "projective" by "injective", gives E_* from T to S, and leaves T and S covariant.

A $\mathscr{P}$-connected pair (T, E^*, S) of contravariant functors consists of two such functors $T, S\colon \mathscr{A}\to\mathscr{R}$ and a function which assigns to each proper short exact $E\colon A \rightarrowtail B \twoheadrightarrow C$ in $\mathscr{A}$ a complex

$$T(C)\to T(B)\to T(A)\xrightarrow{E^*} S(C)\to S(B)\to S(A), \qquad \mathscr{R},$$

which is a functor of E. The pair is *right universal* if and only if to each natural $f\colon T\to T'$ and each connected pair $(T', E^{\#}, S')$, there is a unique $g\colon S\to S'$ such that (f, g) is a morphism of pairs.

Theorem 7.7. *In the presence of enough proper projectives, the contravariant pair (T, E^*, S) is right $\mathscr{P}$-universal if and only if each proper $K\rightarrowtail P\twoheadrightarrow C$ with P proper projective induces an exact sequence*

$$T(P)\to T(K)\to S(C)\to 0, \qquad \mathscr{R}.$$

Example: For D a fixed module, $T(C)=\operatorname{Ext}^n(C, D)$, $S(C)=\operatorname{Ext}^{n+1}(C, D)$.

Proof. This reduces to the previous result if we replace $\mathscr{A}$ by the opposite category $\mathscr{A}^{\mathrm{op}}$. Recall (I.7) that $\mathscr{A}^{\mathrm{op}}$ has an object A^* for each object A of $\mathscr{A}$ and a morphism $\alpha^*\colon B^*\to A^*$ for each $\alpha\colon A\to B$ in $\mathscr{A}$, with $(\alpha\beta)^*=\beta^*\alpha^*$. Thus monics in $\mathscr{A}$ become epic in $\mathscr{A}^{\mathrm{op}}$, the opposite of an abelian category is abelian, and the opposites of a proper class $\mathscr{P}$ of short exact sequences of $\mathscr{A}$ constitute a proper class in $\mathscr{A}^{\mathrm{op}}$. Each covariant $T\colon \mathscr{A}\to\mathscr{R}$ gives a contravariant $T^*\colon \mathscr{A}^{\mathrm{op}}\to\mathscr{R}$ with $T^*(A^*)=T(A)$. Moreover, "enough injectives" becomes "enough projectives". All arrows in $\mathscr{A}$-diagrams are reversed, those in $\mathscr{R}$-diagrams stay put, and Thm. 7.6 becomes Thm. 7.7.

A similar replacement in Thm. 7.2 shows that a contravariant pair (S, E^*, T) is left couniversal if and only if $0\to S(C)\to T(K)\to T(J)$ is exact for each $C\rightarrowtail J\twoheadrightarrow K$. Then S is a left satellite of T.

Exercises

1. Call a diagram $A_1 \rightleftarrows B \leftrightarrows A_2$ "cartesian" if it satisfies the usual direct sum identities $\pi_1\iota_1=1=\pi_2\iota_2$ and $\iota_1\pi_1+\iota_2\pi_2=1$. Prove that an additive functor takes every cartesian diagram into a cartesian diagram, and, conversely, that any functor with this property is additive.

2. Let $T\colon \mathscr{A}\to\mathscr{R}$, not assumed to be additive, be half $\mathscr{P}$-exact. Prove it additive (cf. Ex. 1 and Prop. I.4.2).

3. If T is covariant and left $\mathscr{P}$-exact, show its left satellite zero.

4. If (S, E_*, T) is left couniversal and T half $\mathscr{P}$-exact and covariant, prove (7.3) exact, provided has $\mathscr{A}$ enough proper projectives.

5. Derive Thm. 7.6 from Thm. 7.2 by replacing both $\mathscr{A}$ and $\mathscr{R}$ by their opposites.

8. Connected Sequences of Functors

A *$\mathscr{P}$-connected sequence* $\{T_n, E_n\}$ of covariant functors is a sequence $(\ldots, T_n, E_n, T_{n-1}, E_{n-1}, \ldots)$ of functors $T_n: \mathscr{A} \to \mathscr{R}$ in which each pair (T_n, E_n, T_{n-1}) is $\mathscr{P}$-connected; in other words, such a sequence assigns to each proper short exact E of $\mathscr{A}$ a complex

$$\cdots \to T_{n+1}(C) \xrightarrow{E_{n+1}} T_n(A) \to T_n(B) \to T_n(C) \xrightarrow{E_n} T_{n-1}(A) \to \cdots \qquad (8.1)$$

which is a covariant functor of E. The sequence is *positive* if $T_n = 0$ for $n < 0$ or *negative* if $T_n = 0$ for $n > 0$; in the latter case we usually use upper indices.

Positive connected sequences may be described more directly in terms of graded additive categories. Recall (Thm. 4.4) that $\mathscr{A}$ can be enlarged to a graded additive category $\mathscr{E}_{\mathscr{P}}(\mathscr{A})$ with the same objects and with the elements of $\operatorname{Ext}^n_{\mathscr{P}}(C, A)$ regarded as the morphisms of degree n from C to A. From the range category $\mathscr{R}$ we can construct the category $\mathscr{R}^+$ of graded objects of $\mathscr{R}$, with morphisms of negative degrees. In detail, an object $\mathfrak{R}$ of $\mathscr{R}^+$ is a family $\{R_n\}$ of objects of $\mathscr{R}$, with $R_n = 0'$ for $n < 0$; while an element of $\hom^k(\mathfrak{R}, \mathfrak{R}')$ is a morphism $\mu: \mathfrak{R} \to \mathfrak{R}'$ of degree $-k$; that is, a family of morphisms $\{\mu_n: R_n \to R'_{n-k}\}$ of $\mathscr{R}$, with the evident composition. Then $\mathscr{R}^+$ is a graded additive category. If $\mathscr{R}$ is the category of modules over some ring, $\mathscr{R}^+$ is the category of graded modules over the same ring, with morphisms of negative degrees.

For graded categories, functors are defined as usual, with supplementary attention to the degrees of morphisms. Thus if $\mathscr{G}$ and $\mathscr{H}$ are graded additive categories, a *covariant functor* $\mathfrak{T}: \mathscr{G} \to \mathscr{H}$ assigns to each object G of $\mathscr{G}$ an object $\mathfrak{T}(G)$ of $\mathscr{H}$ and to each morphism $\gamma: G_1 \to G_2$ of degree d in $\mathscr{G}$ a morphism $\mathfrak{T}(\gamma): \mathfrak{T}(G_1) \to \mathfrak{T}(G_2)$ of the same degree in $\mathscr{H}$, with the usual conditions $\mathfrak{T}(1_G) = 1_{\mathfrak{T}(G)}$ and $\mathfrak{T}(\gamma_1\gamma_2) = \mathfrak{T}(\gamma_1)\,\mathfrak{T}(\gamma_2)$ whenever $\gamma_1\gamma_2$ is defined. The functor $\mathfrak{T}$ is additive if $\mathfrak{T}(\gamma_1 + \gamma_2) = \mathfrak{T}(\gamma_1) + \mathfrak{T}(\gamma_2)$ whenever $\gamma_1 + \gamma_2$ is defined. A *natural transformation* $\mathfrak{f}: \mathfrak{T}' \to \mathfrak{T}$ of *degree d* is a function which assigns to each $G \in \mathscr{G}$ a morphism $\mathfrak{f}(G): \mathfrak{T}'(G) \to \mathfrak{T}(G)$ of degree d in $\mathscr{H}$ such that

$$\mathfrak{T}(\gamma)\,\mathfrak{f}(G_1) = (-1)^{(\deg\gamma)(\deg\mathfrak{f})}\,\mathfrak{f}(G_2)\,\mathfrak{T}'(\gamma)$$

for each $\gamma: G_1 \to G_2$ in $\mathscr{G}$.

In particular, consider such functors on $\mathscr{E}_{\mathscr{P}}(\mathscr{A})$ to $\mathscr{R}^+$.

Proposition 8.1. *There is a 1-1 correspondence between covariant additive functors $\mathfrak{T}: \mathscr{E}_{\mathscr{P}}(\mathscr{A}) \to \mathscr{R}^+$ and positive $\mathscr{P}$-connected sequences $\{T_n, E_n\}$ of covariant additive functors $T_n: \mathscr{A} \to \mathscr{R}$.*

Proof. Let $\mathfrak{T}: \mathscr{E}_{\mathscr{P}}(\mathscr{A}) \to \mathscr{R}^+$ be given. The object function of $\mathfrak{T}$ assigns to each object A an object $\{T_n(A)\}$ of $\mathscr{R}^+$. The mapping function of $\mathfrak{T}$ assigns to each morphism of $\mathscr{E}_{\mathscr{P}}(\mathscr{A})$ a morphism of $\mathscr{R}^+$. In particular, each morphism $\alpha: A \to A'$ of $\mathscr{A}$ is a morphism of degree 0 in $\mathscr{E}_{\mathscr{P}}(\mathscr{A})$, so $\mathfrak{T}$ assigns a family of morphisms $\{T_n(\alpha): T_n(A) \to T_n(A')\}$ of $\mathscr{R}$; these make each T_n an additive functor $T_n: \mathscr{A} \to \mathscr{R}$. Moreover, each proper $E: A \rightarrowtail B \twoheadrightarrow C$ in $\mathscr{A}$ is a morphism $E: C \to A$ of degree 1 in $\mathscr{E}_{\mathscr{P}}(\mathscr{A})$, so the mapping function of $\mathfrak{T}$ assigns to E a morphism of degree 1 in $\mathscr{R}^+$; that is, a family of morphisms $\{E_n = T_n(E): T_n(C) \to T_{n-1}(A)\}$ in $\mathscr{R}$. The composition rules $\mathfrak{T}(E\,\gamma) = \mathfrak{T}(E)\,\mathfrak{T}(\gamma)$ and $\mathfrak{T}(\alpha\, E) = \mathfrak{T}(\alpha)\,\mathfrak{T}(E)$ show that these morphisms E_n satisfy the conditions (7.2a), (7.2b), (7.2c) which make (T_n, E_n, T_{n-1}) a connected pair. Thus $\mathfrak{T}$ determines a positive $\mathscr{P}$-connected sequence of functors $\{T_n: \mathscr{A} \to \mathscr{R}\}$.

Conversely, each such connected sequence of functors determines the object function $\mathfrak{T}(A) = \{T_n(A)\}$ and the mapping functions $\mathfrak{T}(\alpha)$, $\mathfrak{T}(E)$ for morphisms of degree 0 and 1 in $\mathscr{E}_{\mathscr{P}}$. Now a morphism of higher degree in $\mathscr{E}_{\mathscr{P}}$ is just a congruence class of long exact sequences S. Each such is the Yoneda composite of short exact sequences E, so the $\mathfrak{T}(E)$ determine each $\mathfrak{T}(S)$; the rules (7.2b) and (7.2c) show that two congruent long exact sequences have the same $\mathfrak{T}(S)$; indeed, this $\mathfrak{T}(S)$ is the "iterated connecting homomorphism" determined by the long exact sequence S. Finally, to show this functor $\mathfrak{T}$ additive we must prove that $\mathfrak{T}(E+E') = \mathfrak{T}(E) + \mathfrak{T}(E')$. This follows from the definition $E+E' = \nabla_A(E \oplus E')\,\Delta_C$ of addition and the rule $(E \oplus E')_n \cong E_n \oplus E_n$ for connecting morphisms, which is a consequence of the condition (7.2) for a connected pair.

This gives the asserted 1-1 correspondence. The same applies to maps:

Proposition 8.2. *If $\mathfrak{T}', \mathfrak{T}: \mathscr{E}_{\mathscr{P}}(\mathscr{A}) \to \mathscr{R}^+$ are two covariant functors, a natural transformation $\mathfrak{f}: \mathfrak{T}' \to \mathfrak{T}$ of degree d is a family of natural transformations $\{f_n: T'_n \to T_{n+d}: \mathscr{A} \to \mathscr{R}\}$ which commute with all connecting morphisms:*

$$T_{n+d}(E)\, f_n(C) = f_{n-1}(A)\, T'_n(E), \qquad E: A \rightarrowtail B \twoheadrightarrow C. \tag{8.2}$$

In other words, for $d=0$, $\mathfrak{f}$ is a chain transformation of the complex (8.1) for $\mathfrak{T}'$ to that for $\mathfrak{T}$.

A covariant functor $\mathfrak{T}: \mathscr{E}_{\mathscr{P}}(\mathscr{A}) \to \mathscr{R}^+$ is called *couniversal* if to each covariant $\mathfrak{T}': \mathscr{E}_{\mathscr{P}}(\mathscr{A}) \to \mathscr{R}^+$ and each natural transformation $f_0: T'_0 \to T_0$ in $\mathscr{A}$ of the components of degree 0, there exists a unique natural transformation $\mathfrak{f}: \mathfrak{T}' \to \mathfrak{T}$ of degree 0 extending f_0. In other words, a couniversal positive connected sequence of covariant functors is such a sequence starting at T_0, extended to the left, and couniversal for all such connected sequences. Thus T_0 uniquely determines $\mathfrak{T}$, up to a natural isomorphism.

Theorem 8.3. *Let $\mathscr{A}$ have enough proper projectives. A covariant functor $\mathfrak{T}$: $\mathscr{E}_{\mathscr{P}}(\mathscr{A})\to\mathscr{R}^+$ is couniversal if and only if, for each proper short exact sequence $K\rightarrowtail P\twoheadrightarrow C$ of $\mathscr{A}$ with P proper projective, the sequence*

$$0\to T_n(C)\to T_{n-1}(K)\to T_{n-1}(P), \qquad \mathscr{R}, \tag{8.3}$$

is exact for every $n>0$.

Proof. Given this condition, some other $\mathfrak{T}'$: $\mathscr{E}_{\mathscr{P}}(\mathscr{A})\to\mathscr{R}^+$, and some $f_0\colon T_0'\to T_0$ we construct by recursion on n the requisite natural transformations $f_n\colon T_n'\to T_n$. If $f_0, \ldots, f_{n-1}$ are already constructed to commute with the connecting homomorphisms, the condition (8.3) shows by Thm. 7.2 that (T_n, E_n, T_{n-1}) is left couniversal, so will construct a unique $f_n\colon T_n'\to T_n$ with $E_n f_n = f_{n-1}E_n'$. Hence $\mathfrak{T}$ is couniversal.

Conversely, suppose that $\mathfrak{T}$ is couniversal. From T_0 we construct the left satellite S_1, and construct in turn each S_n: $\mathscr{A}\to\mathscr{R}$ as the left satellite of S_{n-1}. The resulting connected sequence satisfies (8.3), hence is couniversal, so must agree with the unique couniversal $\mathfrak{T}$ for the given T_0. Therefore any couniversal $\mathfrak{T}$ satisfies (8.3). This argument also proves an existence theorem:

Theorem 8.4. *Let $\mathscr{A}$ have enough proper projectives. Each covariant functor T_0: $\mathscr{A}\to\mathscr{R}$ is the component of degree 0 for a couniversal functor $\mathfrak{T}$: $\mathscr{E}_{\mathscr{P}}(\mathscr{A})\to\mathscr{R}^+$ in which the n-th component T_n is the n-th iterated left satellite of T_0.*

Since $0\rightarrowtail P\twoheadrightarrow P$ is exact for each proper projective, condition (8.3) implies that $T_n(P)=0$ for each $n>0$. Thm. 8.3 includes the weaker result:

Corollary 8.5. *If $\mathfrak{T}$ satisfies $T_n(P)=0$ for each projective P and for each $n>0$, and if the long sequence (8.1) is exact for each proper exact E of $\mathscr{A}$, then $\mathfrak{T}$ is couniversal.*

In particular, if $\mathscr{A}$ is the category of all left modules over some ring R and G is a fixed right R-module, Thm. V.8.5 asserts that the functors $T_n(A)=\operatorname{Tor}_n^R(G, A)$ satisfy this condition.

Corollary 8.6. *If U: $\mathscr{R}\to\mathscr{R}'$ is exact and covariant, while $\{T_n, E_n\}$ is a couniversal positive connected sequence, so is $\{UT_n, UE_n\}$.*

Proof. Since $E_n\colon T_n(C)\to T_{n-1}(A)$ is a morphism of $\mathscr{R}$ while U is a functor, $UE_n\colon UT_n(C)\to UT_{n-1}(A)$ is a morphism of $\mathscr{R}'$. Since U preserves exactness, condition (8.3) for couniversality is preserved.

Note. If U is not exact, the description of the left satellite of the functor UT_0 in terms of U and T_0 involves an important spectral sequence [Cartan-Eilenberg, XVI, § 3; Grothendieck 1957, p. 147].

To handle negative connected sequences

$$\cdots \to T^0(C) \to T^1(A) \to T^1(B) \to T^1(C) \to T^2(A) \to \cdots$$

of covariant functors T^n: $\mathscr{A} \to \mathscr{R}$, use the graded additive category $\mathscr{R}^-$; its objects $\{R^n\}$ are families of objects of $\mathscr{R}$, with $R^n = 0$ for $n < 0$; its morphisms μ of degree $k \geqq 0$ are the families $\{\mu_n: R^n \to R'^{n+k}\}$ of mor phisms of $\mathscr{R}$. A covariant functor $\mathfrak{T}$: $\mathscr{E}_{\mathscr{P}}(\mathscr{A}) \to \mathscr{R}^-$ is then a negative connected sequence of functors T^n: $\mathscr{A} \to \mathscr{R}$, much as in Prop. 8.1.

Contravariant functors require attention as to sign. Thus if $\mathscr{G}$ and $\mathscr{H}$ are graded additive categories, a contravariant $\mathfrak{T}$: $\mathscr{G} \to \mathscr{H}$ assigns to each object G an object $\mathfrak{T}(G)$ in $\mathscr{H}$, and to each morphism $\gamma: G_1 \to G_2$ of $\mathscr{G}$ a morphism $\mathfrak{T}(\gamma)$: $\mathfrak{T}(G_2) \to \mathfrak{T}(G_1)$ of the same degree in $\mathscr{H}$, with $\mathfrak{T}(1_G) = 1_{\mathfrak{T}(G)}$ and

$$\mathfrak{T}(\gamma_1\gamma_2) = (-1)^{(\deg \gamma_1)(\deg \gamma_2)}\, \mathfrak{T}(\gamma_2)\, \mathfrak{T}(\gamma_1), \tag{8.4}$$

in accord with the sign conventions. A natural transformation $\mathfrak{f}$: $\mathfrak{T}' \to \mathfrak{T}$ of degree d is a function which assigns to each object G of $\mathscr{G}$ a morphism $\mathfrak{f}(G)$: $\mathfrak{T}'(G) \to \mathfrak{T}(G)$ of degree d in $\mathscr{H}$ in such wise that

$$\mathfrak{T}(\gamma)\, \mathfrak{f}(G_2) = (-1)^{(\deg \gamma)(\deg \mathfrak{f})}\, \mathfrak{f}(G_1)\, \mathfrak{T}'(\gamma)$$

— just as usual, except for sign.

Exercises

1. Show that the condition $T_n(P) = 0$ cannot be dropped from Cor. 8.5: Use $T'_n(A) = T_n(A) \oplus T_{n-1}(A)$.

2. Describe a contravariant additive functor $\mathfrak{S}$: $\mathscr{E}_{\mathscr{P}}(\mathscr{A}) \to \mathscr{R}^+$ as a suitably connected sequence of functors on $\mathscr{A}$ to $\mathscr{R}$.

9. Derived Functors

A standard method is: Take a resolution, apply a covariant functor T: $\mathscr{A} \to \mathscr{R}$, take the homology of the resulting complex. This gives a connected sequence of functors, called the derived functors of T.

In detail, let $\mathscr{A}$ have enough proper projectives. Each object A thus has a proper projective resolution ε: $X \to A$. If ε': $X' \to A'$ is a second such, the comparison theorem lifts each α: $A \to A'$ to a chain transformation f: $X \to X'$, and any two such are homotopic. Since T is additive, it carries homotopies to homotopies, and so the induced chain transformation $T(f)$: $T(X) \to T(X')$ in $\mathscr{R}$ is determined up to a homotopy. Therefore $L_n(A) = H_n(T(X))$ defines a function of A, independent of the choice of X, and $L_n(\alpha) = H_n(T(f))$: $L_n(A) \to L_n(A')$ makes each L_n a covariant functor $\mathscr{A} \to \mathscr{R}$. It is the n-th *left derived* functor of T.

Now let $E: A \rightarrowtail B \twoheadrightarrow C$ be any proper short exact sequence in $\mathscr{A}$. Take an allowable pro ctive resolution $\varepsilon: K \to E$ in the category of short exact sequences of $\mathscr{A}$, as in Thm. 6.4; this amounts to a short exact sequence $X \rightarrowtail W \twoheadrightarrow Y$ of complexes in $\mathscr{A}$ with $X \to A$, $W \to B$, and $Y \to C$ proper projective resolutions; moreover, $W_n = X_n \oplus Y_n$ for each n. As T is additive, this last shows that $T(X) \rightarrowtail T(W) \twoheadrightarrow T(Y)$ is a short exact sequence of complexes in $\mathscr{R}$, so gives connecting homomorphisms $H_n(T(Y)) \to H_{n-1}(T(X))$ for $n>0$. Since X is a resolution of A and Y one of C, this is a homomorphism $E_* = L_n(E)$: $L_n(C) \to L_{n-1}(A)$. The general comparison theorem for allowable resolutions (Thm. IX.4.3) shows this independent of the choice of the resolution K and shows that $L_n(E)$ is a natural transformation of functors of E.

Theorem 9.1. *For each additive covariant functor $T: \mathscr{A} \to \mathscr{R}$, the left derived functors $L_n: \mathscr{A} \to \mathscr{R}$ and the connecting homomorphisms $L_n(E)$ constitute a positive connected sequence of functors with L_0 right $\mathscr{P}$-exact. This sequence is couniversal for the initial component L_0. If T is right $\mathscr{P}$-exact, $L_0 = T$.*

Proof. If P is proper projective, the resolution $P \to P$ shows $L_n(P) = 0$ for $n>0$. For each proper exact E, the exactness of the long sequence (8.1) for $L_n = T_n$ follows from the usual long exact sequence for the homology of $T(X) \rightarrowtail T(W) \twoheadrightarrow T(Y)$. In particular, L_0 is right exact. The connected sequence $\{L_n, L_n(E)\}$ satisfies the conditions of Cor. 8.5, hence is couniversal.

Suppose that the original T is right $\mathscr{P}$-exact. In any resolution, the portion $X_1 \to X_0 \to A \to 0$ is right exact; hence so is $T(X_1) \to T(X_0) \to T(A) \to 0$. This gives $L_0(A) = H_0(T(X)) \cong T(A)$, a natural isomorphism.

This theorem is of interest when T is right exact. It can then be read either as a characterization of the sequence of left derived functors of T as the couniversal sequence for $L_0 = T$, or as the statement that the left satellites of T and their connecting homomorphisms can be calculated from resolutions.

To have a definite derived functor L_n one must choose a resolution X for each A. This sweeping use of the axiom of choice is legal in small categories $\mathscr{A}$ and possible in all those relevant examples of categories in which there is a canonical way of choosing a projective resolution. If the range category $\mathscr{R}$ is not a category of modules, but any abelian category, the proof above requires that we know the exact homology sequence, with its connecting homomorphisms, for an abelian category. We have indicated only too briefly in § 3 how this could be accomplished, using additive relations.

Let us summarize the properties of the derived functors in this case.

I. A covariant functor $\mathfrak{T}: \mathscr{E}_{\mathscr{P}}(\mathscr{A}) \to \mathscr{R}^+$ is a positive connected sequence $\{T_n, E_n\}$ consisting of covariant functors $T_n: \mathscr{A} \to \mathscr{R}$ and homomorphisms $E_n: T_n(C) \to T_{n-1}(A)$ natural in E. It assigns to each proper $E: A \rightarrowtail B \twoheadrightarrow C$ a complex

$$\cdots \to T_n(A) \to T_n(B) \to T_n(C) \xrightarrow{E_n} T_{n-1}(A) \to \cdots \tag{9.1}$$

in $\mathscr{R}$. Suppose that $\mathscr{A}$ has enough proper projectives. Each right $\mathscr{P}$-exact covariant $T: \mathscr{A} \to \mathscr{R}$ has a left derived functor $\mathfrak{T}: \mathscr{E}_{\mathscr{P}}(\mathscr{A}) \to \mathscr{R}^+$ which is determined by T, up to natural isomorphism, by any one of the following three conditions:

(Ia) $T_0 = T$ and $\mathfrak{T}$ is couniversal,

(Ib) $T_0 = T$, (9.1) is exact, and $T_n(P) = 0$ for $n > 0$ and P proper projective,

(Ic) $T_n(A) = H_n(T(X))$ for some proper projective resolution $\varepsilon: X \to A$, while E_n is similarly calculated from short exact sequences of such resolutions.

These considerations may be dualized: Replace one or both of the categories $\mathscr{A}$ and $\mathscr{R}$ by its opposite. For example, replacing $\mathscr{A}$ by $\mathscr{A}^{\text{op}}$ gives

II. Let $T: \mathscr{A} \to \mathscr{R}$ be a right $\mathscr{P}$-exact contravariant functor and suppose that $\mathscr{A}$ has enough proper injectives. For each object A take a proper injective coresolution $\varepsilon: A \to Y$. This Y is a negative complex $Y^0 \to Y^1 \to \cdots$; application of the contravariant T yields a positive complex $T(Y): T(Y^0) \leftarrow T(Y^1) \leftarrow \cdots$; that is $[T(Y)]_n = T(Y^n)$. Its homology $H_n(T(Y)) = T_n(A)$ is the n-th left derived functor T_n of T. For each proper E, coresolutions of E give a corresponding connecting homomorphism $E_n: T_n(A) \to T_{n-1}(C)$, natural in E. They constitute a positive connected sequence $\{T_n, E_n\}$ of contravariant functors which assigns to each proper $E: A \rightarrowtail B \twoheadrightarrow C$ a complex

$$\cdots \to T_n(C) \to T_n(B) \to T_n(A) \xrightarrow{E_n} T_{n-1}(C) \to \cdots \tag{9.2}$$

in $\mathscr{R}$. This sequence $\{T_n, E_n\}$ may also be described as a contravariant functor $\mathfrak{T}: \mathscr{E}_{\mathscr{P}}(\mathscr{A}) \to \mathscr{R}^+$. Given the right exact functor $T: \mathscr{A} \to \mathscr{R}$, its left derived functors may be characterized by their construction from injective coresolutions or by either of the properties:

(IIa) $T_0 = T$ and $\mathfrak{T}$ is couniversal; that is, given $\mathfrak{T}': \mathscr{E}_{\mathscr{P}}(\mathscr{A}) \to \mathscr{R}^+$, each natural $f_0: T'_0 \to T_0$ extends to a unique natural $\mathfrak{f}: \mathfrak{T}' \to \mathfrak{T}$,

(IIb) $T_0 = T$, (9.2) is always exact, and $T_n(J) = 0$ for $n > 0$ and each proper injective J.

The categorical dual of I (replace $\mathscr{A}$ by $\mathscr{A}^{op}$, $\mathscr{R}$ by $\mathscr{R}^{op}$) is

III. Let $T: \mathscr{A} \to \mathscr{R}$ be left $\mathscr{P}$-exact and covariant (sample, $T(A) = \operatorname{Hom}_R(G, A)$). Its right derived functors are $T^n(A) = H^n(T(Y))$, where $\varepsilon: A \to Y$ is a proper injective coresolution (assume enough injectives). With the corresponding connecting homomorphisms they constitute a negative connected sequence of covariant $T^n: \mathscr{A} \to \mathscr{R}$ which assigns to each E a complex,

$$\cdots \to T^{n-1}(C) \xrightarrow{E^n} T^n(A) \to T^n(B) \to T^n(C) \to \cdots \tag{9.3}$$

in $\mathscr{R}$; that is, a covariant $T: \mathscr{E}_{\mathscr{P}}(\mathscr{A}) \to \mathscr{R}^-$. The T^n are characterized in terms of T by either of the properties

(IIIa) $T^0 = T$ and $\mathfrak{T}$ is universal; that is, given $\mathfrak{T}': \mathscr{E}_{\mathscr{P}}(\mathscr{A}) \to \mathscr{R}^-$, each natural $f^0: T^0 \to T'^0$ extends to a unique natural $\mathfrak{f}: \mathfrak{T} \to \mathfrak{T}'$,

(IIIb) $T^0 = T$, (9.3) is exact, and $T^n(J) = 0$ for $n > 0$ and each injective J.

Finally, replace $\mathscr{R}$ in case I by $\mathscr{R}^{op}$

IV. Let $T: \mathscr{A} \to \mathscr{R}$ be left $\mathscr{P}$-exact and contravariant (sample: $T(A) = \operatorname{Hom}_R(A, G)$). Suppose that $\mathscr{A}$ has enough proper projectives. A projective resolution $\varepsilon: X \to A$ gives a negative complex $T(X)$ in $\mathscr{R}$, hence derived functors $T^n(A) = H^n(T(X))$ and connecting homomorphisms which constitute a negative connected sequence $\{T^n, E^n\}$ and for each E a complex

$$\cdots \to T^{n-1}(A) \xrightarrow{E^n} T^n(C) \to T^n(B) \to T^n(A) \to \cdots; \tag{9.4}$$

that is, a contravariant $\mathfrak{T}: \mathscr{E}_{\mathscr{P}}(\mathscr{A}) \to \mathscr{R}^-$ characterized by either

(IVa) $T^0 = T$ and $\mathfrak{T}$ is universal, or

(IVb) $T^0 = T$, (9.4) is exact, and $T^n(P) = 0$ for $n > 0$ and each projective P.

To summarize (examples with G a fixed module):

	T_0	Variance	Derived	T	Resolution	Type, $T_n(A)$
I	Right exact	Co	Left	couniversal	projective	$\operatorname{Tor}_n(G, A)$
II	Right exact	Contra	Left	couniversal	injective	?
III	Left exact	Co	Right	universal	injective	$\operatorname{Ext}^n(G, A)$
IV	Left exact	Contra	Right	universal	projective	$\operatorname{Ext}^n(A, G)$

Thus a change in variance or a change from left to right switches the type of resolution used.

For example, if Λ is a K-algebra, $\mathscr{A}$ the category of left Λ-modules, $\mathscr{P}$ the class of K-split short exact sequences of Λ-modules, and $\mathscr{R}$ the category of K-modules, then $\mathrm{Hom}_A(C, A)$ is left exact. As a functor of C it is contravariant (case IV); its right $\mathscr{P}$-derived functors are $\mathrm{Ext}^n_{(A,K)}(C, A)$. As a functor of A, Hom_A is covariant (case III); its right $\mathscr{P}$-derived functor is again given by the sequence of functors $\mathrm{Ext}^n_{(A,K)}(C, A)$, this time with the connecting homomorphisms in the second argument A.

Notes. Characterization of functors. For categories of modules, right or left exact additive functors are often given just by the usual functors $\otimes$ and Hom. Specifically (Watts [1960], Eilenberg [1960]), if C is a fixed S-R-bimodule, the tensor product with C gives a covariant functor $T_C(A) = C \otimes_R A$ of $_RA$ which is right exact and carries (infinite) direct sums to direct sums. Any functor T on the category of R-modules to the category of S-modules with these properties has this form for some C; namely, for $C = T(R)$. Again, any left exact contravariant functor T on R-modules to S-modules which converts (infinite) direct sums into direct products is naturally equivalent to the functor $T(A) = \mathrm{Hom}_R(A, C)$ for some left $(R \otimes S)$-module C (to wit, $C = T(R)$). Finally (Watts [1960]) any covariant left exact functor from R-modules to abelian groups which commutes with inverse limits has the form $T(A) = \mathrm{Hom}_R(C, A)$ for a suitable C. Mitchell [1962] has generalized these theorems to suitable abelian categories.

Bifunctors. Let $T_0(C, A)$ be a bifunctor, additive and right exact in each variable separately. Replacing both arguments by projective resolutions, taking the total complex of the resulting bicomplex and its homology gives the left derived functors $T_n(C, A)$ — as for example for $\mathrm{Tor}_n(C, A)$ as a bifunctor (Thm. V.9.3). This and related cases, with difference in variance, are treated in detail in Cartan-Eilenberg. This theory is not needed for $C \otimes A$, because this bifunctor becomes exact when either of the variables is replaced by a projective, so the derived functors can be constructed by the one-variable case. A relevant example is the trifunctor $C \otimes B \otimes A$ for three modules over a commutative ring, which must be treated as a functor of at least two variables. Its derived functors, called Trip_n, occur in the Künneth formulas for the homology of the tensor product of three complexes (MacLane [1960]). At present, there appears to be no way of characterizing derived functors of two or more variables by "universal" properties or by "axioms". For example, a suitable definition of a tensor product of two abelian categories would allow the reduction of bifunctors to functors of one variable.

Other constructions of derived functors. If T_0 is right exact and covariant on the category of all modules, each $S \in \mathrm{Ext}^n(C, A)$ gives an iterated connecting homomorphism $S_*: T_n(C) \to T_0(A)$, so each $t \in T_n(C)$ yields a natural transformation $\mathrm{Ext}^n(C, A) \to T_0(A)$ of functors of A. Indeed, $T_n(C)$ may be defined [Yoneda 1960, Hilton-Rees 1961] as

$$T_n(C) = \mathrm{Nat\ hom}_A(\mathrm{Ext}^n(C, A), T(A)).$$

This provides another definition of the torsion products. We have already remarked that an additive category $\mathscr{A}$ is a "ringoid" (usual ring axioms, but compositions not everywhere defined). In the same sense, each covariant additive functor T on $\mathscr{A}$ to the category of abelian groups is a left "$\mathscr{A}$-moduloid" (axioms for a left module over a ring; compositions not always defined), while a contravariant S is a right $\mathscr{A}$-moduloid. Yoneda [1960] has defined a corresponding tensor product $S \otimes_{\mathscr{A}} T$ and used it to construct satellites. Again, let T be a contravariant additive functor. The short exact sequences $E: A \rightarrowtail B \twoheadrightarrow C$ ending in a fixed C may be partly ordered by $E' \leqq E$ if there is a morphism $(\alpha, \beta, 1_C): E' \to E$; these E then from a "directed" class; the direct limit of the kernels of $T(A) \to T(B)$ taken over the directed class gives the right satellite of T [BUCHSBAUM 1960], defined in this way without assuming that there are enough projectives. This construction has been studied further by AMITSUR [1961]; RÖHRL [1962] gives an existence theorem for satellites of half exact functors, with applications to the theory of sheaves. For any additive functor which is not half-exact one must distinguish the derived functors, the satellites, and the cosatellites; their interrelations are studied in BUTLER-HORROCKS [1961].

Derived functors of non-additive functors have been studied by DOLD-PUPPE [1961] using iterated bar constructions. Indeed, the homology groups $H_{n+k}(\Pi, n; G)$ of Π provide many examples of non-additive functors (EILENBERG-MACLANE [1954a]). The classical example is the functor Γ of J. H. C. WHITEHEAD [1950]. For each abelian group A, $\Gamma(A)$ is the abelian group with generators $[\gamma(a) \mid a \in A]$, relations $\gamma(-a) = \gamma(a)$ and

$$\gamma(a+b+c) - \gamma(a+b) - \gamma(a+c) - \gamma(b+c) + \gamma(a) + \gamma(b) + \gamma(c) = 0.$$

These are the relations valid for a "square" $\gamma(a) = a^2$.

10. Products by Universality

The universal properties of derived functors may often be used to construct homomorphisms, such as the cup product for the cohomology of a group Π. In the notational scheme of § 7, take $\mathcal{R}$ to be the category of abelian groups, $\mathcal{A}$ the category of all left Π-modules, and $\mathcal{P}$ the class of Z-split short exact sequences of Π-modules. We first show that there are enough proper injectives in $\mathcal{A}$.

To each abelian group M construct the Π-module $J_M = \operatorname{Hom}_Z(Z(\Pi), M)$ with left operators defined for each $f \in J_M$ by $(x f) r = f(r x)$, with $x \in \Pi$, $r \in Z(\Pi)$. These are the left operators induced by the right Π-module structure of the group ring $Z(\Pi)$. Define a homomorphism $e = e_M: J_M \dashrightarrow M$ of abelian groups by setting $e(f) = f(1)$ for each $f: Z(\Pi) \dashrightarrow M$. This has the usual couniversal property, dual to that of Prop. VI.8.2:

Lemma 10.1. *If A is a left Π-module and $h: A \dashrightarrow M$ a homomorphism of abelian groups, there exists a unique Π-module homomorphism $\gamma: A \to J_M$ with $e\gamma = h$.*

Proof. Consider $A \overset{h}{\dashrightarrow} M \overset{e}{\dashleftarrow} J_M$. The condition $e\gamma = h$ requires for each $a \in A$ and $x \in \Pi$ that

$$h(x a) = e[\gamma(x a)] = [\gamma(x a)]\, 1 = [x(\gamma a)]\, 1 = (\gamma a) x.$$

Conversely, if one defines γ by $(\gamma a) x = h(x a)$, γ is a Π-map and satisfies $e\gamma = h$.

A standard argument now shows that each J_M is relatively injective. Moreover, if A is any Π-module, the Lemma gives a unique Π-module homomorphism $\gamma: A \to \operatorname{Hom}_Z(Z(\Pi), A) = J_A$ with $e\gamma = 1_A$. Hence γ is proper monic and $\gamma: A \to J_A$ embeds each A into a proper injective. Therefore there are enough proper injectives.

For each Π-module C let C^Π denote the subgroup of Π-invariant elements of C. The covariant functors

$$H^p(C) = H^p(\Pi, C) = \mathrm{Ext}^p_{Z(\Pi), Z}(Z, C)$$

have connecting homomorphisms for each proper E,

$$E_*: H^p(C) \to H^{p+1}(A), \quad E: A \rightarrowtail B \twoheadrightarrow C,$$

defined (say) by Yoneda composition, and giving the usual long exact sequence. Moreover, $H^p(J) = 0$ for $p > 0$ and J proper injective (any extension of a proper injective splits). Hence the $H^p(C)$ are the right derived functors of $H^0(C) = C^\Pi$.

Lemma 10.2. *For each fixed integer q and each fixed Π-module C', the functors $H^p(C) \otimes H^q(C')$ constitute the components of a universal sequence of functors with $\mathscr{P}$-connecting homomorphisms $E_* \otimes 1$.*

Proof. Let $E_0: A \rightarrowtail J \twoheadrightarrow K$ be any Z-split short exact sequence with J proper injective. For $p > 0$, $H^{p-1}(J) \to H^{p-1}(K) \to H^p(A) \to 0 (= H^p(J))$ is exact. As the tensor product over Z is right exact, so is the sequence

$$H^{p-1}(J) \otimes H^q(C') \to H^{p-1}(K) \otimes H^q(C') \to H^p(A) \otimes H^q(C') \to 0.$$

This is the condition parallel to (8.3) in the dual of Thm. 8.3; hence $H^p(C) \otimes H^q(C')$ is the universal sequence for its given initial component $H^0(C) \otimes H^q(C')$.

Lemma 10.3. *If $C \otimes C'$ has the diagonal Π-module structure $[x(c \otimes c') = xc \otimes xc'$ for $x \in \Pi]$, then for fixed q and C' the functors $H^{p+q}(C \otimes C')$ of C constitute a $\mathscr{P}$-connected sequence of functors with connecting homomorphisms $(E \otimes C')_*$.*

Proof. Since E is Z-split and exact, the tensor product

$$E \otimes C': A \otimes C' \rightarrowtail B \otimes C' \twoheadrightarrow C \otimes C'$$

is exact and Z-split, hence gives the required (natural) connecting maps.

Similarly, for p and C fixed, the functors $H^p(C) \otimes H^q(C')$ constitute a universal $\mathscr{P}$-connected sequence, when the connecting homomorphisms $1 \otimes E'_*$ are defined with the usual sign:

$$(1 \otimes E'_*)(\sigma \otimes \sigma') = (-1)^p \sigma \otimes E'_* \sigma', \quad \sigma \in H^p(C), \quad \sigma' \in H^q(C'). \tag{10.1}$$

Moreover, the $H^{p+q}(C \otimes C')$ constitute a $\mathscr{P}$-connected sequence of functors of C' with connecting homomorphisms $(C \otimes E')_*$.

For $p = 0$, $H^0(C) = C^\Pi$ is the subgroup of Π-invariant elements of C. Now $c \in C^\Pi$ and $c' \in C'^\Pi$ give $c \otimes c' \in (C \otimes C')^\Pi$, so the identity induces a homomorphism $C^\Pi \otimes C'^\Pi \to (C \otimes C')^\Pi$.

Theorem 10.4. *There exists a unique family of group homomorphisms*

$$f^{p,q}\colon\ H^p(C)\otimes H^q(C')\to H^{p+q}(C\otimes C') \tag{10.2}$$

defined for all $p\geqq 0$, $q\geqq 0$ and all Π-modules C and C', such that

(i) *$f^{0,0}$ is the map induced by the identity, as above,*

(ii) *$f^{p,q}$ is natural in C and C',* $\quad p\geqq 0,\ q\geqq 0,$

(iii) $f^{p+1,q}(E_*\otimes 1)=(E\otimes C')_* f^{p,q},\quad p\geqq 0,\ q\geqq 0,$

(iv) $f^{p,q+1}(1\otimes E'_*)=(C\otimes E')_* f^{p,q},\quad p\geqq 0,\ q\geqq 0;$

the latter two for all Z-split short exact sequences E and E'.

The last two conditions assert that the maps f commute with the connecting homomorphisms.

Proof. We are given $f^{0,0}$. For $q=0$ and C' fixed, the left hand side of (10.2) is $\mathscr{P}$-universal, while the right hand side is $\mathscr{P}$-connected. Hence the maps $f^{p,0}$, natural in C, exist and are unique subject to (iii) for $q=0$. These maps are also natural in C'. For consider $\gamma\colon C'\to D'$. Then $\gamma f^{p,0}$ and $f^{p,0}\gamma$ are two natural transformations of the $\mathscr{P}$-universal functor $H^p(C)\otimes H^0(C')$ to the $\mathscr{P}$-connected functor $H^p(C\otimes D')$ which agree for $p=0$ and hence for all p.

Now hold p and C fixed. In (10.2), $f^{p,q}$ is given for $q=0$, and by (iv) must be a natural transformation of a universal to a connected sequence. Hence it exists and is unique; as before it is also natural in C.

Our construction gives (iii) only for $q=0$; it remains to prove it for $q>0$. For p fixed, let φ^q be the left-hand side and ψ^q be the right-hand side of (iii). Both are maps

$$\varphi^q,\ \psi^q\colon\ H^p(C)\otimes H^q(C')\to H^{p+q+1}(A\otimes C')$$

of a universal to a connected sequence of functors of C'. They both anticommute with the connecting homomorphisms given by E'. Indeed, by (iv),

$$(A\otimes E')_*\varphi^q=(A\otimes E')_* f^{p+1,q}(E_*\otimes 1)=f^{p+1,q+1}(1\otimes E'_*)(E_*\otimes 1),$$

$$\varphi^{q+1}(1\otimes E'_*)=f^{p+1,q+1}(E_*\otimes 1)(1\otimes E'_*),$$

and $(1\otimes E'_*)(E_*\otimes 1)=-(E_*\otimes 1)(1\otimes E'_*)$ by the definition (10.1). Also

$$(A\otimes E')_*\psi^q=(A\otimes E')_*(E\otimes C')_* f^{p,q},$$

$$\psi^{q+1}(1\otimes E'_*)=(E\otimes A')_* f^{p,q+1}(1\otimes E'_*)=(E\otimes A')_*(C\otimes E')_* f^{p,q},$$

and $(A\otimes E')\circ(E\otimes C')$ is congruent to $-(E\otimes A')\circ(C\otimes E')$ by the 3×3 splice lemma (VIII.3.1). Since $\varphi^0=\psi^0$, the uniqueness of the maps on a universal sequence gives $\varphi^q = \psi^q$ in all dimensions. This completes the proof.

Now the cup product (as defined — say — by Yoneda composites of long exact sequences) for the cohomology of groups satisfies exactly the conditions for the maps $f^{p,q}$ of our theorem. Thus we have still another construction of these cup products (VIII.9). This construction may be used to "calculate" these products for Π cyclic.

A similar argument in $H_n(\Pi, C) = \mathrm{Tor}_n^{(Z(\Pi), Z)}(Z, C)$ will construct a product which agrees with the internal product for the relative torsion functor. If Π is finite, these two products can be combined in a single product [CARTAN-EILENBERG, Chap. XII].

11. Proper Projective Complexes

Let $\mathscr{K}$ be the abelian category of positive complexes K (of left modules over some ring), with morphisms all chain transformations $f: K \to L$. Call a short sequence of complexes $K \xrightarrow{f} L \xrightarrow{g} M$ *proper exact* if, for all n,

(i) $0 \to K_n \to L_n \to M_n \to 0$ is exact, and

(ii) $0 \to C_n(K) \to C_n(L) \to C_n(M) \to 0$ is exact,

where $C_n(K)$ denotes the module of n-cycles of K. Since (i) implies that (ii) is left exact, (ii) may be replaced by

(ii') $C_n(L) \to C_n(M)$ is an epimorphism for all n.

In other words, a chain epimorphism $g: L \to M$ is proper if to each $m \in M$ with $\partial m = 0$ there exists an $l \in L$ with $g\,l = m$ and $\partial l = 0$. Equivalently, a chain monomorphism $f: K \to L$ is proper if to each $l \in L$ with $\partial l \in fK$ there is a $k \in K$ with $\partial l = \partial f\,k$. With these characterizations, the reader may verify that this class of proper short exact sequences satisfies the axioms of § 4 for propriety. Since a long exact sequence is a Yoneda composite of short exact sequences, we have

Lemma 11.1. *A sequence of complexes* $\cdots \to K \to L \to M \to N \to \cdots$ *is proper exact if and only if, for every dimension* $n \geqq 0$, *both* $\cdots \to K_n \to L_n \to M_n \to N_n \to \cdots$ *and* $\cdots \to C_n(K) \to C_n(L) \to C_n(M) \to C_n(N) \to \cdots$ *are exact.*

Proposition 11.2. *If* $K \rightarrowtail L \twoheadrightarrow M$ *is a proper short exact sequence of complexes, then each of the following sequences is exact for all* n:

(iii) $0 \to B_n(K) \to B_n(L) \to B_n(M) \to 0$,

(iv) $0 \to H_n(K) \to H_n(L) \to H_n(M) \to 0$,

(v) $0 \to K_n/B_n(K) \to L_n/B_n(L) \to M_n/B_n(M) \to 0$,

(vi) $0 \to K_n/C_n(K) \to L_n/C_n(L) \to M_n/C_n(M) \to 0$.

Proof. The modules $B_{n-1}(K) = \partial K_n$ of boundaries are defined by the exactness of the short sequence $C_n(K) \rightarrowtail K_n \twoheadrightarrow B_{n-1}(K)$. These sequences for K, L, and M form a 3×3 diagram with rows (ii), (i), and (iii), so the 3×3 lemma gives (iii). The homology $H_n(K)$ is defined by the exactness of $B_n(K) \rightarrowtail C_n(K) \twoheadrightarrow H_n(K)$; the 3×3 lemma gives (iv). The proofs of (v) and (vi) are similar, via $B_n \rightarrowtail K_n \twoheadrightarrow K_n/B_n(K)$ and the dual description of the homology modules by the exact sequences $H_n(K) \rightarrowtail K_n/B_n(K) \twoheadrightarrow K_n/C_n(K)$.

Next construct proper projective complexes. To each module A and each integer n introduce the special complex $U = U(A, n)$ with $U_n = A$ and $U_m = 0$ for $m \neq n$. If K is any complex, each module homomorphism $\alpha\colon A \to C_n(K)$ defines a chain transformation $h = h(\alpha)\colon U(A, n) \to K$ with h_n the composite $A \to C_n(K) \to K_n$; all chain transformations $h\colon U \to K$ have this form.

To each module A and each integer n introduce the special complex $V = V(A, n)$ with $V_n = V_{n+1} = A$, all other chain groups zero, and $\partial\colon V_{n+1} \to V_n$ the identity 1_A. Then $H_m(V) = 0$ for all m. If K is a complex, each module homomorphism $\gamma\colon A \to K_{n+1}$ defines a chain transformation $h = h(\gamma)\colon V(A, n) \to K$ with $h_{n+1} = \gamma$, $h_n = \partial\gamma$; all $h\colon V \to K$ have this form.

Lemma 11.3. *For a projective module P, the special complexes $U(P, n)$ and $V(P, n)$ are proper projective complexes.*

Proof. Let $g\colon L \twoheadrightarrow M$ be a proper epimorphism of complexes and $h = h(\gamma)\colon V(P, n) \to M$ any chain transformation. Now $g_{n+1}\colon L_{n+1} \to M_{n+1}$ is epic, so $\gamma\colon P \to M_{n+1}$ lifts to $\beta\colon P \to L_{n+1}$ with $g_{n+1}\beta = \gamma$. Therefore $h(\gamma)$ lifts to $h(\beta)\colon V \to L$. The corresponding argument for U uses the fact that $C_n(L) \to C_n(M)$ is epic. We then have

Lemma 11.4. *If P_n and Q_n are projective modules, then*

$$S = \sum_{n=0}^{\infty} U(P_n, n) \oplus \sum_{n=0}^{\infty} V(Q_n, n) \tag{11.1}$$

is a proper projective complex with $H_n(S) \cong P_n$, $B_n(S) \cong Q_n$. Any complex K with all $H_n(K)$ and $B_n(K)$ projective has this form.

Proof. The direct sum of proper projectives is proper projective. Set $Q_{-1} = 0$. The complex S has the form

$$\cdots \to Q_{n+1} \oplus P_{n+1} \oplus Q_n \to Q_n \oplus P_n \oplus Q_{n-1} \to \cdots$$

with ∂ induced by the identity $Q_n \to Q_n$, so $H(S)$ and $B(S)$ are as stated. The last assertion follows by induction from the fact that every extension by a projective module splits.

We can now prove that there are enough proper projective complexes.

Lemma 11.5. *For each complex K there exists a proper projective complex S of the form* (11.1) *and a proper epimorphism $h: S \twoheadrightarrow K$ of complexes.*

Proof. For each n, there is a projective module P_n and an epimorphism $\varrho_n: P_n \twoheadrightarrow H_n(K)$; lift ϱ_n to a homomorphism $\alpha_n: P_n \to C_n(K)$. This α_n determines $h(\alpha_n): U(P_n, n) \to K$. For each n, there is a projective module Q_n and an epimorphism $\sigma_n: Q_n \to B_n(K)$; since $K_{n+1} \to B_n$ is epic, lift σ_n to a homomorphism $\gamma_n: Q_n \to K_{n+1}$. This γ_n determines $h(\gamma_n): V(Q_n, n) \to K$. For S as in (11.1), these chain transformations $h(\alpha_n)$ and $h(\gamma_n)$ combine to give $h: S \to K$. If $s_n = q_n + p_n + q_{n-1} \in S_n$, then $h s_n = \partial \gamma_n q_n + \alpha_n p_n + \gamma_{n-1} q_{n-1}$, so h is epic. To show it proper epic, we must show that $\partial h s_n = 0$ implies $\partial s_n = \partial s_n'$ for an s_n' with $h s_n' = 0$. But $\partial h s_n$ is $\partial \gamma_{n-1} q_{n-1}$. Since $\gamma_{n-1} q_{n-1}$ is a cycle of $C_n(K)$ while σ_n and ϱ_n are epic, there are p_n' in P_n and q_n' in Q_n with $\gamma_{n-1} q_{n-1} = \alpha_n p_n' + \partial \gamma_n q_n'$. Then $s_n' = -q_n' - p_n' + q_{n-1} \in S_n$ has $\partial s_n = \partial s_n' = q_{n-1}$ and $h s'_n = 0$, as required.

These results combine to give

Proposition 11.6. *For each (positive) complex L there exists a proper projective resolution*

$$\cdots \to Y_q \xrightarrow{\partial} Y_{q-1} \to \cdots \to Y_1 \xrightarrow{\partial} Y_0 \to L \to 0 \qquad (11.2)$$

in which each Y_q is a proper projective complex of the form (11.1).

Here $Y = \{Y_q\}$ is a complex of complexes; each Y_q is a graded module $\{Y_{q,r}\}$ with a boundary homomorphism $\partial'': Y_{q,r} \to Y_{q,r-1}$ with $\partial''\partial'' = 0$. The resolution itself provides chain transformations ∂ with $\partial''\partial = \partial\,\partial''$. Change the sign of ∂ (just as in the process of condensation, X.9) by setting $\partial' = (-1)^q \partial: Y_{q,r} \to Y_{q-1,r}$. Then $(Y, \partial', \partial'')$ is a positive bicomplex.

For positive complexes K and L of right and left R-modules, respectively, we now introduce certain "hyperhomology" modules. Take a resolution Y of L, as above, and form $K \otimes Y$, where $\otimes$ is $\otimes_R$. This is a trigraded module $\{K_p \otimes Y_{q,r}\}$, with three boundary operators given by $\partial_I = \partial_K: K_p \otimes Y_{q,r} \to K_{p-1} \otimes Y_{q,r}$,

$$\partial_{II}(k \otimes y) = (-1)^{\dim k} k \otimes \partial' y, \quad \partial_{III}(k \otimes y) = (-1)^{\dim k} k \otimes \partial'' y; \qquad (11.3)$$

it is a *tricomplex* (each ∂ of square zero, each pair of ∂'s anticommutative). The corresponding total complex $T = \operatorname{Tot}(K \otimes Y)$ has $T_n = \sum K_p \otimes Y_{q,r}$ for $p+q+r = n$, $\partial = \partial_I + \partial_{II} + \partial_{III}$. An application of the comparison theorem for proper projective resolutions shows $H_n(T)$ independent of the choice of the resolution Y. We define the *hyperhomology* modules of K and L to be

$$\mathfrak{R}_n(K, L) = H_n(\operatorname{Tot}(K \otimes Y)). \tag{11.4}$$

Remark. The often used fact that the tensor product of two complexes is a bicomplex applies to functors other than the tensor product. Let $T(A, B)$ be a covariant bifunctor of modules A and B with values in some additive category $\mathscr{C}$. If K and L are positive complexes of modules, T applies to give a bigraded object $T(K_p, L_q)$ in $\mathscr{C}$ and the boundary homomorphisms of K and L induce morphisms

$$\partial' = T(\partial_K, 1)\colon\ T(K_p, L_q) \to T(K_{p-1}, L_q),$$

$$\partial'' = (-1)^p T(1, \partial_L)\colon\ T(K_p, L_q) \to T(K_p, L_{q-1})$$

which satisfy $\partial'\partial' = 0$, $\partial''\partial'' = 0$, and $\partial'\partial'' = -\partial''\partial'$, the latter because T is a bifunctor. Therefore $T(K, L) = \{T(K_p, L_q), \partial', \partial''\}$ is a bicomplex in $\mathscr{C}$ with an associated total complex $\operatorname{Tot}[T(K, L)]$. If homotopies are to be treated, one assumes T *biadditive*; that is, additive in each variable separately. When T is the tensor product, $T(K, L)$ is the familiar bicomplex $K \otimes L$.

Exercises

1. Let $K \xrightarrow{f} L \xrightarrow{g} M$ be a sequence of complexes with $gf = 0$. Show that it is a proper short exact sequence if an only if both (iii) and (iv) of Prop. 11.2 hold, and also if and only if both (ii) and (iii) hold. Find other sufficient pairs of conditions.

2. Show that every proper projective positive complex has the form given in Lemma 11.4.

3. Show that $\mathfrak{R}_n(K, L)$ is independent of the choice of the resolution of L, and prove that it can also be computed from a proper projective resolution of K, or from resolutions of both K and L.

4. Study proper exact sequences for complexes not necessarily positive.

5. Let $\mathscr{P}$ be a proper class of short exact sequences in an abelian category $\mathscr{A}$. Study the corresponding proper class in the abelian category of positive complexes in $\mathscr{A}$.

6. Each additive functor $T\colon \mathscr{A} \to \mathscr{R}$ induces a functor T on $\mathscr{A}$-complexes K to $\mathscr{R}$-complexes. For S left exact and T right exact, construct natural maps

$$H_n SK \to SH_n K, \qquad TH_n K \to H_n TK.$$

Extend to bifunctors, and obtain the homology product as a special case for $T = \otimes$.

12. The Spectral Künneth Formula

Spectral sequences provide a generalization of the KÜNNETH formula.

Theorem 12.1. *If K and L are positive complexes of right and left R-modules, respectively, and if*

$$H(\operatorname{Tot}[\operatorname{Tor}_m(K, L)]) = 0 \qquad \textit{for all } m > 0, \tag{12.1}$$

there is a first quadrant spectral sequence $\{E^r_{p,q}, d_r\}$ *with*

$$E^2_{p,q} = \sum_{s+t=q} \operatorname{Tor}_p(H_s(K), H_t(L)), \quad E^r_{p,q} \underset{p}{\Rightarrow} H(K \otimes L). \tag{12.2}$$

The hypothesis (12.1) for this theorem requires that each of the complexes $\operatorname{Tor}_m(K, L)$, defined as in the remark of § 11, has zero homology for $m>0$. The stronger hypothesis that each K_n is flat would imply that each $\operatorname{Tor}_m(K, L)=0$ for $m>0$, hence (12.1).

For positive complexes, the previous KÜNNETH Theorem (Thm. V.10.2) is included in this one. In detail, the hypotheses of that theorem required that $C_n(K)$ and $B_n(K)$ be flat; i.e., that $\operatorname{Tor}_p(C_n, G)=0=\operatorname{Tor}_p(B_n, G)$ for all G and $p>0$. Since $C_n(K) \rightarrowtail K_n \twoheadrightarrow B_{n-1}(K)$ is exact, the following portion of the standard exact sequence for the torsion product

$$\operatorname{Tor}_1(C_n, G) \to \operatorname{Tor}_1(K_n, G) \to \operatorname{Tor}_1(B_{n-1}, G)$$

is exact, so each $\operatorname{Tor}_1(K_n, G)=0$, K_n is flat, and (12.1) holds. Moreover, $B_n(K) \rightarrowtail C_n(K) \twoheadrightarrow H_n(K)$ is exact, so $\operatorname{Tor}_p(C_n, G) \to \operatorname{Tor}_p(H_n, G) \to \operatorname{Tor}_{p-1}(B_n, G)$ is exact, and therefore $\operatorname{Tor}_p(H_n(K), G)=0$ for $p>1$. The spectral sequence (12.2) thus has $E^2_{p,q}=0$ for $p \neq 0, 1$, hence consists of two columns only, and so has zero differential. The filtration of $H_n(K \otimes L)$ amounts to an exact sequence with $E^2_{0,n}$ and $E^2_{1,n-1}$, as follows:

$$0 \to \sum_{s+t=n} H_s(K) \otimes H_t(L) \to H_n(K \otimes L) \to \sum_{s+t=n-1} \operatorname{Tor}_1(H_s(K), H_t(L)) \to 0.$$

This is the usual KÜNNETH exact sequence. In other words, the present theorem shows that higher torsion products of $H(K)$, $H(L)$ affect $H(K \otimes L)$ via a suitable spectral sequence.

This theorem will be derived from a more general result.

Theorem 12.2. *If K and L are positive complexes of right and left R-modules, respectively, with hyperhomology $\mathfrak{R}_n(K, L)$ defined as in § 11, there are two first quadrant spectral sequences*

$$E'^r_{p,q} \underset{p}{\Rightarrow} \mathfrak{R}(K, L) \underset{q}{\Leftarrow} E''^r_{p,q}, \tag{12.3}$$

$$E'^2_{p,q} \cong H_p(\operatorname{Tot}[\operatorname{Tor}_q(K, L)]), \quad E''^2_{p,q} \cong \sum_{s+t=q} \operatorname{Tor}_p(H_s(K), H_t(L)). \tag{12.4}$$

Under the previous hypothesis (12.1), the first sequence collapses to the base, gives $\mathfrak{R}_n \cong E'^2_{n,0} \cong H_n(K \otimes L)$, hence yields the result of the first theorem.

Proof. Choose a proper projective resolution Y of L and form the triple complex $K \otimes Y$ of (11.3) with three boundary operators $\partial_I, \partial_{II}, \partial_{III}$.

By totalizing the first and third indices, construct a double complex with

$$X_{p,q}=\sum_{s+t=p} K_s\otimes Y_{q,t}, \quad \partial'=\partial_I+\partial_{III}, \quad \partial''=\partial_{II}.$$

Then $\operatorname{Tot} X=\operatorname{Tot}(K\otimes Y)$ has homology $\mathfrak{R}(K, L)$. The two spectral sequences of this double complex will yield the result.

In the first spectral sequence, $E'^2_{p,q}=H'_p H''_q(X)$. In each dimension t, $\cdots\to Y_{q,t}\to\cdots\to Y_{0,t}\to L_t\to 0$ is a projective resolution of L_t, so the torsion product $\operatorname{Tor}_q(K_s, L_t)$ may be calculated from this resolution as $H''_q(K\otimes Y)$; the remaining boundary $\partial'=\partial_I+\partial_{III}$ is then the boundary operator of the complex $\operatorname{Tor}_q(K, L)$. Hence E'^2 is as stated.

For the second spectral sequence, write X with renamed indices as $X_{q,p}$, so that p is the filtration index for the (second) filtration, and $E''^2_{p,q}=H''_p H'_q(X)$. For fixed p, $X_{q,p}=\sum K_s\otimes Y_{p,t}$ with $s+t=q$ is just the complex $\operatorname{Tot}(K\otimes Y_p)$ with boundary $\partial'=\partial_I+\partial_{III}$. In each complex Y_p the modules of cycles and of homologies are projective, by construction, so the KÜNNETH tensor formula (Thm. V.10.1), with hypotheses on the second factor, applies to give

$$H'_q(X_p)=H_q(K\otimes Y_p)\cong\sum_{s+t=q} H_s(K)\otimes H_t(Y_p).$$

Now each Y_p has the form S of (11.1), so each $H_n(Y_p)$ is projective, while the definition of proper exact sequences of complexes shows that for each t

$$\cdots\to H_t(Y_p)\to H_t(Y_{p-1})\to\cdots\to H_t(Y_0)\to H_t(L)\to 0$$

is a projective resolution of $H_t(L)$. Taking the tensor product with $H_s(K)$ and the homology with respect to ∂'' is the standard method of computation for $\operatorname{Tor}(H_s(K), H_t(L))$. Therefore we get the formula of (12.4) for $E''^2_{p,q}$.

This theorem can be regarded as the formation from K and L of a large collection of "hyperhomology invariants": The modules $\mathfrak{R}(K, L)$, the two filtrations of $\mathfrak{R}$, and the two spectral sequences converging, as above, to the graded modules associated with these filtrations. For example, if the ground ring R is the ring of integers, the result becomes:

Corollary 12.3. *If K and L are positive complexes of abelian groups with hyperhomology groups $\mathfrak{R}(K, L)$, there is a diagram*

$$\begin{array}{ccccccc}
 & & \sum_{p+q=n} H_p(K)\otimes H_q(L) & & & & \sum_{p+q=n-1} H_p(K)\otimes H_q(L) \\
 & & \downarrow & & & & \downarrow \\
\cdots\to H_{n-1}(\operatorname{Tor}_1(K,L)) & \to & \mathfrak{R}_n & \to H_n(K\otimes L)\to & H_{n-2}(\operatorname{Tor}_1(K,L)) & \to & \mathfrak{R}_{n-1}\to\cdots \\
 & & \downarrow & & & & \downarrow \\
 & & \sum_{p+q=n-1}\operatorname{Tor}_1(H_p(K), H_q(L)) & & & & \sum_{p+q=n-2}\operatorname{Tor}_1(H_p(K), H_q(L))
\end{array}$$

with (long) exact row and short exact columns.

Here $\operatorname{Tor}_1(K, L)$ is short for $\operatorname{Tot}[\operatorname{Tor}_1(K, L)]$.

Proof. Over Z, Tor_p vanishes for $p>1$, so the first spectral sequence has only two non-vanishing rows ($q=0$, $q=1$) and only one non-zero differential d^2: $E^2_{n,0}\to E^2_{n-2,1}$; hence the exact sequence

$$0\to E'^{\infty}_{n,0}\to H_n(K\otimes L)\xrightarrow{d^2} H_{n-2}(\operatorname{Tor}_1(K, L))\to E'^{\infty}_{n-2,1}\to 0.$$

Spliced with the exact sequences expressing the filtration of $\mathfrak{R}_n$, this yields the long horizontal exact sequence above. The second spectral sequence has only two non-vanishing columns ($p=0$, $p=1$), hence has all differentials $d^2=d^3=\cdots=0$; this yields the vertical exact sequences.

The reader may show that the composite map

$$H_p(K)\otimes H_q(L)\to\mathfrak{R}_n\to H_n(K\otimes L)$$

in this diagram is the homology product; the composite map

$$H_{n-1}(\operatorname{Tor}_1(K, L))\to\mathfrak{R}_n\to\sum\operatorname{Tor}_1(H_p(K), H_q(L))$$

is a corresponding "product" for the left exact functor Tor_1, as defined in Ex. 11.6.

Note. The hyperhomology modules are due to CARTAN-EILENBERG; the treatment in terms of proper exact sequences is due to EILENBERG (unpublished).

Bibliography

(With most listings the citation X.8 (to Chap. X, § 8) indicates where the article at issue is quoted in this book.)

ADAMS, J. F.: On the Cobar Construction. Proc. NAS USA **42**, 409—412 (1956). X.13

— On the non-Existence of Elements of Hopf Invariant One. Ann. of Math. **72**, 20—104 (1960). VII.6; X.8

AMITSUR, S. A.: Derived Functors in Abelian Categories. J. Math. and Mech. **10**, 971—994 (1961). XII.9

ARTIN, E.: Galois Theory. Notre Dame Math Lectures, No. 2, Notre Dame (Ind.), 2nd ed. 1944. IV.2

—, and J. TATE: Class Field Theory. (Mimeographed notes) Cambridge: Harvard University 1960. IV.11

ASANO, K., and K. SHODA: Zur Theorie der Darstellungen einer endlichen Gruppe durch Kollineationen. Comp. Math. **2**, 230—240 (1935). IV.11

ASSMUS, E. F., Jr.: On the Homology of Local Rings. Ill. J. Math. **3**, 187—199 (1959). VII.7

ATIYAH, M. F.: On the Krull-Schmidt Theorem with Application to Sheaves. Bull. Soc. Math. France **84**, 307—317 (1956). IX.2

— Characters and Cohomology of Finite Groups. Pub. Math. No. 9, Inst. Hautes Etudes, Paris.

AUSLANDER, M.: On the Dimension of Modules and Algebras. III. Nagoya Math. J. **9**, 67—77 (1955). VII.1

—, and D. A. BUCHSBAUM: Homological Dimension in Noetherian Rings. Proc. NAS USA **42**, 36—38 (1956). VII.7

— — Homological Dimension in Local Rings. Trans. AMS **85**, 390—405 (1957).

— — Homological Dimension in Noetherian Rings. II. Trans. AMS **88**, 194—206 (1958). VII.7

— — Codimension and Multiplicity. Ann. of Math. **68**, 625—657 (1958). VII.7

— — Unique Factorization in Regular Local Rings. Proc. NAS USA **45**, 733—734 (1959). VII.7

BAER, R.: Erweiterung von Gruppen und ihren Isomorphismen. Math. Z. **38**, 375—416 (1934). III.11; IV.11

— Automorphismen von Erweiterungsgruppen. Actualités Scientifiques et Industrielles, No. 205. Paris 1935. III.11; IV.11

— Abelian Groups that Are Direct Summands of Every Containing Abelian Group. Bull. AMS **46**, 800—806 (1940). III.11

BARRATT, M. G.: Homotopy Ringoids and Homotopy Groups. Q. J. Math. Oxon (2) **5**, 271—290 (1954). IX.1

—, V. K. A. M. GUGENHEIM and J. C. MOORE: On Semisimplicial Fibre Bundles. Am. J. Math. **81**, 639—657 (1959). VIII.9

BASS, H.: Finitistic Dimension and a Homological Generalization of Semi-Primary Rings. Trans. AMS **95**, 466—488 (1960). V.4; VII.1

BOCKSTEIN, M.: Sur le spectre d'homologie d'un complexe. C. R. Acad. Sci. Paris **247**, 259—261 (1958). V.11

BOCKSTEIN, M.: Sur la formule des coefficients universels pour les groupes d'homologie. C. R. Acad. Sci. Paris **247**, 396—398 (1958). V.11

BOREL, A.: Sur la cohomologie des espaces fibrés principaux et des espaces homogènes de groups de Lie compacts. Ann. of Math. **57**, 115—207 (1953). VI.9

— Nouvelle démonstration d'un théorème de P. A. Smith. Comment. Math. Helv. **29**, 27—39 (1955). XI.11

BOURBAKI, N.: Algèbre Multilinéare. (Chap. III of Book II in Éléments de Mathématique) Hermann, Paris 1948. 2nd ed. 1951. V.11

BRAUER, R.: Untersuchungen über die arithmetischen Eigenschaften von Gruppen linearer Substitutionen. I. Math. Z. **28**, 677—696 (1928). III.11

BROWDER, W.: Torsion in H-Spaces. Ann. of Math. **74**, 24—51 (1961). II.4; XI.5

BROWN, E. H., Jr.: Twisted Tensor Products. I. Ann. of Math. **69**, 223—246 (1959). VIII.9

BUCHSBAUM, D. A.: Exact Categories and Duality. Trans. AMS **80**, 1—34 (1955). IX.2

— A Note on Homology in Categories. Ann. of Math. **69**, 66—74 (1959). XII.4; XII.5

— Satellites and Universal Functors. Ann. of Math. **71**, 199—209 (1960). XII.4; XII.9

BUTLER, M. C. R., and G. HORROCKS: Classes of Extensions and Resolutions. Phil. Trans. Roy. Soc. London **254**, 155—222 (1961). XII.4

CARTAN, H.: Sur la cohomologie des espaces où opère un groupe. C. R. Acad. Sci. Paris **226**, 148—150, 303—305 (1948). XI.7

— Séminaire de Topologie Algébrique, 1950—1951 (École Norm. Sup.). Paris 1951. III.11

— Extension du théorème des "chaînes de syzygies". Rend. Matem. Appl. Roma **11**, 156—166 (1952). VII.6

— Sur les groupes d'Eilenberg-MacLane $H(\Pi, n)$: I. Méthode des constructions; II. Proc. NAS USA **40**, 467—471, 704—707 (1954). X.13

— Séminaire (École Norm. Sup.). Algèbres d'Eilenberg-MacLane et Homotopie. Paris 1955. X.12; XI.11

—, and S. EILENBERG: Homological Algebra. Princeton 1956.

—, and J. LERAY: Relations entre anneaux d'homologie et groupe de Poincaré. Colloque Topologie Algébrique, Paris 1949, p. 83—85. XI.7

ČECH, E.: Les groupes de Betti d'un complex infini. Fund. Math. **25**, 33—44 (1935). V.11

CHEVALLEY, C.: Fundamental Concepts of Algebra. Academic Press, New York 1956. VI.9

—, and S. EILENBERG: Cohomology Theory of Lie Groups and Lie Algebras. Trans. AMS **63**, 85—124 (1948). X.13

COHN, P. M.: On the Free Product of Associative Rings. Math. Z. **71**, 380—398 (1959); **73**, 433—456 (1960). VI.4

DIEUDONNÉ, J.: Remarks on quasi-Frobenius Rings. Ill. J. Math. **2**, 346—354 (1958). V.4

DIXMIER, J.: Homologie des anneaux de Lie. Ann. Sci. École Norm. Sup. **74**, 25—83 (1957). X.13

DOLD, A.: Homology of Symmetric Products and Other Functors of Complexes. Ann. of Math. **68**, 54—80 (1958). VIII.5

— Zur Homotopietheorie der Kettenkomplexe. Math. Ann. **140**, 278—298 (1960). II.4

— Über die Steenrodschen Kohomologieoperationen. Ann. of Math. **73**, 258—294 (1961). VIII.8

DOLD, A., and D. PUPPE: Homologie nicht-additiver Funktoren; Anwendungen. Ann. Inst. Fourier **11**, 201—312 (1961). XII.9

ECKMANN, B.: Der Cohomologie-Ring einer beliebigen Gruppe. Comment. Math. Helv. **18**, 232—282 (1945—1946). IV.11; VIII.9

— Zur Cohomologietheorie von Räumen und Gruppen. Proc. Int. Cong. Math. Amsterdam 1954, III, p. 170—177. VIII.9

—, and P. J. HILTON: On the Homology and Homotopy Decomposition of Continuous Maps. Proc. NAS USA **45**, 372—375 (1959).

— — Homotopy Groups of Maps and Exact Sequences. Comment. Math. Helv. **34**, 271—304 (1960).

—, and A. SCHOPF: Über injektive Moduln. Archiv Math. **4**, 75—78 (1953). III.11; IX.4

EHRESMANN, C.: Gattungen von lokalen Strukturen. Jber. DMV **60**, 49—77 (1957). I.8

EILENBERG, S.: Singular Homology Theory. Ann. of Math. **45**, 407—447 (1944). II.9

— Homology of Spaces with Operators. I. Trans. AMS **61**, 378—417 (1947).

— Homological Dimension and Syzygies. Ann. of Math. **64**, 328—336 (1956). Errata thereto: Ann. of Math. **65**, 593 (1957). VII.6; VII.7

— Abstract Description of some Basic Functors. J. Indian Math. Soc. **24**, 231—234 (1960). XII.9

—, and S. MACLANE: Group Extensions and Homology. Ann. of Math. **43**, 757—831 (1942). III.11; XII.4

— — Relations between Homology and Homotopy Groups. Proc. NAS USA **29**, 155—158 (1943). IV.11

— — General Theory of Natural Equivalences. Trans. AMS **58**, 231—294 (1945). I.8

— — Relations between Homology and Homotopy Groups of Spaces. Ann. of Math. **46**, 480—509 (1945). IV.11

— — Cohomology Theory in Abstract Groups. I. Ann. of Math. **48**, 51—78 (1947). VIII.5; VIII.9

— — Cohomology Theory in Abstract Groups. II. Group Extensions with a non-Abelian Kernel. Ann. of Math. **48**, 326—341 (1947). IV.9

— — Cohomology and Galois Theory. I. Normality of Algebras and Teichmüller's Cocycle. Trans. AMS **64**, 1—20 (1948).

— — Homology of Spaces with Operators. II. Trans. AMS **65**, 49—99 (1949). XI.7

— — Relations between Homology and Homotopy Groups of Spaces. II. Ann. of Math. **51**, 514—533 (1950). XI.7

— — Cohomology Theory of Abelian Groups and Homotopy Theory. I. Proc. NAS USA **36**, 443—447 (1950). X.13

— — Homology Theories for Multiplicative Systems. Trans. AMS **71**, 294—330 (1951). X.12; X.13

— — Acyclic Models. Am. J. Math. **75**, 189—199 (1953). VIII.7; VIII.8

— — On the Groups $H(\Pi, n)$. I. Ann. of Math. **58**, 55—106 (1953). VIII.5; X.12

— — On the Groups $H(\Pi, n)$. II. Methods of Computation. Ann. of Math. **60**, 49—139 (1954). X.11; XII.9

— — On the Groups $H(\Pi, n)$. III. Operations and Obstructions. Ann. of Math. **60**, 513—557 (1954).

— — On the Homology Theory of Abelian Groups. Can. J. Math. **7**, 43—55 (1955). X.12

EILENBERG, S., and J. MOORE: Limits and Spectral Sequences. Topology **1**, 1—23 (1962). XI.11

—, and N. STEENROD: Foundations of Algebraic Topology. Princeton 1952. II.9

—, and J. A. ZILBER: Semi-simplicial Complexes and Singular Homology. Ann. of Math. **51**, 499—513 (1950). VIII.5

— — On Products of Complexes. Am. J. Math. **75**, 200—204 (1953). VIII.5

EVENS, L.: The Cohomology Ring of a Finite Group. Trans. AMS **101**, 224—239 (1961). XI.11

EVERETT, C. J.: An Extension Theory for Rings. Am. J. Math. **64**, 363—370 (1942). X.13

FADELL, E., and W. HUREWICZ: On the Structure of Higher Differential Operators in Spectral Sequences. Ann. of Math. **68**, 314—347 (1958). XI.3; XI.11

FEDERER, H.: A Study of Function Spaces by Spectral Sequences. Trans. AMS **82**, 340—361 (1956). XI.11

FITTING, H.: Beiträge zur Theorie der Gruppen endlicher Ordnung. Jber. DMV **48**, 77—141 (1938). IV.11

FREUDENTHAL, H.: Der Einfluß der Fundamentalgruppe auf die Bettischen Gruppen. Ann. of Math. **47**, 274—316 (1946). IV.11

FREYD, P.: Functor Theory, Dissertation, Princeton University, 1960. XII.3

FRÖHLICH, A.: On Groups over a d.g. Near Ring. II. Categories and Functors. Q. J. Math. Oxon (2) **11**, 211—228 (1960).

— Non Abelian Homological Algebra. I. Derived Functors and Satellites. Proc. London Math. Soc. (3) **11**, 239—275 (1961); II. Varieties. Proc. London Math. Soc. (3) **12**, 1—28 (1961).

FRUCHT, R.: Zur Darstellung endlicher Abelscher Gruppen durch Kollineationen. Math. Z. **63**, 145—155 (1955). IV.11

FUCHS, L.: Abelian Groups, Budapest, 1958. Also published by Pergamon Press, Oxford and New York, 1960.

— Notes on Abelian Groups. I. Ann. Univ. Sci., Budapest, Sect. Math. **2**, 5—23 (1959). XII.4

GABRIEL, P.: Des catégories abéliennes. Bull. Soc. Math. France **90**, 323—448. (1962) IX.2

GODEMENT, R.: Théorie des Faisceaux. Hermann, Paris 1958. I.8; IX.4

GÖDEL, K.: The Consistency of the Continuum Hypothesis. Ann. of Math. Studies 3. Princeton 1940. I.7

GRAY, J. W.: Extensions of Sheaves of Algebras. Ill. J. Math. **5**, 159—174 (1961). X.13

— Extensions of Sheaves of Associative Algebras by non-trivial Kernels. Pac. J. Math. **11**, 909—917 (1961). X.13

GREEN, J. A.: On the Number of Automorphisms of a Finite Group. Proc. Roy. Soc. London, Ser. A **237**, 574—581 (1956). XI.11

GRÖBNER, W.: Über die Syzygien-Theorie der Polynomideale. Monatsh. Math. **53**, 1—16 (1949). VII.6

GROTHENDIECK, A.: Sur quelques Points d'Algèbre Homologique. Tohoku Math. J. **9**, 119—221 (1957). IX.2; IX.4; XII.8

— (with J. DIEUDONNÉ): Éléments de Géometrie Algébrique. I, II. Pub. Math. Inst. des Hautes Etudes. Paris 1960, 1961. Nos. 4 and 8. I.8

GRUENBERG, K. W.: Resolutions by Relations. J. London Math. Soc. **35**, 481—494 (1960). X.5

GUGENHEIM, V. K. A. M.: On a Theorem of E. H. Brown. Ill. J. Math. **4**, 292—311 (1960). VIII.9

GUGENHEIM, V. K. A. M.: On Extensions of Algebras, Co-algebras, and Hopf Algebras I. Am. J. Math. **84**, 349—382 (1962).

—, and J. C. MOORE: Acyclic Models and Fibre Spaces. Trans. AMS **85**, 265—306 (1957). VIII.7; XI.11

HALL, M.: Group Rings and Extensions. I. Ann. of Math. **39**, 220—234 (1938). IV.11

HALPERN, E.: Twisted Polynomial Hyperalgebras. Memoir AMS 29. Providence, 1958. VI.9

HASSE, H., R. BRAUER, and E. NOETHER: Beweis eines Hauptsatzes in der Theorie der Algebren. J. reine angew. Math. **167**, 399—404 (1932). III.11

HARRIS, B.: Cohomology of Lie Triple Systems and Lie Algebras with Involution. Trans. AMS **98**, 148—162 (1961). X.13

HARRISON, D. K.: Infinite Abelian Groups and Homological Methods. Ann. of Math. **69**, 366—391 (1959); **71**, 197 (1960). XII.4

— Commutative Algebras and Cohomology. Trans. AMS **104**, 191—204; (1962). X.13

HATTORI, A.: On Fundamental Exact Sequences. J. Math. Soc. Japan **12**, 65—80 (1960). XI.11, ex. 4

HELLER, A.: Homological Algebra in Abelian Categories. Ann. of Math. **68**, 484—525 (1958). XII.4; XII.6

HILBERT, D.: Über die Theorie der Algebraischen Formen. Math. Ann. **36**, 473—534 (1890). VII.6

HILTON, P. J., and W. LEDERMANN: Homology and Ringoids. I. Proc. Camb. Phil. Soc. **54**, 152—167 (1958). IX.1

—, and D. REES: Natural Maps of Extension Functors and a Theorem of R. G. Swan. Proc. Camb. Phil. Soc. **57**, 489—502 (1961). XII.9

—, and S. WYLIE: Homology Theory. Cambridge 1960. II.9; VIII.9; XI.7

HOCHSCHILD, G.: On the Cohomology Groups of an Associative Algebra. Ann. of Math. **46**, 58—67 (1945). VII.4; VII.7

— On the Cohomology Theory for Associative Algebras. Ann. of Math. **47**, 568—579 (1946). VII.7

— Cohomology and Representations of Associative Algebras. Duke M. J. **14**, 921—948 (1947). VII.4; IX.3; X.13

— Local Class Field Theory. Ann. of Math. **51**, 331—347 (1950). IV.11

— Cohomology Classes of Finite Type and Finite Dimensional Kernels for Lie Algebras. Am. J. Math. **76**, 763—778 (1954). X.13

— Lie Algebra Kernels and Cohomology. Am. J. Math. **76**, 698—716 (1954). X.13

— Relative Homological Algebra. Trans. AMS **82**, 246—269 (1956). IX.8; XII.4

—, and J.-P. SERRE: Cohomology of Group Extensions. Trans. AMS **74**, 110—134 (1953). XI.10; XI.11

HÖLDER, O.: Die Gruppen der Ordnungen p^3, $p\,q^2$, $p\,q\,r$, p^4. Math. Ann. **43**, 301—412 (1893) (especially § 18). III.11

HOPF, H.: Über die Abbildungen der dreidimensionalen Sphäre auf die Kugelfläche. Math. Ann. **104**, 637—665 (1931). XI.2

— Über die Abbildungen von Sphären auf Sphären niedrigerer Dimension. Fund. Math. **25**, 427—440 (1935). XI.2

— Fundamentalgruppe und zweite Bettische Gruppe. Comment. Math. Helv. **14**, 257—309 (1941/42). IV.11

— Relations between the Fundamental Group and the Second Betti Group. Lectures in Topology. Ann. Arbor 1942. IV.11

— Über die Bettischen Gruppen, die zu einer beliebigen Gruppe gehören. Comment. Math. Helv. **17**, 39—79 (1944/45). III.11; IV.11

HU, S.-T.: Homotopy Theory. Academic Press, New York and London 1959. XI.2; XI.7

HUBER, P. J.: Homotopy Theory in General Categories. Math. Ann. **144**, 361—385 (1961).

HUREWICZ, W.: Beiträge zur Topologie der Deformationen. Proc. Akad. Amsterdam **38**, 112—119, 521—538 (1935); **39**, 117—125, 215—224 (1936). IV.11

— On Duality Theorems. Bull. AMS, abstract 47-7-329 **47**, 562—563 (1941). I.8

JACOBSON, N.: Structure of Rings. AMS, Providence, 1956.

— Lie Algebras. New York: [Interscience] John Wiley & Sons 1962. X.13

JANS, J. P.: Duality in Noetherian Rings. Proc. AMS 829—835 (1961). V.4; VII.7

KAN, D. M.: Adjoint Functors. Trans. AMS **87**, 294—329 (1958). IX.6

— A Combinatorial Definition of Homotopy Groups. Ann. of Math. **67**, 282—312 (1958). VIII.5

KAPLANSKY, I.: On the Dimension of Modules and Algebras. X. Nagoya Math. J. **13**, 85—88 (1958). VII.1

KASCH, F.: Dualitätseigenschaften von Frobenius-Erweiterungen. Math. Z. **77**, 219—227 (1961). VII.7

KELLEY, J. L., and E. PITCHER: Exact Homomorphism Sequences in Homology Theory. Ann. of Math. **48**, 682—709 (1947). II.9

KLEISLI, H.: Homotopy Theory in Abelian Categories. Can. J. Math. **14**, 139—169 (1962). XII.4

KOCHENDÖRFFER, R.: Über den Multiplikator einer Gruppe. Math. Z. **63**, 507—513 (1956). IV.11

KOSZUL, J. L.: Sur les opérateurs de dérivation dans un anneau. C. R. Acad. Sci. Paris **225**, 217—219 (1947). XI.11

— Sur un type d'algèbres différentielles en rapport avec la transgression. Colloque de Topologie, Brussels 1950, p. 73—81. VII.2

— Homologie et cohomologie des algèbres de Lie. Bull. Soc. Math. France **78**, 65—127 (1950). X.13

KUDO, T., and S. ARAKI: Topology of H_n-Spaces and H-Squaring Operations. Mem. Fac. Sci. Kyusyu Univ., Ser. A **10**, 85—120 (1956). XI.11

KÜNNETH, H.: Über die Bettischen Zahlen einer Produktmannigfaltigkeit. Math. Ann. **90**, 65—85 (1923). V.10

— Über die Torsionszahlen von Produktmannigfaltigkeiten. Math. Ann. **91**, 125—134 (1924). V.10

LAMBEK, J.: Goursat's Theorem and the Zassenhaus Lemma. Can. J. Math. **10**, 45—56 (1958). II.9

LANG, S.: Review of "Éléments de géométrie algébrique". Bull. AMS **67**, 239—246 (1961). I.8

LEFSCHETZ, S.: Algebraic Topology. AMS, New York 1942.

LERAY, J.: Structure de l'anneau d'homologie d'une représentation. C. R. Acad. Sci. Paris **222**, 1419—1422 (1946). XI.2; XI.11

— L'anneau spectral et l'anneau filtré d'homologie d'un espace localement compact et d'une application continue. J. Math. Pures Appl. **29**, 1—139 (1950). XI.2; XI.11

— L'homologie d'un espace fibré dont la fibre est connexe. J. Math. Pures Appl. **29**, 169—213 (1950). XI.2; XI.11

LIULEVICIUS, A.: The Factorization of Cyclic Reduced Powers by Secondary Cohomology Operations. Proc. NAS USA **46**, 978—981 (1960). X.8

LORENZEN, P.: Über die Korrespondenzen einer Struktur. Math. Z. **60**, 61—65 (1954). II.9

LUBKIN, S.: Imbedding of Abelian Categories. Trans. AMS **97**, 410—417 (1960). II.9, XII.3

LYNDON, R. C.: The Cohomology Theory of Group Extensions. Harvard University Thesis, 1946. XI.10; XI.11

— The Cohomology Theory of Group Extensions. Duke Math. J. **15**, 271—292 (1948). XI.10; XI.11

— Cohomology Theory of Groups with a Single Defining Relation. Ann. of Math. **52**, 650—665 (1950). X.5

MACAULEY, R. A.: Analytic Group Kernels and Lie Algebra Kernels. Trans. AMS **95**, 530—553 (1960). X.13

MACKEY, G. W.: Unitary Representations of Group Extensions. I. Acta Math. **99**, 265—311 (1958). IV.11

MAC LANE, S.: Duality for Groups. Bull. AMS **56**, 485—516 (1950). IX.2

— Slide and Torsion Products for Modules. Rend. del Sem. Math. Torino **15**, 281—309 (1955/56). V.11

— Homologie des anneaux et des modules. Colloque de Topologie algébrique, Louvain, 1956, p. 55—80. X.13

— Extensions and Obstructions for Rings. Ill. J. Math. **2**, 316—345 (1958). X.13

— Group Extensions by Primary Abelian Groups. Trans. AMS **95**, 1—16 (1960). II.4; XII.4

— Triple Torsion Products and Multiple Künneth Formulas. Math. Ann. **140**, 51—64 (1960). XII.9

— An Algebra of Additive Relations. Proc. NAS USA **47**, 1043—1051 (1961). II.9

— Locally Small Categories and the Foundations of Set Theory. Infinitistic Methods. (Warsaw symposium, 1959.) Oxford 1961. I.8; IX.2

MASSEY, W. S.: Exact Couples in Algebraic Topology. Ann. of Math. **56**, 363—396 (1952). XI.5; XI.11

— On the Universal Coefficient Theorem of Eilenberg and MacLane. Bol. Soc. Mat. Mex. (2) **3**, 1—12 (1958). III.11

MATLIS, E.: Injective Modules over Noetherian Rings. Pac. J. Math. **8**, 511—528 (1958). III.11

— Applications of Duality. Proc. AMS **10**, 659—662 (1959). VII.1

— Modules with Descending Chain Condition. Trans. AMS **97**, 495—508 (1960). VII.7

MAYER, W.: Über Abstrakte Topologie. I und II. Monatsh. Math. u. Physik **36**, 1—42, 219—258 (1929). II.9

— Topologische Gruppensysteme. Monatsh. Math. u. Physik **47**, 40—86 (1938/39). II.9

MILNOR, J.: The Steenrod Algebra and its Dual. Ann. of Math. **67**, 150—171 (1958). VIII.8

—, and J. C. MOORE: On the Structure of Hopf Algebras. Forthcoming. VI.9

MITCHELL, B.: The full embedding Theorem. Forthcoming in Am. J. Math. **86**, 619—637 (1964).

MORITA, K.: Duality for Modules and its Applications to the Theory of Rings with Minimum Condition. Sci. Rep. Tokyo Kyoiku Daigaku, Sec. A **6**, 83—142 (1958). V.4

NAGATA, M.: A General Theory of Algebraic Geometry over Dedekind Rings. II. Am. J. Math. **80**, 382—420 (1958). VII.7

NAKAYAMA, T.: On the Complete Cohomology Theory of Frobenius Algebras. Osaka Math. J. **9**, 165—187 (1957). VII.7

NAKAYAMA, T., and T. TSUZUKU: On Frobenius Extensions. I, II. Nagoya Math. J. 17, 89—110 (1960); 19, 127—148 (1961). VII.7
NEUMANN, B. H.: An Essay on Free Products of Groups with Amalgamations. Phil. Trans. Roy. Soc. London A 246, 503—554 (1954). XII.1
NORTHCOTT, D. G.: Ideal Theory. Cambridge, 1953.
— An Introduction to Homological Algebra. Cambridge, 1960. VII.1
NUNKE, R. J.: Modules of Extensions over Dedekind Rings. Ill. J. Math. 3, 222—241 (1959). XII.4
PALERMO, F. P.: The Cohomology Ring of Product Complexes. Trans. AMS 86, 174—196 (1957). V.11
PUPPE, D.: Homotopie und Homologie in abelschen Gruppen und Monoidkomplexen. I, II. Math. Z. 68, 367—406, 407—421 (1958).
— Korrespondenzen in Abelschen Kategorien. Math. Ann. 148, 1—30 (1962). II.9, XII.3
REDEI, L.: Die Verallgemeinerung der Schreierschen Erweiterungstheorie. Acta Sci. Math. Szeged 14, 252—273 (1952). X.13
REE, R.: Lie Elements and an Algebra Associated with Shuffles. Ann. of Math. 68, 210—220 (1958). X.13
RIM, D. S.: Modules over Finite Groups. Ann. of Math. 69, 700—712 (1959).
RÖHRL, H.: Über Satelliten halbexakter Funktoren. Math. Z. 79, 193—223 (1962). XII.9
ROSE, I. H.: On the Cohomology Theory for Associative Algebras. Am. J. Math. 74, 531—546 (1952). X.7
ROSENBERG, A., and D. ZELINSKY: Cohomology of Infinite Algebras. Trans. AMS 82, 85—98 (1956). X.3
SCHMID, J.: Zu den Reduktionssätzen in der homologischen Theorie der Gruppen. Archiv der Math. 15, 28—32 (1964). IX.7
SCHREIER, O.: Über die Erweiterungen von Gruppen. I. Monatsh. Math. u. Phys. 34, 165—180 (1926); II. Abh. Math. Sem. Hamburg 4, 321—346 (1926). III.11; IV.11
SERRE, J.-P.: Homologie Singulière des Espaces Fibrés. Applications. Ann. of Math. 54, 425—505 (1951). XI.2; XI.11
— Sur la dimension homologique des anneaux et des modules noethériens. Proceedings. Symposium on Algebraic Number Theory, Tokyo, 1956, p. 175—189. VII.7
— Algèbre Locale — Multiplicités. (Notes written by P. GABRIEL) Paris 1957/58. VII.7
SHUKLA, U.: Cohomologie des algèbres associatives. Ann. Sci. École Norm., Sup. 78, 163—209 (1961). X.13
SPECHT, W.: Gruppentheorie. Berlin-Göttingen-Heidelberg: Springer 1956. XII.1
STEENROD, N. E.: The Topology of Fibre Bundles. Princeton 1951. XI.2
— Cyclic Reduced Powers of Cohomology Classes. Proc. NAS USA 39, 217—223 (1953). IV.7; VIII.8
— The Cohomology Algebra of a Space. L'Ens. Math., II. ser. 7, 153—178 (1961). VIII.8
—, and D. B. A. EPSTEIN: Cohomology Operations. Lectures by N. E. STEENROD, written and revised by D. B. A. EPSTEIN. Annals of Math. Studies 50, Princeton, 1962. VIII.8
SWAN, R. G.: A Simple Proof of the Cup Product Reduction Theorem. Proc. NAS USA 46, 114—117 (1960). VIII.8; VIII.9
— Induced Representations and Projective Modules. Ann. of Math. 71, 552—578 (1960).

SZCZARBA, R. H.: The Homology of Twisted Cartesian Products. Trans. AMS **100**, 197—216 (1961). VIII.9

SZENDREI, J.: On Schreier Extension of Rings without Zero-divisors. Publ. Math. Debrecen **2**, 276—280 (1952). X.13

TATE, J.: The Higher Dimensional Cohomology Groups of Class Field Theory. Ann. of Math. **56**, 294—297 (1952). IV.11

— Homology of Noetherian Rings and Local Rings. Ill. J. Math. **1**, 14—27 (1957). VII.7

TEICHMÜLLER, O.: Über die sogenannte nicht kommutative Galoissche Theorie und die Relation .... Deutsche Math. **5**, 138—149 (1940). IV.11

VENKOV, B. B.: Cohomology Algebras for some Classifying Spaces [Russian]. Dokl. Akad. Nauk. SSSR **127**, 943—944 (1959). XI.11

WALL, C. T. C.: Resolutions for Extensions of Groups. Proc. Camb. Phil. Soc. **57**, 251—255 (1961). X.5

— On the Cohomology of Certain Groups. Proc. Camb. Phil. Soc. **57**, 731—733 (1961). X.5

WALLACE, A. H.: An Introduction to Algebraic Topology. Pergamon, London, 1957. II.9

WATTS, C. E.: Intrinsic Characterizations of some Additive Functors. Proc. AMS **11**, 5—8 (1960). XII.9

WEDDERBURN, J. H. M.: Homomorphisms of Groups. Ann. of Math. **42**, 486—487 (1941). II.9

WHITEHEAD, J. H. C.: A Certain Exact Sequence. Ann. of Math. **52**, 51—110 (1950). XII.9

— On Group Extensions with Operators. Q. J. Math. Oxon (2) **1**, 219—228 (1950).

WHITNEY, H.: Tensor Products of Abelian Groups. Duke Math. J. **4**, 495—528 (1938). V.11

YAMAGUTI, K.: On the Cohomology Space of a Lie Triple System. Kumamoto J. Sci. A **5**, 44—51 (1960). X.13

YONEDA, N.: On the Homology Theory of Modules. J. Fac. Sci. Tokyo, Sec. I **7**, 193—227 (1954). III.11

— Notes on Products in Ext. Proc. AMS **9**, 873—875 (1958). VIII.4

— On Ext and Exact Sequences. J. Fac. Sci. Tokyo, Sec. I **8**, 507—526 (1960). III.11; XII.4; XII.9

ZASSENHAUS, H.: The Theory of Groups. 2nd ed., New York, 1958 (especially p. 237: Morphisms). II.9

ZEEMAN, E. C.: A Proof of the Comparison Theorem for Spectral Sequences. Proc. Camb. Phil. Soc. **53**, 57—62 (1957). XI.11

List of Standard Symbols

$\rightarrow$	Homomorphism
$\rightarrowtail$	Monomorphism
$\twoheadrightarrow$	Epimorphism
$\dashrightarrow$	Homomorphism after "neglect" of some structure (IX.5.2)
$\dashrightarrow$	Homomorphism to be constructed
$\rightharpoonup$	Additive relation (II.6)
$\Rightarrow$	Implies; convergence of a spectral sequence (XI.3)
$\Leftrightarrow$	If and only if
$\exists$	There exists
&	And
∂	Boundary
δ	Coboundary (II.3.1)
$*$, #	Induced homomorphism (subscript or superscript); (I.2.4); (I.6.1); (III.6.8)
$\sim$	Homologous (II.1.2); (II.2)
$\simeq$	Homotopic (II.2.3)
$\cong$	Isomorphic
$\leqq$	Contained in (XII.2)
$\equiv$	Congruent (of extensions); (III.5.2)
$\circ$	Composition
$\in$	Member of
$\emptyset$	Null set
$\subset$	Inclusion
[]	Base element of bar resolution (IV.5.1); (X.2)
$\frown$, $\cap$	Intersection
$\smile$, $\cup$	Union
$\smile$	Cup product (VIII.9.5)
$\vee$	Wedge product (VIII.4.1)
$\times$	Cartesian product
$\otimes$	Tensor product
Σ, $\oplus$	Direct sum (I.4)
Π	Direct product (I.4)
$\Vert$	Short exact sequence (IX.1.12)
$\square$	Functor neglecting some structure (IX.5)

∇, ∇_A	Codiagonal map (III.2.1'); (IX.1.6)
0	Zero element or mapping
$0'$	Zero object of a category (IX.1)
1, 1_A	Identity element of a group, ring, or algebra
1, 1_A	Identity mapping $A \to A$
$\mathscr{A}$, $\mathscr{C}$, $\mathscr{R}$	Category
$\mathscr{A}^{\mathrm{op}}$	Opposite category
$\mathscr{M}_R$	Category of right R-modules
E	Exact sequence; exterior algebra
E_{K}	Exterior algebra over K
F	Free module; Field
H	Homology or cohomology
I	"Identity element" map $\mathsf{K} \to \Lambda$
K	Complex
L	Complex; projective module; Left ideal (I.2)
P	Projective module
P_{K}	Polynomial algebra over K
Q	Field of rational numbers
R	Ring
R^{op}	Opposite ring
S	Long exact sequence; Ring
$S(X)$	Singular complex of X
T	Functor; Singular simplex; Ring
U	DGA-algebra
V	Vector space; Hopf algebra
X, Y	Complex; Topological space
Z	Ring of integers
a, r	Elements $a \in A$, $r \in R$, etc.
d_i	Face operator (simplicial set); (VIII.5)
p, p_A, p_H	Homology products (VIII.1)
s, t	Homotopy
s_i	Degeneracy operator (simplicial set); (VIII.5)

Abbreviations (Caps for modules, lower case for categories)

bidim	Dimension as a bimodule (VII.5)
cls	(Homology) class of
Coim, coim	Coimage
Coker, coker	Cokernel
Def	Domain of definition (of an additive relation); (II.6)
deg	Degree of an element or morphism
Ext, ext	Group of extensions

h.dim	Homological dimension (VII.1)
Hom, hom	Group of homomorphisms
Im, im	Image
Ind	Indeterminacy (of an additive relation); (II.6)
Ker, ker	Kernel
l.gl.dim.	Left global dimension of a ring (VII.1)
Tor	Torsion product

Γ	Morphism of diagrams (I.7); (XII.6)
Δ, Δ^n	Simplex, n-dimensional (affine) simplex (II.7)
Δ, Δ_A	Diagonal map (III.2.1); (IX.1.6)
K	Commutative ground ring
Λ	Algebra
Π	Multiplicative group
Σ	Algebra
Ω	Algebra
α	Associative map for tensor product (VI.2.3); (VI.8.3)
η	Adjoint associativity (V.3.5); (VI.8.7)
θ	Isomorphism; Equivalence in a category (IX.1)
ι	Injection of direct sum (I.4.1)
$\varkappa$	Monomorphism
λ	Monomorphism
π_i	Projection of a direct sum (I.4.1)
π, π_A	Product map of an algebra (VI.1.2)
ϱ	Epimorphism
σ, τ	Epimorphism
τ	Middle four exchange (VI.2.4); (VI.8.4)
ψ	Codiagonal map of coalgebra or Hopf algebra (VI.9)

Index

∞ CIM

M. Aigner Combinatorial Theory ISBN 978-3-540-61787-7
A. L. Besse Einstein Manifolds ISBN 978-3-540-74120-6
N. P. Bhatia, G. P. Szegő Stability Theory of Dynamical Systems ISBN 978-3-540-42748-3
J. W. S. Cassels An Introduction to the Geometry of Numbers ISBN 978-3-540-61788-4
R. Courant, F. John Introduction to Calculus and Analysis I ISBN 978-3-540-65058-4
R. Courant, F. John Introduction to Calculus and Analysis II/1 ISBN 978-3-540-66569-4
R. Courant, F. John Introduction to Calculus and Analysis II/2 ISBN 978-3-540-66570-0
P. Dembowski Finite Geometries ISBN 978-3-540-61786-0
A. Dold Lectures on Algebraic Topology ISBN 978-3-540-58660-9
J. L. Doob Classical Potential Theory and Its Probabilistic Counterpart ISBN 978-3-540-41206-9
R. S. Ellis Entropy, Large Deviations, and Statistical Mechanics ISBN 978-3-540-29059-9
H. Federer Geometric Measure Theory ISBN 978-3-540-60656-7
S. Flügge Practical Quantum Mechanics ISBN 978-3-540-65035-5
L. D. Faddeev, L. A. Takhtajan Hamiltonian Methods in the Theory of Solitons ISBN 978-3-540-69843-2
I. I. Gikhman, A. V. Skorokhod The Theory of Stochastic Processes I ISBN 978-3-540-20284-4
I. I. Gikhman, A. V. Skorokhod The Theory of Stochastic Processes II ISBN 978-3-540-20285-1
I. I. Gikhman, A. V. Skorokhod The Theory of Stochastic Processes III ISBN 978-3-540-49940-4
D. Gilbarg, N. S. Trudinger Elliptic Partial Differential Equations of Second Order ISBN 978-3-540-41160-4
H. Grauert, R. Remmert Theory of Stein Spaces ISBN 978-3-540-00373-1
H. Hasse Number Theory ISBN 978-3-540-42749-0
F. Hirzebruch Topological Methods in Algebraic Geometry ISBN 978-3-540-58663-0
L. Hörmander The Analysis of Linear Partial Differential Operators I – Distribution Theory and Fourier Analysis ISBN 978-3-540-00662-6
L. Hörmander The Analysis of Linear Partial Differential Operators II – Differential Operators with Constant Coefficients ISBN 978-3-540-22516-4
L. Hörmander The Analysis of Linear Partial Differential Operators III – Pseudo-Differential Operators ISBN 978-3-540-49937-4
L. Hörmander The Analysis of Linear Partial Differential Operators IV – Fourier Integral Operators ISBN 978-3-642-00117-8
K. Itô, H. P. McKean, Jr. Diffusion Processes and Their Sample Paths ISBN 978-3-540-60629-1
T. Kato Perturbation Theory for Linear Operators ISBN 978-3-540-58661-6
S. Kobayashi Transformation Groups in Differential Geometry ISBN 978-3-540-58659-3
K. Kodaira Complex Manifolds and Deformation of Complex Structures ISBN 978-3-540-22614-7
Th. M. Liggett Interacting Particle Systems ISBN 978-3-540-22617-8
J. Lindenstrauss, L. Tzafriri Classical Banach Spaces I and II ISBN 978-3-540-60628-4
R. C. Lyndon, P. E Schupp Combinatorial Group Theory ISBN 978-3-540-41158-1
S. Mac Lane Homology ISBN 978-3-540-58662-3
C. B. Morrey Jr. Multiple Integrals in the Calculus of Variations ISBN 978-3-540-69915-6
D. Mumford Algebraic Geometry I – Complex Projective Varieties ISBN 978-3-540-58657-9
O. T. O'Meara Introduction to Quadratic Forms ISBN 978-3-540-66564-9
G. Pólya, G. Szegő Problems and Theorems in Analysis I – Series. Integral Calculus. Theory of Functions ISBN 978-3-540-63640-3
G. Pólya, G. Szegő Problems and Theorems in Analysis II – Theory of Functions. Zeros. Polynomials. Determinants. Number Theory. Geometry ISBN 978-3-540-63686-1
W. Rudin Function Theory in the Unit Ball of $\mathbb{C}^n$ ISBN 978-3-540-68272-1
S. Sakai C*-Algebras and W*-Algebras ISBN 978-3-540-63633-5
C. L. Siegel, J. K. Moser Lectures on Celestial Mechanics ISBN 978-3-540-58656-2
T. A. Springer Jordan Algebras and Algebraic Groups ISBN 978-3-540-63632-8
D. W. Stroock, S. R. S. Varadhan Multidimensional Diffusion Processes ISBN 978-3-540-28998-2
R. R. Switzer Algebraic Topology: Homology and Homotopy ISBN 978-3-540-42750-6
A. Weil Basic Number Theory ISBN 978-3-540-58655-5
A. Weil Elliptic Functions According to Eisenstein and Kronecker ISBN 978-3-540-65036-2
K. Yosida Functional Analysis ISBN 978-3-540-58654-8
O. Zariski Algebraic Surfaces ISBN 978-3-540-58658-6